14th South African Conference on Computational and Applied Mechanics (SACAM 2025)

Johannesburg, South Africa
21-23 January 2025

ISBN: 979-8-3313-3299-0

Printed from e-media with permission by:

Curran Associates, Inc.
57 Morehouse Lane
Red Hook, NY 12571

Some format issues inherent in the e-media version may also appear in this print version.

South African Association for Theoretical and Applied Mechanics (SAAM)
c/o Schalk Kok
9-21 Engineering I
Department of Mechanical Engineering
University of Pretoria
South Africa

Phone: 27 21 808 3554
Fax: 27 86 615 5206

Schalk.Kok@uj.ac.za

Additional copies of this publication are available from:

Curran Associates, Inc.
57 Morehouse Lane
Red Hook, NY 12571 USA
Phone: 845-758-0400
Fax: 845-758-2633
Email: curran@proceedings.com
Web: www.proceedings.com

Proceedings of the 14th South African Conference on Computational and Applied Mechanics

SACAM 2025

21-23 January 2025

Editors: P.W. Loveday and M. Sharifpur

on behalf of the

South African Association for Theoretical and Applied Mechanics

(SAAM)

SACAM 2025

14th South African Conference on Computational and Applied Mechanics

SAAM

Copyright

Conference Title

14th South African Conference on Computational and Applied Mechanics.

Organising Institute

University of the Witwatersrand, Johannesburg, South Africa.

Date and Location

Held on 21-23 January 2025, at the University of the Witwatersrand, Johannesburg, South Africa.

Editors

P.W. Loveday

M. Sharifpur

Table of Contents

Aerodynamics

AI/ML in Mechanics

Biomechanics

Computational Methods in Mechanics

Dynamics & Vehicle Dynamics

Fluid Dynamics & Computational Fluid Dynamics

Heat Transfer & Thermodynamics

Robotics & Mechatronics

Structural Mechanics, Materials & Manufacturing

Ultrasonics

Vibration & Condition Monitoring

Preface

The SACAM conference is held every second year under the auspices of the South African Association for Computational and Applied Mechanics (SAAM).

SACAM provides a forum for presenting and discussing the latest developments in applied mechanics. We aim to bring together scientists, engineers and applied mathematicians from academia and industry to encourage scientific exchange of ideas and opportunities in applied mechanics.

The conference provides a unique opportunity for postgraduate students to engage with the community, and postgraduate students' participation is especially welcome.

SACAM 2025 was the 14th conference in this series and was held at the University of the Witwatersrand, Johannesburg, on 21-23 January 2025.

The conference was attended by 83 delegates. Authors could choose to submit a conference paper for review and possible inclusion in the proceedings or a 2-page extended abstract. These proceedings contain 56 papers, which were accepted after the review process.

The ISBN number of conference proceedings is 978-1-0492-4021-3. The proceedings are available from the SAAM website https://saam.africa/ .

Peer Review Statement

Full-length manuscripts presented at the conference were subjected to peer review after the conference, and accepted papers are included in the Conference Proceedings. Two independent reviewers per manuscript were tasked to ensure that:

- The editorial standard of the paper is acceptable for an international conference.
- The contents of the paper are of interest to the target audience.
- The standard, methodology and quality of the work is appropriate.

The conference programme included 91 presentations, including four keynote presentations. Following the peer review process, the editors accepted 56 papers, which are included in these proceedings.

Authors of selected accepted papers were invited to submit extended papers to special issues of the MPDI journals, "Applied Mechanics" and "Mathematical and Computational Applications".

Organising Committee

Chair

Prof. Philip Loveday
University of the Witwatersrand, Johannesburg

Members

Dr Alfred Bogaers
EX MENTE

Mr Wesley Dharmalingum
University of the Witwatersrand, Johannesburg

Mr Markus Erwee
Samancor

Mr John Jones
University of the Witwatersrand, Johannesburg

Ms Makhosazana Moyo
University of the Witwatersrand, Johannesburg

Dr Randall Paton
University of the Witwatersrand, Johannesburg

Dr Kevin Pereira
University of the Witwatersrand, Johannesburg

Mr Francesco Pietra
University of the Witwatersrand, Johannesburg

Prof. Claudia Polese
University of the Witwatersrand, Johannesburg

Prof. Robert Reid
University of the Witwatersrand, Johannesburg

Dr David Reinecke
CSIR

Prof. Quinn Reynolds
Mintek

Dr Willem Roos
EX MENTE

Prof. Mohsen Sharifpur
University of the Witwatersrand, Johannesburg

Scientific Committee

Chair

Prof. Mohsen Sharifpur, University of the Witwatersrand, Johannesburg

Members

Prof. Thorsten Becker, University of Cape Town

Prof. Tunde Bello Ochende, University of Cape Town

Dr Alfred Bogaers, EX MENTE

Dr Natasha Botha, CSIR

Prof. Steeve Chung Kim Yuen, University of Cape Town

Prof. Corne Coetzee, Stellenbosch University

Dr Danie De Kock, QFINSOFT

Dr Andie De Villiers, Stellenbosch University

Prof. Dawood Desai, Tshwane University of Technology

Prof. Jat Du Toit, North West University

Dr Regan Dunne Tshwane, University of Technology

Prof. Thomas Franz, University of Cape Town

Prof. Igle Gledhill, University of the Witwatersrand

Dr Reuben Govender, University of Cape Town

Prof. Kamran Hassani, University of the Witwatersrand

Prof. Stephan Heyns, University of Pretoria

Prof. Weihua Ho, University of Cape Town

Dr Helen Inglis, University of Pretoria

Prof. Schalk Kok, University of Pretoria

Dr Jan-Hendrik Kruger, North West University

Dr Chioniso Kuchwa-Dube, University of the Witwatersrand

Prof. Philip Loveday, University of the Witwatersrand

Prof. Daniel Madyira, University of Johannesburg

Prof. Arnaud Malan, University of Cape Town

Prof. Josua Meyer, Stellenbosch University

Dr Dithoto Modungwa, CSIR

Prof. Velaphi Msomi, UNISA

Dr Tiyamike Ngonda, Cape Peninsular University of Technology

Prof. Harry Ngwangwa, UNISA

Prof. Graeme Oliver, Cape Peninsular University of Technology

Dr Aarti Panday, University of the Witwatersrand

Prof. Thanyani Pandelani, UNISA

Dr Randall Paton, University of the Witwatersrand

Dr Kevin Pereira, University of the Witwatersrand

Dr Mohammad Pourbehi, Cape Peninsula University of Technology

Prof. Daya Reddy, University of Cape Town

Prof. Robert Reid, University of the Witwatersrand

Dr David Reineke, CSIR

Prof. Quinn Reynolds, Mintek

Prof. Pieter Rousseau, Stellenbosch University

Prof. Sebastian Skatulla, University of Cape Town

Dr Bevan Smith, University of the Witwatersrand

Prof. Glen Snedden, University of Kwazulu-Natal

Dr Andre Storm, University of the Witwatersrand

Prof. Gerhard Venter, Stellenbosch University

Prof. Martin Venter, Stellenbosch University

Prof. Nico Wilke, University of Pretoria

Prof. Alphose Zingoni, University of Cape Town

Sponsors

We gratefully acknowledge the support of our sponsors.

Platinum Sponsor

QFINSOFT

Qfinsoft provides technical support services to Ansys, Ansys Rocky and CFTurbo users in Southern Africa, Ghana, Rwanda, Kenya and Ethiopia through a team of highly qualified and experienced engineers.

Gold Sponsor

AEROTHERM

Improving efficiency, reliability, and maintainability within a wide scope of industries.

Silver Sponsors

ULTIMATE PARTNER

TANDM

SIMTEQ

MPDI

Keynote Speakers

Prof. Schalk Kok, University of Pretoria, "Recent advances in surrogate modelling"

Abstract

Prof. Schalk Kok will present a personal perspective on surrogate modelling. He has worked on and off on surrogate modelling for almost three decades. His first exposure to surrogate modelling occurred during his masters studies in 1996, when he used polynomial surrogates to replace a transient thermoelastic finite element model. He next encountered surrogates in 2009, where he was involved with a mesh movement project. Radial basis functions were used to move the fluid mesh in a fluid-structure interaction (FSI) solver. More recently (2022-2024), Prof. Kok and Prof. Nico Wilke supervised doctoral student Johann Bouwer to develop near-optimal gradient enhanced surrogates. Specifically noteworthy is the development of a data pre-processing step that transforms the dataset using scaling and rotation. The intent is to transform the dataset to be more isotropic, making it more likely that a radial basis function surrogate (consisting of the summation of isotropic basis functions) will approximate the data accurately.

Bio

Prof. Schalk Kok is an experienced academic in the field of mechanical engineering, currently serving as Professor and Head of the Department of Mechanical and Aeronautical Engineering at the University of Pretoria. He is also currently appointed as the Chair of the School of Engineering, in the EBIT Faculty. Prof. Kok completed his B.Eng. and M.Eng. degrees at the University of Pretoria, followed by a Ph.D. at the University of Illinois at Urbana-Champaign, supported by notable scholarships including a Fulbright award.

Prof. Kok's professional journey has spanned academia and applied research. He worked at the University of Pretoria from 2003 to 2009, at the Council for Scientific and Industrial Research (CSIR) from 2009 to 2013, and he returned to the University of Pretoria in 2013. Since returning to UP, his research has concentrated on computational solid mechanics and material modeling, contributing to areas such as finite element analysis and material parameter identification.

His contributions have also extended to professional service, including leadership roles in the South African Association of Theoretical and Applied Mechanics (SAAM). He is a past president of SAAM, serving in the role for three consecutive 2-year terms (2010-2016).

Prof. Claudia Polese, University of the Witwatersrand, "Shaping South Africa's Future: Innovations in Advanced Manufacturing"

Bio

Claudia Polese is an Associate Professor in Aerospace Manufacturing and Design at the School of Mechanical, Industrial and Aeronautical Engineering at the University of the Witwatersrand, Johannesburg, South Africa, and she also serves as the Advisory Aerospace Research Director of the National Aerospace Centre. A National Research Foundation-rated researcher since 2011 and a Fellow of the Royal Aeronautical Society, Prof. Polese leads as Head of the Aeronautical Engineering Stream and Deputy Director of the ARUA Centre of Excellence in Materials, Energy and Nanotechnology. Prof. Polese has played a pivotal role in advancing South African research in cutting-edge aerospace technologies through collaborations with the Council for Scientific and Industrial Research, Airbus and various international institutions. In recognition of her contributions to fostering industrial and scientific partnerships between Italy and South Africa, she was honoured with the Knight of the Order of the Star of Italy and the NSTF-South32 Science Oscar for her work in advancing research, infrastructure and human capital in advanced manufacturing technologies.

Dr Alfred Bogaers, Ex-Mente & Prof. Quinn Reynolds, Mintek, "Life at 1500°C - a multiphysics modelling journey through pyrometallurgy"

Bio

Quinn Reynolds is 48 years old and holds degrees in Engineering and Applied Mathematics. He has worked in the Pyrometallurgy Division at Mintek for over 25 years. Mintek is a state-owned research institute serving the extensive mineral and metallurgical processing sector in South Africa. Quinn is also an extraordinary professor in the department of Chemical Engineering at Stellenbosch University.
His expertise includes mathematical and computational modelling of complex coupled phenomena in high temperature processes, and in particular the application of high-performance computing and open source software platforms to pyrometallurgy. His current areas of research include magnetohydrodynamic modelling of electric arcs, xiiiultiphysics fluid flow problems in furnace tapping and phase separation, combustion modelling for metallurgical processing, and reduced order modelling for digital twin applications.

Dr Mehdi Safari, Mintek, "Applications of Computational Fluid Dynamics for Mineral Processing"

Abstract

Computational Fluid Dynamics (CFD) is a vital tool in mineral processing, providing detailed insights into fluid dynamics and multiphase interactions essential for various operations. By solving the Navier-Stokes equations, CFD enables precise simulation and analysis of fluid behavior, enhancing equipment design, process optimization, and efficiency. A key application of CFD is in optimizing hydrocyclones, crucial for particle classification and separation. Simulations allow visualization of flow patterns and particle trajectories, facilitating design improvements that boost separation efficiency while reducing energy consumption. Similarly, CFD plays a pivotal role in flotation processes by modeling interactions between air bubbles and mineral particles, leading to better cell designs and operational parameters that improve mineral recovery. Furthermore, CFD is instrumental in optimizing mixing tanks and reactors, where uniform mixing is essential. It analyzes fluid flow and mixing patterns, guiding impeller design and improving efficiency. In slurry transport systems and thickeners, CFD models particle settling and slurry flow through pipelines, minimizing energy use and preventing blockages for more efficient transport. Beyond process optimization, CFD contributes to environmental management in mining. It models pollutant dispersion in air and water, aiding in the design of effective ventilation systems and effluent treatment processes, thus reducing environmental impacts and ensuring regulatory compliance. Applications in mine ventilation also enhance workplace safety. In summary, CFD fosters a deeper understanding of fluid flow phenomena, enabling the development of more efficient, sustainable, and environmentally friendly mineral processing technologies. Its role in improving productivity and minimizing environmental impact continues to grow, positioning CFD as a cornerstone of innovation in the mineral processing industry.

Bio

As an industrial research leader, Dr. Safari has over 17 years of experience applying cutting-edge research to practical industry solutions, specializing in novel mineral processing technologies. He is an NRF C1-rated researcher by the National Research Foundation. His expertise lies in scaling up laboratory to pilot-scale technologies and processes to achieve profitability. Mehdi has extensive knowledge in the classification, physical separation, and flotation of various minerals and has collaborated with leading mining houses such as Rio Tinto, Anglo American, BHP, Glencore, Vale, Fortescue, Sibanye-Stillwater, Impala Platinum, Assore, and Vedanta. Additionally, he has worked with renowned equipment suppliers including FLSmidth, Eriez, and Metso:Outotec on global projects involving REE, PGM, gold, copper, cobalt, nickel, lead, zinc, iron, barite, apatite, and phosphates. Throughout his career, Mehdi has consistently delivered technical, financial, and reputational success. Currently, he is responsible for providing development, engineering, and support services, focusing on enhancing productivity, managing risks, and delivering sustainable, safe, and efficient solutions to clients in the mining industry.

Fluid structure interaction in flapping wings: how important is flexibility in performance predictions?

Uma P. Krige[a], Kim Boon Lua[b,*], Wei Hua Ho[a,c]

[a]*University of the Witwatersrand, 1 Jan Smuts Ave, Braamfontein, Johannesburg, 2017, South Africa*

[b]*National Yang Ming Chiao Tung University, 1001, Daxue Rd, East District, Hsinchu City, 30010, Taiwan*

[c]*University of Cape Town, Rondebosch, Cape Town, 7700, South Africa*

**corresponding author email: krige.uma@gmail.com*

Abstract

Flapping propulsion in nature inherently incorporates flexibility which has inspired advancements in biomimetic propulsion systems. A deeper understanding is needed to optimise the design and performance of these systems.

Tests were conducted in a continuous water tunnel using a geometrically stepped flexible sample and a rigid sample both made of the same acrylic. They were subjected to different flapping parameters using a sinusoidal pitching motion.

It was found that for a low Strouhal (St) number (0.1) and a small amplitude (A) of 20°, the difference in performance between the two systems is negligible. For low St (St = 0.1) and higher A of 30°, the rigid system outperforms the flexible one in both lift and thrust. However, once the St aligns with what is optimal in nature (St = 0.2), the flexible systems outperform the rigid ones aligning with what is expected.

Keywords: Fluid-Structure Interaction, Flapping Wings, Wing Flexibility, Unsteady Flows, Flow-Visualization, Biomimicry

1. Background and Motivation

Flapping propulsion exists in birds, fish, insects and other animals. Inspired by nature's time-tested propulsion method, micro aerial vehicles (MAVs) and underwater autonomous vehicles (UAVs) are now starting to implement this harmonic oscillation into their designs. Flapping propulsion is less prone to the cavitation effects corroding traditional propellors in undersea propulsion [1]. Unsteady wings outperform stationary wings of the same geometry as a result of vortices being shed from the leading edge (LE) and trailing edge (TE) of the wing [2, 3].

In nature, most of these flapping-propelled creatures exhibit some form of flexibility on their propulsors, which begs the question: when taking inspiration from nature, should the component of flexibility be included?

The flexibility seen in nature may not be due to material limitations, but due to performance enhancements leading to an evolutionary advantage. Research has shown that flexible flapping wings tuned with the correct inputs can yield increased thrust, reduced drag and improved efficiency [4, 5, 6]. This therefore presents an opportunity for not only performance improvements in MAVs and UAVs, but also potential material reduction in their manufacture if flexibility is embraced. These improvements can only be realised through the full understanding of the effects of flexibility on the flow through the study of fluid-structure interactions.

1.1. Flapping wing motion explained

As a wing moves through an arc of pitching up, a high-pressure region develops on the underside of the wing and a low-pressure region above the wing this pressure difference generates a lift force. A stationary wing in this orientation would eventually stall but a harmonically pitching wing delays this stall through the induced flow tripping up the boundary layer on the upper half of the wing. Because swimming and flying typically happen in low Reynolds number (Re) regimes (1), instead of a wake fully developing, the disturbed boundary layer rolls up into a vortex, named the leading-edge vortex (LEV). The velocity difference of the flow on the upper and lower surfaces of the wing causes a shear layer to form into a vortex sheet at the TE, the trailing edge vortex (TEV). This sheet delays the mixing of the high- and low-pressure fluid on the lower and upper surfaces respectively. The LEV is associated with high drag but also high lift due to its enhancement of the pressure gradient.

$$Re = \frac{UL}{\nu} = \frac{\rho UL}{\mu} \tag{1}$$

Eventually the LEV can no longer remain attached to the shear layer, and it is shed off into the surrounding flow. This shedding happens at regular intervals described by the St (2).

$$St = \frac{f \cdot h}{U} = \frac{3k \cdot sin(A)}{2\pi} \quad where\ k = \frac{\pi \cdot f \cdot c}{U} \tag{2}$$

As the wing reaches its peak angle of attack (AoA), it reverses direction and eventually sheds another vortex that rotates in the opposite direction. This eventually forms a wake with counter-rotating vortices. Should the vortices align in such a way that a jet is created pushing fluid backwards, the wing will be pushed forward and thrust (negative drag) will be generated; this is known as a reverse von Karman vortex street (RvKVS). The opposite is true for a von Karman vortex street (VKVS) wherein the jet pushes fluid forward and hence results in a net-drag effect rather than thrust. The flapping frequency (f) and A determine when the LE vortices are shed and hence where they are positioned in the wake. The wake shape and width determine the overall performance of the system [7]. This overall performance is measured through the force coefficients of lift (C_L) and drag (C_D) normalized with dynamic pressure using (4) and (5):

$$C_L = \frac{2F_y}{\rho \cdot U^2 \cdot S} \tag{4}$$

$$C_D = \frac{2F_x}{\rho \cdot U^2 \cdot S} = -C_T \tag{5}$$

1.2. Flexibility in flapping wings

As the wing pitches upward in its motion, the TE moves in a downward arc, the fluid resists this by pushing upward on the TE. In a sufficiently soft wing, this bending force due to fluidic pressure causes an elastic deformation and a concave-up curvature on the wing. This shape reduces drag, especially in the case of wings with a heaving component allowing less power to be put into the actuation motion [8].

When a flexible wing is oscillated, as the wing is rotated through peak amplitude, the elastic energy stored in the bent wing is released allowing the TE to travel further than the actuated motion. This additional "flick" further widens the wake and can thus enhance the thrust [8].

2. Objectives and Scope

Flapping wings undergo harmonic motion, which in nature involves complex three-dimensional movements. Typically, the root of the wing remains nearly fixed, while the most pronounced motions—pitching and heaving—occur near the tip. This three-dimensional motion presents significant challenges in both experimental and computational studies. To simplify the analysis, the motion can be approximated as two-dimensional (2D), assuming spanwise uniformity with a combination of pure pitching and heaving. Since the heaving component effectively alters the AoA, the problem can be further simplified by isolating the pitching motion, wherein an idealised wing of infinite span rotates harmonically about a fixed point over time (t), as described by (6):

$$AoA\ (^\circ) = A \cdot \sin\left(2\pi \cdot f \cdot t\right) \tag{6}$$

The scope of this paper is limited to this such motion using a 2D wing. With this, the following cases were tested with the objective of comparing the stiff wing to the flexible equivalent:

Table 2. Cases and corresponding parameters

Case	U ($m.s^{-1}$)	A (°)	f (Hz)	Re	St
2	0.20	20	0.38	7970	0.1
4	0.20	30	0.26	7970	0.1
6	0.14	30	0.36	5580	0.2

3. Methodology

Two wings were manufactured out of the same material, acrylic, with the same 4mm LE thickness, however one steps down to 1mm thickness for the 60 mm of the 90 mm chord both with a span of 180 mm. These wings were inserted (cross section into the water)into a recirculating water tunnel that is able to achieve speeds (U) of up to 0.6 $m.s^{-1}$ without waves or free surface effects, which is well above the velocities used in this study. The model is oriented such that the wing root is at the top of the 0.45 x 0.5 x 2m test section, so that it can be actuated by MDrive 23 Plus MDM5FSD23C7 stepper motors above the water. The wing is connected to custom built force sensors.

The flow is seeded with 20µm spherical DANTEC PSP-20 glass particles with a similar enough density to the water to allow neutral buoyancy and a small enough diameter to not affect the flow properties. The 532 nm single laser sheet, emitted by a Microvec Pte Ltd SMSEMI-5W DPSS laser, is aligned parallel to the flow and perpendicular to the wingspan, and the seeded glass particles refract the laser light. The particle images are then captured using a high-speed Mikrotron-GmbH MotionBLITZ EoSens® camera. The camera has a resolution of 1280x1024 pixels and a maximum frame rate of 506FPS allowing it to capture the moving particles as dots rather than streaks to allow for post-processing using PIVlab on Matlab.

4. Key Results and Discussion

In Case 2, which features the smallest flapping A, the difference in performance between the stiff and flexible wings is minimal, with the stiff wing outperforming the flexible one in both lift and thrust, as illustrated in Figure 1a. This outcome suggests that a significant portion of the input energy is dissipated through elastic deformation, rather than being effectively recovered through the anticipated "flick" effect. This observation is further supported by the hysteresis curve in Figure 1b, which is the narrowest among all cases examined.

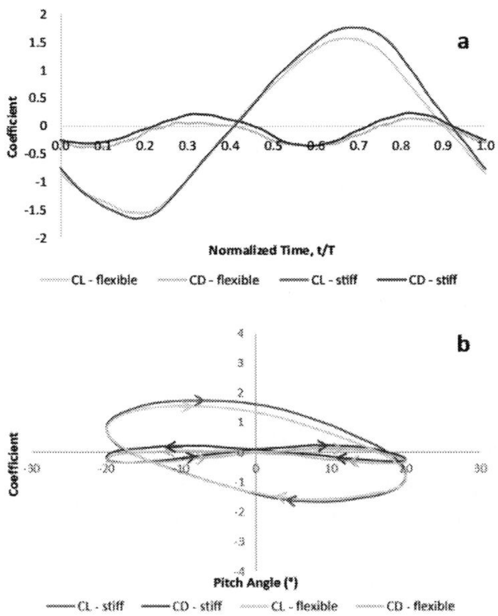

Fig. 1. Case 2 force coefficients (a) over time, (b) hysteresis curve

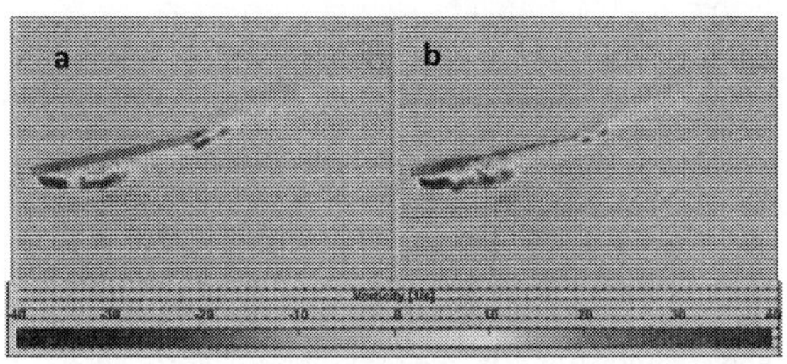

Fig. 2. Case 2 flow field at peak lift (t/T=0.7) for (a) rigid wing, (b) flexible wing

Based on these findings, it can be concluded that for low A and low St cases, the influence of wing flexibility is negligible and may not need to be considered in design decisions. This negligible effect is further corroborated by the particle image velocimetry (PIV) flow fields at t/T=0.7, as shown in Figure 2, where the flow characteristics of both wing types exhibit minimal differences.

Case 4 exhibits a significant difference in C_L and C_D characteristics between the stiff and flexible wings, as shown in Figure 3a. Notably, the flexible wing demonstrates a delay in both C_L and C_D peaks compared to the stiff wing. While the thrust peaks of the flexible wing are slightly lower than those of the stiff wing, the difference is not substantial, reinforcing the conclusion that at low St, stiff wings generally outperform flexible wings in thrust generation.

Despite these performance differences, the PIV flow fields at this stage, shown in Figure 4, do not reveal any major variations between the two wing types. However, the hysteresis curve in Figure 3b for the flexible wing is the widest among the cases studied, with a gentler slope at the lift peaks, forming a wider and less rotated oval shape. This indicates a modification in the stall characteristics, where the flexible wing experiences a more gradual loss of lift after reaching its peak. At the point of peak lift, the region of positive vorticity, as depicted in Figure 3a, is stronger and more sustained in the flexible wing compared to the stiff wing, leading to a slower dissipation of the vortex structures.

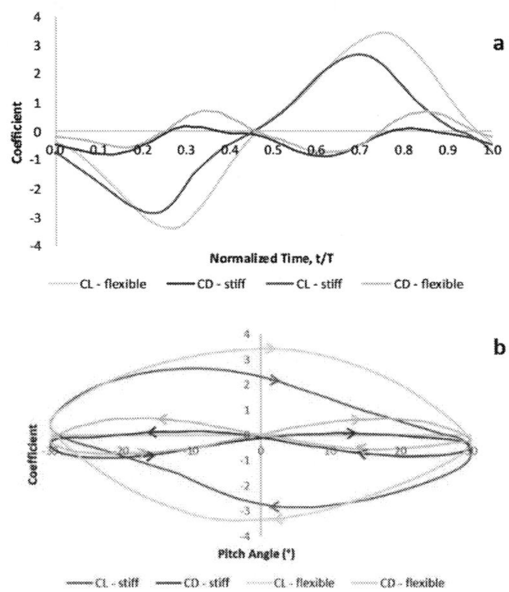

Fig. 3. Case 4 force coefficients (a) over time, (b) hysteresis curve

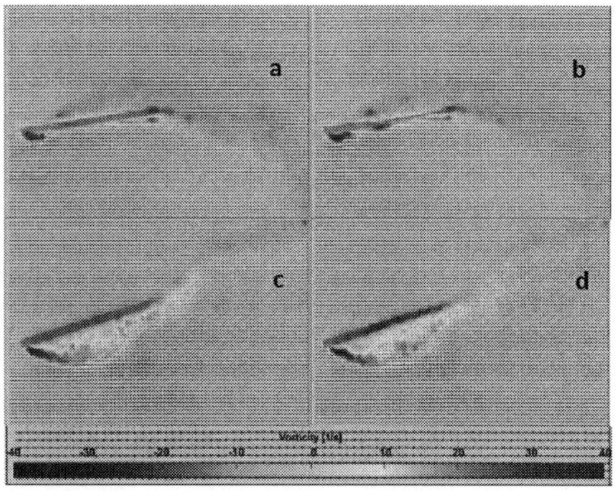

Fig. 4. Case 4 flow field at peak thrust (t/T=0.35) for (a) rigid wing, (b) flexible wing and peak lift (t/T=0.7) for (c) rigid wing and (d) flexible wing.

Case 6 is the only test conducted within the St range typical of natural fliers and swimmers (~0.2–0.4). In this case, the flexible wing begins to outperform the stiff counterpart in thrust generation. Consistent with the observations from Case 4, which had the same amplitude, Case 6 exhibits a wider hysteresis curve (Figure 5b) for both lift and drag, along with a noticeable delay in the lift and thrust peaks (Figure 5a).

The most pronounced difference between the stiff and flexible wings is observed in the PIV flow fields shown in Figure 6. The flexible wing sheds discrete, strong vortices at the TE, in contrast to the stiff wing, which produces a more diffused vortex sheet that dissipates more rapidly.

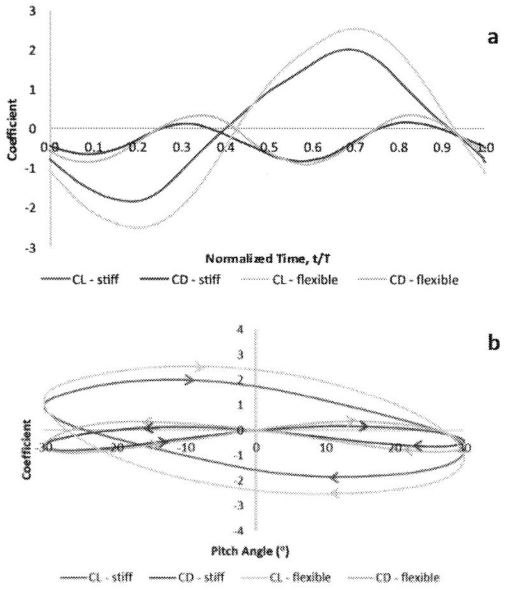

Fig. 5. Case 6 force coefficients (a) over time, (b) hysteresis curve

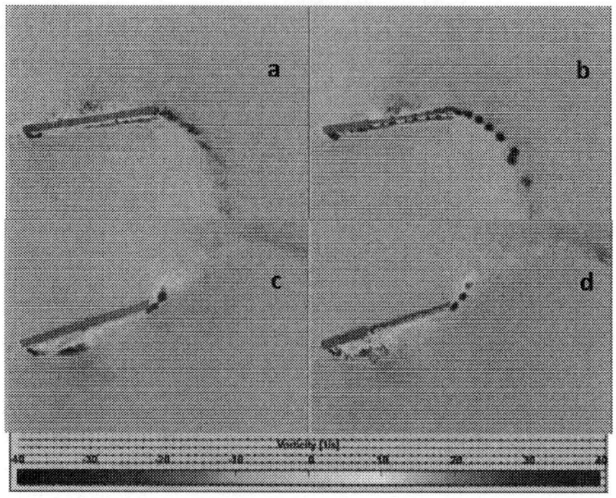

Fig. 9. Case 6 flow field at peak thrust (t/T=0.35) for (a) rigid wing, (b) flexible wing and peak lift (t/T=0.7) for (c) rigid wing and (d) flexible wing.

Although the work of Shinde and Arakeri is in the absence of a freestream with higher frequencies, they also investigate the effects of flexibility. The results in this study agree with the findings in that flexibility only improves performance within certain conditions. [9] The work of Ferreira de Sousa et al also agrees that Strouhal number controls wake topology and thrust. [4]

5. Conclusion

The results from Cases 2 and 4, which operate at St below 0.1, indicate that the flexible wing underperforms compared to its rigid counterpart in both lift and thrust. This suggests that flexibility may not be beneficial in such non-optimal conditions. However, practical applications are unlikely to operate in this low St range, reinforcing the conclusion that flexible wings should be designed to function within the optimal range of St = 0.2–0.4, rather than assuming rigidity as a superior design choice.

Case 6, conducted within the optimal St range, confirms that flexible flapping propulsors outperform their rigid counterparts, as expected. These findings highlight the importance of incorporating fluid-structure interaction (FSI) effects in performance predictions and design considerations for bio-inspired propulsion systems.

Additionally, the observation of discrete vortices shedding from the TE of the flexible wing in Case 6 has raised further interest in understanding the impact of the wing's stepped shape. Future work will focus on investigating this effect in greater detail through 2D LES computational fluid dynamics (CFD) simulations to gain deeper insights into the role of flexibility and wake dynamics.

Acknowledgements

The author would like to thank National Yang Ming Chiao Tung University in Taiwan for the opportunity to use their recirculating wind tunnel and their PIV setup. This was realised through the short-term exchange research programme at NYCU and the NRF South Africa-Taiwan Joint Science and Technology Research Collaboration Grant (TAI190212416022) and the Wits Postgraduate Scholarship.

6. References

[1] W. S. Vorus, "SHIP VIBRATIONS," in Encyclopedia of Vibration, Elsevier, 2001, pp. 1167-1173.

[2] T. J. Flint, M. C. Jermy, T. H. New and W. H. Ho, "Computational study of a pitching bio-inspired corrugated airfoil," International Journal of Heat and Fluid Flow, pp. 328-341, 2017.

[3] R. A. Norberg, "Hovering flight of the dragonfly Aeschna Juncea L. kinematics and aerodynamics," Swimming and Flying in Nature, p. 763–781, 1975.

[4] P. J. S. A. Ferreira de Sousa and J. J. Allen, "Thrust efficiency of harmonically oscillating flexible flat plates," Journal of Fluid Mechanics, vol. 674, pp. 43-66, 2011.

[5] M. Nicholas and J. Moore, "Analytical results on the role of flexibility in flapping propulsion," Journal of Fluid Mechanics, vol. 757, pp. 599-612, 2014.

[6] J. Pederzani and H. Haj-Hariri, "Numerical Analysis of Heaving Flexible Airfoils in a Viscous Flow," AIAA Journal, vol. 44, pp. 2773-2779, 2006.

[7] J. Young and J. C. S. Lai, "Oscillation Frequency and Amplitude Effects on the Wake of a Plunging Airfoil," AIAA Journal, vol. 42, no. 10, pp. 2042-2052, 2004.

[8] D. B. Quinn, G. V. Lauder and A. J. Smits, "Maximizing the efficiency of a flexible propulsor using experimental optimisation," Journal of Mechanics of Fluids, vol. 767, pp. 430-448, 2015.

[9] S. Y. Shinde and J. H. Arakeri, "Physics of unsteady thrust and flow generation by a flexible surface flapping the the absence of a free stream," in Royal Society Publishing, 2018.

Numerical simulations of induced flow asymmetry of truncated aerospike nozzles

Claude Hohls[a,*], Craig Law[a], Lara MacLucas[a]

[a]*School of Mechanical, Industrial and Aeronautical Engineering, University of the Witwatersrand, Johannesburg*

**2090258@students.wits.ac.za*

Abstract

Aerospike nozzles are a type of nozzle that allows for efficient thrust at all atmospheric conditions, from sea level to vacuum due to its altitude adaption properties [1]. This study investigates the effects of secondary flow injection at the point of truncation of an aerospike on the flow field and the line of action of the thrust vector produced by the nozzle.

To this end, steady-state Reynolds-averaged Navier-Stokes (RANS) simulations were performed to analyse the flow structures upstream and downstream of the injection point.

This secondary flow injection introduces complex interactions among the high-Mach-number exhaust stream, the injected secondary flow, and the low-speed recirculating flow at the nozzle base. Upstream of the injection point, a series of shocks form to divert the incoming flow away from the nozzle, creating a transverse momentum component in the exhaust. This coincides with an asymmetric pressure distribution that forms across the nozzle. Additionally, a shock-induced flow separation occurs upstream of the injection which has a strong influence on the pressure distribution along the nozzle. These asymmetries are then accompanied by the general deflection of the downstream nozzle exhaust.

Keywords: CFD, Aerospike, Nozzles, Secondary fluid injection

1. Background and Motivation

Aerospike nozzles offer efficiency improvements over conventional bell nozzles, but their adoption presents several challenges. One potential challenge is implementing thrust vector control (TVC) on aerospikes to enable dynamic vehicle control. The most common method of TVC in bell nozzles involves tilting the entire nozzle, deflecting the line of action of the thrust vector from the original nozzle axis. However, this approach is not always practical for aerospikes due to the wide base required by their unique shape [2]. As an alternative, fluidic thrust vectoring (FTV) can be employed. FTV operates by injecting secondary flow from within the aerospike nozzle wall into the main exhaust stream, generating a series of shocks that deflect the flow vector and produce a lateral force on the nozzle. The nozzle in Fig. 1.(a) was a prototype aerospike for the now cancelled X-33 program. The case analysed in this paper is shown in Fig. 1.(b).

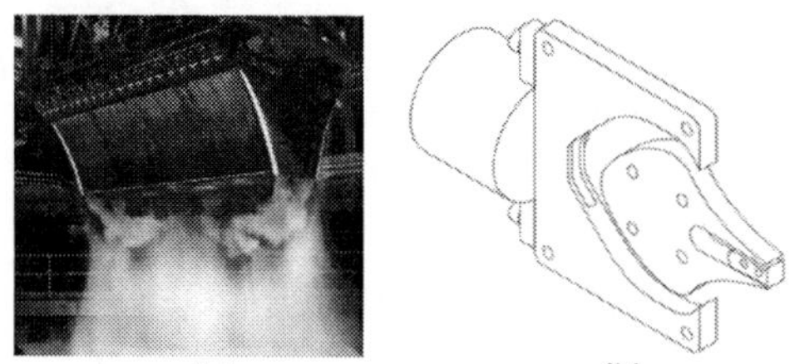

(a) (b)

Fig. 1. (a) The XRS-2200 truncated aerospike [3]; (b) Nozzle used in this work

Secondary injection along the spike has been the most recent subject of investigation [1], [4], [5], [6]. Studies, such as those by Eilers [7], have shown that there are two mechanisms that cause a side-force. The momentum flux of the injected flow and, secondly, the modified pressure distribution that is caused by the interaction between the primary and secondary flow. The first mechanism, momentum flux, is analogous to a second nozzle pointing sideways.

The pressure distribution that forms due to the injection can be seen in Fig. 2. The flow injection causes shocks to form upstream, causing slower flow and a corresponding higher static pressure along the wall. Conversely, downstream of the injection point, the flow accelerates, leading to lower pressure and inducing a net moment on the wall.

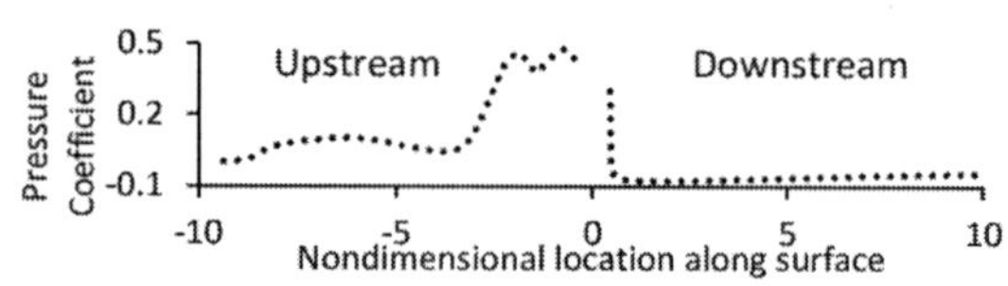

Fig. 2. Surface pressure distribution [8]. Injection is at 0.

The high-pressure upstream acts to increase the side-force produced which acts in the same direction as the mass flux vector while the low pressure subtracts from it. To enhance the side force produced by injection, this study investigates placing the injection point at the point of truncation, which has not been investigated previously. This should increase the net side force as it removes any surface with a horizontal component in which the downstream low pressure can influence the side force.

2. Objectives and Scope

This study aims to numerically model the flow field and use computational fluid dynamic simulations to allow for the identification of general flow structures that form when there is secondary flow injection at the point of truncation and how that influences the line of action of the thrust vector.

Structures such as shocks, expansion fans, and shear layers are of particular interest as they are the primary factors that influence the total deflection of the flow.

Key quantitative measurements are thrust and deflection angle (primarily influenced by flow structures such as shocks, expansion fans, and shear layers).

3. Methodology

Utilizing computational fluid dynamics (CFD) within Simcenter STAR-CCM+, this work investigates how secondary injection at the point of truncation of aerospike nozzles achieves flow deflection. Numerical results will be qualitatively validated against experimental schlieren images. As an initial step, 2-D steady-state Reynolds-Averaged Navier-Stokes (RANS) simulations were performed due to their relative simplicity and reliability in capturing general flow trends [9], despite potential inaccuracies particularly regarding the unsteady wake. However, these inaccuracies should be isolated to just the wake.

3.1. Setup and boundary conditions

A density-based coupled solver was used with the Spalart-Allmaras (S-A) turbulence model. S-A was selected as the RANS model as it has shown to have equal or better predictions of the pressure distribution along the wall when compared to other RANS models such as κ-ε and κ-ω for the specific case of secondary flow injection [9], [10]. The simulations used the ratio of specific heats for air of 1.4. Fig. 3 shows overall geometry used in the simulation.

The nozzle contour was created using a method that was developed by C.C. Lee [11] but adapted to a two dimensional nozzle. Using Lee's method, the shape of a nozzle with a throat area of 150 mm^2 was created such that a single expansion fan at the throat turns the velocity vector of the flow to align axially at a nozzle pressure ratio of 45.

For this work the nozzle has complete asymmetric flow injection through a slot with an area of 12% of the throat area and a stagnation pressure of 7 bar. This is the same as the stagnation pressure for the primary flow and the mass flow rate ratio between the two flows is 11%.

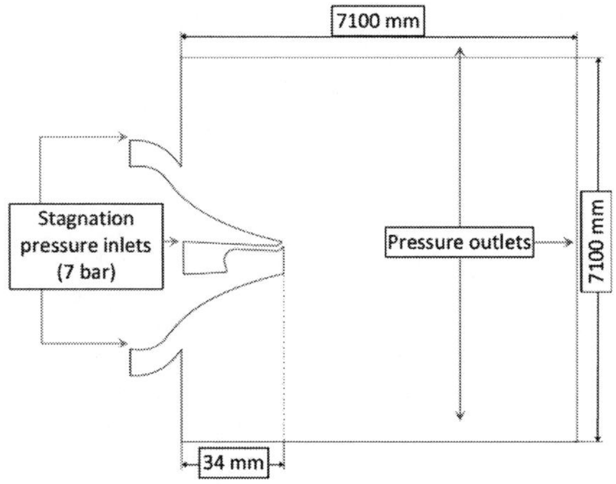

Fig. 3. Geometric layout and boundary conditions

3.2. Meshing

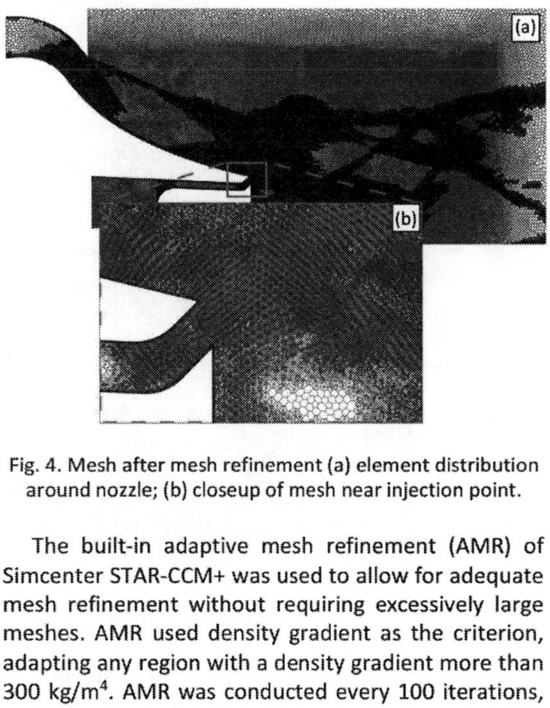

Fig. 4. Mesh after mesh refinement (a) element distribution around nozzle; (b) closeup of mesh near injection point.

The built-in adaptive mesh refinement (AMR) of Simcenter STAR-CCM+ was used to allow for adequate mesh refinement without requiring excessively large meshes. AMR used density gradient as the criterion, adapting any region with a density gradient more than 300 kg/m^4. AMR was conducted every 100 iterations, causing refinement in regions that contained shocks and boundaries of the jet flow such as the outer viscous layer of the jet and the wall boundary layer along the spike. Fig. 4. Shows an example of a mesh that has run to completion. Regions which had AMR correspond with regions of high mesh density, which is easily seen in Fig. 4.(a). Three levels of refinement were found to be adequate without causing significant slowdown caused by memory requirements of the AMR process.

3.3. Mesh Independence

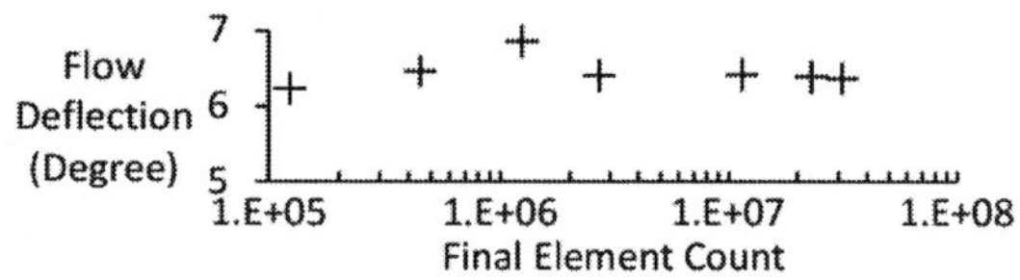

Fig. 5. Mesh independence study

The mesh independence study indicates that the simulations are mesh independent due to the convergence of the flow deflection to a value of 6.4°. There are still slight fluctuations, visible in Fig. 5, but they are within the expected accuracy of measurements taken during the planned experimental runs.

3.4. Post Processing

The primary outputs of the simulations are numerical schlieren and Mach contour plots. The plots allow for qualitative validation and identification of pertinent flow features such as shock waves and expansion fans.

4. Discussion

The general flow field with asymmetric injection can be seen in Fig. 6. It has 7 bar stagnation for both primary and secondary inlets.

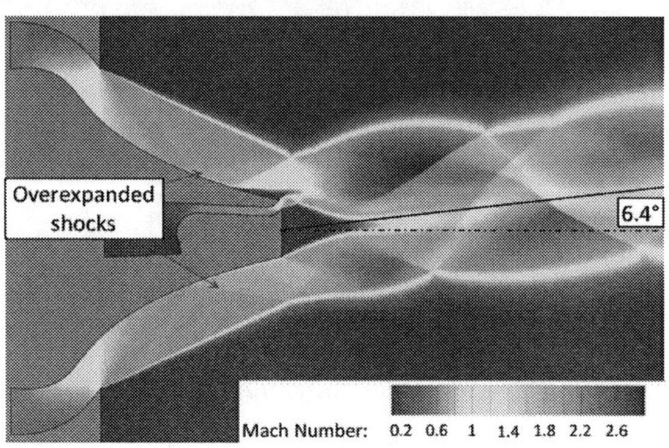

Fig. 6. Mach number plot of the flow field

The total deflection of the flow in this case is 6.4°. This deflection is caused by the top flow interacting with the injected flow while the bottom flow expands unobstructed around the truncation corner. An overexpanding shock is formed symmetrically on both the top and bottom flow due to the nozzle operating in the overexpanded state.

Downstream of the nozzle, shock waves separate regions of relatively lower and higher pressure. These shocks are labelled as trailing shocks in Fig. 7, also labelled are expansions fans. These expansion fans occur at the corner of the throat and in the design condition (NPR = 45) would turn the flow to be completely axial but because the nozzle is overexpanded, the angle in which the flow turns is smaller, thereby causing the overexpanded shock patterns.

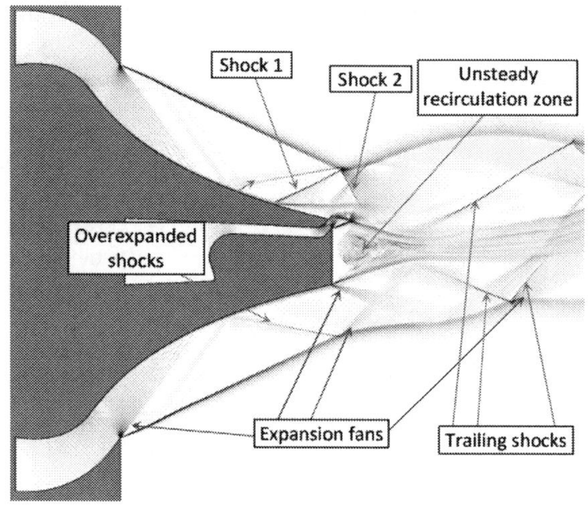

Fig. 7. Numerical schlieren of aerospike

Further expansion fans exist at the point of truncation for both sides of the nozzle. The expansion fan at the point of injection is of similar structure to the one that forms at the throat of the nozzle. Expansion fans also exist wherever trailing shocks meet the shear layers at the boundary of the jet.

The recirculation zone, evident in Fig. 7, acts as a primary source of flow fluctuations, vortex shedding, and overall flow instability. Due to its inherently unsteady nature, capturing the full dynamics of this region necessitates transient simulations, as steady-state simulations are insufficient to capture all the intricate details. Examples of unsteady regions would be the shear layer between both the jet and wake and the jet and ambient air which would cause vortices through various mechanisms such as the Kelvin-Helmholtz instability.

Fig. 8 shows a numerical schlieren closeup at the point of secondary injection. The very strong shock upstream of the injection point causes flow separation and a general turning of the primary flow of 34°, creating fully axial flow until it reaches shock 2 which turns it further. Once the flow passes through shock 2, it passes through the expansion fan that originates from where shock 2 meets with the upper shear layer. Shock 2 has slight curving near the middle, this is probably caused by complex interactions between shock 2, the trailing shock downstream of it and the expansion fan that is also downstream.

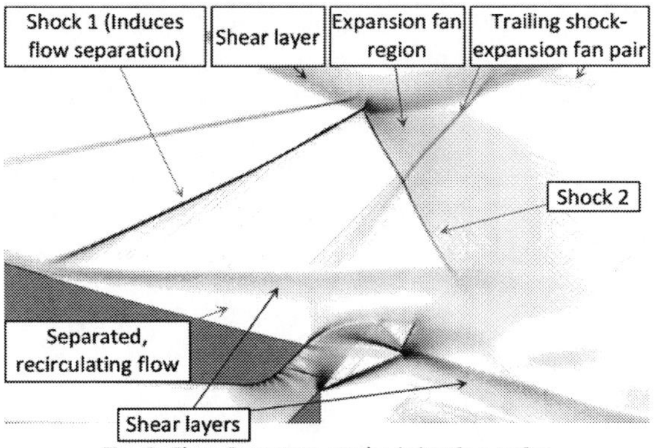

Fig. 8. Flow Structure at the injection point

The double shock structure consists of the two shocks meeting in a shear layer, turning the flow towards that shear layer. Fig. 9(a) shows the structure and the general trend of the flow. Fig. 9(b) shows a mirrored structure that forms in the expanding region of the injected flow.

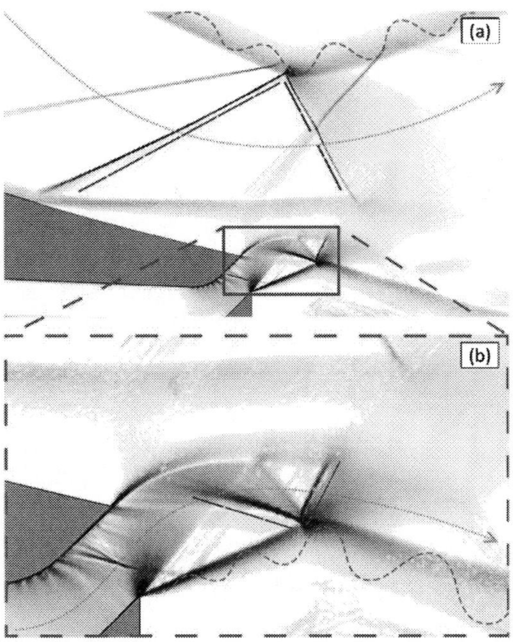

Fig. 9. Comparison of the turning of the (a) main flow and; (b) the secondary flow

Fig. 10. shows the pressure distribution along both sides of the aerospike. A strong positive gradient in this plot indicates the position of a shock along the wall. There are two distinct shocks, namely, the overexpanded shock, which is symmetrical on both sides of the spike, and shock 1 which is the shock causing flow separation along the top wall of the spike.

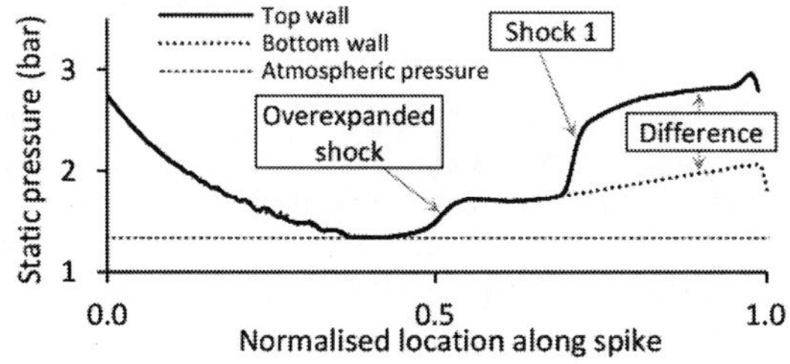

Fig. 10. Pressure distribution along spike wall

The integral of the difference in pressure between the two walls is another quantity to judge the effectiveness of the injection. In this case the net side-force is 15 N and the specific impulse (I_{sp}) of the side-force is 60 seconds.

This is a preliminary study with a focus on the qualitative measures of the flow field. To know the true effectiveness of the fluid injection scheme a full comparison to alternative thrust vector control system, need to be done. The source of fluid for the injection is of great importance. If it comes from gas cycle exhaust it will most likely increase overall thrust however in other engine cycles, it may decrease overall thrust and then would become be a trade-off between thrust and controllability.

5. Conclusion

Secondary flow injection allows for a potential method of control when applied to truncated aerospikes. Steady-state RANS simulations were conducted to analyse the effect secondary flow injection, at the point of truncation, has on the general flow field of the nozzle. The general flow field prediction is one where there is a single strong shock leading into a region of shock-induced flow separation causing a moderate amount of flow diversion, subsequently followed by a weaker secondary shock that further turns the flow.

By injecting the flow at the point of truncation, several degrees of exhaust flow deflection can be achieved through the control of shock-induced flow separation and the series of shocks that form downstream of the point of separation. Such results would be important in the design of a system that aims to incorporate thrust vectoring into a design without adding many moving parts or by degrading thrust performance by an unacceptable amount.

Experimental cold-flow tests need to be conducted in the coming months to provide both qualitative and potentially quantitative validation for the simulation results. These future experiments are required to allow for the evaluation of the reliability of the RANS simulations, particularly in accurately predicting the shock-induced flow separation.

Acknowledgements

The authors acknowledge the Centre for High Performance Computing (CHPC), South Africa, for providing computational resources to this research project. The 'Aerothermodynamics with vortices, vorticity and shear' group (MECH1532) was used for this research.

References

[1] G. Hagemann, H. Immich, and M. Terhardt, “Flow phenomena in advanced rocket nozzles - The plug nozzle,” in *34th AIAA/ASME/SAE/ASEE Joint Propulsion Conference and Exhibit*, Cleveland,OH,U.S.A.: American Institute of Aeronautics and Astronautics, Jul. 1998. doi: 10.2514/6.1998-3522.

[2] M. Onofri *et al.*, “Plug nozzles: summary of flow features and engine performance - Overview of RTO/AVT WG 10 subgroup 1,” in *40th AIAA Aerospace Sciences Meeting & Exhibit*, Reno,NV,U.S.A.: American Institute of Aeronautics and Astronautics, Jan. 2002. doi: 10.2514/6.2002-584.

[3] NASA/MarshallSpaceFlightCenter, “XRS-2200,” Testing of Twin Linear Aerospike XRS-2200 Engine. Accessed: Jan. 06, 2025. [Online]. Available: https://archive.org/details/MSFC-0103149

[4] H. Greer, “Rapid Method for Plug Nozzle Design,” *ARS J.*, vol. 31, no. 4, pp. 542–568, Apr. 1961, doi: 10.2514/8.5551.

[5] T. Ito and K. Fujii, “Numerical analysis of the base bleed effect on the aerospike nozzles”.

[6] P. P. Nair, A. Suryan, and H. D. Kim, “Study of Conical Aerospike Nozzles with Base-Bleed and Freestream Effects,” *JSR*, vol. 56, no. 4, pp. 990–1005, Jul. 2019, doi: 10.2514/1.A34256.

[7] S. D. Eilers, M. D. Wilson, S. A. Whitmore, and Z. W. Peterson, “Side-Force Amplification on an Aerodynamically Thrust-Vectored Aerospike Nozzle,” *J. Propuls. Power*, vol. 28, no. 4, pp. 811–819, Jul. 2012, doi: 10.2514/1.B34381.

[8] V. Viti, R. Neel, and J. A. Schetz, “Detailed flow physics of the supersonic jet interaction flow field,” *Phys. Fluids*, vol. 21, no. 4, p. 046101, Apr. 2009, doi: 10.1063/1.3112736.

[9] M. Propst *et al.*, “ACTIVE – Optimisation of a fluidic thrust vector control on aerospike nozzles,” 2021. [Online]. Available: https://www.researchgate.net/publication/359092036_ACTIVE_-_Optimisation_of_a_Fluidic_Thrust_Vector_Control_on_Aerospike_Nozzles

[10] M. Propst, V. Liebmann, J. Sieder-Katzmann, C. Bach, and M. Tajmar, “Maximizing side force generation in aerospike nozzles for attitude and trajectory control,” Oct. 2018. [Online]. Available: https://www.researchgate.net/publication/328145959_Maximizing_side_force_generation_in_aerospike_nozzles_for_attitude_and_trajectory_control

[11] C.C. Lee, “Fortran programs for plug nozzle design,” 1963, Accessed: Jun. 01, 2025. [Online]. Available: https://archive.org/details/nasa_techdoc_19630012259

Thermal effects on accelerating bodies

T. Makomva[a], H. Roohani[a], I.M.A Gledhill[a]

[a]*University of Witwatersrand, School of Mechanical, Industrial and Aeronautical Engineering, Johannesburg, South Africa*

1037222@students.wits.ac.za, Hamed.Roohani@wits.ac.za, igle.gledhill@wits.ac.za

Abstract

This study investigates the thermal effects in the region ahead of accelerating and decelerating blunt bodies, focusing on the area between the blunt body, detached shock wave, and sonic lines. In contrast to steady flight, aerodynamic equilibrium is not reached during acceleration or deceleration. This research investigates objects with adiabatic wall boundaries in the transonic and supersonic regimes. Numerical simulations were conducted using ANSYS Fluent, with numerical validation performed against STAR-CCM+. For experimental validation, numerical results for constant velocity flight were compared against previously published experimental data. The differences in stagnation temperature during constant velocity, acceleration, and deceleration were computed. Increased acceleration increases the temperature in the range of interest for flight Mach numbers M less than 1, and decreases it for M > 1. Deceleration reverses the temperature effect. In the region of interest, close to the upstream stagnation point, temperature is predominately influenced by adiabatic compression, while flow history plays a dominant role in bow shock position, and in transonic and supersonic regions. The study also examines the impact of varying levels of acceleration, revealing that both the stagnation temperature and pressure increase with higher acceleration values, and the stand-off distance changes accordingly.

Keywords: Blunt bodies, stagnation point, thermal effects, acceleration and deceleration, subsonic and supersonic flows, numerical simulation

1. Introduction

Much research has been carried out in the field of thermal effects in compressible flow focuses on objects moving at constant velocity in air. The study needs to be expanded to include bodies in acceleration because flow behavior and aerodynamic loads at the same Mach number differ significantly for accelerating, decelerating, and constant velocity motion. Aerodynamic equilibrium is achieved in steady flight, whereas equilibrium is not achieved at any Mach number during acceleration or deceleration [1].

The aim of this study is to investigate the thermal effects of air in the region in front of accelerating and decelerating bluff bodies. This study focuses on objects that operate predominantly in the transonic and supersonic regimes, including fighter jets, missiles, and re-entry vehicles, among others. The study of thermal effects is critical in designing missiles and fighter jets operating at high speeds. Detached shock waves form ahead of these vehicles, creating intense pressure and temperature variations. The thermal effects directly influence material selection and internal component layout for high-speed systems. The temperatures produced by acceleration are insufficiently characterized for accelerating bodies. Understanding these effects is essential to prevent structural failure or interference with infrared imaging in extreme flight conditions. Ongoing research has resulted in the development of missiles capable of reaching linear accelerations of up to 400g (where g = 9.81 m/s^2).

Lilley et al. [2] conducted some of the earliest significant research on the geometry of shock behaviour in acceleration. Roohani et al. [1] have performed notable aerodynamics analyses of accelerating bodies using Computational Fluid Dynamics (CFD).

In this study, CFD was used. Experiments cannot be performed in wind tunnels, because acceleration of air in the tunnel relies on pressure gradients that do not accurately represent the in-flight behavior of accelerating bodies.

2. Methodology

Two computational fluid dynamics (CFD) techniques were implemented: a non-inertial relative-frame method, adopted as the primary approach, and an inertial moving-reference-frame (MRF) method, used as a validation benchmark. In the relative (body-fixed) frame, a coordinate system is attached to the body while the surrounding flow field is accelerated to the desired Mach number [1]. Transforming the Navier–Stokes equations into this non-inertial frame introduces explicit acceleration terms, causing all fluid elements to accelerate simultaneously; the far-field boundaries are also time-dependent in both pressure and velocity. In contrast, the MRF technique involving the translation of the body and mesh through quiescent air, with computations carried out in the inertial frame of the stationary fluid [3]. Here, the governing equations remain unaltered, and the distant fluid is nominally at rest. Both methods employed time-dependent solvers, density-based boundary conditions, and the Spalart–Allmaras turbulence model, selected for its computational efficiency and proven performance in external aerodynamics [1].

Preliminary comparisons showed that the two techniques produced stagnation temperature differences of less than 1%, and therefore the relative-frame formulation was adopted for the main study. Flows over a 150 mm two-dimensional cylinder, a 30 mm axisymmetric sphere, and a NACA 0012 aerofoil were simulated in Fluent under constant-velocity, accelerating, and decelerating regimes. Accelerations of ±106g, ±212g, and ±318g were applied while the Mach number increased from 0.1 to 1.6. A time-step sensitivity study yielded a final physical step of 1×10^{-5} s; second-order MUSCL spatial discretisation was applied, and the Courant number was maintained near unity to preserve temporal accuracy. At the far field, a pressure of 101 325 Pa and a temperature of 300 K were prescribed, while all solid walls were treated as adiabatic.

Verification involved simulating accelerating air in an empty domain, yielding a residual pressure gradient of less than 1 Pa, and cross-checking inertial vs non-inertial frames in Fluent. Validation comprised code-to-code agreement between Fluent's inertial frame and Star-CCM+'s non-inertial counterpart, plus comparison with constant-velocity experiments, Abbott's aerofoil data [4] and Achenbach's cylinder measurements [5] and with Saito et al.'s decelerating-sphere results [6].

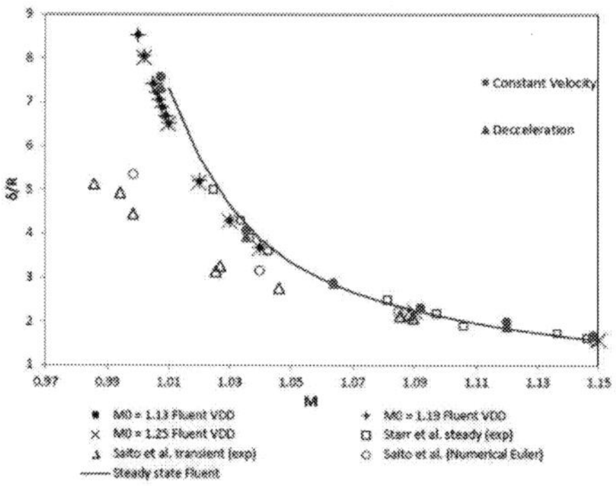

Figure 1: Flow over a sphere from literature (black), constant velocity and deceleration cases from Fluent (blue) [6]

Fluent results and results from the literature in Figure 1 follow the same trend. Fluent results match the experimental results reasonably well. Mahomed et al. [7] validated their numerical models on the study of flow over an accelerating hemisphere-flare-cylinder configuration using ballistic range experiments. No experimental data are known to exist for thermal effects on transonic accelerating bodies, and experimental validation of those effects lies beyond the present scope.

3. Results and Discussion

The stagnation point, defined as the location where the streamline normal to the body is brought to rest relative to the body surface, was the primary focus of this study. This region is of particular interest because it experiences the highest thermal and pressure loads during high-speed flow as shown in Figure 2.

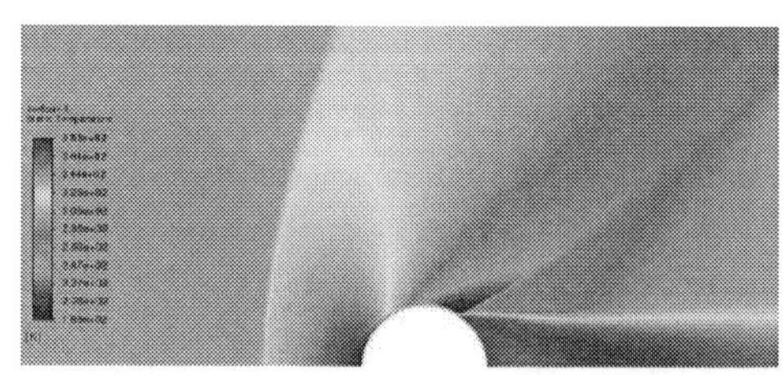

Figure 2: Static temperature flood plot around an axisymmetric sphere at Mach 1.2 during acceleration of 212g

The variation in static temperature ahead of the objects was analyzed for different flow conditions. As illustrated in Figure 3, at a given Mach number, the bow shock is positioned closer to the axisymmetric sphere during deceleration compared to the acceleration and constant velocity cases. This behavior aligns with findings by Gledhill et al., who reported comparable differences in stand-off distances during their study of stagnation pressure on an accelerating airfoil [8]. Pressure profiles show similar trends.

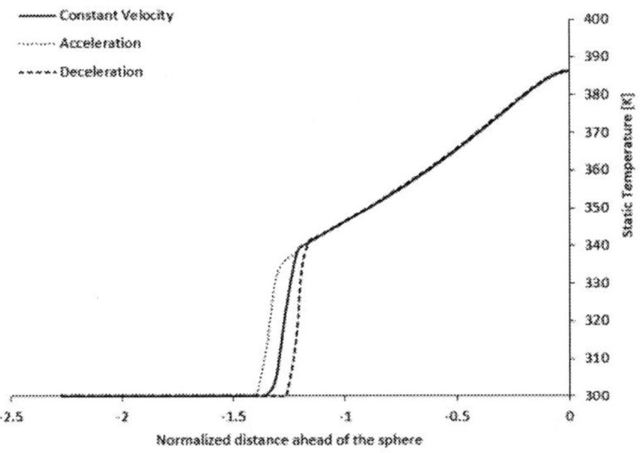

Figure 3: Constant Velocity, Acceleration, and Deceleration comparison for the axisymmetric sphere at Mach 1.2 for an acceleration of 106*g*

The temperature at the stagnation point on a 2D cylinder was analysed in Fluent for cases involving constant velocity, acceleration, and deceleration. The total temperature was obtained from Fluent (Tf), while the theoretical stagnation temperature (Tt) was calculated using the Rankine-Hugoniot equations.

Figure 4 illustrates that constant-velocity results closely match the theoretical values across the Mach number range. However, in the cases of acceleration and deceleration, there is a small deviation in the subsonic region, with the difference increasing significantly in the transonic and supersonic regions.

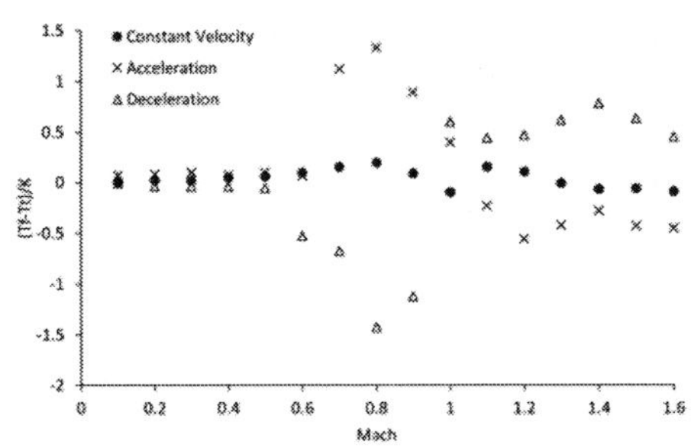

Figure 4: Stagnation temperature difference of Fluent and theory against Mach number for an acceleration of 106*g*

An analysis of the variation in the difference between the stagnation temperatures from Fluent and the theoretical stagnation temperature as the Mach number changes at different acceleration values was conducted. As shown in Figure 5, at the same Mach number, the absolute difference in stagnation temperature increases with higher acceleration values.

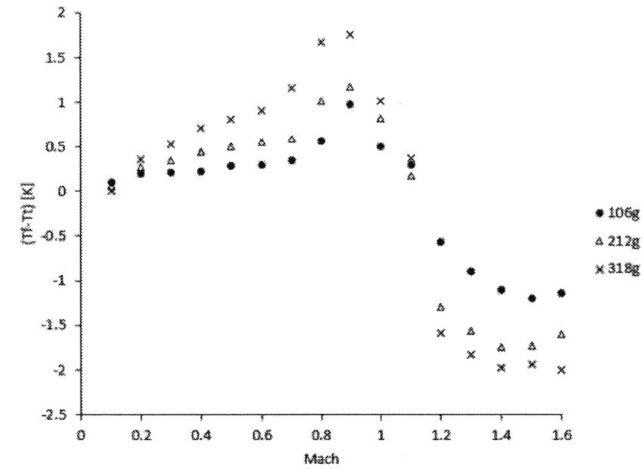

Figure 5: Stagnation temperature against Mach number for different positive acceleration values

A comparison of the stagnation temperature on the different types of blunt objects was conducted as shown in

Figure 6. There are qualitatively similar effects on all bodies, with the exception in the range of M > 1.2. It is observed that the difference in stagnation temperatures is in the range of ± 2 K for the acceleration cases.

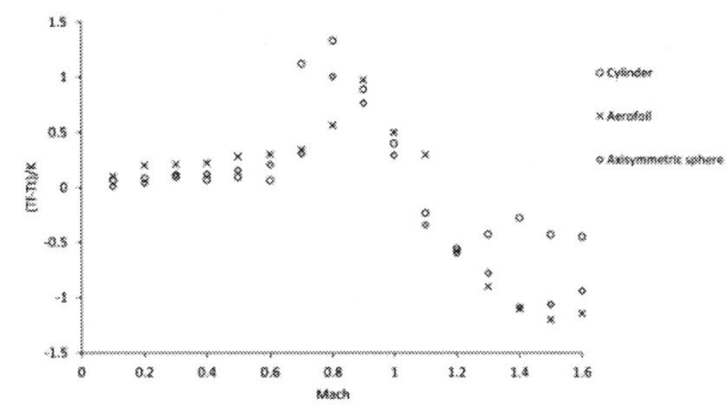

Figure 6: The stagnation temperature difference between constant velocity theoretical values and Fluent cases at an acceleration of +106*g*

The study explored the thermal effects of blunt bodies moving through the air from subsonic to supersonic speeds, with a specific focus on stagnation properties where the highest temperatures occur.

Constant velocity results closely aligned with theoretical predictions across the entire Mach number range of 0.1 to 1.6. In subsonic conditions, increased acceleration resulted in higher stagnation values, while deceleration produced lower values; this trend reversed in the supersonic regime. These variations were attributed to compression, inertial, and flow history effects. Inertial effects occur as the surrounding air is compressed by rapid changes in the body's motion. Flow history describes delays in the adjustment of the flow field when a body accelerates through a compressible medium, the unsteady flow field that develops does not have sufficient time to establish equilibrium at any given Mach number. This results in transient flow fields being influenced by their past characteristics [9].

As the body moves, the air ahead is compressed. The compression increases pressure and temperature as the work done on the air elevates its internal energy, consistent with the conservation of energy. The static temperature elevation for $M < 1$ is explained by this effect.

Since the bow shock standoff distance decreases with increasing Mach number [6] in all cases (constant velocity, acceleration and deceleration), flow history effects, which preserve earlier flow field properties, explain the bow shock stand-off distance being larger under acceleration and smaller under deceleration.

In subsonic regions, the observed differences in flow properties were primarily influenced by compression effects, whereas in supersonic regions, flow history effects became the dominant factor.

4. Conclusion

This study investigated thermal effects on blunt bodies: a two-dimensional cylinder, an axisymmetric sphere, and a NACA 0012 aerofoil, during acceleration and deceleration in air, extending prior research on constant-velocity flows. Results showed that in the range of 160*g* to 318*g*, stagnation temperatures differ from constant velocity flight significantly under unsteady motion. In the subsonic regime (Mach 0.1–1.0), stagnation temperature increased during acceleration and decreased during deceleration due to compression and inertial effects. In the supersonic regime, flow history effects reversed this trend, with higher subsonic stagnation values during deceleration. Deviations from steady-flow predictions grew with increasing acceleration levels (±106*g* to ±318*g*), and stand-off distances expanded under acceleration but contracted during deceleration. Above Mach 1.2, the axisymmetric sphere displayed distinct trends attributed to wake and separation-line variations. These insights are directly relevant to the design of thermal protection systems and material selection for aerospace vehicles operating under rapid transients, helping engineers anticipate localized heating during maneuvers or re-entry. Future studies should include experimental validation, consider walls with finite thermal conductivity, and investigate advanced turbulence models to better resolve separation under high accelerations.

References

[1] H. Roohani et al., *Shock Waves*, vol. 30, pp. 115-129, 2020.

[2] G. Lilley et al., *Journal of the Royal Aeronautical Society College of Aeronautics*, vol. 71, pp. 396-414, 1953.

[3] I. Mahomed et al., in the *23rd International Shock Interaction Symposium*, Kruger National Park, South Africa, 2018.

[4] I. H. Abbott and A. E. von Doenhoff, General Publishing Company, Ontario, 1959.

[5] E. Achenbach, *Journal of Fluid Mechanics*, vol. 34, p. 625–639, 1968.

[6] T. Saito et al., *Shock waves*, vol. 21, pp. 483-489, 2011.

[7] I. Mahomed et al., *International Journal of Aeronautical and Space Sciences*, vol. 22, no. 6, pp. 1293-1301, 2021

[8] I. Gledhill et al., in the *23rd International Shock Interaction Symposium*, Kruger National Park, South Africa, 2018.

[9] P. Stathoulis et al., in the *23rd International Shock Interaction Symposium*, Kruger National Park, South Africa, 2018.

Aerodynamic loading for transonic flight with acceleration

Mahomed I.[a,*], Roohani H.[a], Skews B.W.[a], Gledhill I.M.A.[a]

[a] *University of the Witwatersrand, School of Mechanical, Industrial and Aeronautical Engineering*

1 Jan Smuts Avenue, Johannesburg, South Africa

**irshaad.mahomed@wits.ac.za*

Abstract

Flight performance calculations are usually performed with a constant velocity approximation, but the effects of acceleration, which are still being explored, can be significant. Acceleration effects for a cone-cylinder at 10° angle of attack are investigated in the transonic Mach regime using ANSYS Fluent (V19.0 series) with the Moving Reference Frame acceleration technique. Axial force plots show acceleration effects at certain flight Mach numbers within the range of interest 0.80<M(t)<1.20 at 400g. The transonic drag rise is reduced by approximately 10% for the acceleration case, primarily due to the lag in the development of the shoulder shock system and the bow shock. The transonic drag is increased by approximately 15% for the deceleration case, also due to the flow history effect, but overtaking wave propagation leads to rapid variations in axial force near Mach number 0.8.

***Keywords*: Transonic flow, Shock waves, Acceleration aerodynamics, Computational Fluid Dynamics**

1. Introduction

This work is part of a programme to establish the fundamental flow physics of the relatively new field of flight with significant acceleration and deceleration. Studies of acceleration aerodynamics have identified that in the transonic Mach number range the compressible flow field exhibits considerable differences of the shock-expansion wave structures compared to constant velocity flight [1,2]. This fundamental investigation considers transonic aerodynamic forces for a cone-cylinder at 10° angle of attack, and at constant 400g magnitude acceleration (where g is the gravity constant = 9.81 $m.s^{-2}$). The body vortices developed at this angle of attack are characterized as a symmetric and steady vortex pair based on the observation of [3] and it can be assumed that flow a three-dimensional domain with a symmetry plane is a reasonable model. The results were obtained using a validated acceleration technique in the inertial frame in ANSYS Fluent (V.19 series) with turbulence modeling. The flow domain was a half 3D-model with a symmetry plane.

1.1. Objectives and Scope

This fundamental investigation contrasts the aerodynamic force between three motion types, acceleration, deceleration and constant velocity for a cone-cylinder at 10° angle of attack. Wave propagation and flow history are shown to be dominant acceleration effects, in the transonic Mach number range, which modifies the body loading. This causes the aerodynamic forces to differ between the three types of motion.

The objectives are to identify dominant acceleration effects and to contrast significant differences in the flow field between constant velocity and acceleration flight. This paper, although part of broader investigation (which considered angles of attack 5°,10°,15°), will describe a subset of these results and will focus on the axial force component. The data selected for this paper was to illustrate the shock wave propagation and flow history effect caused by acceleration.

Fig. 1 illustrates the orientation of the cone-cylinder with respect to the free-stream velocity components, and the axial orientation. Symbols in this figure have the following meaning, U_x axial velocity, U_y normal velocity, U_∞ velocity magnitude, L_N nose length, L_c cylinder length, D base diameter, Θ cone half-angle, and α angle of attack.

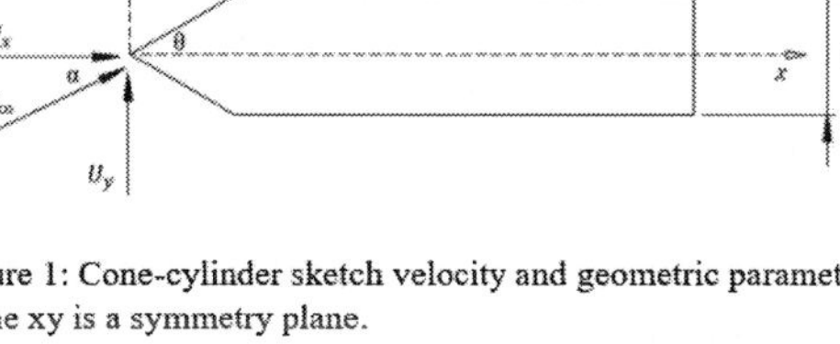

Figure 1: Cone-cylinder sketch velocity and geometric parameters. Plane xy is a symmetry plane.

The cone-cylinder (with a slender ratio of ~10, D=25, L_N=70.6mm, L_c=200, Θ=10°), accelerates from an initial flight Mach number 0.80 until 1.20 approximately and decelerates from an initial flight Mach number 1.10 until 0.80 approximately. An acceleration magnitude of 400g constant was selected to observe the acceleration effects. The flight Reynolds number based on the cylinder diameter varied from $1x10^6$ to $1.4x10^6$. Simulations were performed with ANSYS Fluent (V.19.0 series) and implementation of the Moving Reference Frame (MRF) acceleration technique with a User Defined Function [4]. This technique solves the Navier-Stokes equations in an inertial frame with fluid velocities and accelerations transformed to the non-inertial frame for analysis, the thermodynamic properties are frame-invariant, and background theory of the Navier-Stokes transformation between non-inertial and inertial reference frames is covered in [4,5,6,7].

Results of the grid and time-step convergent studies are shown in Figs. 2 and 3, noting that grid adaptation was used. Menter's k-ω SST turbulence model with curvature correction was selected to resolve the boundary layer and viscous flows. The solver is URANS (Unsteady Reynolds Averaged Navier-Stokes), second order and implicit.

The flux scheme for spatial discretization is Roe Flux Difference Splitting scheme. Temporal discretization is implicit dual-time-stepping. The flux schemes selected are for transonic dominant flow and to capture wave propagation. The grid was adapted per time-step on parameters pressure, density, relative velocity magnitude, and vorticity magnitude.

The validation results with constant velocity and deceleration cases are reported in [7], which are based on wind-tunnel experiments by Hsieh (1975) [8] for transonic flight, and Hsieh (1977) [9] for the supersonic Mach regime; the free-flight segment from a ballistic range test by Yamashita et al. (2018) [10] was repeated numerically using the MRF acceleration technique in ANSYS Fluent. The ballistic range validation result is shown in Fig. 2 by an overlay plot of schlieren and density contours.

The unstructured grid had pressure-far-field as the boundary condition and a symmetry plane as a half-cone-cylinder model was simulated. This is three-dimensional flow domain with one symmetrical plane. The body boundary is a no-slip wall. Prism cells were used to capture boundary layer flow with y^+ approximately less than 0.6. The unadapted grid cell-count was $30x10^6$ cells. For the constant velocity cases the cell-count was ~$40x10^6$ cells, ~$100x10^6$ cells for the acceleration case, and ~$200x10^6$ cells for the deceleration case.

The reason for the increased cell-count during acceleration and deceleration is caused by wave propagation and the persistence of flow structures formed earlier in time that will convect through the flow field, reducing or increasing in strength, depending on the nature of the local flow field interactions.

3. Results and Discussion

Axial force plots and symmetry plane contours are presented for the acceleration case first, then followed by the deceleration case. The force plots aid in identifying regions where acceleration effects dominate in comparison to the transonic flight Mach number range of interest.

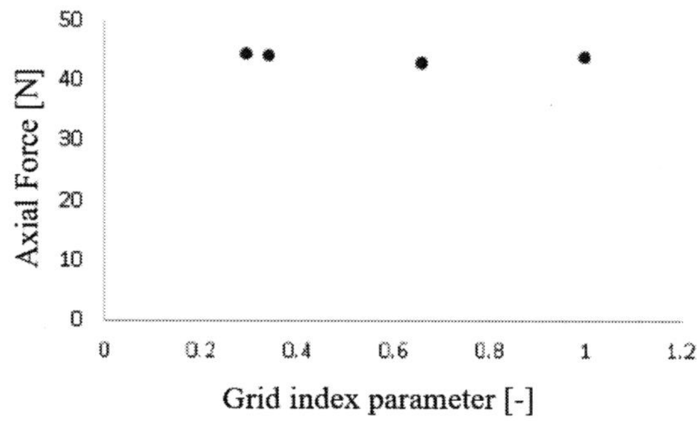

Figure 2: Grid convergence plot at Mach number 0.95 steady state. Vertical axis is axial force, Fa [N], and horizontal axis is a non-dimensional grid index parameter. Angle of attack 10°.

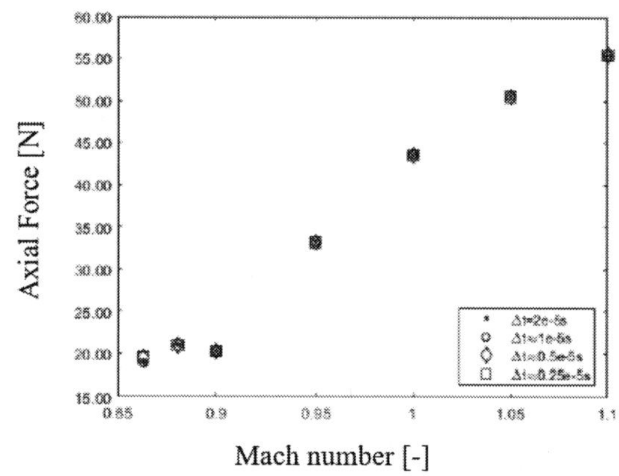

Figure 3: Time-step convergence plot at 400g deceleration from flight Mach number 1.10 until 0.85. Vertical axis is transient axial force, Fa(t), [N], and horizontal axis is the flight Mach number, M(t). Angle of attack 15°.

4. Results and Discussion

4.1. Acceleration

The axial force during transonic acceleration is shown in Fig. 5 and this is contrasted with the axial force obtained for constant velocity flight. Variations occur from high subsonic flight, M(t) near 0.95, and continue until M(t) near 1.15.

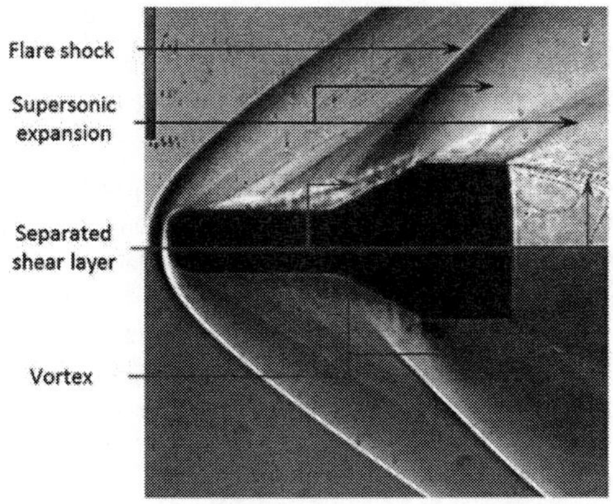

Figure 4: Schlieren image at M(t) 1.98, overlaid with a density contour plot deceleration shown at M(t) ≈ 1.98. [7].

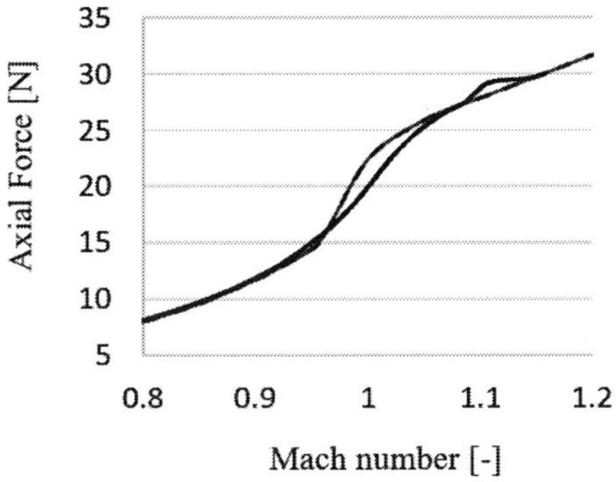

Figure 5: Axial force plot at 400g acceleration, shown for flight Mach numbers 0.8 until 1.2. Vertical axis is transient axial force, Fa(t), [N], and horizontal axis is the flight Mach number, M(t). Angle of attack 10°. Solid curve: Acceleration, Dotted curve: Constant velocity.

A reduced transonic drag rise is evident in the vicinity of M(t)=1. The transonic drag rise is associated with shocks that occur near the shoulder, the asymmetric bow shock, and wake shock. The reduced drag rise is an observation of an acceleration effect where the shoulder and asymmetric bow shocks develop gradually as the cone-cylinder accelerates from subsonic to supersonic speeds. Flow field results are shown on the symmetry plane. This is inferred from the contour plots in Fig. 6 and 7, where the shoulder expansion and shock system for the acceleration result (Fig. 7) has a smaller extent than in the constant velocity case (Fig. 6), reducing wave drag. This is due to the lag of in the development of the shock system in the transient case. In the constant velocity case, a fully developed equilibrium is reached; but under acceleration, the transient flow features resemble those of an earlier Mach number at which the shoulder shock extends to a smaller distance from the body. This is a flow history effect.

This gradual shock development modifies the cone-surface pressure in the axial direction. Similarly, the wake shock will modify the base pressure load in the axial direction and the wake shock's development depends on the upstream flow.

There is a noticeable bump near flight Mach number 1.1. This behavior of the axial force coincides with propagation of an asymmetric compression wave from the shoulder to the base, and although the surface pressure change from this wave is negligible in the axial direction, there is a change in base pressure and surface shear as this wave propagates away from the cone-cylinder. The bump in axial force near flight Mach number 1.1 is likely to be caused by overpressure as the bow shock develops, a reduced base pressure, and a marginal increase in shear force. This overpressure peak is attained at M(t)≈1.1 and tends towards the constant velocity solution at M(t)>1.15.

The origin of the asymmetric compression wave is the foot of the lambda shock structure, formed during high subsonic flight and aft of the shoulder vertex. The downstream propagation of this wave represents an example of a flow history acceleration effect. This wave is illustrated in Fig. 8, shown on the leeside only. There is no corresponding wave in Fig. 9, for the constant velocity case.

4.2. Deceleration

The axial force during transonic deceleration is shown in Fig. 10 and this is contrasted with the axial force obtained for constant velocity flight. Variations are observed from supersonic flight, M(t) near 1.0 and continue until M(t) near 0.85.

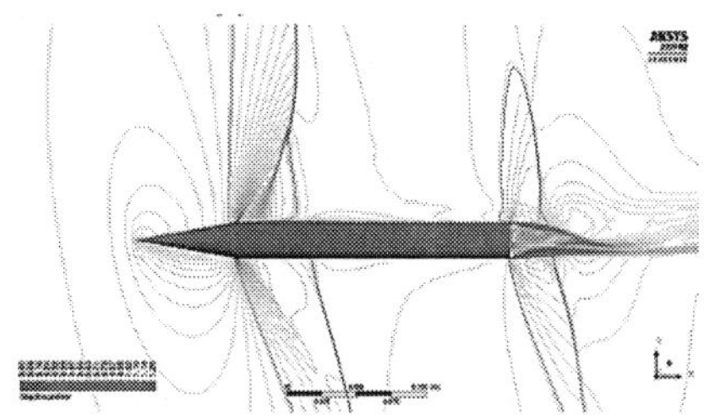

Figure 6: Cone-cylinder at M=0.98, constant velocity.

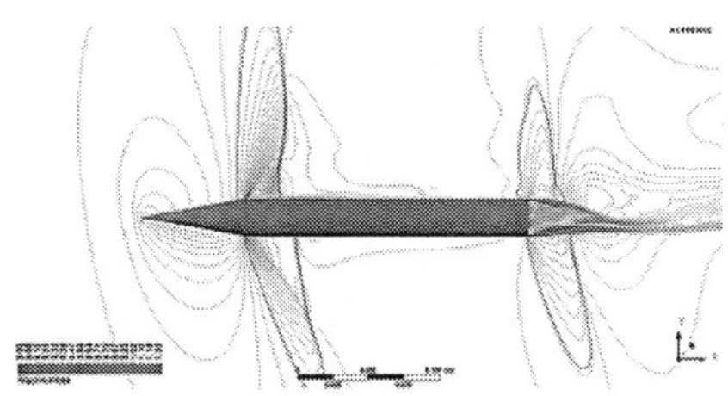

Figure 7: Cone-cylinder at M(t)=0.98, 400g acceleration magnitude.

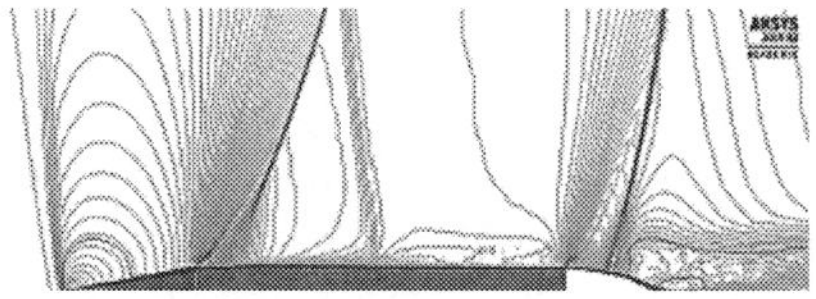

Figure 8: Cone-cylinder at M(t)=1.06, 400g acceleration magnitude, highlighting the asymmetric compression wave.

The drag loading is elevated in the vicinity of flight Mach number M(t)=0.95. This is caused by upstream propagation of the asymmetric bow shock, whereby the post-shock pressure modifies the cone-cylinder compression field, contributing to increased axial loading. Propagation of the bow shock is illustrated in Fig. 11 at flight Mach number M(t)=1.05 and Fig. 12. At M(t)=0.93 approximately. The bow shock has traversed a distance upstream, and away from the cone apex.

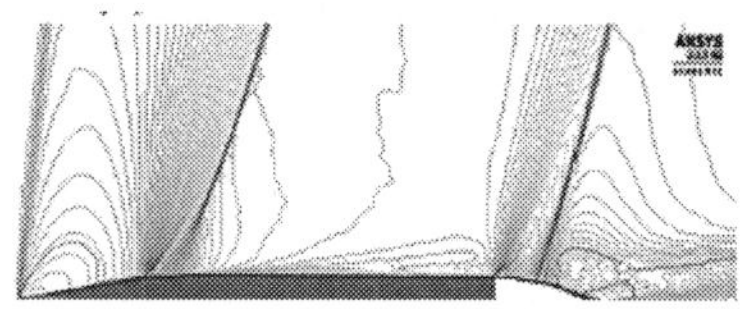

Figure 9: Cone-cylinder at M(t)=1.06, constant velocity, showing no presen

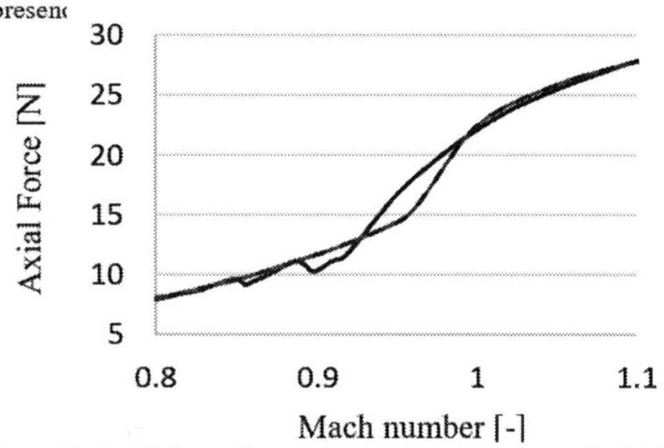

Figure 10: Axial force plot at 400g deceleration, shown for flight Mach numbers 1.1 until 0.8. Vertical axis is transient axial force, Fa(t), [N], and horizontal axis is the flight Mach number, M(t). Angle of attack 10°. Solid curve: Acceleration, Dotted curve: Constant velocity.

The distinct change in gradient near M(t)=0.90 and M(t)=0.85 is associated with upstream propagation of the terminal and wake shocks upon the cone-surface. Post shock compression on the cone surface raises the pressure load on the cone surface contributing to an increased axial force.

5. Conclusion and Future Work

Acceleration effects in the transonic Mach regime were identified for a cone-cylinder at 10° angle of attack, during accelerating and decelerating flight at an acceleration magnitude of 400g constant. This represents a subset of results from a broader investigation that focused on shock dynamics up to 15° angle of attack. To emphasize the flight Mach numbers where acceleration effects dominate relative to the Mach number range of interest, plots of the axial force were considered.

During accelerating flight, the transonic drag rise is gradual between approximate flight Mach numbers 1.05 to 0.95, with an approximately 10% reduction compared to constant velocity at a flight Mach number of 1. The reduced transonic drag rise is caused by the gradual development of the asymmetric bow shock, shoulder shock system and wake shock. The downstream propagation of the asymmetric compression is clear in the acceleration results, with the origin of this wave as the rear-foot of the lambda shock at the aft shoulder. Gradual wave development and downstream propagation represent acceleration effects for positive acceleration.

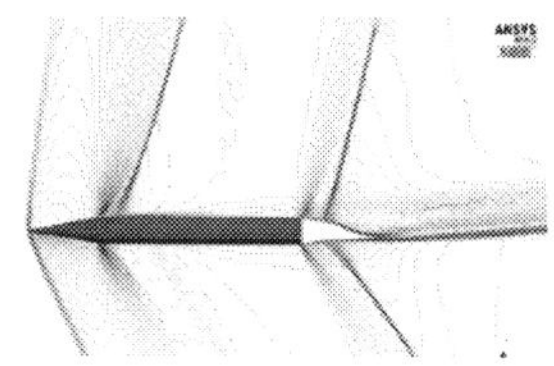

Figure 11: Cone-cylinder at M(t)=1.05, 400g deceleration magnitude, asymmetric bow shock is attached at the cone-apex.

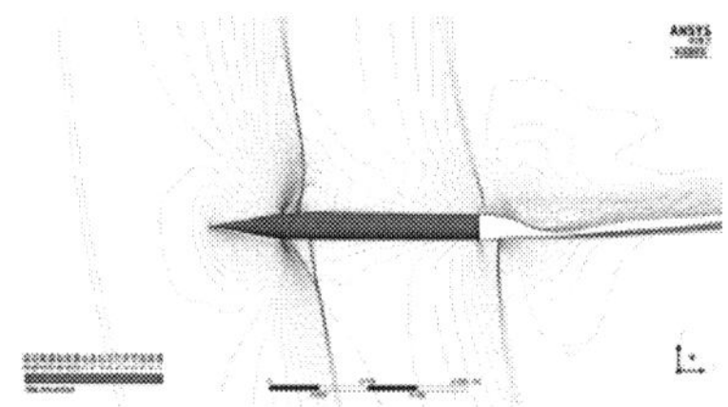

Figure 12: Cone-cylinder at M(t)=0.93, 400g deceleration magnitude, asymmetric bow shock is detached and has propagates upstream, away from the cone-apex.

The deceleration case shows increased drag between approximate flight Mach numbers 1.0 to 0.91, with a maximum reduction of approximately 15% that occurs in the subsonic Mach regime near flight Mach number 0.95. The increased drag is caused by upstream propagation of the asymmetric bow shock, the post-shock pressure is like that of a blast wave profile, and the cone-surface pressure is elevated while the bow shock is sufficiently close to the cone-apex. Below flight Mach number 0.91, upstream propagation of the wake and terminal shocks cause step-changes in the axial force. This is an area of further investigation.

The observed change in aerodynamic drag due to acceleration effects must be well understood in order to enhance control law effectiveness during rapid air vehicle manoeuvre.

Acknowledgements

The financial assistance of the National Research Foundation (NRF) towards this research is hereby acknowledged. Opinions expressed and conclusions arrived at, are those of the author and are not necessarily to be attributed to the NRF.

References

[1] Roohani, H. and Skews, B.W. (2009). *The influence of acceleration and deceleration on shockwave movement on and around aerofoils in transonic flight*. Shock Waves J., **19**, 297–305.

[2] Saito, T. et al. (2011). *Shock stand-off distance of a solid sphere decelerating in transonic velocity range*. Shock Waves J., **21**, 483–489.

[3] Reding, J.P. and Ericsson, L.E. (1986). *Asymmetric Vortex Shedding from Bodies of Revolution.*, Tech. rep. Sunnyvale, California: Lockheed Missiles and Space Company Inc.

[4] ANSYS (2015). *Flows with Moving Reference Frames* in ANSYS Fluent Theory Guide Release 17.0. ANSYS, 17–21.

[5] I.M.A. Gledhill et al. (2016). *Theoretical treatment of fluid flow for accelerating bodies.,* In: Theoretical and Computational Fluid Dynamics **30**, 449 - 467.

[6] N.M. Musehane, M.L. Combrinck and L.N. Dala. (2021). *Eulerian Derivation of the Conservation Equation for Energy in a Non-inertial Frame of Reference in Arbitrary Motion.,* In: Applied Mathematics and Computation 399, 126002.

[7] Mahomed I., Roohani H., Skews B.W., and Gledhill I.M.A. (2021), *Unsteady Shock Wave Dynamics in Accelerating and Decelerating Flight.,* The Aeronautical Journal, **126**, 401 – 423.

[8] Hsieh T. (1975). *Flow field about a hemisphere cylinder in the transonic and low supersonic Mach number range.,* Arnold Engineering Development Center, Tennessee, report AEDC-TR-75-114.

[9] Hsieh T. (1977). *Low supersonic flow over a hemisphere cylinder at incidence.*, J. Spacecraft and Rockets, **14**, 11.

[10] Yamashita Y., Fujiwara H., Suzuki T., Iwakawa A., Sasoh A., Mahomed I., Roohani H., Gledhill I.M.A., Skews B.W. (2019), *Supersonic flow fields with various deceleration levels.* In: Proc. 32nd International Symposium on Shock Waves, Singapore, 0496.

Flow topology around a supersonic aircraft body

Robert C. Clark[a,*], Beric W. Skews[a], Lara J. MacLucas[a]

[a]School of Mechanical, Industrial, and Aeronautical Engineering, University of the Witwatersrand, 1 Jan Smuts Ave., Braamfontein, Johannesburg, 2017, South Africa

*1108315@students.wits.ac.za

Abstract

The principle of methodological reductionism, whereby a complex system is divided into its constituent parts, is widely employed by gas dynamics researchers. This approach is used for the purpose of simplification, as well as isolation of principal effects resulting from certain individual components. Reductionism is insufficient, however, for gaining a comprehensive understanding of the flow topology around a supersonic aircraft, as it eliminates higher-order shock wave—expansion interaction effects that are unique to the overall fuselage-wing combination. Thus, the three-dimensional flow topology present around an analogue delta wing aircraft body, at zero incidence to the oncoming Mach 3.0 flow, was investigated using computational and experimental techniques.

Implicit Computational Fluid Dynamics (CFD) models of the aircraft body were created, with corresponding experimental models studied in a supersonic wind tunnel facility for qualitative validation purposes. Schlieren photography was used to capture the shock wave profiles around the body.

Complex, multi-order interactions between the three-dimensional shock wave and expansion profiles of the fuselage and wing were observed. The interaction and interference effects identified were characteristic of the combination of elements, rather than the constituents. Whilst not primarily aimed at modelling aerodynamic effects, this study aids in the fundamental understanding of the shock and expansion fan topologies resulting from common aircraft configurations. This could contribute to future work in optimising aerodynamic forces resulting from complex aircraft configurations, or with multiple aircraft in formation.

***Keywords*:** supersonic flow, delta wing aircraft, shock wave, expansion fan.

1. Background and Motivation

Methodological reductionism is an approach used to isolate constituent parts of a complex geometry set, in order to understand the effects of each on the overall factor of interest. In the field of aeronautical engineering specifically, this entails examining the flow field surrounding the wing and fuselage separately, in order to determine the effects on lift and drag. The shortcoming of this approach is that any interference effects on the flow field are neglected. The present research aims to examine the flow topology around an aircraft wing-body geometry at supersonic speeds, with a focus on the flow topology arising from interference effects resulting from the wing-body combination.

Aircraft design companies will often test full aircraft configurations, however, the results of these tests are often proprietary or classified (in the case of military aircraft). Additionally, the tests conducted on full aircraft configurations during development are primarily concerned with extracting the aerodynamic load data, rather than examining the complexities of the flow field that results from the fuselage-wing geometry. Thus, the present research aims to address this gap in knowledge by examining the flow topology around a delta wing—fuselage aircraft analogue at supersonic speeds.

Since this research is concerned with the flow topology around an aircraft analogue, it is helpful to consider the previous research conducted on the constituent parts, namely the fuselage and the wing.

1.1. Supersonic Flow around Cones and Ogives

Aircraft fuselages are commonly approximated by cone-cylinder or ogive-cylinder geometries. When supersonic flow flows over a cone at zero incidence, a conical shock wave forms at the tip of the cone. The freestream flow is deflected by the shock wave and curves continuously towards the cone surface; this is due to an effective expansion at the cone vertex causing the streamlines to disperse over the cone surface [1, 2, 3]. Consequently, the flow properties will be constant along coaxial cones [1, 3].

Supersonic flow around an ogival body is similar to that of the conical case, with the difference being that the effective cone angle changes along the length of the body, introducing weak expansion fans into the flow field. These expansion fans interact with the shock wave and curve the shock wave towards the ogival surface. Thus, the shock wave will not be perfectly conical, nor will the flow properties be constant along coaxial cones, as was the case with the conical body.

1.2. Supersonic Flow around Swept Wings

Swept wings are used on supersonic aircraft due to their favourable wave drag characteristics. The wing leading edge can be classified as either subsonic, where the leading edge falls within the Mach cone emanating from the apex (and thus the velocity component normal to the leading edge is subsonic), or supersonic, where the leading edge is placed outside the Mach cone (and the velocity component normal to the leading edge is supersonic) [4,5]. In the case of a subsonic leading edge, for low angles of attack the flow over the leeward side of the wing remains attached, whereas at larger angles of attack the flow separates at the leading edge and forms a spiral vortex, as for subsonic freestream speeds [5]. For the case of a supersonic leading edge, the flow is expanded around the leading edge and deflected towards the wing centreline; the deflected flow is then realigned with the freestream direction by an inboard shock wave due to symmetry requirements [5]. If the inboard shock wave is sufficiently strong (i.e. at higher angles of attack) the spanwise adverse pressure gradient causes boundary layer separation and subsequent formation of spiral vortices [5].

2. Objectives

The aim of this research is to investigate the flow topology surrounding an aircraft in supersonic flow, with particular focus on the flow features arising due to interference from the wing-body combination. The emphasis of the current study is to provide a fundamental understanding of the flow physics, rather than determining the aerodynamic load data.

3. Methodology

A wing-fuselage aircraft analogue was modelled in a Mach 3 freestream at zero incidence both experimentally and numerically using computational fluid dynamics (CFD) software, to investigate the resulting flow topology. The experimental results were used to validate the CFD results, thus allowing the CFD results to provide insight into the resulting complex flow field.

3.1. Experimental Methods

A representative aircraft model was designed and manufactured from brass for wind tunnel testing. The model had an ogival fuselage, with flat surfaces machined on the sides and bottom, and a delta wing with a 5% thickness-to-chord ratio, double-diamond aerofoil cross-section, to mimic the general shape of a supersonic aircraft. The model is shown in Figure 2 below. The model had a fuselage length of 128 mm, a span of 68.5 mm, a root chord of 80 mm, a taper ratio of 0.2, and a wing reference area of 3599 mm^2.

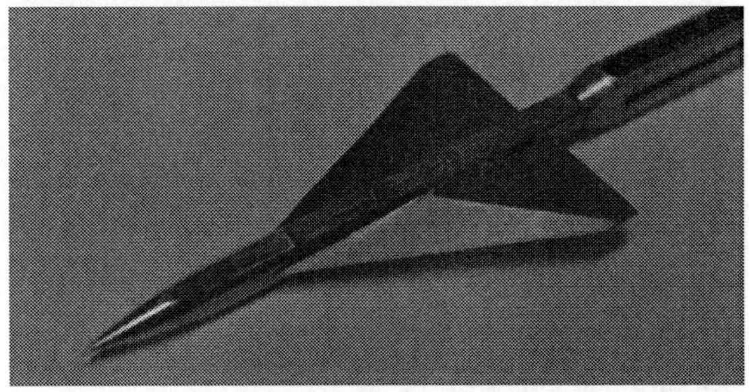

Fig. 2. Delta wing aircraft model.

The wind tunnel testing was conducted in the University of the Witwatersrand's blowdown supersonic wind tunnel, which uses air as a working fluid and has a test section cross-section of dimensions 101.6 mm x 101.6 mm and a freestream speed range of Mach 2.5 - Mach 3.5. The flow was visualised using a standard black and white schlieren system and the images were captured using an SLR camera

3.2. Computational Methods

The computational results were obtained by use of the ANSYS Fluent solver. The computational freestream conditions were set to be representative of the experimental wind tunnel environment, with a Mach 3 freestream and stagnation pressure of 380kPa. All models used a coupled, density-based, steady-state solver, approximating air as an ideal gas, and the flow as inviscid. The flow domain was made large enough to resolve any internal features, with the shock waves moving out through pressure far-field boundaries placed around the aircraft body. The vertical symmetry plane was exploited to halve the computational domain. The upstream far-field boundary was placed at a quarter of the characteristic length forward of the body. The downstream boundary was terminated before the end of the body, in order to prevent the high flow gradients that a sudden termination of the model would create. The fuselage was required to be uniformly extended in order to allow for a more fully developed downstream flow-field.

Several different tetrahedral meshes were constructed, and a mesh independence study conducted, the results of which are presented in Table 1. Adaptive mesh refinement techniques were used to provide better shock wave resolution, with curvature of the flow density, Mach number, static pressure, and temperature used as refinement criteria. An example of the mesh structure post-refinement is shown in Figure 3. Wing surface pressure, a measure of mesh independence, differed by a maximum of 0.802%, indicating suitable independence.

Table 1. Mesh independence results

	Coarse Mesh	Fine Mesh
Initial Cell Count	8 069 765	8 965 821
Post-adaption Cell Count	26 667 148	39 308 973
Fuselage Surface Pressure Avg. (Pa)	10 289.0	10 289.6
Fuselage Surface Pressure Avg. (% Diff.)	0.0058%	-
Wing Surface Pressure Avg. (Pa)	12 205.4	12 203.3
Wing Surface Pressure Avg. (% Diff.)	0.802%	-

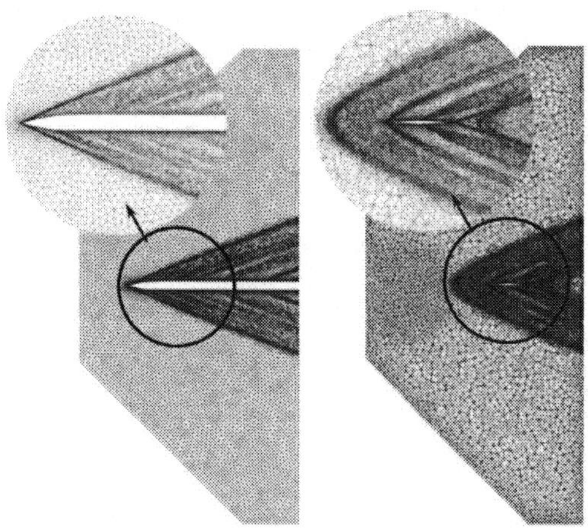

Fig. 3. Longitudinal sections of the flow domain mesh after adaptive mesh refinement.

4. Key Results and Discussion

4.1. Validation

Schlieren images were captured for the aircraft in a Mach 3 freestream at both a 0° and 90° roll angle and the resultant images were compared with the CFD results for validation. Figure 4 shows the stitched experimental schlieren image and the CFD results for the aircraft at a 90° roll angle. It can be observed that the CFD resolves all the flow features of interest, namely the bow shock wave, the leading edge shock wave outboard of the wingtip (A in Figure 4), the flow feature labelled B (discussed further below), and the expansion fans at the corners of the fuselage (denoted C). The CFD results correlate well with the experimental results and thus can be considered valid.

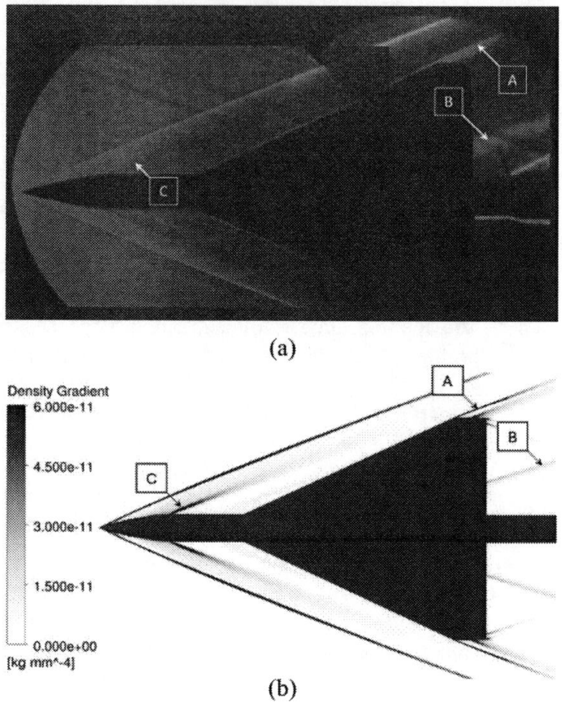

Fig. 4. Experimental (a) and computational (b) planforms of the density gradient topology around the aircraft body.

4.2. Wing-Fuselage Interference Effects

The constituent flow topologies of the wing and fuselage were found to interact, producing several higher-order interference effects that did not result from either body alone. These effects are best understood by taking cross sections of the flow around the aircraft, a legend of which is shown in Figure 5, which shows the density gradient topology of the flow.

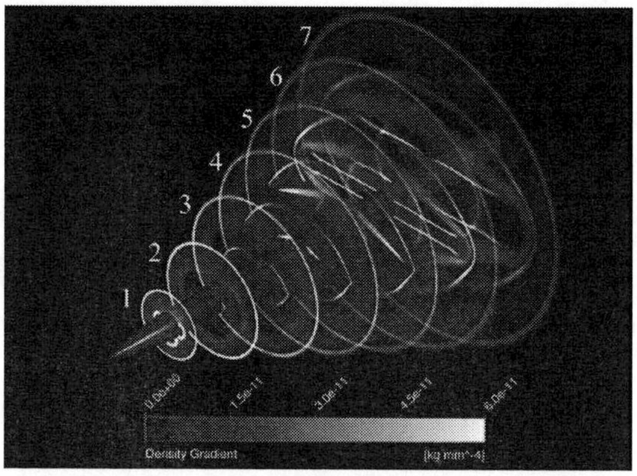

Fig. 5. Transverse cross-sections of the density gradient topology around the aircraft body.

The density profile of the third cross-section in Figure 5 is shown in Figure 6. The expansion fans resulting from the longitudinal fuselage 'corners' cause a region of low density marked by the points A in Figure 6. These low-density regions are asymmetrical about the planform plane of the aircraft, since the ogival surface of the fuselage is flattened on the underneath and sides. This asymmetry results in a slightly stronger leading edge shock wave on the upper surface of the wing, as indicated by point B. The presence of the fuselage, as well as the expansion fans resulting from it, produces a complex leading edge shock wave shape, initially planar near to the leading edge corner, then diffracted inwards toward the fuselage (denoted C).

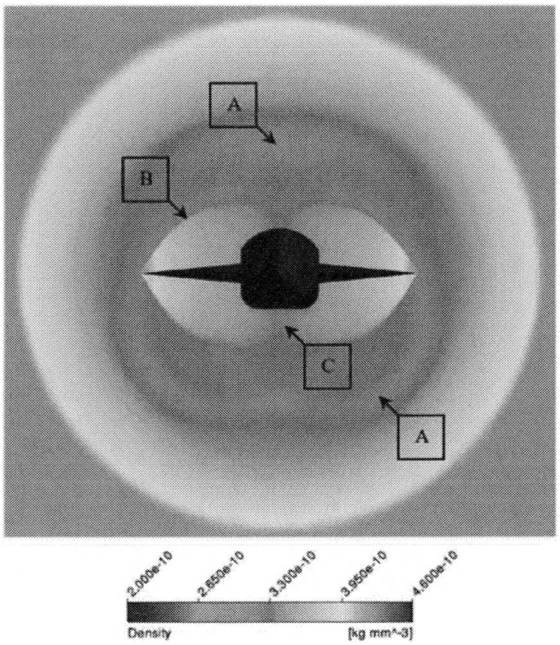

Fig. 6. Density profile of cross-section 3.

Further downstream, the interference effects between the wing and fuselage are evident, as shown by the complex shock wave and expansion profiles seen in Figure 7. This is the seventh and most downstream section shown in Figure 5. The leading edge shock wave is diminished in strength at the centre, due to interaction with the fuselage expansion fans (point A). The trailing edge shock wave (denoted B) exhibits similar characteristics, with diffraction inwards towards the fuselage centreline, and a regular reflection off the symmetry plane. The wing-vertex expansion fan also interacts with the fuselage topology, being diffracted around the body in a circular pattern as shown by indicator C. The expansion fan interferes with the trailing edge shock wave, which diminishes the strength of each as they are diffracted around the wingtip. The semi-circular shock wave, indicated by label D, was identified as a secondary realignment shock wave, which will be discussed in the following section.

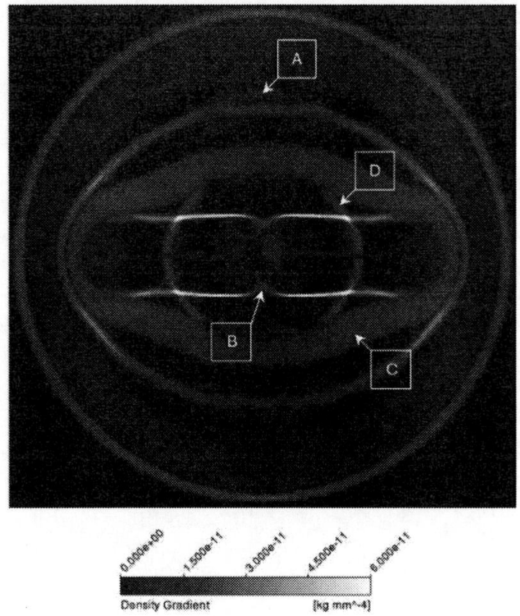

Fig. 7. Density gradient profile of cross-section 7.

4.3. Secondary Vertex Realignment Shock Wave

A circular realignment shock wave was identified in both the experimental and computational results, as denoted by label B in Figure 4. This shock wave is centred on the intersection of the vertex of the diamond wing profile with the fuselage. It originates from the three-dimensional diversion of the streamlines towards the body. The flow aft of the wing mid-vertex, after turning through the convex corner, converges towards the fuselage due to the expansion around the rearward-swept corner. The shock wave exists to realign the flow tangent to the fuselage body. Its transverse profile is semi-circular, mirroring the profile of the fuselage, as shown by D in Figure 7. This results in a semi-conical region of higher density air on the inboard sections of the aft part of the wings.

5. Conclusion

The flow features identified in this study are distinct to the wing-fuselage combination, and the complexity of the shock wave and expansion profiles demonstrate that consideration of higher-order effects can be beneficial to understanding the overall flow topology around typical aircraft bodies. This research could contribute to future studies on the optimisation of aerodynamic loads of complex aircraft configurations, or with multiple aircraft in formation.

Acknowledgements

The authors would like to acknowledge the use of the Centre for High Performance Computing (CHPC), for providing the computational resources used to conduct the CFD simulations.

References

[1] Anderson JD (2003). *Modern Compressible Flow: with Historical Perspective.* 3rd ed. New York: McGraw-Hill.

[2] Zucker R, Biblarz O (2002). *Fundamentals of Gas Dynamics.* 2nd ed. New Jersey: John Wiley & Sons, Inc.

[3] Courant R, Friedrichs K (1976). *Supersonic Flow and Shock Waves.* New York: Springer-Verlag.

[4] Bertin JJ (2009). *Aerodynamics for Engineers.* 5th ed. Upper Saddle River: Pearson Prentice Hall.

[5] Szodruch JG, Peake DJ (1980). *Leeward Flow over Delta Wings at Supersonic Speeds.* NASA Technical Memorandum 81187.

Transient Effects of Acceleration and Deceleration on the Formation and Location of the Bow Shock on an RAE 2822 Aerofoil in Ground Effect

Sean R. Morrow[a,*], Hamed Roohani[a], Beric W. Skews[a] & Irvy M. A. Gledhill[a]

[a]*School of Mechanical, Industrial, and Aeronautical Engineering, University of the Witwatersrand, 1 Jan Smuts Ave, Braamfontein, Johannesburg, 2017, South Africa*

**morrosean@gmail.com*

Abstract

Inspired by the world land speed record projects, the Thrust Supersonic Car (Thrust SSC) and the Bloodhound Land Speed Record (LSR), this investigation focuses on the transient effects of acceleration and deceleration on the formation of the bow shock on a simple test case, an RAE 2822 aerofoil. The aerofoil has a ground clearance of one chord length and is inverted to produce down force. An extreme acceleration and deceleration of 176 g was applied to the aerofoil in order to highlight the transient effects. In this investigation, a novel exception was identified, as the extreme acceleration of the aerofoil led to the inertial effects initially dominating the flow history effects even after the formation of the bow shock.

Flow history refers to when the flow does not instantly adjust to changing conditions; for example, time-dependent phenomena, such as the shedding and propagation of vortices and shock waves, evolve over time. With continued acceleration, the flow history effects became more prominent as the δ_{acc} reduced at a much lower rate than that observed in steady state analysis. By Mach 1.30, the stand-off distances increase in the following order: $\delta_{dec} < \delta_{steady} < \delta_{acc}$, which aligns with the stand-off distances that occur from flow history effects. It is deduced that, the increased inertia of the accelerated case impacted on both the formation and location of the bow shock. Flow history effects were observed: the accelerating flow field lagged the steady state conditions at a lower Mach number, while the decelerating flow field lagged conditions at a higher Mach number. At a much lower acceleration of 3 g, the relationship between the steady state and transient stand-off distances was qualitatively similar to that in free-stream flow. It is shown that at the mid-range example of 40 g, the total effect on the down force produced by the inverted aerofoil is reversed and becomes lift near Mach number 1.30 and that this effect is shifted down to Mach number 1.20 in deceleration. For surfaces on an LSR vehicle either lift or positive pitching moment is catastrophic. The indication is that the effects of deceleration should be investigated in the transonic phase for LSR vehicles.

Keywords: accelerating aerodynamics, transonic flow, inertial effects, flow history, CFD

1. Introduction

In 1997, the Thrust SSC achieved the current world land speed record of 763 mph (1,228 km/h) [1]. To surpass this record, the Bloodhound LSR must accelerate and decelerate at rates of up to 3 g and -3 g, respectively, to attain its top speed and come to a safe stop. This raises the critical question of how rapid acceleration and deceleration influence shock formation on lifting bodies within transonic and supersonic ground effect conditions. This investigation focuses on the transient bow shock formation of an RAE 2822 aerofoil accelerated to and decelerated from Mach 2.00, while in ground effect and contributes to understanding of the risk of unexpected lift developing on vehicles in the transonic range in ground effect, due to acceleration or deceleration. Ground effect has been shown, through different mechanisms, to amplify both the lift and the downforce generated by an aerofoil [2]. Two key considerations are the transient effects of acceleration and the influence of ground effect. Transient flow history and fluid inertia effects have been demonstrated to reduce subsonic lift and increase subsonic drag on an aerofoil during acceleration [3]. Acceleration and deceleration lead to a delay of the formation and dissipation of shock waves surrounding an aerofoil travelling at transonic and supersonic speeds. The transient shock waves are thus located further upstream of the constant velocity shock positions during acceleration and further downstream during deceleration. Consequently, the developing flow field of an accelerating aerofoil reflects the flow field and aerodynamic performance of the aerofoil at a lower constant velocity, with the converse effect being true for a decelerating aerofoil. Furthermore, the observed delay or flow history effect increases as the magnitude of acceleration and deceleration increases. [4].

In general, inertial effects were found to dominate the flow field below the Mach number when the bow shock first forms, defined as M_B, while flow history effects dominated after the formation of the bow shock $M > M_B$. In the case of the free stream, comparing the bow shock stand-off distance, δ, of constant velocity flow, δ_{steady}, against the results, the theoretical stand-off distances increase in the following order: $\delta_{dec} < \delta_{steady} < \delta_{acc}$ [5].

2. Methodology

The computational fluid dynamics (CFD) software chosen for flow simulation was Ansys® Fluent, Release 18.2, previously validated and used on multiple acceleration cases [4, 6, 7]. The pressure far-field boundary condition was employed to define the free-stream velocity for both steady-state and transient numerical simulations. To accurately capture shock formation and location, a transient, second-order, AUSM, Green-Gauss Node Based spatial scheme, viscous, density-based solver was chosen, accounting for the compressibility and viscosity of the fluid. The fluid was modelled as an ideal gas, with viscosity variations determined using Sutherland's law and the Spalart-Allmaras turbulence model applied.

The primary velocity range of interest corresponds to that of the Thrust SSC and Bloodhound LSR, spanning Mach 0.05 to Mach 1.40. However, to further investigate the influence of flow history and ground effect, this range was extended to include velocities up to Mach 2.00. For the purposes of the present study a two-dimensional model of the RAE 2822 aerofoil was used, and an example of the computational domain setup is illustrated in Fig 1. Grid adaption was applied in order to further refine the shock waves. To allow local flow field effects to dissipate and achieve far-field conditions, the domain dimensions were defined as 15 chord lengths, c, above the aerofoil and 20 chord lengths upstream and downstream of the aerofoil. The ground clearance, h, was set to one chord length from the lowest point of the aerofoil, and a moving ground wall was established beneath the aerofoil to simulate ground effect. Inflation layers were applied to the walls of the aerofoil and ground wall immediately under the aerofoil, to ensure y-plus values less than one.

Constant velocity simulations were conducted at Mach number intervals of 0.05, ranging from Mach 0.05 (17.35 m/s) to Mach 2.00 (694.18 m/s), to investigate the development of supersonic flow around the aerofoil at a clearance of h/c = 1.00. Acceleration was modelled in the non-inertial frame fixed to the aerofoil. The Navier-Stokes equations have been transformed into non-inertial frames [4, 6] in which additional terms in the momentum and energy equations take simple forms with additional acceleration terms, which effectively accelerate flow elements uniformly. These are coded as time-dependent, user-defined source terms. The flow velocity was prescribed by defining the time-dependent Mach number function at the pressure far-field boundary and a velocity function at the moving ground wall. The flow field was initialised from a converged constant velocity simulation at the initial Mach number. The method has been validated against ballistic range data [7].

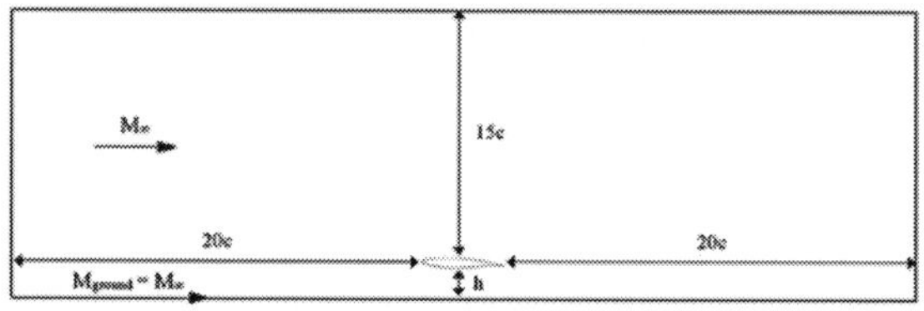

Fig. 1: Domain, boundary settings and meshing technique used for the RAE 2822 aerofoil model.

An acceleration rate of 40 g was chosen to correspond with an experimental study conducted on the TSSC geometry. The higher value of 176 g (1 726 m/s^2) was employed to evaluate the extreme effects of acceleration. The transient effects are determined by comparing the acceleration cases against the equivalent constant velocity case or cases when the analysis spans over a set speed range. Density gradients and contour plots were used to evaluate the location and development of the bow shock wave in each case. Each case was set with an angle of attack of -2.79° and subjected to a free-stream temperature of 300 K and an atmospheric pressure of 101 325 Pa. A chord length of 3.05 m was used in order that the Reynolds numbers reflect real-world flight conditions.

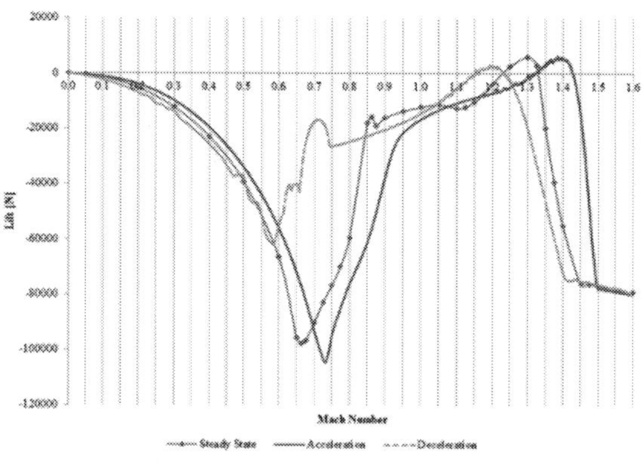

Fig. 2: Inverted RAE 2822 aerofoil, h/c = 1.00 and α = -2.79° for constant velocity and 40 g acceleration and 40 g deceleration: downforce performance comparison.

3. Results and Discussion

In Fig 2, the flow history effects on the downforce can be seen by the lateral shifts in the down force. The minimum downforce value at Mach 1.30 occurs at Mach 1.40 under acceleration and then at Mach 1.20 under acceleration. The same effect is seen for the maximum downforce, with additional variations being introduced by the downstream flow field overtaking the aerofoil.

In Fig 3, the bow shock at a constant aerofoil velocity of Mach 1.10 is compared with the developing flow fields of the transient cases. Density profiles are shown in Fig 4. It is immediately apparent that the shock wave has not yet developed at M = 1.10 for accelerations of 40g and 176g but that there is a significant compression wave close to the aerofoil leading edge. A typical stand-off distance for this wave is of interest and can be defined as δ_w for ease of reference. It may also be noted that ground reflections change the regular reflection of the tail shock to a Mach reflection at 176g.

In the two high acceleration cases the formation of the transient bow shock is delayed to a higher critical Mach number, $M_{B\,steady} < M_{B\,acc}$. Below the critical Mach number, the compression wave distance from the leading edge is smaller than the constant velocity shock standoff distance, $\delta_{acc} < \delta_{steady}$. Both shock and compression wave standoff distances are shown in Table 1. As the flow continues accelerating, the stand-off distance of the steady-state bow shock decreases more rapidly than that of the transient case. By Mach 1.30, $\delta_{acc} > \delta_{steady}$. The same transient effects occur during the 40 g acceleration, but to a lesser degree and the transient stand-off distance of the bow shock exceeds that of the steady state stand-off at a slightly lower velocity of Mach 1.20.

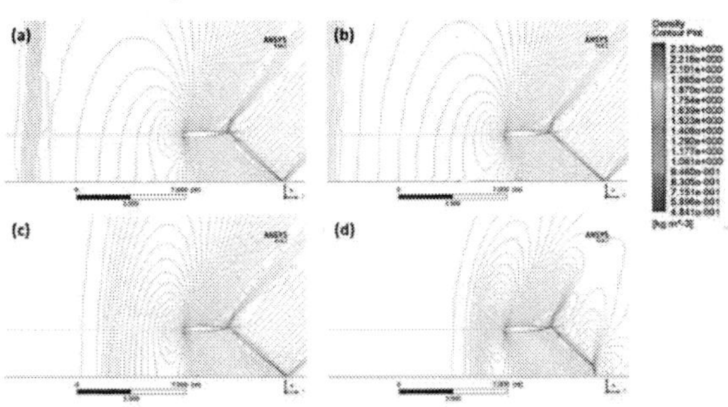

Fig. 3. Density contour plots of the bow shock wave development on an RAE 2822 aerofoil traveling at Mach 1.10: (a) steady-state, and accelerations of (b) 3 g, (c) 40 g and (d) 176 g.

The density gradients presented in Fig 4 further indicate how the magnitude of the acceleration impacts on where the bow shock wave forms. In the 3 g case, flow history effects are dominant and the stand-off distance of the accelerating aerofoil is consistently above that of the steady state case, $\delta_{acc} < \delta_{steady}$.

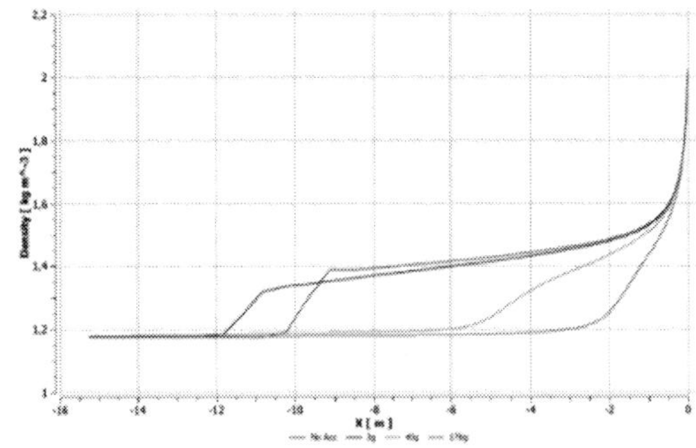

Fig. 4. Steady state and transient comparison of density plots of the Mach 1.10 flow field directly upstream of the leading edge of the RAE 2822 aerofoil

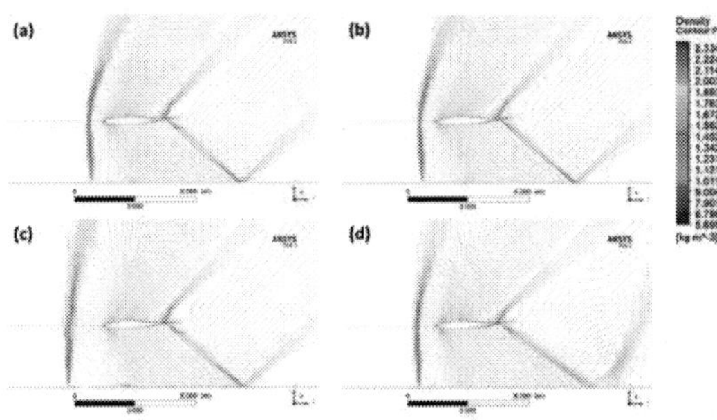

Fig. 5. Density contour plots of the bow shock wave development on an RAE 2822 aerofoil traveling at Mach 1.30: (a) steady-state, and accelerations of (b) 3 g, (c) 40 g and (d) 176 g.

From Fig 5 and Table 1, only the 176 g case features a lower stand-off distance of the bow shock at Mach 1.20. By Mach 1.30, the differences between the transient cases and steady state reduce as the bow shock fully develops and becomes more stable in each case. As expected, the stand-off distance gradually decreases as the Mach number increases. For each Mach number the trend as the acceleration increases is less clear due to the inertial and flow history effects simultaneously affecting the development and position of the bow shock. This is illustrated at Mach 1.25 and 1.30, where the 40 g case has the largest stand-off out of all the cases.

With the later development of the bow shock in each acceleration case the bow shock Mach number M_B increases from approximately Mach 1.00 at steady state and 3 g case to Mach 1.20 and Mach 1.30 for the 40 g and 176 g cases respectively.

During deceleration, the transient flow field resembles the steady state flow field of a higher velocity and thus, the transient stand-off distance steadily increases. Table 1, also provides a comparison of the stand-off distances of the 176 g deceleration cases between Mach 1.10 and 1.40. With the bow shock already developed the transient stand-off distance for the decelerating aerofoil is consistently less than that of the steady state and acceleration cases: $\delta_{dec} < \delta_{steady} < \delta_{acc}$.

Table 1. Normalised bow shock standoff distances, x/c, of an RAE 2822 aerofoil travelling at constant velocity and accelerating at different magnitudes: 3 g, 40 g, and 176 g. In case marked with shading a compression wave is observed.

g	δ/c at various Mach numbers					
	1.10	**1.15**	**1.20**	**1.25**	**1.30**	**1.40**
0	-3.24	-1.59	-0.86	-0.59	-0.27	-0.10
3	-3.86	-2.41	-1.24	-0.67	-0.34	-0.13
40	-1.55	-1.28	-1.15	-0.81	-0.61	-0.20
176	-0.63	-0.53	-0.47	-0.43	-0.33	-0.23
-176	-0.37		-0.25		-0.20	-0.09

At high subsonic Mach numbers and constant acceleration, each increase in velocity a compression wave is propagated ahead. As the body reaches Mach 1.00 relative to reference sound speed it is moving into a compression field having higher sound speed and thus is essentially locally subsonic. As it accelerates further it will engulf this compression field changing its relative Mach number until it develops a normal bow shock.

4. Conclusions

The formation of the bow shock on an accelerating aerofoil in ground effect is delayed in comparison with constant velocity. The greater the acceleration, the greater the delay and therefore the higher the free stream Mach number needs to be for the bow shock to first form: $M_B > M_{B\,steady}$. The underdeveloped flow ahead of the aerofoils accelerating at 40 g and 176 g are significantly smaller than the 3 g and constant velocity cases.

As the flow becomes supersonic, the delayed development and propagation of the bow shock in the 40 g and 176 g acceleration cases led to significantly smaller subsonic region ahead of the leading edge of the aerofoil, lower stand-off distances and steeper pressure gradients. In contrast, the 3 g results feature a larger subsonic region and consistently higher stand-off distances.

The combined inertial and flow history effects result in a series of compression waves that propagate ahead of the aerofoil and delay the formation of the bow shock. As the aerofoil continues to accelerate the compression waves eventually combine to form the bow shock.

References

[1] Evans B, Rose C (2014), *Simulating the aerodynamic characteristics of the Land Speed Record vehicle BLOODHOUND SSC*, Proc. IMechE Part D, J. Automobile Engineering, 228, 10, pp. 1127-1141.

[2] Vogt et al. (2007) *Flow Field Phenomena about Lift and Downforce Generating Cambered Aerofoils in Ground Effect*, Presented at the 16th Australasian Fluid Mechanics Conference (AFMC), 1, pp. 328-335.

[3] Roohani H, Skews BW (2008), *Unsteady aero-dynamic effects experienced by aerofoils during acceleration and retardation*, Proc. IMechE, Part G: J. Aerospace Engineering, 222(G5), pp. 631-636.

[4] Roohani H, Skews BW (2009), *The influence of acceleration and deceleration on shock wave movement on and around aerofoils in transonic flight.*, Shock Waves, 19, 4.

[5] Gledhill IMA, Roohani H, and Skews BW (2018), *Stagnation Pressure on a Uniformly Accelerating Aerofoil*, South Africa, p. 7.

[6] Roohani H, Gledhill IMA, Mahomed I and Skews BW (2020), *Bow shock stand-off distance for subsonic decelerating bodies*, Shock Waves 30, 115–129, ISSN 14322153, 09381287, https://doi.org/10.1007/s00193-019-00921-3

[7] Mahomed I, Roohani H, Skews BW, Gledhill IMA, Yamashita Y, Fujiwara H, Suzuki T, Iwakaka A, Sasoh A (2021), *Numerical Investigation of a Ballistic Range Free Flight Model*, International Journal of Aeronautical and Space Sciences, 22(6), 1293-1301, https://doi.org/10.1007/s42405-021-00405-8

A systematic literature review of techniques used for remaining useful life prediction

Tiaan Mare[a], Jacoba H. Bührmann[a,*]

[a]*School of Industrial Engineering, North-West University, Potchefstroom, 2531, South Africa*

**Joke.Buhrmann@nwu.ac.za*

Abstract

Predictive maintenance (PdM) consists of timely corrective actions before the failure of industry equipment. When successfully implemented, this approach reduces downtime, extends equipment lifetime, and minimizes maintenance costs by performing repairs only when necessary. A key aspect of PdM is Remaining Useful Life (RUL) prediction to forecast the time equipment will function before failure occurs. This paper presents a systematic literature review to investigate the types of datasets used, common models/algorithms employed for feature extraction, prediction, the selected optimization algorithms, and the evaluation metrics used in these studies. Lastly, it presents the recommendations of future work proposed by the research papers.

Keywords: Predictive maintenance, Remaining useful life, RUL, Systematic literature review, Data-driven approaches, Time series data

1. Introduction

The concept of equipment maintenance has been prevalent for thousands of years. There are records of maintenance documents that have been found to date back to 600 BC in ancient Egypt [1]. These examples, referred to as corrective maintenance (CM), involve the repair of equipment after a breakdown or fault has occurred. However, CM is a reactive strategy and comes at a risk of incurring expenses such as unavailability of equipment and high maintenance expenses due to critical equipment damage [2]

Preventive maintenance addresses this challenge by averting failures through periodic maintenance schedules, essentially preventing equipment breakdowns [2]. This technique, however, does not prevent unexpected failures from occurring, as scheduled maintenance is a preset plan and may not cover certain defects/ faults [3]. Additionally, this approach often results in unnecessary expenses due to maintenance actions that may be deemed superfluous [2].

These expenses sparked the popularity of investigating the maintenance approach called Predictive Maintenance (PdM), which involves the action of maintenance only when necessary. Zonta et al. [4] describe PdM as historical data-based models that prevent equipment breakdowns by predicting their health status. Typically, this approach utilizes machine learning (ML) or statistical models to analyze patterns and trends that can identify pending failures proactively [4].

Depending on the availability of historical data, equipment type, prediction model used, and other factors, PdM can be used to plan proactive maintenance schedules or order replacement parts well in advance, even weeks before a failure occurs [5].

Health prognostics of industrial equipment is a fundamental aspect of PdM. Among PdM methodologies, Remaining Useful Life (RUL) prediction is emerging as a prominent focus in literature, presenting a significant challenge and garnering considerable attention [6]. RUL refers to the estimated time an asset/ system will function before it no longer meets its required operational performance standards [7]. For RUL prediction to be effective, a Prognostics and Health Management (PHM) infrastructure with sensors that capture degradation indicators is required [8].

RUL more broadly falls under the Condition-Based-Monitoring maintenance strategy that involves the use of these sensors to monitor parameters such as vibration, temperature, electrical current, and acoustic emissions, among others, in real time. This enables the extraction of statistical degradation trends, health indicators, anomalies to predict impending failures [9]. The goal of health condition prognostics is to estimate future equipment health and predict its RUL. Fault detection forms a foundation for which many RUL prognostic pipelines are built. This includes the signal processing of the raw sensor signals, feature extraction and health assessment before any RUL predictions occur. Health condition prognostics, in this context, uses periodic condition monitoring data to estimate both future health states and the remaining useful life of equipment. When applied operationally, these prognostic insights serve as the foundation for Predictive Maintenance (PdM), allowing maintenance actions to be scheduled just before failure is likely to occur [10]. Improving the accuracy of machinery's RUL predictions facilitates proactive maintenance, consequently mitigating the expenses associated with unplanned machinery downtime [6].

This paper provides a systematic literature review (SLR) of state-of-the-art RUL prediction methods for predictive maintenance, focusing on data-driven approaches. The SLR focused on investigating PdM using data-driven techniques applied to the time series data of industrial assets and equipment. The aim was to investigate the trends and techniques used in current studies as well as future recommended research opportunities.

The SLR focuses on PdM using data-driven techniques applied to time series data of industrial assets and equipment. The aim is to answer the following research questions:

Q1 - What are the annual trends of the publications?

Q2 -What is the geographical distribution of the publications?

Q3 -What articles are most prominent in the field?

Q4 - What type of time series data is used?

Q5 - What feature extraction techniques are used?

Q6 - What prediction algorithms are used?

Q7 - What optimizers are used?

Q8 - What hyperparameter tuning techniques are used?

Q9 - What performance evaluation metrics are used?

Q10 - What future research opportunities are commonly proposed amongst the publications?

2. Methodology

To identify relevant articles that address these questions, the PRISMA 2020 approach was utilized [11]. The approach consists of three parts: identification of records from the databases, screening of the identified works, and selection of the publications for the review.

The search query included all records using the following query: ("Predictive Maintenance" OR "PdM") AND ("time series" OR "temporal data") AND ("Machine Learning" OR "ML" OR "data-driven") AND ("Remaining useful life" OR " RUL"). The search was conducted on September 2, 2024, and included all records up to this date from the databases: IEEE Xplore, Science Direct, Scopus, and Web of Science.

After the removal of duplicates, 186 records were identified. Next, the records were automatically screened to only include published peer-reviewed journal articles written in English, including surveys, reviews and conference proceedings (referred to as E1 – exclusion criteria 1) by using filters on the databases. This resulted in only 58 being deemed relevant.

Next, the records were screened for the following exclusion criteria, resulting in a remaining 30 records:

- Works that do not include evaluation metrics (E2),
- Works that do not explain how hyperparameters were found (E3),
- Works that do not discuss RUL prediction techniques/ algorithms used (E4); and
- Works that do not reference (or explain the origin) of the datasets used (E5).

A list of the relevant publications is available upon request.

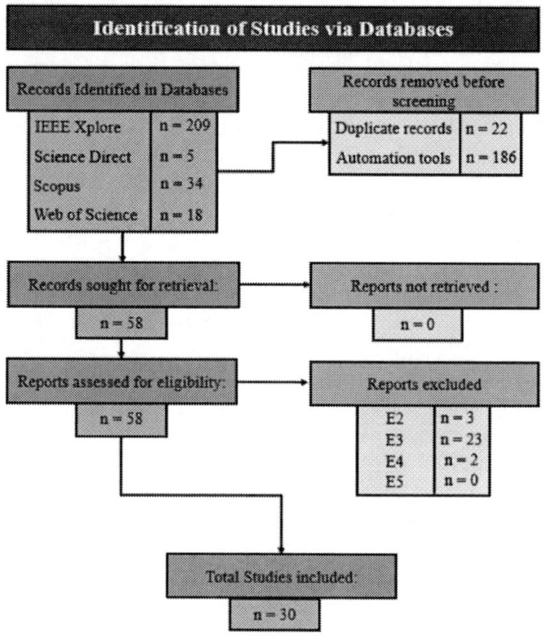

Fig. 1. PRISMA 2020 flow diagram of methodology for the systematic literature review, adapted from [11].

3. Results

The following trends were found in the remaining 30 publications.

3.1. Annual trends of the publications

Fig. 2. shows a line graph illustrating the annual trend of the number of publications.

A general increase overtime is evident, peaking in 2022 and 2024, with 7 and 6 publications, respectively, up to the date of the query.

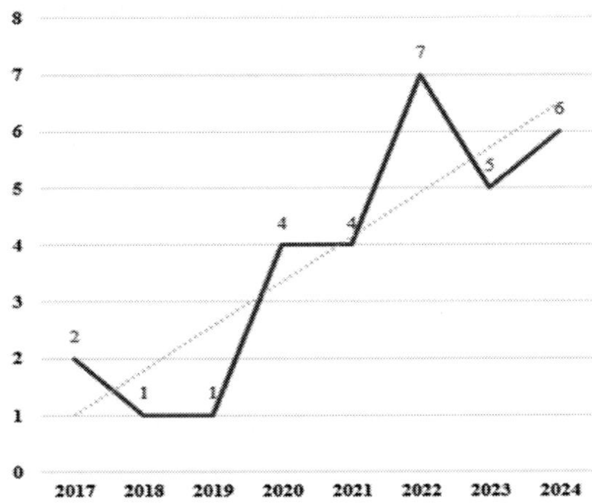

Fig. 2. (a) The number of publications identified per year.

3.2. Geographical distribution

China has the highest contribution with 13 publications, while research in this field shows global interest, with smaller contributions from 13 other countries.

An analysis of the citation counts for the selected works as of October 25, 2024, indicated two publications had a majority of the total 2255 citations: Zhao et al. (2018) [12] had 808 citations (35.8%) and Khelif et al. (2017) [13] had 414 (18.4%).

3.3. Types of time series data used

Data can either be real, meaning it is collected from actual equipment working in real-world environments, or synthetic, meaning it is generated artificially through simulations or algorithms to mimic these real-world environments. An analysis showed that 55% of the publications made use of synthetic data and 45% of real-world data.

Most synthetic datasets (95%) were from the well-known C-MAPSS dataset [14]. This dataset consists of four subsets (FD001, FD002, FD003, FD004), each simulating time-series turbofan engine data from NASA's Prognostics Center of Excellence.

The range of real-world data sources included 5 studies using bearing data, 2 using data from ion mill etching tools and 2 sourced from lithium-ion batteries. The rest of the sources were from various other types of industrial equipment. Attached in Appendix A is a table of all the publicly available datasets used amongst the studies

One reason studies tend to favour real data over synthetic data is that synthetic data often fails to capture real-world complexities, such as waveform source signatures, realistic noise, and accurate reflectivity. This can result in poor performance of trained neural networks during inference [15]. The C-MAPSS dataset, however, is well-established in the RUL prediction industry [14]. It consists of four subsets (FD001, FD002, FD003, FD004), each simulating time-series turbofan engine data from NASA's Prognostics Center of Excellence. These subsets consist of data collected from 21 sensors where the subsets differ in operational conditions and fault types, making them ideal for model development when intra-domain generalization is important.

3.4. Feature extraction techniques

Feature extraction techniques are described as a methodology used to minimize the number of features to improve computation efficiency [16]. The most used technique was Convolutional Neural Network (CNN)-based algorithms [5, 17], used in 16% of the publications. This was followed by Temporal Convolutional Networks (TCN) [18] applied 11%; Auto Encoder (AE) [19] techniques applied 8%. The rest used diverse other feature extraction techniques or basic statistical methods like Root Mean Square (RMS) [20].

CNNs are frequently employed to extract complex degradation features from condition monitoring data [17]. These algorithms are, according to Zhang et al. (2020) [5], one of the most successfully applied ML algorithms that are known for their ability to learn and express features. It involves the use of multiple layers of filters to slide over input sequences to extract hierarchical features through localized data patterns [5]. One key advantage CNNs has is that it has local FE capabilities and fast computational speed [14]. In addition, they are useful in extracting the spatial features of data.

3.5. Prediction model techniques

Prediction models on the other hand are the identified ML model used during training and will directly impact the performance of the final RUL prediction [21].

The most popular approach found for RUL-related predictions was Long Short-Term Memory (LSTM) [22] and dense Artificial Neural Networks (ANN) [23] based models (both found in 23% of the papers). This is followed by the Gaussian Process Regression (GPR) [24] Meng et al., 2023 which was used in 10% of the studies.

LSTM is a variant of RNN-based structures. RNNs are limited to short-term memory, whereas LSTM networks consist of gate control mechanisms that can manage both short-term and long-term memory [22]. The typical structure of LSTM consists of three gates: a forget gate, an input gate, and an output gate. The forget gate decides which information from the previous cell to discard. LSTM is known for addressing the training difficulties that are associated with long sequence data. A key benefit of LSTM is that they are designed to address the vanishing gradient problems that often occur in traditional RNNs [25]. This is because LSTM has cells that allow the network to learn long-range temporal dependencies while having a forget gate to remove hidden states [26]. Therefore, this model can better handle long-term sequential data and address the problem of long-term dependency through its memory function [22].

3.6. Optimizers applied in the selected publications

Optimization is central to many ML approaches. These algorithms revolve around minimizing or maximizing certain functions to improve model performance [27].

22 studies explicitly mentioned the use of optimizers. The Adam Optimizer [16] was the most used optimizer, used in 62% of the relevant papers. This was followed by Stochastic Gradient Descent (SGD) [26] which was used in 15% of the applicable studies.

Adam optimizer proposed by Kingma and Ba (2014) [28], is a stochastic optimization approach that relies only on first-order gradients and has low memory requirements. It involves calculating individual adaptive learning rates for different parameters based on estimates of the first and second moments of the gradients [28]. The Adam optimizer combines the parts of RMSProp and AdaGrad (two well-known optimizers) to estimate the mean (first moment) and uncentered variance (second moment) of the gradients [29].

3.7. Hyperparameter tuning techniques

Most ML algorithms are set up by a collection of hyperparameters, the values of which frequently have a significant impact on performance and must be carefully selected [30]. A hyperparameter tuning algorithm can be used to find the best performance for the specific ML model.

The results show that the Grid Search (GS) [18] method is the most popular (40%). Genetic algorithms (GA) [16] were used in a further 10% of the studies. Other diverse methods made up the rest of the publications, including 10% reporting that hyperparameter tuning was done manually with experimentation without using a specific method.

GS involves thorough searching over a manually selected subset of the hyperparameter space used by the ML model [31]. It works by dividing the range of each parameter into smaller intervals, calculating the target value, which is usually an error, for every parameter combination, and selecting the best one to minimize the target value [32]. Belete and Huchaiah (2022) [31], contend that this method's ease of use, parallelization capabilities, and resilience in low-dimensional areas make it the cutting edge of hypermeter optimization. However, traditional GS methods are time-consuming which limits its application involving large datasets [33].

According to Wang et al. (2024) [34], the Bayesian optimization algorithm (BOA) is a method that has been proven to be effective for hyperparameter optimization. This approach involves combining the GPR model with an acquisition function [34]. The GPR model estimates posterior probability distributions based on observed data, while the acquisition function selects the next point to evaluate.

3.8. Evaluation performance metrics

Performance evaluation metrics are used to determine the performance of the model. Traditionally RUL prediction involves regression since the RUL of equipment involves a continuous value. Out of the 30 publications, one study converted the problem into a classification task, while two other used both classification and regression models. The remaining publications only used regression models.

93.33% of the studies used accuracy / error measurements like Root Mean Squared Error (RMSE) and Mean Absolute Error (MAE) to measure performance. A third of the studies also reported on computational time efficiency as a performance metric. With diverse other metrics, like fitness functions reported only in five studies.

3.9. Future recommendations of studies

Future work sections in papers identify potential areas for improvement or further research in the field. Of the 30 studies, 26 studies proposed a total of 46 future works. The future works were categorized into eight different groups based on the nature of the research focus. Fig. 3. shows the distribution of the proposed future works. The eight categories made up 64% of the mentioned future works and included: subdomain adaptation (13%), application to new domains (13%), hyperparameter tuning 9%), model uncertainty (9%), practical application (7%), transfer learning (4%), anomaly detection (4%), and equipment-specific research (4%). 37% of the studies also made diverse other recommendations applicable to their specific applications and/or datasets.

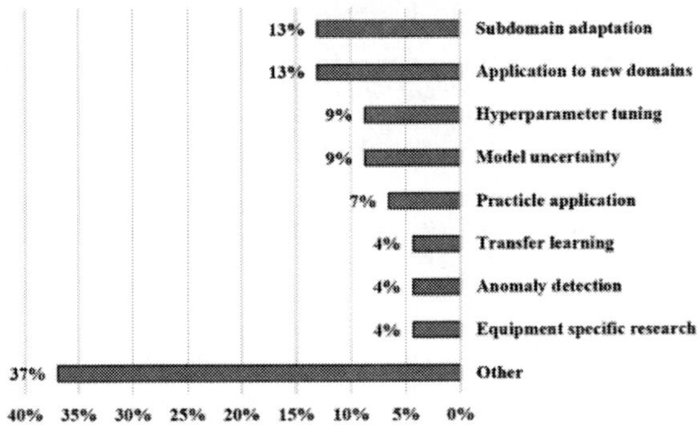

Fig. 3. Distribution of the proposed future works.

4. Conclusion

The results show that the synthetic turbofan engine dataset C-MAPSS was the most popular RUL dataset among the identified publications. The most popular feature extraction techniques were CNN and TCN, while LSTM and dense artificial neural networks were the most popular prediction models. The preferred optimizer was dominated by the Adam optimizer. Hyperparameter tuning techniques were majority of the studies the brute force GS method or manual tuning being used. This suggests that there is much potential for improvement in terms of using more sophisticated techniques like GA (only used in 10% of the publications). As expected, accuracy/error-related evaluation metrics were used in majority of the studies, but other metrics like time efficiency were used in a third of the studies. Evaluation metrics like model efficiency, generalization or interpretability, were notably underutilized. Many studies propose future research focused on subdomain adaptation, expanding to new domains, and uncertainty quantification.

References

[1] Poór P, Ženíšek D., Basl J.(2019). *Historical Overview of Maintenance Management Strategies: Development from Breakdown Maintenance to Predictive Maintenance in Accordance with Four Industrial Revolutions*, Proceedings of the International Conference on Industrial Engineering and Operations Management, 495–504.

[2] Molęda M, Małysiak-Mrozek B, Ding W, Sunderam V, Mrozek D (2023). *From Corrective to Predictive Maintenance—A Review of Maintenance Approaches for the Power Industry.* Sensors **23**:5970.

[3] Carvalho TP, Soares FAAMN, Vita R, Francisco RDP, Basto JP, Alcalá SGS (2019). *A systematic literature review of machine learning methods applied to predictive maintenance.* Computers & Industrial Engineering **137**: 106024.

[4] Zonta T, Da Costa CA, Da Rosa Righi R, De Lima MJ, Da Trindade ES, Li GP (2020). *Predictive maintenance in the Industry 4.0: A systematic literature review.* Computers & Industrial Engineering **150**: 106889.

[5] Zhang H, Zhang Q, Shao S, Niu T, Yang X (2020). *Attention-Based LSTM Network for Rotatory Machine Remaining Useful Life Prediction.* IEEE Access **8**, 132188–132199.

[6] Lei Y, Li N, Gontarz S, Lin J, Radkowski S, Dybala J (2016). *A Model-Based Method for Remaining Useful Life Prediction of Machinery.* IEEE Trans. Rel. **65**, 1314–1326.

[7] Liu Q, Zhang Z, Guo P, Wang Y, Liang J (2023). *Enhancing aircraft engine remaining useful life prediction via multiscale deep transfer learning with limited data.* Journal of Computational Design and Engineering **11**, 343–355.

[8] L Zhou and H Wang (2024), *MST-GAT: A multi-perspective spatial-temporal graph attention network for multi-sensor equipment remaining useful life prediction*, Information Fusion, **110**, 102462

[9] M Nuruzzaman , G Limon , A Chowdhury ,M Kahn (2025), *MAINTENANCE IN POWER TRANSFORMERS: A SYSTEMATIC REVIEW OF AI AND IOT APPLICATIONS*, ASRC Procedia, **1**:1, 34–47

[10] Z Tian, T Jin, B Wu, and F Ding, *Condition based maintenance optimization for wind power generation systems under continuous monitoring*, Renewable Energy, **36**:5, 1502–1509

[11] Page MJ, McKenzie JE, Bossuyt PM, et al. (2021). *The PRISMA 2020 statement: an updated guideline for reporting systematic reviews*. BMJ, **372**: n71.

[12] Zhao R, Wang D, Yan R, Mao K, Shen F, Wang J (2018). *Machine Health Monitoring Using Local Feature-Based Gated Recurrent Unit Networks.* IEEE Trans. Ind. Electron. **65**: 1539–1548.

[13] Khelif R, Chebel-Morello B, Malinowski S, Laajili E, Fnaiech F, Zerhouni N (2017). *Direct Remaining Useful Life Estimation Based on Support Vector*

Regression. IEEE Trans. Ind. Electron. **64**: 2276–2285.

[14] Gao H, Li Y, Zhao Y, Song Y (2023). *Dual Channel Feature Attention-Based Approach for RUL Prediction Considering the Spatiotemporal Difference of Multisensor Data*. IEEE Sensors J. **23**: 8514–8525.

[15] Alkhalifah T., Wang H., Ovcharenko O (2022). *MLReal: Bridging the gap between training on synthetic data and real data applications in machine learning*, Artificial Intelligence in Geosciences **3**, 101–114.

[16] Sharma DK, Brahmachari S, Singhal K, Gupta D (2022). *Data driven predictive maintenance applications for industrial systems with temporal convolutional networks*. Computers & Industrial Engineering **169**: 108213.

[17] Liu Y, Pan D, Liu Y (2024). *Structural-Guided Interaction and Attention-Enhancing Compensation Network for Machine Remaining Useful Life Prediction.* IEEE Trans. Instrum. Meas. **73**: 1–11.

[18] Hsu CY, Lu YW, Yan JH (2022). *Temporal Convolution-Based Long-Short Term Memory Network With Attention Mechanism for Remaining Useful Life Prediction*. IEEE Trans. Semicond. Manufact. **35**: 220–228.

[19] Aydemir G, Paynabar K (2020). *Image-Based Prognostics Using Deep Learning Approach.* IEEE Trans. Ind. Inf. **16**: 5956–5964.

[20] Zhao Q, Zhang X (2024). *HMDM: A hybrid GM model-based and MHSA-GRU data-driven method for remaining useful life prediction of rolling bearings.* Journal of Vibration and Control doi:10775463241273080.

[21] Si XS, Wang W, Hu CH, Zhou DH (2011). *Remaining useful life estimation – A review on the statistical data driven approaches.* European Journal of Operational Research. **213**: 1–14.

[22] Chen JC, Chen TL, Liu WJ, Cheng CC, Li MG (2021). *Combining empirical mode decomposition and deep recurrent neural networks for predictive maintenance of lithium-ion battery.* Advanced Engineering Informatics. **50**: 101405.

[23] He J, Wu C, Luo W, Qian C, Liu S (2024). *Remaining Useful Life Prediction and Uncertainty Quantification for Bearings Based on Cascaded Multiscale Convolutional Neural Network.* IEEE Trans. Instrum. Meas. **73**: 1–13.

[24] Meng J, Yue M, Diallo D (2023). *A Degradation Empirical-Model-Free Battery End-of-Life Prediction Framework Based on Gaussian Process Regression and Kalman Filter.* IEEE Trans. Transp. Electrific. **9**: 4898–4908.

[25] Ullah S, Li S, Khan K, Khan S, Khan I, Eldin S.M, (2023). *An Investigation of Exhaust Gas Temperature of Aircraft Engine Using LSTM,* IEEE Access, **11**, 5168–5177.

[26] Lu BL, Liu ZH, Wei HL, Chen L, Zhang H, Li XH (2021). A Deep Adversarial Learning Prognostics Model for Remaining Useful Life Prediction of Rolling Bearing. IEEE Trans. Artif. Intell. **2**: 329–340.

[27] Smola A, Vishwanathan SVN (2010). *Introduction to Machine Learning*. Cambridge University Press, Cambridge; United Kingdom.

[28] Kingma D.P, Ba J, (2014), *Adam: A Method for Stochastic Optimization* **9**, 1-15

[29] Nwankpa C.E, (2020), *Advances in Optimisation Algorithms and Techniques for Deep Learning*. Adv. sci. technol, 563–577

[30] Bischl, B, Binder M, Lang M, et al. (2023), *Hyperparameter optimization: Foundations, algorithms, best practices, and open challenges,* WIREs Data Min & Knowl, **13**:2, e1484.

[31] Belete D.M, Huchaiah M.D, (2022), *Grid search in hyperparameter optimization of machine learning models for prediction of HIV/AIDS test results. International,* Journal of Computers and Applications, **44**, 875–886.

[32] Hu K, Cheng Y, Wu J, Zhu H, Shao X (2023). *Deep Bidirectional Recurrent Neural Networks Ensemble for Remaining Useful Life Prediction of Aircraft Engine,* IEEE Trans. Cybern. **53**, 2531–2543.

[33] Beltrami M, Da Silva A.C.L, (2020), *A grid-quadtree model selection method for support vector machines.* Expert Systems with Applications **146**, 113172.

[34] Wang L, Chen Y, Zhao X, Xiang J (2024). *Predictive Maintenance Scheduling for Aircraft Engines Based on Remaining Useful Life Prediction*, IEEE Internet Things J, **11**, 23020–23031.

Appendix

A. Table of Publicly Available Datasets

Dataset Name	Data Source	Creators
C-MAPSS	Aircraft Turbofan engine	(Saxena et al., 2008)
IEEE PHM 2012 Prognostic Challenge	Bearings	(Nectoux et al., 2012)
Dataset1_BatteryAgingARC-FY08Q4	Lithium-Ion Batteries	(Saha and Goebel, 2007)
XJTU-SY bearing datasets	Bearings	(Wang et al., 2020)
Incipient Fault Diagnosis of Rolling Element Bearings	Bearings	(Wei et al., 2020)
2010 PHM data challenge	CNC milling machine cutters	(PHM Society, 2010)
PHME20	Filtration system	(PHM Society, 2020)
Phoenix Contact Relay Dataset	Electromechanical relays	(Winkel and Deuse-Kleinsteuber, 2022)
Backblaze dataset	Hard disk drives	(Bauer, 2018)
S.M.A.R.T dataset	Hard disk drives	(Baidu Inc, 2013)
2018 PHM Data Challenge	Ion mill etching tool	(PHM Society, 2018)
Oxford Battery Degradation Dataset	Lithium-ion batteries	(Birkl, 2017)
WDC data	Lubrication systems	(Fan et al., 2017)

Inter-transfer learning approach for the RUL prediction of different industrial equipment types.

Tiaan Mare[a], Jacoba H. Bührmann[a,*]

[a]School of Industrial Engineering, North-West University, Potchefstroom, 2531, South Africa

*Joke.Buhrmann@nwu.ac.za

Abstract

Predictive maintenance (PdM) strategies have enabled industries to optimize operational efficiency, minimize unplanned downtime, and extend the lifespan of industrial assets. A key part of many PdM approaches is the estimation of Remaining Useful Life (RUL), which involves using historical asset degradation data to predict the future usability of equipment. However, the availability and quality of the data are often equipment specific and limited.

To address this limitation, transfer learning (TL) is a promising technique that utilizes knowledge from well-documented equipment domains to those with limited data availability. This research investigates the feasibility of transferring knowledge from turbofan engines, using the publicly available C-MAPSS dataset, to a lithium-ion batteries dataset. The feasibility assessment utilizes the Maximum Mean Discrepancy (MMD) metric to quantify domain similarity. Results indicate that the MMD between the source (turbofan engines) and target (lithium-ion batteries) datasets is 0.5654. This closely aligned and was notably slightly smaller than the baseline intra-domain value of 0.5995, suggesting promising feasibility for cross-domain knowledge transfer. The research then tested this by applying several TL approaches. A naïve partial freezing approach resulted in negative transfer, but with a dedicated adapter layer the negative transfer was significantly mitigated. The best TL model extended the adapter only approach by unfreezing the penultimate layer and improved the model accuracy by 17% over a battery-only-trained baseline model.

Keywords: Predictive maintenance, Transfer Learning, Cross-Domain, Remaining useful life, RUL, Data-driven approaches, Time series data

1. Introduction

RUL prediction has garnered significant attention in literature with the number of publications increasing every year. When implemented effectively, it can prevent downtime, minimize costs, and extend the lifespan of equipment, among other benefits. However, the field faces many challenges limiting its applicability. One major challenge is data source limitations [1]. In the review of the challenges in PdM, Nunes et al. [2] discovered that despite significant progress in PdM research, current approaches are tailored to specific parts or equipment, making these solutions highly specialized and not easily adaptable to other applications. This presents a research opportunity to investigate how pre-trained knowledge can be incorporated for RUL prediction in cross-domain PdM applications.

A promising technique to address this challenge is Transfer Learning (TL) [1]. TL can provide several benefits including accuracy improvement and efficient use of scarce data [3]. The review by Iman M et al. [4] emphasizes the increasing relevance of transfer learning within the field of ML, particularly due to its capability to reduce the reliance on large volumes of training data and the associated high costs of model training. Moreover, it addresses the challenge of building high-performance models using data that is more readily available from diverse domains [5]. This approach can be categorized in two areas namely intra-domain and cross-domain TL. Intra-domain involves transferring knowledge from one data set to another from a shared or similar source and cross-domain TL involves transferring knowledge across different and often unrelated domains [6].

While intra-domain transfer learning offers several benefits such as maximizing the utility of data within a specific domain, reducing training time and improving model generalization, it inherently limits itself to scenarios where substantial domain-specific datasets are available [6]. However, many industries have specialized equipment that might not have widely available datasets and thus they are restricted to their own confined dataset which may be limited in size. Thus, cross-domain TL is particularly useful as it addresses this problem by enabling the application of learned capabilities across fundamentally different types of datasets, increasing the versatility of models [6].

Several studies have investigated the application of TL for RUL prediction, however, typically they fall under the intra-domain category. In research done by Sun et al. [7], a deep transfer learning based on a sparse autoencoder was used to predict the RUL of cutting tools operating under different conditions.

Although some studies have claimed to explore cross-domain transfer learning, they often rely on datasets sourced from similar equipment types. For example, Xiang et al [8] introduced a Micro Transfer Learning Mechanism for RUL prediction. The experiments were conducted separately on turbo-engines and wind turbine gearbox datasets, where TL was performed within the same equipment type under varying operating conditions and degradation trends. While the method successfully reduced the impact of time-varying data distributions and outperformed other techniques, it focused on related domains rather than achieving true cross-domain transfer.

Only one study could be found that has truly investigated cross-domain transfer learning in the strict sense of the term. Na et al [9] proposed a heterogeneous transfer learning framework to transfer knowledge from aircraft engine RUL prediction tasks to railway track circuit fault prediction. The feasibility of transfer was assessed using Maximum Mean Discrepancy (MMD) to compare the data distributions of the source (aircraft engines) and target (track circuits) domains. By transforming the track circuit (target) data into a classification task through discretizing RUL values into fault classes and using a Data Distribution Fitting Network (DDFN) effective knowledge transfer was achieved. The model attained over 95% accuracy in track circuit fault prediction, significantly outperforming non-transfer learning models, which only reached 70%. This highlights a significant gap in RUL prediction-related research vast majority of transfer learning studies remain confined to intra-domain scenarios.

This study investigates the feasibility of applying true cross-domain transfer learning between two fundamentally different datasets: turbofan engines as the source dataset $D_s = \{\boldsymbol{x}_i^s, \boldsymbol{y}_i^s\}_{i=1}^{n_s}$, representing a mechanical system, and lithium-ion batteries as the , target dataset $D_t = \{\boldsymbol{x}_i^s, \boldsymbol{y}_i^s\}_{i=1}^{n_s}$, representing an electrochemical system.

2. Overview of Datasets and Preprocessing Steps

The datasets selected for this study include the C-MAPSS dataset [10], which provides degradation data for turbofan engines, and two lithium-ion battery datasets. Where one battery dataset is used as the target domain [12] and the other as a baseline for the transferability assessment [13].

2.1. C-MAPSS Dataset

C-MAPSS, or Commercial Modular Aero-Propulsion System Simulation, is a software simulation tool that models the behavior of a large commercial turbofan engine with a control system that closely mirrors actual engine operations [10]. NASA's Prognostics Center of Excellence have used this simulation environment to create a set of four datasets that are well-established benchmarks for PdM-related research.

This study utilizes the FD001 dataset, which comprises 100 training trajectories and 100 test trajectories, representing observations of aircraft engines operating until failure. The dataset includes several types of simulated faults, such as the degradation of the high-pressure compressor (HPC) and faults in the fan. Fig 1. illustrates the turbofan engine components.

Fig. 1. Turbofan Engine Components adapted from [11].

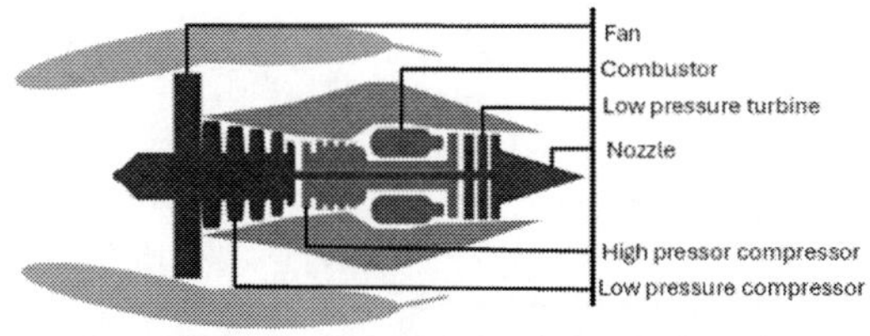

2.2. Hawaii Natural Energy Institute Relaxation Data for Commercial NMC and LFP Cells Dataset $(\boldsymbol{D_T})$

For the first target dataset, a dataset derived from data collected by the Hawaii Natural Energy Institute (HNEI), which examined the cycling behavior of 14 NMC-LCO 18650 lithium-ion batteries, was used. This processed dataset, hosted on Kaggle [12], includes features that characterize voltage and current behavior over 1,000 charge-discharge cycles at a controlled temperature of 25°C, using a constant current-constant voltage (CC-CV) charging method at a C/2 rate and discharging at 1.5C. Key features include discharge time, charging time, voltage behavior, and other metrics that reflect battery performance degradation over cycles.

2.3. NASA's Li-ion Battery Aging Datasets $(\boldsymbol{D_B})$

The second dataset was a baseline dataset, the NASA's Li-ion Battery Aging Datasets [13]. The dataset consists of operational data from four Li-ion batteries (b005, b006, b007, and b0018) subjected to accelerated ageing through repeated charge-discharge cycles at room temperature. Charging was performed in a constant current (CC) mode at 1.5A until the voltage reached 4.2V, followed by a constant voltage (CV) mode until the current dropped to 20mA. Discharge operations were conducted at 2A until specific cutoff voltages (ranging from 2.2V to 2.7V) depending on the battery. The batteries were cycled until reaching an end-of-life (EOL) threshold, defined as a 30% reduction in capacity from the rated 2Ah to 1.4Ah. The dataset includes structured information on voltage, current, temperature, and capacity for charge and discharge cycles, as well as impedance characteristics. Each cycle represented a complete charge-to-discharge activity, aligning the structure with the target dataset (D_T).

3. RUL Labeling

To ensure that the datasets under investigation are interpreted in the context of RUL degradation, they must be appropriately labelled. RUL refers to the estimated time an asset or system will function before it no longer meets its required operational performance standards, as illustrated in Fig 2. [14].

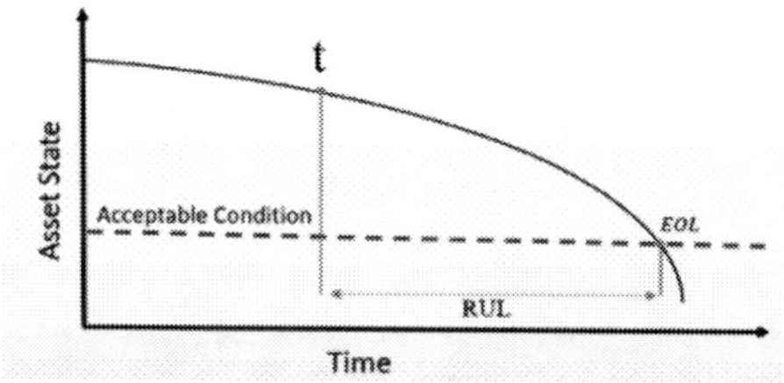

Fig. 2. Asset Remaining Useful Life Visualisation [14].

Subsequently, the labels of the datasets can be inferred through Eq. (1) for the CMAPSS data and Eq. (2) for the battery data [14].

$$RUL_{ij} = C_{i,max} - C_{i,j} \tag{1}$$

$$RUL_{ij} = \begin{cases} C_{i,EOL} - C_{i,j}, & C_{i,j} \leq C_{i,EOL} \\ 0, & C_{i,j} \geq C_{i,EOL} \end{cases} \tag{2}$$

Where, $C_{i,max}$ represents the last cycle, $C_{i,EOL}$ is the cycle in which the battery reaches its end-of-life (EOL) threshold and $C_{i,j}$ is the currently observed cycle.

4. Autoencoders

An autoencoder is used to extract features from D_s and D_t by mapping the input data into a 10-dimensional latent space. performs non-linear transformations of input data into a latent space and reconstructs it back to the original domain using encoder and decoder neural networks. This allows the model to capture implicit non-linear relationships and temporal dependencies [15].

Using this architecture separate autoencoders were trained for the target (D_T) and source (D_S) datasets to extract meaningful and compatible latent representations. The autoencoders comprise of two components the encoder, see Eq. (3), and the decoder, see Eq. (4).

$$\boldsymbol{z} = f_{enc}(\boldsymbol{x}; \boldsymbol{\theta_{enc}}) \tag{3}$$
$$\hat{\boldsymbol{x}} = f_{dec}(\boldsymbol{z}; \boldsymbol{\theta_{dec}}) \tag{4}$$

Where, θ_{enc} and θ_{dec} are the parameters of the encoder and decoder networks respectively. z represents the latent representation of the datasets captured at the latent layer of the autoencoder. $\hat{x}$ is the reconstruction input.

Guided by the architecture proposed in [9], the Autoencoder (AE) was initialized with, with 128 units and a ReLU activation function. The model is trained with batch size = 16, epochs = 50, and learning rate = 0.001. Refer to Appendix A for the tabulated architecture of the AE. The visualization and summary of the AE architecture can be referred to in Appendix A and B respectively

The objective of the autoencoder is to minimize the reconstruction loss. The R^2 and root mean square error measures were used to calculate the autoencoder performance [16]. After training, the encoder f_{enc} was used to map the input features into a latent space size of 10 [9]. The Performance of the AE is shown Table 1

Table 1. Reconstruction Performance of the Autoencoder

Performance metric	R^2	MSE
(D_S)	0.9829	0.1181 Type equation h
(D_T)	0.8770	0.2066
(D_B)	0.9848	0.1226

The results demonstrate that the autoencoder effectively captures the underlying structure and features of both the source and target datasets. Each AE yielded reconstruction that captures more than 85% of the variance and yielded low reconstruction losses. Fig 3. further demonstrates the autoencoder's performance by visualizing the correlation matrices of the raw datasets compared to the extracted latent space datasets.

As shown in Fig 3., the transformation applied to the source (D_S) and target (D_T) datasets significantly improve the correlation structure in the latent space.

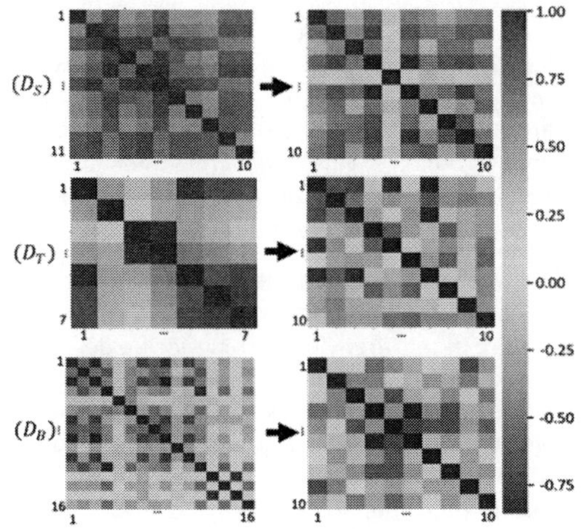

Fig. 3. Correlations Matrices of the Source (D_S) and Target (D_T) Datasets (Raw vs Encoded Data).

5. Maximum Mean Discrepancy Transferability Assessment

The Maximum Mean Discrepancy (MMD) is a kernel-based metric. This metric measures the difference between two probability distributions by comparing their embeddings in a reproducing kernel Hilbert space [16]. Consequently, MMD is used to assess whether knowledge can be reliably shared between different datasets and can be used as a transferability assessment by quantifying the embedded feature gaps [9].

5.1. Maximum Mean Discrepancy results

The results of the MMD calculations between the three datasets are given in Table 2.

Table 2. MMD Results

MMD calculation	Value
$(D_S) \rightarrow (D_T)$	0.5654
$(D_B) \rightarrow (D_T)$	0.5995
$(D_{S\rightarrow T}) : (D_{B\rightarrow T})$	1.0601

The MMD results indicate only moderate distribution shifts, reinforcing that transfer learning is potentially feasible. Notably, the learned latent-space gap between the turbofan engine (D_S) and the target battery (D_T) is very close, even slightly smaller, than the baseline gap. This providing strong evidence that cross-domain learning could be feasible.

6. Transfer Learning Experiment

Section 5 indicated strong evidence that knowledge sharing between (D_S) and (D_T) is plausible. This section carries out transfer learning to prove its feasibility.

The experiment design is illustrated below

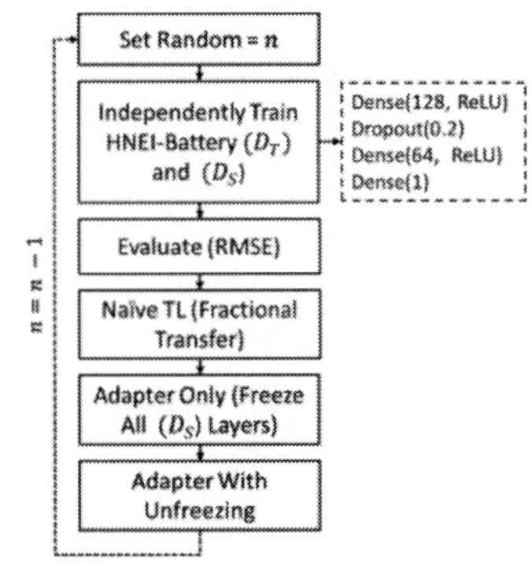

Fig. 4. TL Experimental Design

The implementation is built using TensorFlow and Keras (Version 2.10.0). The experiment was carried out 10 times, setting a different seed for each trial.

A simple two layered dense neural network was selected as the baseline model. The hyperparameters and training settings used are detailed in Appendix C. The hyperparameters were selected through trial and error based on validation set performance. To conduct the trials a small, predefined search space was used. Specifically, layer 1 was fixed to the input size (11 for D_S and 16 for D_T. Layers 2 and 3 were tested within the ranges [32, 64, 128]. Dropout rates were also explored between 0.1 and 0.3. The best-performing (D_T). baseline architecture (selected via validation) was then used to independently train FD001.

Next phase involved experimenting with three transfer learning approaches. The first naively applied fractional transfer learning by copying the (D_S) weights, freezing the first 70% of the layers, and then fine-tuning the remaining 30% on (D_T). Another approach was to freeze all the core FD001 layers and only train on the Dense (128 unit) adapter on (D_T). The third model used an adapter and unfreezing layers. Specifically, all layers of the model trained on FD001 were frozen while only the Dense adapter was trained, and then afterward, the penultimate Dense (64 unit) layer was unfrozen and fine-tuned to (D_S).

7. Results and Discussion

The results of the trials are shown in Fig 5. Across the 10 random seeds, the baseline model achieved an RMSE of 113 ± 1.9. While, Negative transfer learning occurred in the Naïve TL approach, scoring scored an RMSE of 291 ± 20.2. The adapter only approach also incurred negative transfer (notably less dramatically) with an RMSE score of 145.4 ± 4.2. Notably, the adapter with unfreezing strategy delivered the best accuracy of 93.8 ± 4.6 (17% improvement over baseline).

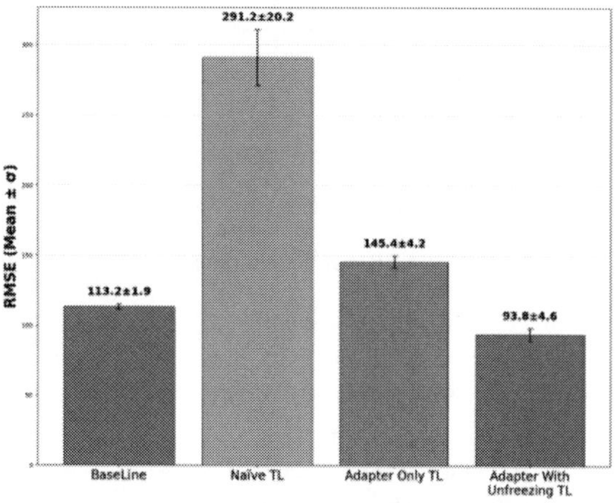

Fig. 5. TL Model RMSE Scores

The Naïve TL model highlights that even with only moderate a domain distance, the feature misalignment is still to large to simply blindly freeze and fine tune pretrained weights. As a result, the error rose by 157% and variance significantly increased from 1.9 to 20.2. The second approach achieved a notable improvement over the Naïve TL approach by training only the 128-Unit adapter. However, it still resulted in negative transfer (where inherited weights harm rather than help [18]) underperforming the baseline by 100%.

The Adapter-with-Unfreezing strategy extended the adapter-only approach by unfreezing the penultimate 64-unit layer after adapter training. This approach allowed the model to retain the general embedded patterns captured in (D_S) while still giving it the flexibility to realign its parameters toward the (D_S) representations. This resulted in the lowest RMSE score of 93.8 (with ± 4.6 variance), improving accuracy by roughly 17% over the baseline. Thus, showcasing that the turbofan engine (D_S) dataset could effectively be used to transfer knowledge to help predictions on a cross-domain lithium-ion battery dataset.

8. Conclusion

The research presented evidence supporting the feasibility of shared knowledge transfer between datasets from heterogeneous domains.

First, MMD was used to quantify the statistical distance of distributions in the reproducing kernel Hilbert space between two cross domain datasets (aircraft turbofan engines and lithium-ion batteries). The results demonstrated only a moderate domain gap between the source and target sets.

To investigate transferability feasibility several TL fine-tuning-based approaches were used. It was found that "naïve based" transfer learning can be detrimental due to domain mismatch (manifesting as negative transfer). However, it can be mitigated through more thought-out TL strategies. More, specifically the study introduced a trainable adapter layer with penultimate unfreezing (fine tuning) and found it to be effective. The approach achieved roughly a 17% lower RMSE in RUL prediction compared to a baseline model trained from scratch on the battery data. This serves as a demonstration that cross-domain TL between fundamentally different systems (Thermo-fluid-mechanical and electrochemical) can improve RUL prediction accuracy of ML models. These findings have important implications for the field of PdM. They suggest that even markedly different industrial assets can share predictive models when a common structure in their degradation patterns exists (as reflected by a moderate MMD distance).

Future Work should explore other more advanced domain adaptation techniques on this dataset pair to further improve performance (such as DANN [18] or DDFN [9]). Moreover, other heterogeneous domain pairs should be investigated. Additionally, investigating the conditions that contribute to negative transfer would be valuable to develop guidelines on how to implement TL in PdM tasks.

9. References

[1] Achouch M, Dimitrova M, Ziane K, Sattarpanah Karganroudi S, Dhouib R, Ibrahim H, Adda M (2022). *On Predictive Maintenance in Industry 4.0: Overview, Models, and Challenges.* Applied Sciences. 2022; **12(**16):8081.

[2] Nunes P, Santos J, Rocha E (2023). *Challenges in predictive maintenance - A review.* CIRP Journal of Manufacturing Science and Technology. **40**:53–67.

[3] Hattula E, Zhu L, Raninen J, Oksanen J, Hyyppä J (2023). *Advantages of Using Transfer Learning Technology with a Quantative Measurement.* Remote Sensing. **15**(17):4278.

[4] Iman M, Arabnia HR, Rasheed K (2023). *A Review of Deep Transfer Learning and Recent Advancements*. Technologies. **11**(2):40. .

[5] Weiss K, Khoshgoftaar TM, Wang D (2016). A survey of transfer learning. Journal of Big Data **3**:9.

[6] Otović E, Njirjak M, Jozinović D, Mauša G, Michelini A, Štajduhar I (2022), *Intra-domain and cross-domain transfer learning for time series data -- How transferable are the features?* Knowledge-Based Systems, **239:**107976.

[7] Sun C, Ma M, Zhao Z, Tian S, Yan R, Chen X (2019). *Deep Transfer Learning Based on Sparse Autoencoder for Remaining Useful Life Prediction of Tool in Manufacturing.* IEEE Trans. Ind. Inf., 15:4; p. 2416-2425.

[8] Xiang S, Li P, Luo J, Qin Y (2024). *Micro Transfer Learning Mechanism for Cross-Domain Equipment RUL Prediction*. IEEE Trans. Automat. Sci. Eng., p. 1460-1470.

[9] Na L, Cai B, Zhang C, Liu J, Li J (2025). *A heterogeneous transfer learning method for fault prediction of railway track circuit*, Engineering Applications of Artificial Intelligence; **140**:109740.

[10] Frederick DK, DeCastro JA, Litt JS, *User's Guide for the Commercial Modular Aero-Propulsion System Simulation (C-MAPSS).* NTRS – NASA Techinical Report. Accessed: Jan. 30, 2025. [Online]. Available: https://ntrs.nasa.gov/citations/20070034949

[11] Saxena A, Goebel K, Simon D, Eklund N (2008). *Damage propagation modeling for aircraft engine run-to-failure simulation*. In: 2008 International Conference on Prognostics and Health Management, Denver, CO, USA: IEEE, p. 1–9.

[12] Vinuales I, *Battery Remaining Useful Life (RUL)*. Kaggle. Accessed: Jan. 01, 2025. [Online]. Available: https://www.kaggle.com/datasets/ignaciovinuales/battery-remaining-useful-life-rul

[13] Saha B, Goebel K (2007). *Battery Data Set*, NASA Prognostics Data Repository, NASA Ames Research Center, Moffett Field, CA. Accessed: Jan. 01, 2025. [Online]. Available: https://phm-datasets.s3.amazonaws.com/NASA/5.+Battery+Data+Set.zip

[14] Wang Y, Zhao Y, Addepalli S (2020). *Remaining Useful Life Prediction using Deep Learning Approaches: A Review.* Procedia Manufacturing, **49;** p. 81–88.

[15] Nguyen N, Quanz B (2021). *Temporal Latent Auto-Encoder: A Method for Probabilistic Multivariate Time Series Forecasting*, AAAI, 35:10, p. 9117–9125.

[16] Arthur G, Borgwardt KM, Rasch MJ, Scholkopf B, Smola A (2012). *A Kernel Two-Sample Test.* Journal of Machine Learning Research, **13**, p. 723–773.

[17] Ganin Y, Ustinova E, Ajakan H, Germain P, Larochelle H, Laviolette F, March M, Lempitsky V (2016). *Domain-adversarial training of neural networks.* Journal of machine learning research, **17**:59, p.1-35.

[18] Wang Z, Dai Z, Poczos B, Carbonell J (2019) *Characterizing and Avoiding Negative Transfer.* Proceedings of the IEEE/CVF Conference on Computer Vision and Pattern Recognition (CVPR), p.11293-11302

10. Appendices

10.1. Appendix A Autoencoder High Level Architecture

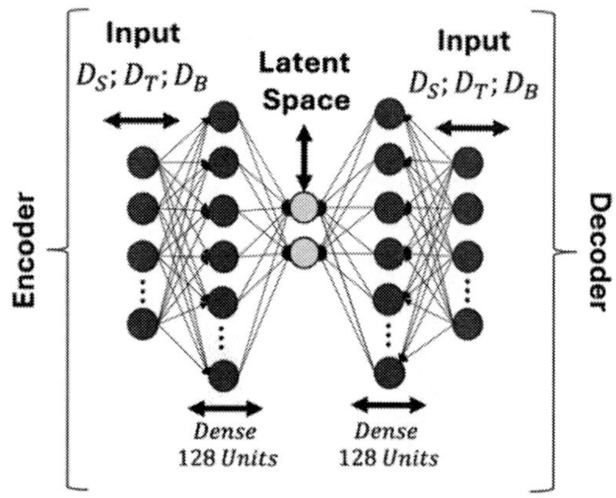

10.2. Appendix B Autoencoder: Summary of Hyper Parameters

Parameter	Value
Input Layer	11 (D_S); 7 (D_T)
Encoder Layer	128 Units
Latent Layer	10 Units
Activation	ReLU
Decoder Layer	128
Activation	ReLU
Training Settings	
Optimizer	Adam
Learning Rate	0.001
Loss Function	MSE

10.3. Appendix C: Baseline Model Architecture

Type	Value
Input Layer	11 (D_S); 7 (D_T)
Dense Layer 1	128 Units
Drop Out Layer	20%
Activation	ReLU
Dense Layer 2	64
Activation	ReLU
Training Settings	
Optimizer	Adam
Learning Rate	0.001
Loss Function	MSE
Seeds	1, 2, 3, 4, 5 (. set_random_seed)

The challenges of applying hybrid modelling to real-world engineering systems

Karl G. Hellberg[a,*], P.Stephan Heyns[a], Johann Wannenburg[a]

[a]*Centre for Asset Integrity Management, Department of Mechanical and Aeronautical Engineering, University of Pretoria, Pretoria, 0002, South Africa*

**corresponding author email: karl.hellberg@tuks.co.za*

Abstract

The application of hybrid modelling to real-world complex engineering systems can face problems related to the computational expense of the physics-based models involved, and to the partial observation of the system state. Reduced-order modelling and data assimilation can successfully address these issues, but result in their own challenges. Using simplified problems, we investigate whether reduced-order models can retain sufficient degrees of freedom to represent system states from the real process, and whether the predictions made by hybrid models can be relied on for system state variables that are never observed. For reduced order models, taking anticipated model uncertainty into account when reducing the number of degrees of freedom of the model is found to be critical to achieving an improvement relative to the physics-based model, and the physics-based component of hybrid models gives them a clear advantage relative to data-driven models when it comes to making predictions for unobserved state variables.

Keywords: hybrid models, reduced-order modelling, proper orthogonal decomposition, data assimilation

1. Background and motivation

Complex real-world engineering systems must often be operated without a full understanding of their condition and of their response to changing process conditions. Two sources of information are typically available to help combat this problem: sensors installed on the system, and mathematical models of the system. The digital twin framework proposes using both these sources of information to construct a digital representation of the system [1], so that real-time information about the system becomes available for better decision-making.

Mathematical models form the backbone of digital twins, and must be of the highest possible fidelity to maximize the value of the information available from the digital representation. Classically, mathematical models would be constructed using either physics-based or data-driven modelling techniques. Both have significant drawbacks when used in isolation [2]. Hybrid models combine these two modelling techniques, leveraging each of their strengths while avoiding each of their drawbacks. The hybrid modelling concept encompasses a wide variety of ways in which physics-based and data-driven models can be combined [3]. We will focus here on delta learning, in which the error made by a physics-based model relative to a real process is corrected by a data-driven model.

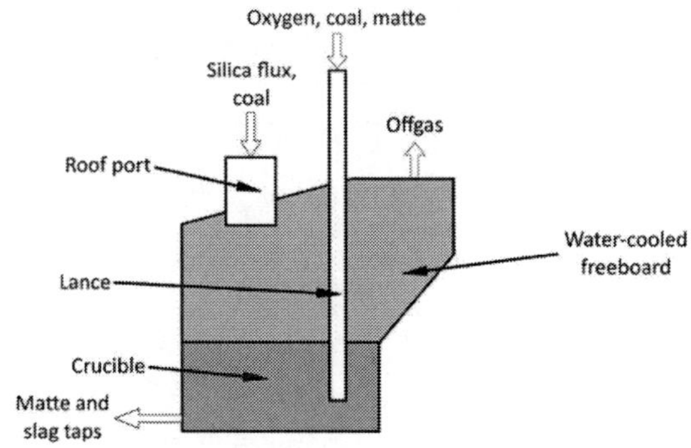

Fig. 1. A schematic representation of the process converter.

We study the application of hybrid modelling to an example of a complex real-world engineering system in the form of the water-cooled freeboard of a process converter. The process converter is illustrated schematically in Fig. 1. As shown, the water-cooled freeboard is situated above the crucible in which molten matte is contained. It consists of a repeating pattern of cooling tubes connected to each other by webs, as is shown in Fig. 3. Of particular interest is that the fireside of the freeboard interacts with the gases in the process converter, as well as the molten material in the crucible. These complex processes are modelled using simple convection boundary conditions, and we apply a hybrid modelling methodology to address the resulting deficiencies in the physics-based model.

2. Objectives and Scope

In this paper, we concern ourselves with two challenges that arise from the two starting points of the hybrid model: the available physics-based model, and the available data.

The physics-based model is a transient finite element model, the computational cost of which is prohibitive. We address this challenge by making use of reduced-order modelling [4-10]. In doing so, we will discuss some pitfalls that the process of constructing a reduced-order model creates, and describe steps that we have taken to mitigate these.

The available data originate from thermocouples installed on the freeboard. Compared to the number of system state variables, which correspond to the degrees of freedom of the finite element model, the number of thermocouples is small. As a result, learning the correct dynamics of the system with the data-driven component of the hybrid model is challenging, especially for system state variables that are never observed.

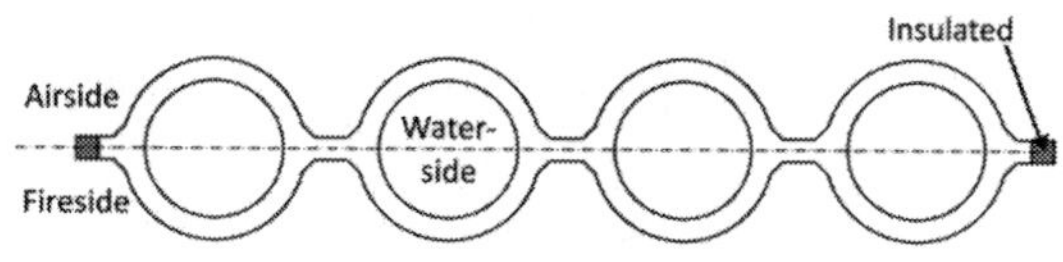

Fig. 3. The geometry for problem 1.

We make use of simplified problems with which we can recreate and study these two challenges of the real problem, while not being ultimately limited by them. Section 3 describes the simplified models which we use, our approach to reduced-order and hybrid modelling, and the numerical experiments which we have conducted. The results we have obtained are presented and discussed in section 4. Finally, conclusions and future work are presented in section 5.

3. Methodology and Computational Approach

3.1. Simplified problems

We use two different simplified problems in our study. The geometry of problem 1 is shown in Fig. 3, and that of problem 2 is shown in Fig. 2. Across the two problems, there are three different sets of convection boundary conditions: the fireside, waterside and airside boundary conditions. Problem 1 is intended to be representative of a small section of the water-cooled freeboard, with the fireside exposed to the process conditions, the airside exposed to the air outside the process converter, and the waterside exposed to the flow of cooling water inside the cooling tubes. For problem 2, there is no such intention behind the naming of the boundary conditions, but their parameters have similar values and temporal and spatial behaviour to those in problem 1.

For each of the two simplified problems, we construct two transient thermal finite element models. The first of these models is meant to represent the simulation model that we would have available in the real problem. For this model, all three boundary condition types are assumed to be spatially constant across the faces to which they are applied, though the fireside convection coefficient varies temporally. The second model represents the "real process", from which our simulated measurements are obtained. It differs from the first model in that the fireside convection coefficient is assumed to vary both spatially and temporally. The fireside boundary condition parameters are sampled from a Gaussian process prior [11] with different length scales in the spatial and temporal dimensions. The error of the simulation model relative to the "real process" is therefore introduced as insufficient fidelity in modelling the boundary conditions, which is representative of the error which we expect to be present in the model of the water-cooled freeboard.

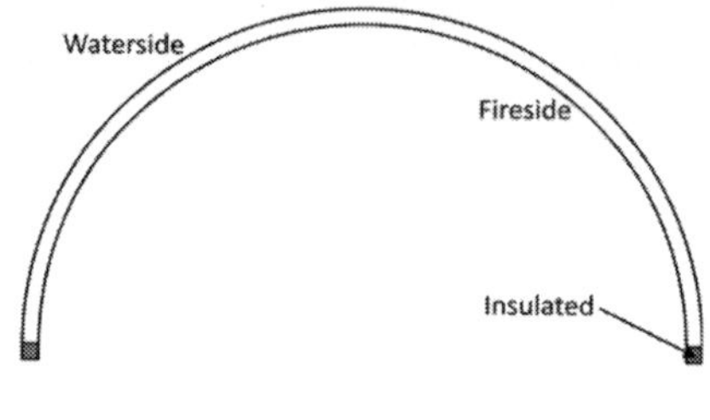

Fig. 2. The geometry for problem 2.

3.2. Reduced-order modelling

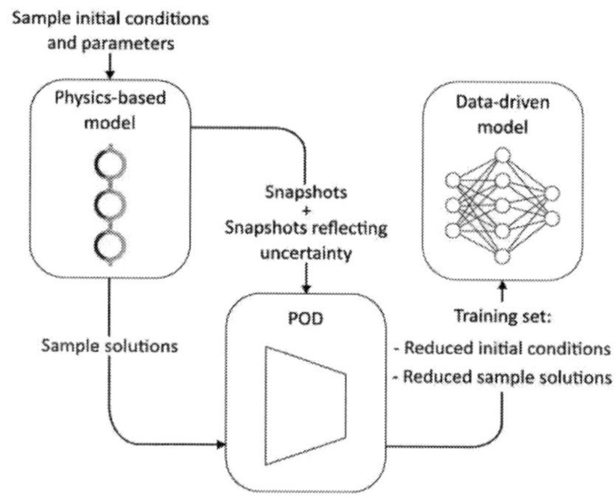

Fig. 4. An illustration of the reduced-order modelling process.

A reduced-order model (ROM) provides approximately the same solutions as the full-order finite element model, but at reduced computational cost. The approach which we follow is illustrated schematically in Fig. 4. As a first step, the dimensionality of the system state is reduced using proper orthogonal decomposition (POD) [12] and the method of snapshots [13]. Normally, snapshots are generated from the best-knowledge physics-based model only [14-17]. Because the system states that can be represented using POD are in the span of the snapshots, this may make it impossible to adequately represent system states resulting from the real process. As a result, hybrid modelling may be ineffective. This is a significant pitfall, which we combat by generating additional snapshots to reflect anticipated model uncertainty.

After the POD subspace has been defined, sample solutions are generated from the physics-based model using sample initial conditions and parameters. The initial conditions and sample solutions are projected into the POD subspace to assemble the training set for a data-driven model. The reduced-order modelling approach described here is similar to the approach followed in [8-10], which we have extended for the purpose of hybrid modelling by taking anticipated model uncertainty into account when reducing the dimensionality of the system state.

3.3. Hybrid model architecture

We create a hybrid model by using a data-driven model to correct the output of a physics-based model. We use a feed-forward neural network (FFNN) with the architecture shown in Fig. 5. The FFNN has three inputs: a set of parameters, the initial condition in the POD subspace, and the output, in the POD subspace, of the physics-based model corresponding to the initial condition and parameters. This set of inputs was found in [18] to work well for delta learning. The inputs are then passed through two hidden layers with *tanh* activation functions. Finally, there is a skip connection from the ROM output to the hybrid model output, so that the model can easily be initialized to give the same output as the physics-based model.

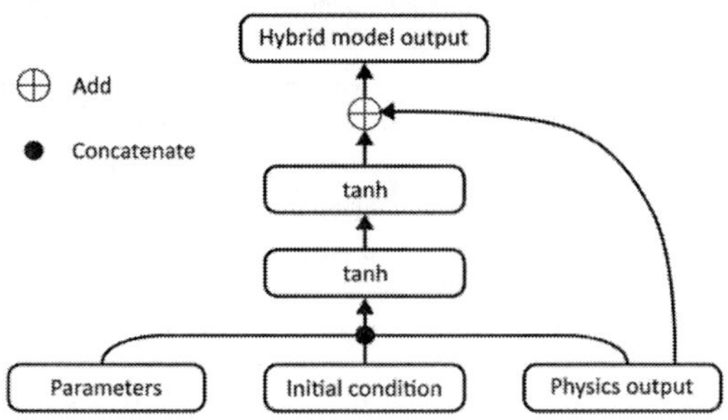

Fig. 5. The FFNN architecture used for the hybrid model.

3.4. Data assimilation

The training set of the hybrid model described in section 3.3 must contain the parameters and the initial system state to generate the inputs of the FFNN, as well as the final system state to evaluate the error made by the FFNN. Neither of these system states are available, because the limited number of sensors installed on the system provide only partial information about the system state at any time. We use data assimilation to estimate the full system state by combining the information available from the sensors and from the best-knowledge model. While numerous data assimilation algorithms exist [19-21], we make use of a merging particle filter (MPF) [22].

Once a training set has been obtained from the data assimilation algorithm, the hybrid model is trained, and a new best-knowledge model becomes available. As a result, the full system states must be re-estimated using data assimilation. The resulting iterative process with which we train our hybrid models is known as machine learning-data assimilation (MLDA) [23-25].

When following the MLDA algorithm, the hybrid model is trained using system states that are estimated based on the best-knowledge model, which could be the physics-based model or the previous iteration of the hybrid model. For unobserved system state variables, it is therefore possible that the hybrid model learns only quantities that are useful for predicting the evolution of the observed system states, rather than a physically meaningful quantity like temperature in our case. The physics-based model, as the starting model, is important for ensuring that physically meaningful quantities are learned, as one of the numerical experiments in section 3.5 is designed to show.

3.5. Numerical experiments

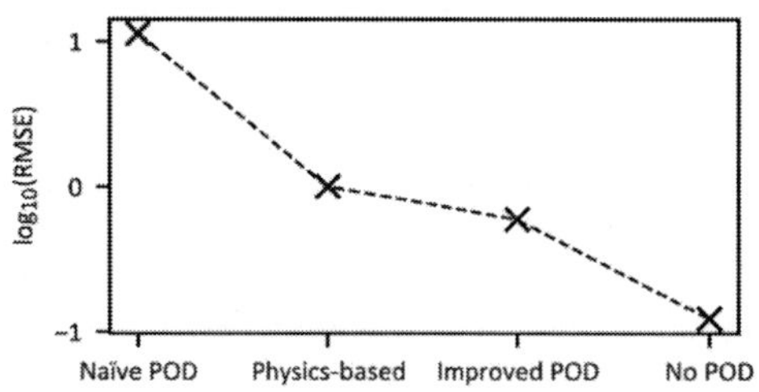

Fig. 6. The performance in terms of $\log_{10}$(RMSE) of different models in numerical experiment 1.

We conduct two different numerical experiments. Experiment 1 aims to investigate the effect of adding snapshots reflecting model uncertainty into the set of snapshots used for POD when constructing reduced-order models as shown in Fig. 4. We use problem 1 from section 3.1, and to simplify hybrid model training, we assume that all state variables (i.e. all nodes of the finite element model) are observed. This means that data assimilation and the use of the MLDA algorithm as described in section 3.4 are not necessary. We use as physics-based inputs into the hybrid model in Fig. 5 the following: a "naïve" ROM constructed with only snapshots from the best-knowledge physics-based model, an improved ROM constructed with additional snapshots reflecting model uncertainty, and the full-order physics-based model.

Experiment 2 aims to investigate the performance of hybrid models for state variables that are never observed. For this, we use problem 2 from section 3.1. Two different hybrid models are trained. One uses as the physics-based input an improved ROM of the simulation model. The second uses as the physics-based input a version of this ROM that has been manipulated to retain the same output where the system state is observed, while giving a different output where the system state is not observed, resulting in a worst-case starting point. Unlike for experiment 1, use is made of data assimilation and the MLDA algorithm as discussed in section 3.4.

4. Key Results and Discussion

4.1. Experiment 1

The results of experiment 1 are shown in Fig. 6 in terms of the logarithm of the root mean square error ($\log_{10}$(RMSE)) averaged over all degrees of freedom of the full-order finite element model, and over all time steps. All models, except the physics-based model, are hybrid models. The naïve POD approach is limited by a poor representation of the system states that result from the real process, causing a large model error. The physics-based model, despite the inadequate modelling of the convection boundary condition, can perform better. Improved representation of the real system states allows the hybrid modelling approach to make improvements relative to the original physics-based model. The best results are achieved when the full-order physics-based model is used in the hybrid model. Importantly, neither the "physics-based" nor the "no-POD" approaches are viable for process converter problem due to computational limitations.

4.2. Experiment 2

In Fig. 7, the hybrid model error in terms of the $\log_{10}$(RMSE) is shown for different MLDA iterations. Again, it has been averaged over all degrees of freedom of the full-order finite element model, and over all time steps. Both starting points result in similar performance where the system is observed, with the modified model even resulting in better performance. Where it is not observed, however, the good base starting point gives better result, and no improvement is observed from the poor modified starting point. For this scenario, a hybrid modelling strategy is not able to overcome the shortcomings of the initial model. These results underline the importance of the physics-based model

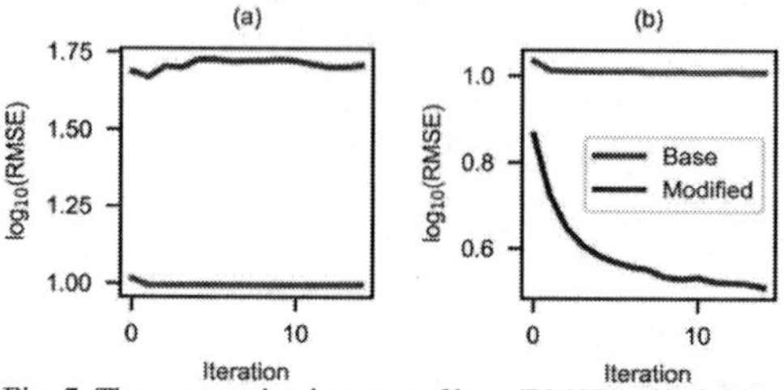

Fig. 7. The progression in terms of $\log_{10}$(RMSE) of hybrid models trained from different starting points for numerical experiment 2. Results for all state variables are shown in (a), and results for measured state variables only are shown in (b).

in the hybrid modelling strategy. The two cases are indistinguishable in a purely data-driven approach. In a hybrid modelling approach, the good initial model is consistent with the known physics of the system, while the poor initial model is not. The hybrid model based on the poor starting point can model the system well where it is observed because the unobserved state variables still contain valuable information to do so. Importantly, however, the unobserved state variables only become interpretable as temperature when starting from a good physics-based model, which constrains their relationship to the observed state variables.

5. Conclusion and Future Work

When we apply a hybrid modelling strategy to a real-world complex engineering system in the form of a process converter freeboard, we are immediately faced with two problems. Firstly, the finite element models that are available are prohibitively computationally expensive, and, secondly, only a few of the state variables of the system are observed by sensors. These two problems can be addressed through the application of reduced-order modelling and data assimilation respectively.

Nevertheless, as we have discussed, a naïve approach to reduced-order modelling, or total reliance on the measurements can both result in poor models. Reduced-order models that are to be used in hybrid models must take anticipated model uncertainty into account during their construction, and a proper physics-based model as a starting point is a critical requirement for obtaining interpretable results for unobserved system variables.

In summary therefore, the primary obstacles to creating hybrid models of the process converter freeboard can be addressed. In doing so, additional issues could be created. We have successfully reduced the extent of these issues on simplified problems. Future work will apply the lessons learned to the real water-cooled freeboard problem, making use of an existing finite element model and of thermocouples installed on the freeboard.

References

[1] Grieves M, Vickers J (2017). *Digital Twin: Mitigating Unpredictable, Undesirable Emergent Behavior in Complex Systems*. In: Transdisciplinary Perspectives on Complex Systems, Springer International Publishing; pp. 85–113.

[2] Rasheed A, San O, Kvamsdal T (2020). *Digital Twin: Values, Challenges and Enablers From a Modeling Perspective*. IEEE Access, **8**: pp. 21980–22012.

[3] Willard J, Jia X, Xu S, Steinbach M, Kumar V (2023). *Integrating Scientific Knowledge with Machine Learning for Engineering and Environmental Systems*. ACM Computing Surveys, **55**: pp. 1–37.

[4] Bialecki RA, Kassab AJ, Fic A (2005). *Proper orthogonal decomposition and modal analysis for acceleration of transient FEM thermal analysis*. International Journal for Numerical Methods in Engineering, **62**: pp. 774–797.

[5] Ito K, Ravindran SS (1998). *A Reduced-Order Method for Simulation and Control of Fluid Flows*. Journal of Computational Physics, **143**: pp. 403–425.

[6] Paul-Dubois-Taine A, Amsallem D (2015). *An adaptive and efficient greedy procedure for the optimal training of parametric reduced-order models*. International Journal for Numerical Methods in Engineering, **102**: pp. 1262–1292.

[7] Cicci L, Fresca S, Manzoni A (2022). *Deep-HyROMnet: A Deep Learning-Based Operator Approximation for Hyper-Reduction of Nonlinear Parametrized PDEs*. Journal of Scientific Computing, **93**: p. 57.

[8] Xiao D, Fang F, Buchan AG, Pain CC, Navon IM, Muggeridge A (2015). *Non-intrusive reduced order modelling of the Navier–Stokes equations*. Computer Methods in Applied Mechanics and Engineering, **293**: pp. 522–541.

[9] Xiao D, Fang F, Pain C, Hu G (2015). *Non-intrusive reduced-order modelling of the Navier–Stokes equations based on RBF interpolation*. International Journal for Numerical Methods in Fluids, **79**: pp. 580–595.

[10] Walton S, Hassan O, Morgan K (2013). *Reduced order modelling for unsteady fluid flow using proper orthogonal decomposition and radial basis functions*. Applied Mathematical Modelling, **37**: pp. 8930–8945.

[11] Williams CKI, Rasmussen CE (2006). *Gaussian Processes for Machine Learning*. Cambridge: MIT Press.

[12] Rozza G, Stabile G, Ballarin F (2022). *Advanced Reduced Order Methods and Applications in Computational Fluid Dynamics*. Philadelphia, PA: Society for Industrial and Applied Mathematics.

[13] Sirovich L (1987). *Turbulence and the dynamics of coherent structures. I. Coherent structures*. Quarterly of Applied Mathematics, **45**: pp. 561–571.

[14] Pawar S, Ahmed SE, San O, Rasheed A (2020). *Data-driven recovery of hidden physics in reduced order modeling of fluid flows*. Physics of Fluids, **32**

[15] San O, Maulik R (2018). *Machine learning closures for model order reduction of thermal fluids*. Applied Mathematical Modelling, **60**: pp. 681–710.

[16] San O, Maulik R (2018). *Neural network closures for nonlinear model order reduction*. Advances in Computational Mathematics, **44**: pp. 1717–1750.

[17] Wan ZY, Vlachas P, Koumoutsakos P, Sapsis T (2018). *Data-assisted reduced-order modeling of extreme events in complex dynamical systems*. PLOS ONE, **13**: p. e0197704.

[18] Daw A, Karpatne A, Watkins W, Read J, Kumar V (2017). *Physics-guided Neural Networks (PGNN): An Application in Lake Temperature Modeling*. arXiv:1710.11431.

[19] Carrassi A, Bocquet M, Bertino L, Evensen G (2018). *Data assimilation in the geosciences: An overview of methods, issues, and perspectives*. WIREs Climate Change, **9**.

[20] van Leeuwen PJ, Künsch HR, Nerger L, Potthast R, Reich S (2019). *Particle filters for high-dimensional geoscience applications: A review*. Quarterly Journal of the Royal Meteorological Society, **145**: pp. 2335–2365.

[21] Arulampalam MS, Maskell S, Gordon N, Clapp T (2002). *A tutorial on particle filters for online nonlinear/non-Gaussian Bayesian tracking*. IEEE Transactions on Signal Processing, **50**: pp. 174–188.

[22] Nakano S, Ueno G, Higuchi T (2007). *Merging particle filter for sequential data assimilation*. Nonlinear Processes in Geophysics, **14**: pp. 395–408.

[23] Brajard J, Carrassi A, Bocquet M, Bertino L (2020). *Combining data assimilation and machine learning to emulate a dynamical model from sparse and noisy observations: A case study with the Lorenz 96 model*. Journal of Computational Science, **44**: p. 101171.

[24] Farchi A, Laloyaux P, Bonavita M, Bocquet M (2021). *Using machine learning to correct model error in data assimilation and forecast applications*. Quarterly Journal of the Royal Meteorological Society, **147**: pp. 3067–3084.

[25] Wikner A, Pathak J, Hunt BR, Szunyogh I, Girvan M, Ott E (2021). *Using data assimilation to train a hybrid forecast system that combines machine-learning and knowledge-based components*. Chaos: An Interdisciplinary Journal of Nonlinear Science, **31**.

Graph Structure Learning in Engineering: A Study on Temporal Domains

Ruben Grobler[a,b] and Daniel N. Wilke[c]

[a]University of Pretoria, Lynnwood Rd, Hatfield, Pretoria, 0002, South Africa
[b]Kriterion AI Research, 121 Boshoff St, Nieuw Muckleneuk, Pretoria, 0081
[c]University of Witwatersrand, 1 Jan Smuts Ave, Braamfontein, Johannesburg 2017, South Africa

** Corresponding author email: ruben.grobler@tuks.co.za*

Abstract

In recent years, the use of graphs to model complex data relationships has seen a surge in popularity across various domains [1]. In particular, directed graphs have gained prominence in engineering applications, which represent processes and the underlying cause-and-effect relationships between components or subsystems. These graphs provide a powerful framework to model the dependencies and interactions in complex systems. However, one significant challenge arises when the underlying physics or governing equations of a problem are unknown, making it difficult to derive the graph's structure directly. Graph structure learning algorithms aim to infer the relationships between variables by analyzing data, and numerous state-of-the-art algorithms are now available for this purpose [2]. This study aimed to investigate and compare the performance of a wide range of these graph structure learning algorithms, particularly in the context of temporal data where causal identifiability is not guaranteed [3], commonly encountered in mechanical engineering problems. This study evaluated nine general-purpose, static graph discovery algorithms, and three were specifically designed for time-series data. The algorithms were tested on a simulated spring-mass-damper system for the time-series (temporal) data. This system was chosen because it represents a canonical system in mechanical engineering. A ground truth graph was generated, which served as a reference point for evaluating the accuracy of the algorithms. The results from the algorithms were compared to the ground truth. This study provides key insights into the performance of various graph structure learning algorithms in the temporal domain, with implications for their application in real-world engineering problems.

Keywords: Dynamics, Causality, Directed Acyclic Graphs, Graph Discovery, Optimization, Deep Learning

1 Background and Motivation

The practical application of engineering has always relied heavily on the understanding of underlying cause and effect relationships that govern the interactions between parameters of an engineering problem. Traditional engineering relies on physics-based models that encode causal relationships in an analytical solution based on rigorous mathematics and well-chosen assumptions. These models usually manifest in the form of one or more mathematical relationships that describe the outcome as some function of one or more inputs; the causes.

There have been lots of successful data-driven approaches for representing physical systems as functions. Modern data-driven approaches traditionally make use of artificial neural networks and their Universal Function Approximation property [4] to accurately represent physical systems. These models have found great success, especially in modern engineering applications with rapid advancement of computing power allowing for models to be implemented faster than ever. However, neural networks consist of many weights and nodes in the hidden layers that connect the input and output layers of these networks. Thus, it's very difficult, if not impossible, for humans to understand exactly what the neural network is doing to transform the input to output. This makes it very difficult to interpret neural networks and understand the underlying cause and effect relationships of inputs and outputs without intervention approaches like conditioning on specific inputs to determine their effect.

Graphical models, directed graphs in particular, show potential as a data-driven method of representing physical systems due to their clearly defined directional flow of information as well as their interpretability. It is thus useful to explore their application in the representation of physical systems, particularly dynamic ones. This allows for models to encode the reasoning of a problem [5].

2 Objectives and Scope

This study focuses on the data driven discovery of a binary directed acyclic graph that most accurately represents the physical system that generated said data. The study focuses on comparing popular static and dynamic graph structure learning algorithms. The study aims to determine how well graph structure learning algorithms perform when applied to a typical dynamics problem encountered in engineering where causal identifiability is not guaranteed. In doing this, it aims to determine how successfully static graph discovery algorithms that have been shown to perform well on purely static data [2, 3, 6–12] generalize to a temporal setting when the problem is cast as static, as well as, determining how well all the algorithms perform when some of their underlying assumptions are violated. These algorithms are compared with the performance of dynamic algorithms in an engineering context. The binary adjacency matrices representing the graphs recovered by the graph discovery algorithms were compared to the ground truth graphs generated from the sample datasets and critically discussed.

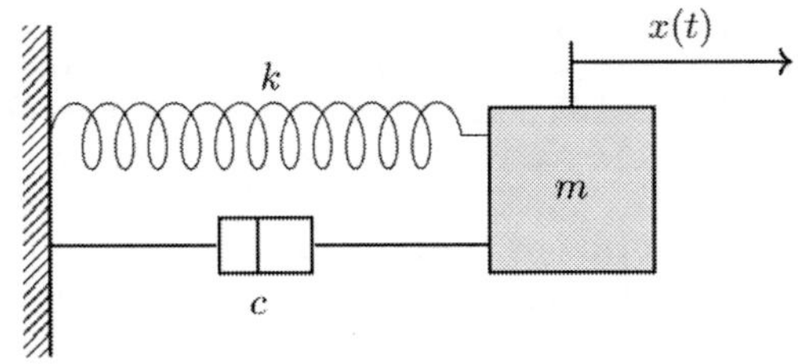

Figure 1: Free vibrating spring mass damper system

3 Methodology and Computational Approach

To generate time series data, a generalized 1D single degree of freedom spring mass damper system was used. Spring mass damper systems are well suited for causal datasets since it is reasonably intuitive to infer causal relationships between various properties from the data generation process. A zero forcing function vector was used to eliminate the forcing function as a potential latent confounder and yields the following governing equation:

$$m\ddot{x} + c\dot{x} + kx = 0 \tag{1}$$

The mass, damping, stiffness as well as the initial conditions were randomly initialized. Displacement and velocity data for each mass was generated by numerically integrating the systems of governing equations over a discrete time interval of 20 seconds, with $s = 1000$ evenly spaced samples using Runge-Kutta numerical integration [13]. Acceleration data was then generated by plugging the displacement and velocity data into the governing equation. Properties such as spring force, kx, and damping force, $c\dot{x}$, were also added to the dataset to represent the complete physics of the system. Gaussian multiplicative noise was then applied to the generated data.

The use of Runge-Kutta integration allowed for the generation of a ground truth DAG that was the only member in its Markov equivalence class [5]. Although Runge-Kutta could be used to describe the full structural causal model (SCM) of the data [14], for comparison between algorithms that do and do not learn the underlying SCM, a binary ground truth was used instead. The ground truth adjacency matrix is shown in fig. 2.

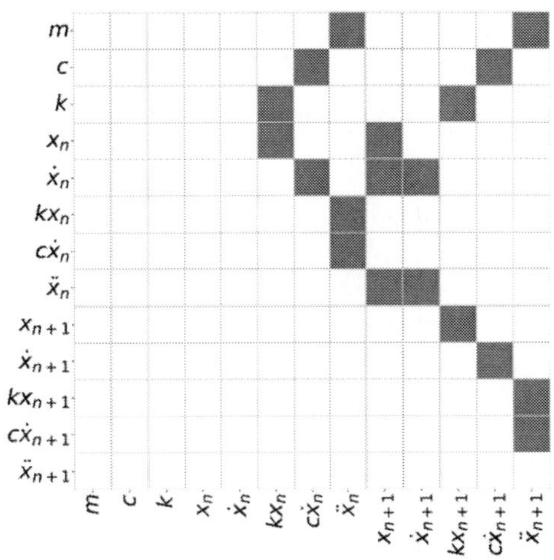

Figure 2: Ground truth adjacency matrix of the spring mass damper system. Edge direction convention is from row index to column index.

The tests were done on data generated by the numerical integration of eq. (1). For the static cases, the mean of the analytical signal envelope [15, 16] was used to represent the dynamic responses parameterized by a specific combination of mass, damping, stiffness, and initial conditions. This was done to eliminate the need for auto-correlated samples during static testing which violates the i.i.d. assumption made by static algorithms [2, 3, 6–8, 10–12]. For the dynamic cases, the system parameters such as mass, damping, and stiffness were introduced as constant columns. These modifications allowed for the discovered graphs to be compared to the same ground truth for a more even comparison. Test case data was generated by keeping specific system input parameters constant whilst varying others. This resulted in 31 total test cases. Each algorithm was run five times on each test case and the average of these runs used for comparison against the ground truth. The data and ground truth generation for both cases is discussed in more detail in Appendix A.

A variety of popular algorithms were chosen to be compared. These algorithms included statistical algorithms as well as gradient based algorithms and general static as well as dynamic algorithms were selected for comparison. All algorithms were implemented using existing graph discovery libraries and the default hyperparameters presented by the original authors. Most libraries were modified to ensure correct implementation of included parallelization functionality. The graph discovery algorithms compared are shown in table 1.

Table 1: Graph discovery algorithms compared

Algorithm	Type
DirectLiNGAM [6]	Static
LiNGAM [2]	Static
GAE [7]	Static
GOLEM [8]	Static
GraNDAG [9]	Static
NOTEARS [10]	Static
NOTEARS-MLP [11]	Static
MCSL [12]	Static
PC [3]	Static
DyNOTEARS [17]	Dynamic
PCMCI [18]	Dynamic
VAR-LiNGAM [19]	Dynamic

Although graph learning is an unsupervised task, a supervised performance evaluation of the models was done by comparing results to the ground truth graphs generated from each respective dataset. The average discovered graphs were compared to the ground truth and discussed. The average F1 score of each algorithm was also compared.

4 Key Results and Discussion

The average learned adjacency matrices were compared to the ground truth adjacency matrices in fig. 3 and the average F1-scores are shown in table 2.

Table 2: Average F1-scores across all test cases

Algorithm	F1-score
DirectLiNGAM	0.346
DyNOTEARS	0.268
GAE	0.097
GOLEM	0.155
GraNDAG	0.165
LiNGAM	0.293
MCSL	0.222
NOTEARS	0.245
NOTEARS-MLP	0.171
PC	0.229
PCMCI	0.345
VAR-LiNGAM	0.273

The two top performing algorithms were DirectLiNGAM and PCMCI, achieving the highest F1-scores. These scores are similar to the F1-scores achieved in literature on real-world dynamic datasets [17, 19]. Dynamic algorithms favoured learning dense graphs, most likely due to their increased sensitivity to auto-correlated as can be seen by the increased number edges predicted that involve dynamic parameters. Poor sensitivity to static parameters resulted in lower scores. On average, the static algorithms performed poorly except for the LiNGAM models, achieving the highest and 3rd highest F1-scores respectively. This suggests that there is potential for using static graph discovery algorithms on dynamic data through sensible pre-processing and sufficient hyperparameter tuning beyond the scope of this study.

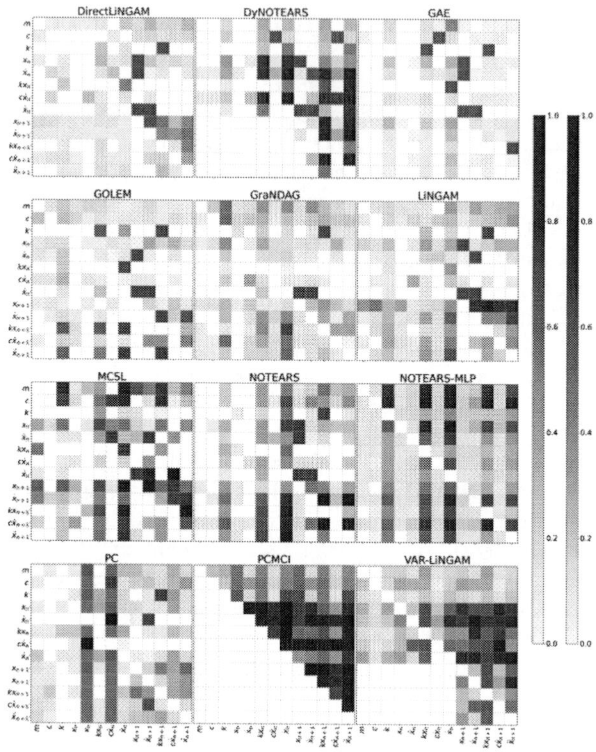

Figure 3: Average discovered graphs vs ground truth. Ground truth edges shown in blue. False positives and true positives shown in red and green respectively.

5 Conclusion and Future Work

Majority of static algorithms performed poorly on the dynamic dataset, even when cast as a static problem. The promising performance of the LiNGAM models suggests potential for static algorithms on augmented datasets, given sufficient tuning. Dynamic algorithms showed poor sensitivity to static parameters and favoured learning dense graphs. The best performing algorithms showed competitive performance when compared to other real-world datasets however, hyperparameter tuning is recommended for future study. Additionally, it is recommended that future studies investigate quantifying performance on similar datasets beyond binary DAGs by using learned structural causal models and reconstruction loss. The emerging field of graph neural networks is also recommended for investigation as a possible alternative to discovering graph structure of a dynamic system since they have shown promising results in the realm of representation learning [20].

Acknowledgements

R.G. acknowledges the support of Kriterion AI Research in the preparation of this document.

References

[1] Aayush Chudiwal. Advancements in probabilistic graphical models: Representation, inference, and learning in complex uncertain domains. August 2024.

[2] Shohei Shimizu, Patrik O. Hoyer, Aapo Hyvärinen, and Antti Kerminen. A linear non-gaussian acyclic model for causal discovery. *Journal of Machine Learning Research*, 7:2003–2030, December 2006.

[3] Peter Spirtes, Clark N. Glymour, and Richard Scheines. *Causation, prediction, and search*. Adaptive computation and machine learning. The MIT Press, Cambridge, Massachussetts London, second edition, 2000.

[4] Kurt Hornik, Maxwell Stinchcombe, and Halbert White. Multilayer feedforward networks are universal approximators. *Neural Networks*, 2(5):359–366, 1989.

[5] Daphne Koller and Nir Friedman. *Probabilistic graphical models: principles and techniques*. Adaptive computation and machine learning. MIT Press, Cambridge, MA, 2009.

[6] Shohei Shimizu, Takanori Inazumi, Yasuhiro Sogawa, Aapo Hyvärinen, Yoshinobu Kawahara, Takashi Washio, Patrik O. Hoyer, and Kenneth Bollen. Directlingam: A direct method for learning a linear non-gaussian structural equation model. April 2011.

[7] Ignavier Ng, Shengyu Zhu, Zhitang Chen, and Zhuangyan Fang. A graph autoencoder approach to causal structure learning. November 2019.

[8] Ignavier Ng, AmirEmad Ghassami, and Kun Zhang. On the role of sparsity and dag constraints for learning linear dags. January 2021.

[9] Sébastien Lachapelle, Philippe Brouillard, Tristan Deleu, and Simon Lacoste-Julien. Gradient-based neural dag learning. February 2020.

[10] Xun Zheng, Bryon Aragam, Pradeep Ravikumar, and Eric P. Xing. Dags with no tears: Continuous optimization for structure learning. November 2018.

[11] Xun Zheng, Chen Dan, Bryon Aragam, Pradeep Ravikumar, and Eric P. Xing. Learning sparse nonparametric dags. March 2020.

[12] Ignavier Ng, Shengyu Zhu, Zhuangyan Fang, Haoyang Li, Zhitang Chen, and Jun Wang. Masked gradient-based causal structure learning. January 2022.

[13] J.R. Dormand and P.J. Prince. A family of embedded runge-kutta formulae. *Journal of Computational and Applied Mathematics*, 6(1):19–26, 1980. ISSN 0377-0427. doi: https://doi.org/10.1016/0771-050X(80)90013-3. URL `https://www.sciencedirect.com/science/article/pii/0771050X80900133`.

[14] Judea Pearl. An introduction to causal inference. *Int. J. Biostat.*, 6(2):Article 7, February 2010.

[15] Frank R. Kschischang. The hilbert transform. Lecture notes / technical report, The Edward S. Rogers Sr. Department of Electrical and Computer Engineering, University of Toronto, Toronto, ON, Canada, October 2006. Originally released October 22, 2006; updated March 10, 2015.

[16] Timothy Ulrich. Envelope calculation from the hilbert transform. 03 2006.

[17] Roxana Pamfil, Nisara Sriwattanaworachai, Shaan Desai, Philip Pilgerstorfer, Paul Beaumont, Konstantinos Georgatzis, and Bryon Aragam. Dynotears: Structure learning from time-series data. April 2020.

[18] Jakob Runge, Peer Nowack, Marlene Kretschmer, Seth Flaxman, and Dino Sejdinovic. Detecting and quantifying causal associations in large nonlinear time series datasets. *Science Advances*, 5 (11):eaau4996, November 2019.

[19] Aapo Hyvärinen, Kun Zhang, Shohei Shimizu, and Patrik O. Hoyer. Estimation of a structural vector autoregression model using non-gaussianity. *Journal of Machine Learning Research*, 11(56):1709–1731, 2010.

[20] William L. Hamilton. Graph representation learning. *Synthesis Lectures on Artificial Intelligence and Machine Learning*, 14(3):1–159, 2020.

A Data and ground truth generation

A.1 Spring-mass-damper system

This study aims to focus on graph discovery tasks in an engineering setting and thus the data had to reflect that need. A spring-mass-damper system response was used as it represents a canonical engineering problem. Spring mass damper systems are also well suited for causal datasets, since it is reasonably intuitive to infer causal relationships between various properties. Synthetic data was generated using simple free vibrating spring mass damper system as shown in Figure 4.

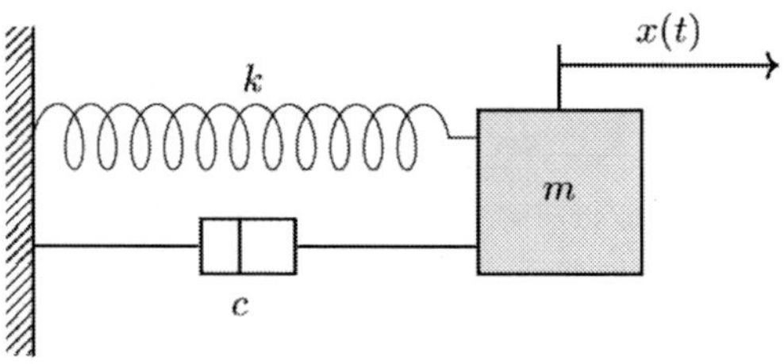

Figure 4: Free vibrating spring mass damper system

For a single body spring mass damper system undergoing free vibration, the general governing equation for the response $x(t)$ of the system given as

$$m\ddot{x}(t) + c\dot{x}(t) + kx(t) = 0 \tag{2}$$

where m, c, k are the mass, damping constant, and spring stiffness respectively. Equation (2) was rearranged to obtain

$$\ddot{x} = -\frac{(c\dot{x} + kx)}{m} \tag{3}$$

which was then used to calculate the acceleration response of the spring-mass-damper system. Numerical integration was used to generate displacement and velocity responses.

A.2 Dynamic properties and numerical integration

Numerical integration plays a key role in defining the underlying graph of the system and thus it is important to discuss the method of numerical integration used and the implications thereof on the ground truth graph of the system. The Python package used to numerically integrate this system makes use of the Dormand-Price method [13] for Runge-Kutta numerical integration. To make use of this method, the second order governing equation had to be rewritten as a system of two first order differential equations. To achieve this we define $\mathbf{x}$ with two components:

$$\mathbf{x} = \begin{bmatrix} x \\ \dot{x} \end{bmatrix} \tag{4}$$

such the derivative $\dot{\mathbf{x}}$ is defined as:

$$\dot{\mathbf{x}} = \begin{bmatrix} \dot{x} \\ \ddot{x} \end{bmatrix} \tag{5}$$

and when substituting Equation (3) into Equation (5) it yields:

$$\dot{\mathbf{x}} = \begin{bmatrix} \dot{x} \\ -\frac{(c\dot{x}+kx)}{m} \end{bmatrix} \tag{6}$$

which allows for the solution of $\dot{\mathbf{x}} = f(t, \mathbf{x})$ using a first order method. The Dormand-Prince method uses seven stages of intermediate derivatives to calculate the value of $\mathbf{x}$ at the next timestep. The exact coefficients used to calculate these value are not of concern to us, since we are only concerned with the flow of causality during this process. These intermediate derivatives are then used to calculate the response vector $\mathbf{x}_{n+1}$. Once $\mathbf{x}_{n+1}$ has been calculated, it's values $x, \dot{x}$ can be used to calculate $\ddot{x}$ using Equation (3).

Since the weighted sum of the intermediate gradients $\mathbf{k}_i$ can be interpreted as the instantaneous estimate for the derivative of $\mathbf{x}$, it is reasonable to substitute out all the intermediate gradients in the graph and replace then with the instantaneous derivative $\dot{\mathbf{x}}$, which yields the graph shown in Figure 5. When the process is repeated over many timesteps, a dynamic Bayesian network emerges, that captures the dynamics of the system through time. Finally, to represent the relationships of the individual components of the response, the nodes $\mathbf{x}_n$, $\dot{\mathbf{x}}_n$, and their counterparts in following timesteps, can be decomposed as per Equations (4) and (5) resulting in a graph showing how displacement, velocity and acceleration interact through time as shown in Figure 6.

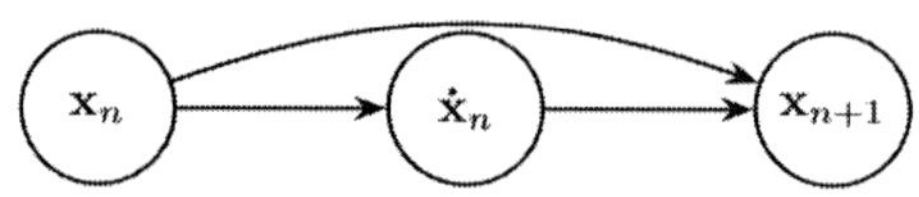

Figure 5: Simplified graph of the Runge-Kutta process

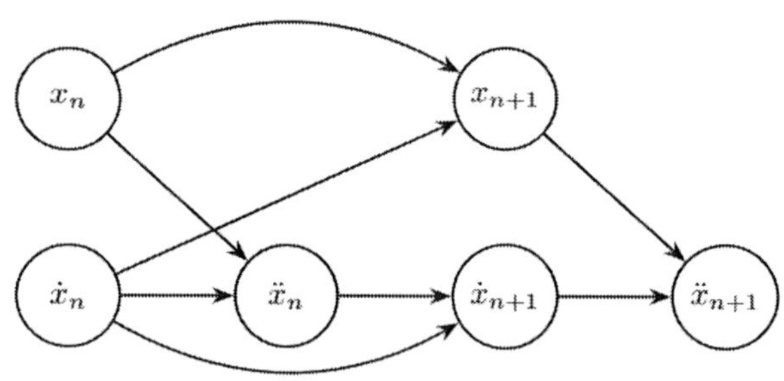

Figure 6: Dynamic graph of individual dynamic properties during the Runge-Kutta process

A.3 Static Properties

The dynamic characteristics of a spring mass damper arise from its specific static parameters like spring stiffness and damping. Thus, there is a clear causal relationship between these static parameters that are fixed for a specific system and its dynamic response. Considering these parameters minimizes the potential latent confounders in the graph.

A.3.1 System parameters

Consider once again the recursive graph of the dynamics in Figure 7, this time with red edges portraying nodes that are inputs to evaluations of $f(t, \mathbf{x})$. These red edges represent processes that involve system parameters m, c, k. These edges can then be expanded to include the influence of these static parameters as shown in Figure 8 as per Equations (4) and (5).

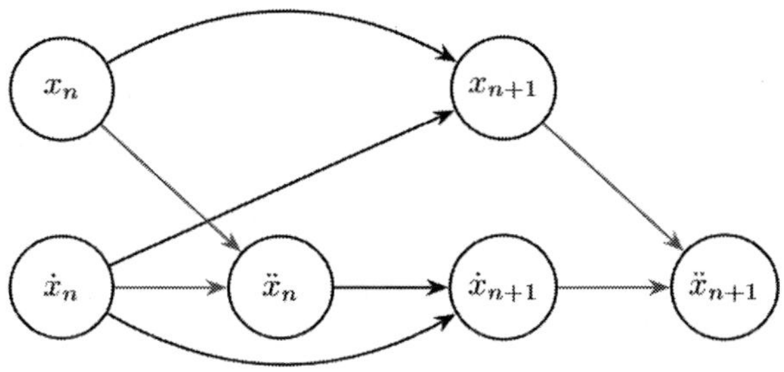

Figure 7: Dynamic graph of individual dynamic properties during the Runge-Kutta process. Evaluations of $f(t, \mathbf{x})$ are shown in red.

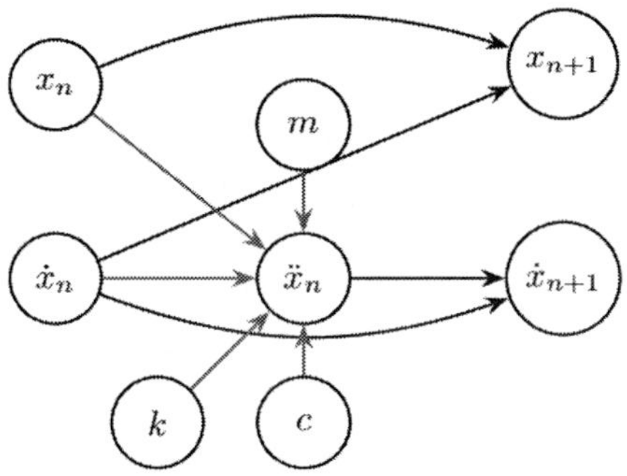

Figure 8: Dynamic graph including influence of static properties. Evaluations of $f(t, \mathbf{x})$ are shown in red.

The inclusion of static parameters also allowed for the inclusion of the forces involved in a spring mass damper oscillation, so the spring force, kx, and damping force, cv were also included in the ground truth. This was done to increase visibility of causal flow by showing a more verbose path of causal flow. When considering that the system parameters m, c, k are constant through time, the graph can be simplified by only using one node for each parameter. This results in the graph as shown in Figure 9.

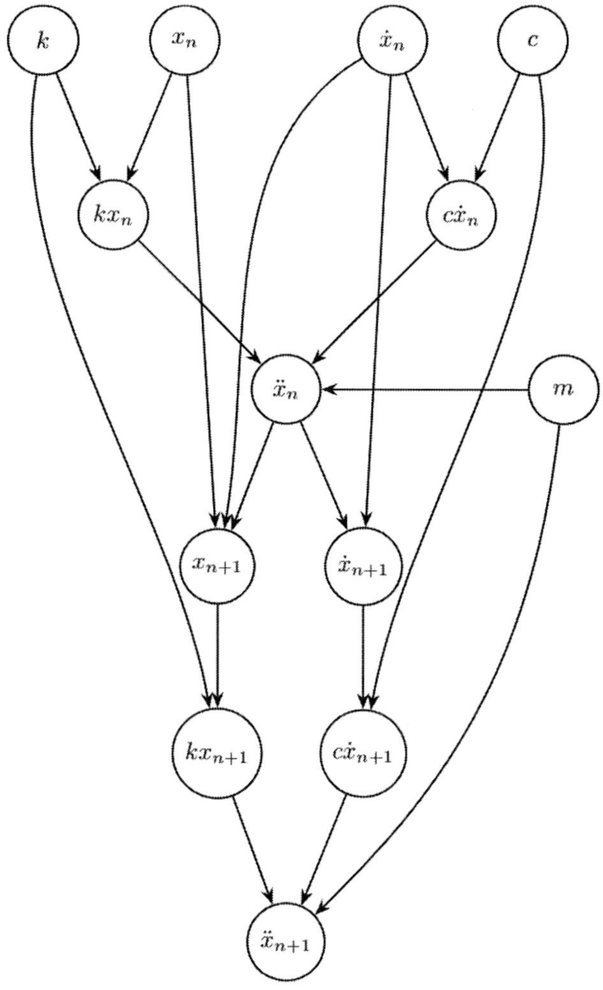

Figure 9: Dynamic graph with single static parameter nodes

A.3.2 Initial conditions

The final static parameters to consider when casting a dynamic system as a DAG are the initial conditions, namely, x_0 and $\dot{x}_0$. These are the first values used in the numerical integration that takes place during data generation. Fortunately, it is trivial to add these to a dynamic graph since they represent the first timestep in the response and correspond to all of the nodes where $n = 0$. As a result, the graph in Figure 9 only needs to be slightly modified to include the parameters $x_0, \dot{x}_0$. This results in the graph shown in Figure 10. This version of the graph was taken as the base of the ground truth throughout the remainder of the study.

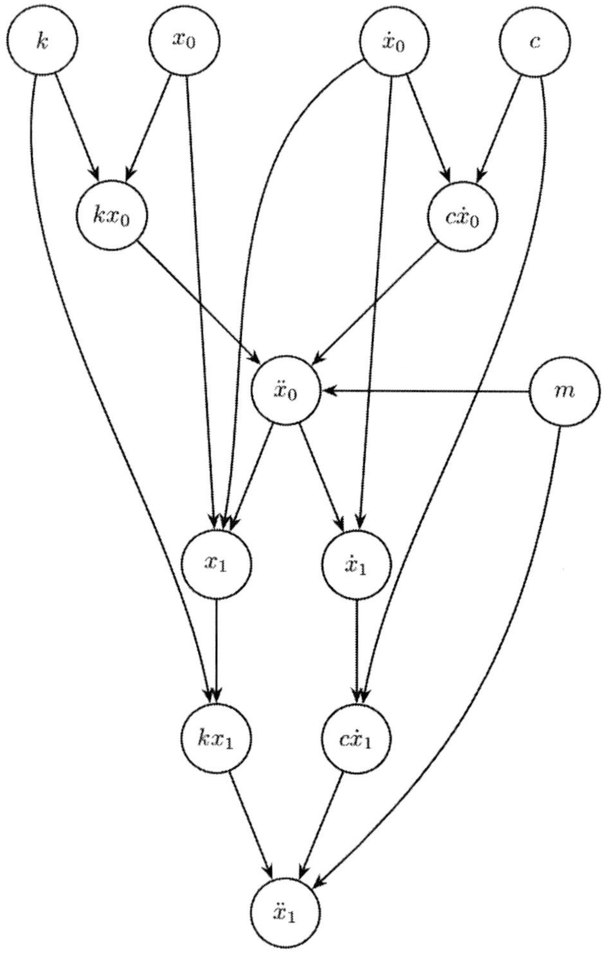

Figure 10: Dynamic graph with initial conditions

A.4 Dataset structure

The simulated spring mass damper system response dataset contains parameters $x, \dot{x}, \ddot{x}$ which can be written as matrix $\mathbf{X} \in \mathbb{R}^{n \times 3}$ for a response with n samples. However, system parameters as well as initial conditions need to be included in the data matrix. Thus, we expand data matrix $\mathbf{X}$ to include $m, c, k, x_0, \dot{x}_0, \ddot{x}_0$. The response data was then relabeled as $\mathbf{x_n}, \dot{\mathbf{x}}_\mathbf{n}, \ddot{\mathbf{x}}_\mathbf{n}$ and initial restorative forces kx_0 and cv_0 as well as dynamic restorative forces $k\mathbf{x_n}$ and $c\mathbf{v_n}$ were added. The new data matrix contains both static and dynamic parameters. To account for this, the static parameters are treated as steady-state dynamic parameters. Thus, for an expanded data matrix $\mathbf{X}$, representing one system's response, the data vectors $\mathbf{m}, \mathbf{c}, \mathbf{k}, \mathbf{x_0}, \dot{\mathbf{x}}_\mathbf{0}, \mathbf{kx_0}, \mathbf{c\dot{x}_0}, \ddot{\mathbf{x}}_\mathbf{0}$ are columns vectors with one repeating value equivalent to a scalar multiple of a vector of ones like $b\mathbf{1}$ where b is the static value of the parameter and $\mathbf{1} \in \mathbb{R}^t$. Thus, the new data matrix can be written as $\mathbf{X} = [\mathbf{m}, \mathbf{c}, \mathbf{k}, \mathbf{x_0}, \dot{\mathbf{x}}_\mathbf{0}, \mathbf{kx_0}, \mathbf{c\dot{x}_0}, \ddot{\mathbf{x}}_\mathbf{0}, \mathbf{x_n}, \mathbf{v_n}, k\mathbf{x_n}, c\dot{\mathbf{x}}_\mathbf{n}, \ddot{\mathbf{x}}_\mathbf{n}]$ where the columns in the entries are vectors with length number of samples t through time. Finally, multiplicative Gaussian noise to ensure consistent signal to noise ratio between parameters. This was achieved by simply multiplying a scalar sample $\epsilon \sim \mathcal{N}(1, 0.1)$ with each element in the data matrix $\mathbf{X}$ to achieve scaled noisy data. In our case multiple system responses were used, which was achieved by stacking multiple data matrices column-wise such that the final data matrix could be written as $\mathbf{X}_{stacked} = col(\mathbf{X_1}, ..., \mathbf{X_m})$ where m is the number of system response datasets to be stacked since we assume i.i.d. row samples.

A.4.1 Signal response enveloping

A challenge faced when attempting causal graph discovery on dynamic systems is the time dependency between data samples of the same feature, such as the time dependency between $\mathbf{x}(t)$ and $\mathbf{x}(t-1)$ in the spring mass damper system. To facilitate the use of general causal graph discovery algorithms on dynamic data, the data had to be cast to a form where the effects of the time dependency between samples of the same feature could be minimized as to preserve the i.i.d. assumption made by many algorithms as well as possible. To mitigate temporal dependencies, a static representation of the response had to be constructed. This study is only concerned with recovering binary causal dependencies between parameters and not reconstructing the response, this is left for future work. We are only concerned with determining whether a dependency between the response and its parameters can be recovered in an unsupervised setting, and not the nature of the response itself. As such, a lower information representation of the data was acceptable if it was still able to sufficiently characterize the response. The average of a signal envelope was chosen as a sufficient static representation of the signal. To obtain this envelope, the Hilbert transform [15] of the signal was used to find the analytic signal. The magnitude of the analytic signal then returns the envelope of the signal [16] as is shown in Figure 11. The mean of that envelope was then taken to reduce the transient responses to singular values. This process was repeated for all dynamic response parameters $\mathbf{x_n}, \dot{\mathbf{x}}_\mathbf{n}, \ddot{\mathbf{x}}_\mathbf{n}$ in $\mathbf{X}$ to yield new dynamic response parameter scalars $\bar{x}, \bar{\dot{x}}, \bar{\ddot{x}}$ and consequently, new restorative force parameter scalars $k\bar{x}, c\bar{\dot{x}}$. This results in one dynamic response sample per system, and as a result the shape of $\mathbf{X}$ reduces to $\mathbb{R}^{m \times 13}$ where m is the number of system's for which data was generated. This removes the need for stacking different responses and having scalar multiples of ones vectors as discussed previously, simplifying the data structure. This also condenses the dynamic ground truth to a finite ground truth graph as shown in Figure 12. A useful property of this ground truth is that it is the only member of its Markov equivalence class and as a result should be fully recoverable by algorithms that are only able to recover a CPDAG.

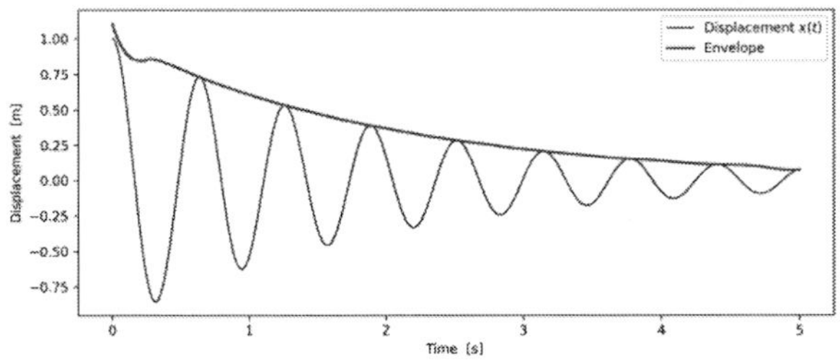

Figure 11: Example of displacement response with envelope

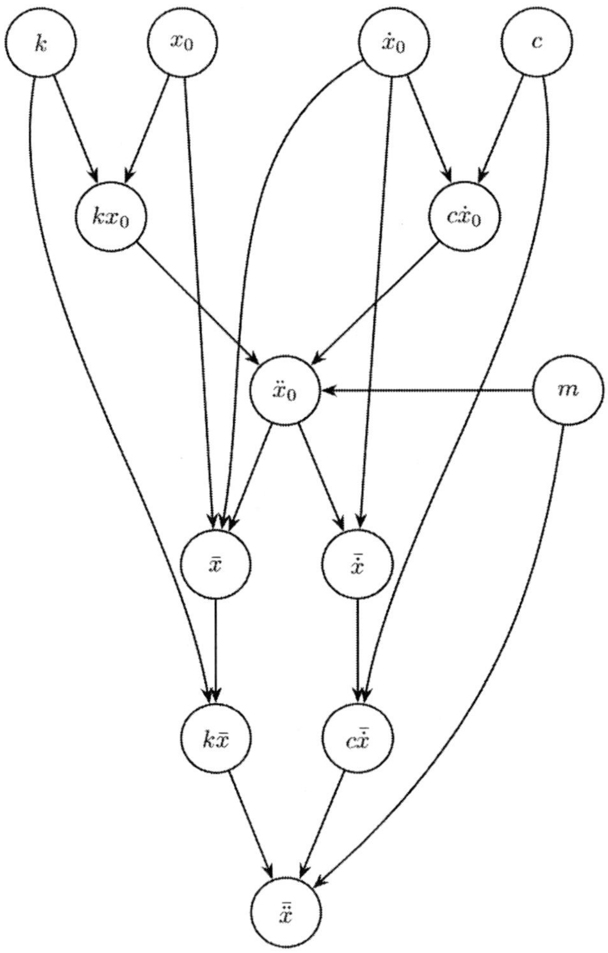

Figure 12: Finite ground truth containing enveloped representations of dynamic responses

A.4.2 Special case: Dynamic algorithms

Dynamic causal graph discovery algorithms are robust to time series signals and do not rely on the i.i.d. sampling assumption of most classical graph discovery algorithms [17–19]. As a result, the strategies discussed above for mitigating the i.i.d. assumption were not necessary, and full system responses could be used as input data for the algorithms. Although row ordering is important in dynamic algorithms, different model responses can still be stacked as in Pamfil et al. [17] since it is assumed that the algorithms should identify independence between the start and endpoints of response datasets. Thus, for the dynamic case, less unique system responses can be used since an individual response will add m rows where m is the number of samples in time, as opposed to the one row in the enveloping case. Dynamic causal graph discovery algorithms are based on autoregression and thus also require a lag parameter. To determine this lag order, partial autocorrelation function (PACF) plots were used. The partial autocorrelation of data generated. A 5% confidence interval was used for order selection. For spring-mass-damper response data generated, an autoregressive order of three selected as shown in Figure 13.

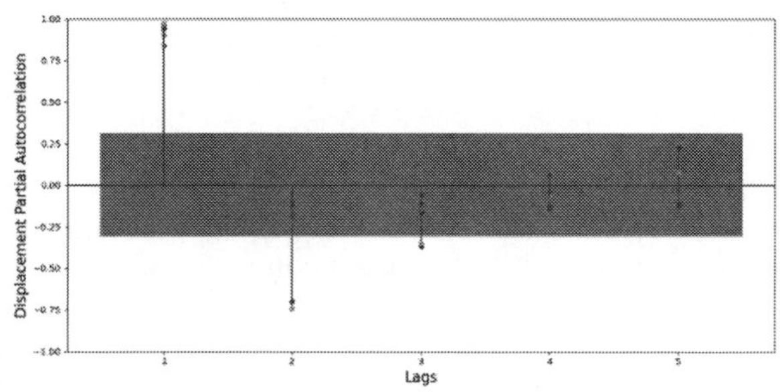

Figure 13: PACF plot of the spring-mass-damper system response with a confidence interval of 5%.

Taking into account the lag, the resulting shape of $\mathbf{X}$ becomes $\mathbb{R}^{m(t+1-l)\times 13}$. From this shape it becomes clear that this approach results in dynamic algorithms being exposed to less varying data responses since the autoregressive sample size $n_a = m_e(t + 1 - l)$ has to be equal to the envelope sample size $n_e = m_e$. The effects of this are discussed in the results chapter. Due to structure of the ground truth graph, it does not need to be drastically changed for the autoregressive case, but there are some additional considerations as to what the ground truth represents. The initial conditions in this case are simply the value of the response at the longest lag t_{-3}, and any parameter with subscript $_n$ is no longer a general representation of the data. Instead the ground truth is now represented by a dynamic graph. It consists of an intra-slice graph shown in Figure 14 connected by inter-slice edges representing temporal causation. It is assumed that the intra-slice graph does not change through time [17–19] and thus an autoregressive graph is simply a group of static graphs joined through time where each static graph represents instantaneous causal dependencies.

To equate the dynamic graph to the static equivalent, we only consider the nodes present in static graph in an attempt to recover the same graphs. This is done by mapping the dynamic graph to a static equivalent graph. This graph incorporates both intra- and inter-slice edges in the graph, but is equivalent to the static graph, allowing the use of a common ground truth between envelope and dynamic datasets. An important consideration is the assumptions made around self-causality, the case where the value of a feature at a previous timestep has an effect on the value of the feature at the current timestep. As with the envelope graph, it is assumed that this self causal influence flows through other variables that arise as a result of a parameter changing. Take, for example, x_{t-3} that influences displacement one timestep ahead, x_{t-2}. Instead of having an edge between x_{t-3} and x_{t-2}, the assumption is made that x_{t-3} influences x_{t-2} through $x_{t-3} \rightarrow kx_{t-3} \rightarrow a_{t-3} \rightarrow x_{t-2}$. It is noted that the direct edges between variables in features in different timesteps are preserved in the reasonable edges graph. The only parameters where this isn't applied are the system parameters m, c, k which remain constant through time, so they are coupled through time. The implications of these assumptions are discussed in the results chapter. Since the system parameters remain constant, when the dynamic graph was mapped, k_t, c_t, m_t were merged with their lagged counterparts $k_{t-1}, c_{t-1}, m_{t-1}$ to form k, c, m. This was done by adding all the rows of corresponding system parameters in the discovered adjacency matrix and masking all nonzero value to one. The mapping between static and dynamic cases is shown in Figures 15 and 16.

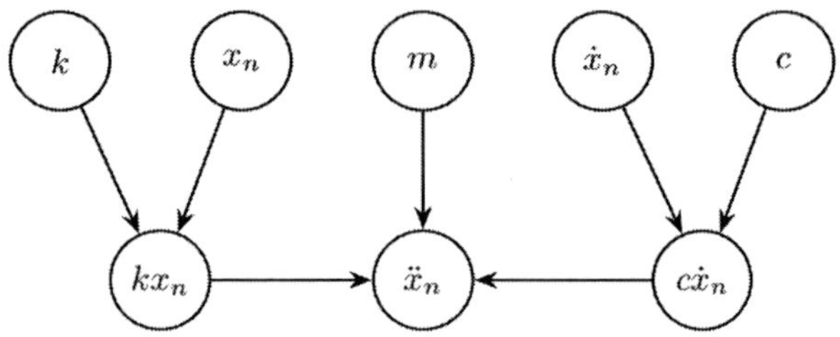

Figure 14: Intra-slice subgraph of the dynamic ground truth.

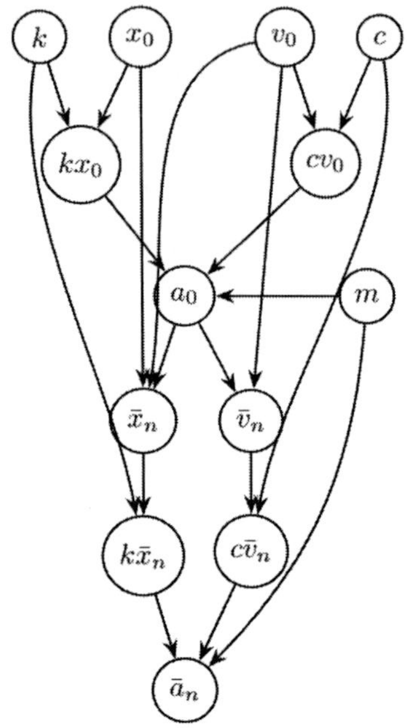

Figure 15: Static ground truth graph using enveloped representations.

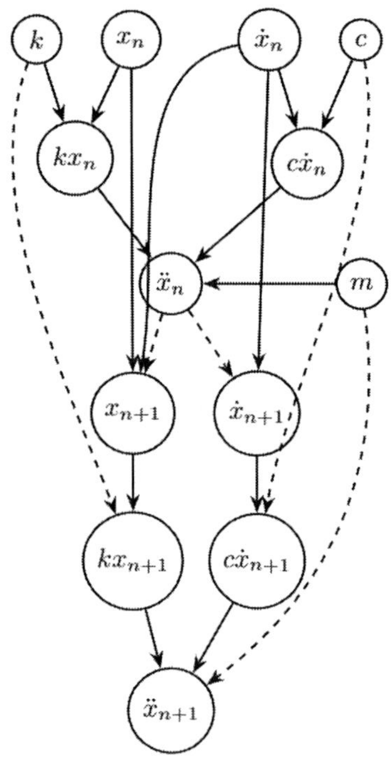

Figure 16: Dynamic ground truth graph. Intra-slice edges are solid and inter-slice edges are dashed.

AI-assisted strategy optimisation for plant operators in power generation

Gabriel Wolfson[a,*], Randall T. Paton[a]

[a]*School of Mechanical, Industrial, and Aeronautical Engineering, University of the Witwatersrand, Johannesburg, 1 Jan Smuts Avenue, Johannesburg 2001, Republic of South Africa*

**corresponding author email: GabrielWolfson8@gmail.com*

Abstract

Concentrated Solar Power (CSP) plant operation requires balancing immediate power generation with managing thermal energy storage (TES). This challenge is exacerbated by the variability of solar irradiation, weather conditions, and seasonal effects which frequently leads to sub-optimal strategies from human plant operators. A machine learning (ML) tool was developed to determine higher-yield strategies using reinforcement learning to aid real-time decision-making. This tool consists of an environment model built with a 3-layer LSTM artificial neural network containing 1,260,400 parameters trained on several years of plant data from a parabolic mirror-trough CSP plant. In addition, a control agent determines optimal turbine and thermal storage feed rate settings using stochastic hill-climbing. Testing on recent plant data demonstrated that the ML tool outperformed human strategies, particularly in winter, boosting revenue by as much as 10–15%.

Keywords: *CSP; solar power; machine learning; reinforcement learning; power plant optimisation.*

Nomenclature:

CSP	Concentrated solar power
DNI	Direct normal irradiation
SF	Solar field
SFA	Solar field availability
TES	Thermal energy storage
HTF	Heat transfer fluid
MS	Molten salt
ML	Machine learning
RL	Reinforcement learning
ANN	Artificial neural network

1. Background and Motivation

As of the date of publication South Africa has 700 MW of installed CSP capacity [1], of which the predominant technology is parabolic mirror trough plants. These systems use specially-shaped mirrors, using single-axis tracking, that focus sunlight onto receiver tubes filled with heat transfer fluid (HTF), which absorb and carry thermal energy to the turbine and TES systems [2]. The sunlight received in the solar field (SF) is a function of the direct normal irradiance (DNI), the incidence angle of the sun in the non-tracking axis, and the solar field availability (percentage of functional mirror trough assemblies). Figure 1 is a schematic of a parabolic mirror trough CSP plant.

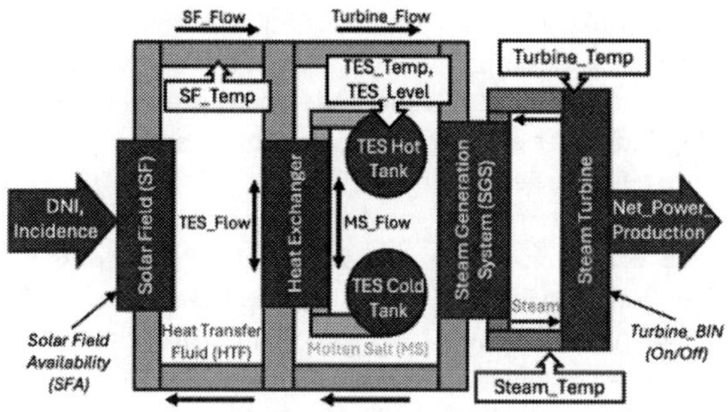

Fig. 1. CSP plant schematic

Plant operators control rates of power production and thermal energy storage by adjusting the HTF flow rates through the system. Effective CSP plant operation involves balancing thermal energy use between live power generation and TES to optimise electricity production and revenue throughout the day. A crucial consideration in this process is the production window and corresponding tariff, as power generated during the peak production window is remunerated at significantly higher rates compared to the standard production window. To capitalise on this, operators prioritise maximising power production during peak hours, often sacrificing standard production in favour of charging TES for use during the peak period.

Determining the optimal balance between live power generation and charging/discharging TES is challenging, especially without the ability to forecast the impact of operating decisions amid changing external factors. As a result, operators frequently adopt conservative strategies to minimise risks, which can lead to suboptimal energy output and revenue.

A data-driven tool powered by Machine Learning (ML) could address this challenge by leveraging real-time data to provide continuous predictive insights into potential strategies and evaluate their risk-reward trade-offs, ultimately recommending higher-yield approaches. This research evaluates how an ML-based tool could support plant operators by recommending optimal strategies to maximise energy production and boost revenue. By enhancing operational decision-making, such a tool could help CSP plants meet growing renewable energy demand while improving financial performance through increased energy output and reduced operational uncertainty.

2. Methodology

Reinforcement learning (RL) is an ideal approach for determining higher-yield strategies in diverse conditions because it can experiment with numerous approaches until it identifies the best-performing policy. To facilitate this exploration, a virtual model of the plant is required. For a complex thermodynamic system like CSP, an artificial neural network is well-suited for mapping the intricate relationships between input and output variables. To capture temporal dependencies and retain information over time, a Long Short-Term Memory (LSTM) network is used [3]. The architecture of this RL environment ANN is included in Figure 2. Each of the relevant inputs and outputs are labelled as in Figure 1. The ANN was trained on 4 years of data from a 100 MW CSP plant using the ADAM optimiser, which is particularly well-suited for LSTMs due to its adaptive learning rates and momentum. Benchmarking of the 1,260,400 parameter ANN showed a 94.6% mean accuracy for predicting net power production during stable turbine operation.

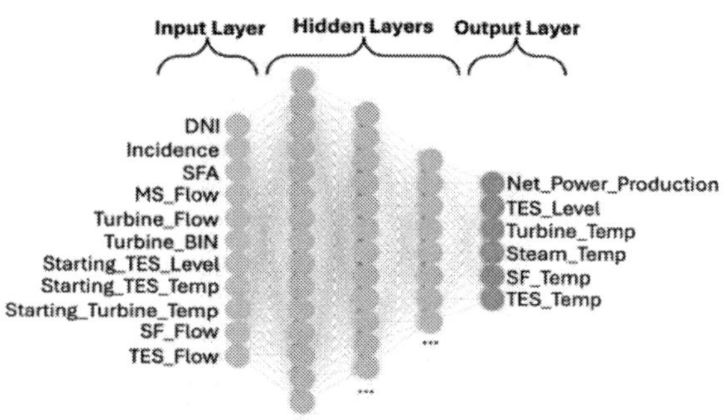

Fig. 2. RL Environment ANN

The ML tool facilitates interaction between the agent and the environment, where the agent generates and tests policies that are evaluated using a value function, as illustrated in Figure 3. To determine how real-time decisions affect an entire day's power production and revenue, the environment requires full-day inputs to produce full-day outputs. This necessitates forecasting the day's DNI based on prior time steps.

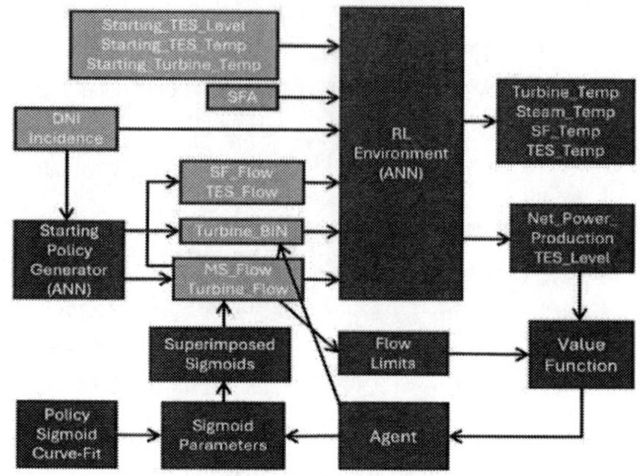

Fig. 3. ML Tool Schematic

The agent proposes full-day policies for the allocation of thermal energy between the turbine and the TES, as well as decisions on when the turbine should be activated or deactivated. These policies are functionally represented as: Turbine_Flow (the HTF flow rate to the turbine), MS_Flow (the molten salt flow rate between tanks), and Turbine_BIN (a binary variable representing the turbine's on/off state). Other inputs to the environment include solar incidence, which is predetermined and seasonally dependent under the assumption of perfect tracking, the plant's initial conditions, and additional HTF flow rates derived from the policy inputs using predefined ratios and continuity equations.

Policy suggestion of arbitrary values at each timestep is inefficient and fails to provide the smoothness necessary for effective flow rate control. A more robust approach involves generating policies as superimposed smooth functions, such as sigmoids [4]. By combining multiple sigmoid functions, this method produces a complex, continuous curve that ensures smoother transitions and improved control. A single sigmoid function is typically defined by the equation:

$$y(x) = \frac{A}{1+e^{-k(x-c)}} \tag{1}$$

In this expression A is the amplitude, k is the steepness of the curve, and c is the centre point where the sigmoid transitions from low to high values. In the context of superimposed sigmoids, several of these individual sigmoid curves are summed together to form a composite function that can capture more complex behaviours over time. Mathematically, this is represented as:

$$y_{total}(x) = \sum_{i=1}^{n} \frac{A_i}{1+e^{-k_i(x-c_i)}} \tag{2}$$

where n is the number of sigmoids being combined, and each sigmoid (i = 1:n) has its own set of parameters A_i, k_i, and c_i. Figure 4 demonstrates the replication of an historical MS_Flow policy using eight superimposed sigmoids.

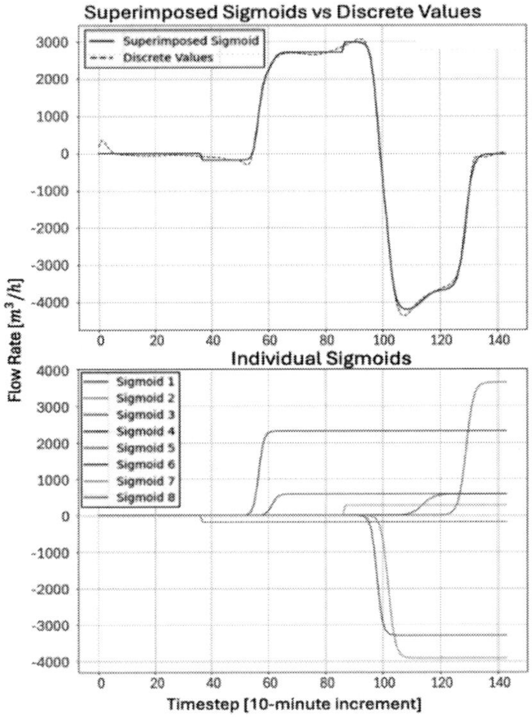

Fig. 4. Policy generation using superimposed sigmoids

Giving the agent control of sigmoid parameters instead of arbitrary value selection per timestep reduces the total policy parameters to 24 per flow policy i.e. 48 total. The Turbine_BIN policy includes 102 binary timesteps (covering production hours) with a switch limit set to six, reflecting the turbine's typical daily operation of no more than three on/off cycles.

The agent improves its policy suggestions using stochastic hill climbing [5]. This method adjusts sigmoid parameters through small, random changes, controlled by an exploration rate of 0.001, to enable a broad search beyond local optima. These adjustments are applied to the policies, which are then tested in the environment to assess their impact on production. Policies are evaluated based on their accumulated reward – from maximising favourable outcomes such as net power production and revenue – while avoiding large penalties for exceeding flow limits or depleting TES levels below 5%. An example of iterative policy improvement using this approach is shown in Figure 5.

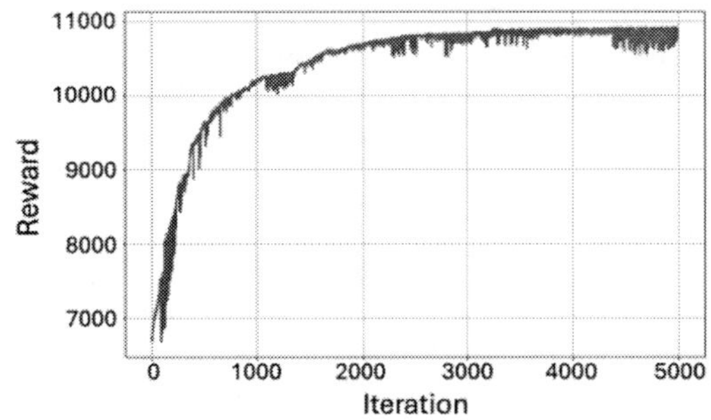

Fig. 5. Reward function maximisation during policy optimisation using stochastic hill-climbing approach

3. Key Results and Discussion

To demonstrate the advantage offered by the ML tool over the conventional strategies used by plant operators, two production cases will be discussed. Firstly, greater standard production is unlocked by superior TES management. Figures 6 and 7 show the net power production and TES management comparisons between the human operators and the ML tool respectively.

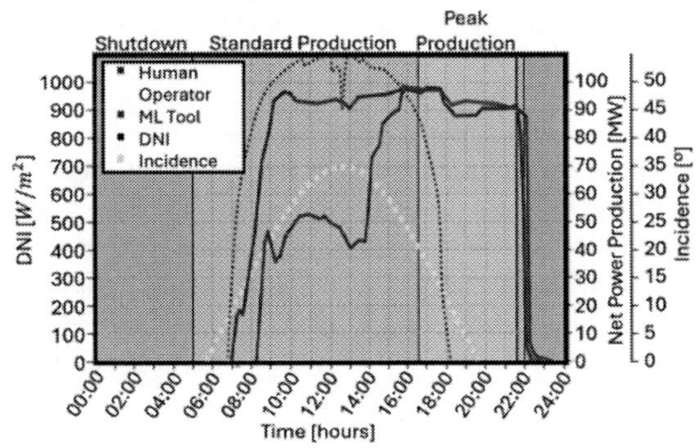

Fig. 6. Net power production comparison for TES management optimisation case

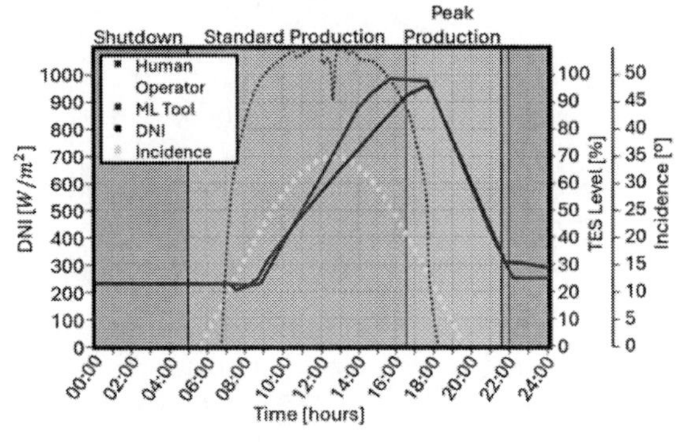

Fig. 7. TES management comparison for TES management optimisation case

In the case shown, the human operator adopted a conservative approach, charging the TES to its maximum level two hours before it was needed for the evening discharge. This limited live production during the standard production window. In contrast, the ML tool, leveraging its ability to forecast and iteratively determine optimal strategies, charged the TES to maximum without a buffer, enabling sustained high production throughout the day and an earlier turbine start to extend live production. A key distinction was the ML tool's slower TES charging rate, supported by computational certainty that the TES would charge in time, unlike the human operator's risk-averse approach. Despite a 5.5% decrease in peak production from a slightly lower TES hot-tank temperature, the ML tool's strategy resulted in a 64.3% increase in standard production, boosting daily revenue by 12.6%.

Between late autumn and early spring, the higher solar incidence angle reduces energy availability from the solar field (SF), particularly during the DNI peak. This decline in thermal energy input limits the plant's capacity to maintain high turbine output within the standard production window while simultaneously charging the TES sufficiently to exploit higher tariffs during the peak production window. Consequently, operators are faced with a strategic trade-off. Shutting down the turbine during high incidence accelerates TES charging and maintains higher temperatures – enabling a longer discharge during the peak production window with enhanced thermal efficiency. However, this approach sacrifices power production during the standard production window. Figures 8 and 9 show the net power production and TES management comparisons between the human operators and the ML tool when approaching this decision.

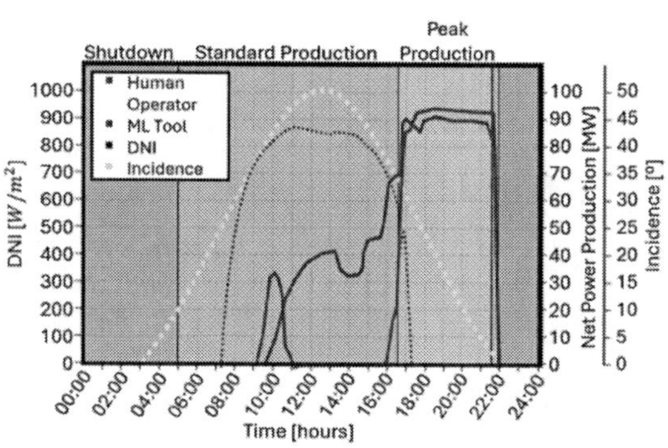

Fig. 8. Net power production comparison for turbine on/off decision optimisation case

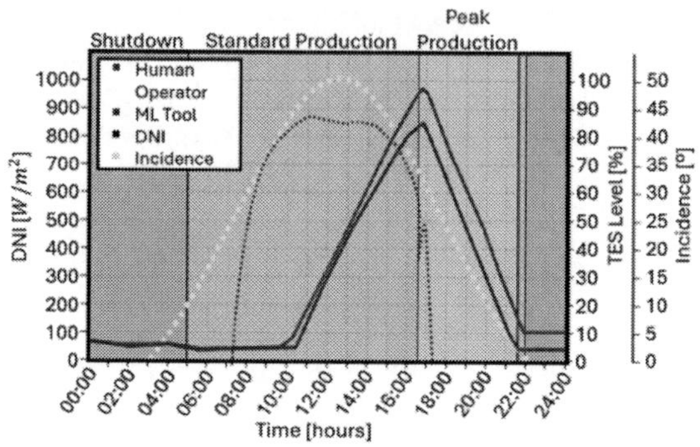

Fig. 9. TES management comparison for turbine on/off decision optimisation case

In this example, the human operator ran the turbine for 2.5 hours before switching it off to prioritise charging, while the ML tool delayed turbine startup but kept it running throughout the day. The human operator's strategy allowed for a more substantial TES charge, but resulted in lower standard production. In contrast, the ML tool's approach sacrificed a small amount of peak power (2.0%) for a 198.5% increase in standard production, generating an additional 168.1 MWh and boosting revenue by 10.4%. By quantifying and leveraging greater risk, the ML tool kept the turbine running for longer, trading off a small peak loss for a much greater standard production gain, whereas the human operator typically avoids this risk, prioritising peak power.

4. Conclusion and Future Work

This research demonstrates the significant potential of a machine learning (ML) tool to optimise CSP plant operations, outperforming human operators by leveraging precise data-driven strategies to maximise power production and revenue. The ML tool's integration of a three-layer LSTM artificial neural network and stochastic hill-climbing reinforcement learning enabled accurate forecasting and iterative policy improvement, achieving as much as 10–15% revenue gains in certain cases by effectively balancing risk and reward. Future work should focus on expanding the tool's functionality by enhancing input resolution, incorporating DNI prediction, and enabling dynamic hyperparameter adjustment for improved adaptability. Additionally, embedding real-time result validation and exploring advanced policy optimisation methods will further refine the tool's capabilities. This will ensure that it can provide robust decision-support for CSP plant operators as our increasing dependence on renewables makes optimisation and efficiency ever more critical.

Acknowledgements

ChatGPT was used as a writing aid in the preparation of this manuscript.

References

1. Zyl NP van. The global value of CSP plants in Africa [Internet]. ESI-Africa.com. 2023 [cited 2025 Jan 13]. Available from: https://www.esi-africa.com/renewable-energy/solar/the-global-value-of-csp-plants-in-africa/

2. Wei S, Liang X, Mohsin T, Wu X, Li Y. A simplified dynamic model of integrated parabolic trough concentrating solar power plants: Modeling and validation. Appl Therm Eng. 2020 Mar;169:114982.

3. www.javatpoint.com [Internet]. [cited 2025 Jan 13]. What are LSTM Networks - Javatpoint. Available from: https://www.javatpoint.com/what-are-lstm-networks

4. Approximation by series of sigmoidal functions with applications to neural networks [Internet]. [cited 2025 Jan 13]. Available from: https://www.researchgate.net/publication/268870507_Approximation_by_series_of_sigmoidal_functions_with_applications_to_neural_networks

5. Stochastic Hill Climbing | Algorithm Afternoon [Internet]. [cited 2025 Jan 13]. Available from: https://algorithmafternoon.com/stochastic/stochastic_hill_climbing/

Design for Sensing and Digitalisation (DSD): A Modern Approach to Engineering Design

Daniel N. Wilke[a,b]

[a]Department of Mechanical Engineering and Aeronautical Engineering, University of Pretoria, Pretoria, South Africa

*corresponding author email: nico.wilke@up.ac.za

Abstract

This paper proposes Design for Sensing and Digitalisation (DSD) as a novel engineering design paradigm that integrates sensor technology for digitisation and digitalisation from the earliest stages of the design process. Unlike traditional methodologies that treat sensing as an afterthought, DSD emphasises sensor integration, signal path optimisation, and real-time data utilisation as core design principles. The objectives are to address gaps in current design approaches by enabling comprehensive system observability, facilitating the implementation of digital twins, and supporting adaptive systems, while also serving as a framework for modern engineering education. Motivated by the demands of Industry 4.0 [4][5], where real-time data is essential for operational optimisation and predictive maintenance, this proposal outlines DSD's key principles, operational process, and educational implications. By adopting DSD, engineers can create more intelligent and adaptable systems that leverage real-time data for continuous design iteration, operational optimisation, and data-driven predictive maintenance.

***Keywords*:** Digitisation, Digitalisation, Sensing, Design, Design for Sensing and Digitalisation (DSD), Industry 4.0, Digital Twins, Sensor Integration, Signal Path Optimisation, Optimal Sensor Placement, Engineering Design, Predictive Maintenance, Real-time Monitoring, Engineering Education, Smart Manufacturing

1. INTRODUCTION

The digital transformation of industry has fundamentally changed how engineered systems are designed, operated, and maintained. While traditional design methodologies, such as Design for Manufacturing (DFM) [1] and Design for Maintenance (DFMaint) [2], remain valuable, they fail to address a critical requirement of modern engineering: the planning, design, and integration of sensing capabilities and data acquisition systems. This paper proposes Design for Sensing and Digitalisation (DSD), a new design methodology that places sensor integration and data acquisition at the core of the engineering design process.

The motivation for DSD stems from the limitations of existing paradigms in supporting Industry 4.0 technologies, such as the Internet of Things (IoT) [4] and digital twins [5], which require high-quality, real-time data for effective implementation. Traditional methods often result in suboptimal sensor integration, leading to incomplete observability and inefficient digitalisation [3][6]. DSD addresses three key objectives: (1) to ensure system observability by design through optimised sensor networks; (2) to enable seamless digital twin integration for adaptive and predictive capabilities; and (3) to provide a curriculum framework that integrates digital competencies with traditional engineering education.

This proposal is grounded in a constructivist epistemological framework, where knowledge of system states is constructed through designed observability and data interpretation, extending beyond empirical observation to proactive design for inference [8].

1.1 The Imperative of Digitalisation

The Internet of Things (IoT) [4] and digital twins [5] demand a fundamental shift in how we approach the engineering design of physical assets and processes. While digital twins and digital processes offer unprecedented insights through real-time monitoring and analysis, their effectiveness hinges entirely on the quality and completeness of the sensor data they receive. This creates a critical challenge: sensor integration must be considered from the earliest design stages, not as an afterthought. Furthermore, modern systems require data collection and optimised sensor networks that ensure comprehensive observability, defined here as the ability to infer the complete internal state of a system from its outputs [8], while minimising noise and interference.

The challenge extends beyond new designs. Legacy systems often require sensor retrofitting, particularly in environments where existing structures limit the optimal placement of sensors. These constraints frequently result in compromised data quality and incomplete system observability, limiting the potential of digital twin implementation. This underscores the need for a design methodology that considers sensing capabilities as fundamental to the design rather than an add-on feature.

1.2 Limitations of Traditional Design Methodologies

Traditional design methodologies, while effective in their respective domains, often fall short when confronted with the complexities of digitisation and digitalisation. For instance, DFM [1] focuses on optimising designs for ease of manufacturing, and DFMaint [2] emphasises reducing maintenance costs and downtime. In comparison, other recognised methods, such as Concurrent Engineering [11], integrate multiple disciplines early but still overlook sensing as a core element. Similarly, Systems Engineering [12] addresses holistic system design but does not explicitly prioritise sensor networks for observability.

These approaches share common shortcomings:

1. Sensor placement and signal path optimisation are typically afterthoughts, leading to suboptimal data collection and potential blind spots in system monitoring.
2. System observability is rarely considered during initial design phases, resulting in situations where critical parameters cannot be effectively measured or monitored.
3. The lack of integrated sensor simulation and virtual calibration capabilities [7][8] means that sensor network effectiveness cannot be verified before physical implementation.

These limitations create a significant gap between traditional design approaches and the requirements of modern digital twin implementations. When sensing and digitalisation are not core design considerations, the resulting systems often require costly modifications or deliver incomplete data, compromising the effectiveness of digital twin applications and limiting opportunities for data-driven operational optimisation, design iteration, and predictive maintenance.

1.2 The DSD Paradigm: Overview and Operational Process

DSD is grounded in a constructivist epistemological framework, where knowledge of system states is constructed through designed observability and data interpretation, extending beyond empirical observation to proactive design for inference [8]. The paradigm around sensing and digitisation can be operationalised as a structured process, illustrated in Figure 1 and briefly discussed.

The Design for Sensing and Digitalisation (DSD) process begins with the definition of requirements, identifying critical parameters for monitoring that encompass manufacturing, maintenance, and operational needs [1][2]. This is followed by an observability analysis, which employs mathematical modelling to ensure complete system state inference through virtual sensing and inference as required [8]. Subsequent sensor network optimisation utilises simulation-driven placement to maximise informational yield while minimising the quantity of sensors [3]. Signal path design then simulates and refines layouts to mitigate interference and preserve signal integrity [6]. Digital twin integration establishes models for real-time behavioural comparison and adaptive adjustments [5,9]. Finally, validation and iteration leverage virtual calibration alongside prototypes to verify and iteratively refine the design [7,8].

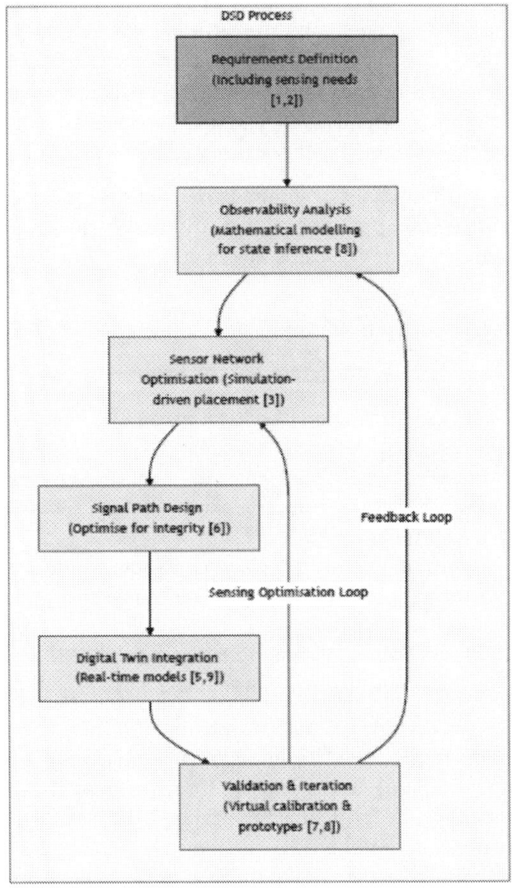

Figure 1 Outline of the DSD Sensing Process.

This process contrasts with traditional methods (e.g., DFM [1]) by embedding sensing early, rather than sequentially.

2. ADDRESSING MODERN DESIGN CHALLENGES

DSD extends virtual calibration [7][8] beyond its traditional role of quantifying parameter estimation uncertainty from experimental data. Instead, DSD empowers designers to proactively manage this uncertainty by incorporating sensor networks and system observability as fundamental components of system design. Through simulation-driven sensor placement [3] and sensor network optimisation [3], DSD ensures that critical system parameters, states, and conditions are observable by design rather than leaving observability to chance or post-design modifications.

This proactive approach naturally incorporates traditional design considerations like manufacturing [1], maintenance [2], and reliability, as the designer can now ensure that key parameters affecting these aspects are continuously monitored and quantified. By putting designers in control of system observability from the outset, DSD transforms the relationship between physical systems and their digital twins, ensuring that the data needed for effective system monitoring, optimisation, and maintenance is available by design rather than by circumstance.

DSD fundamentally reimagines the design process by integrating sensing and digitisation requirements from conception. This paradigm addresses four critical aspects:

2.1 System State Observability by Design

DSD ensures system state observability through:

1. Early identification of critical parameters requiring monitoring.
2. Mathematical modelling to confirm complete system state observability before physical implementation [8].
3. Optimise sensor placement to maximise information capture while minimising sensor count [3].
4. Integration of virtual sensing capabilities where physical sensors are impractical or cost-prohibitive.

DSD makes system state observability a fundamental design requirement by identifying critical parameters and validating observability through mathematical modelling before implementation. This approach ensures comprehensive system monitoring while minimising sensor counts through optimisation-driven sensor placement and virtual sensing strategies. The integration of observability analysis in the design phase enables efficient monitoring of system states throughout the product lifecycle, supporting immediate operational needs and long-term system optimisation.

2.2 Signal Path Optimisation

The paradigm prioritises signal integrity through:

1. Design-stage simulation of sensor networks to identify potential interference patterns [6].
2. Optimisation of physical layouts to minimise signal degradation.
3. Strategic placement of signal conditioning and processing elements [6].
4. Implementation of noise reduction strategies as core design features.

Signal path optimisation in DSD treats signal quality as a primary design constraint through simulation-driven design, sensor network optimisation, and strategic component placement. Addressing signal integrity and interference issues during design ensures high-fidelity data collection while reducing implementation costs and system complexity. The resulting framework provides reliable data for effective system monitoring throughout the product lifecycle.

2.3 Digital Twins for Smart Design & Redesign

DSD integrates digital twins for design and redesign by:

1. Naturally extending redesign through collected data to enable smart redesign.
2. Creating high-fidelity sensor networks that ensure comprehensive data collection.
3. Establishing verified and validated simulation models during the design phase.
4. Enabling real-time operational comparisons between predicted and actual system behaviour.
5. Facilitating continuous model refinement through operational data [9].

DSD creates a natural feedback loop where operational data directly informs design modifications and redesign iterations, providing a foundation for data-driven smart design & redesign, predictive maintenance, and system optimisation throughout the product and process lifecycle [5].

2.4 Adaptive Systems

DSD naturally enables the design of adaptive systems through real-time system identification for continuous model updating and refinement based on operational data, allowing adaptation to changing conditions and ageing effects; optimised sensor networks for automated detection of system changes and anomalies, facilitating rapid responses to evolving requirements; digital twin integration for virtual testing of adaptation strategies before physical implementation, thereby reducing associated risks; and multi-fidelity modelling frameworks to capture rapid dynamic changes and long-term evolution, enabling adaptive responses across varying time scales. Consequently, DSD supports the creation of systems that adjust to shifting operating conditions and requirements in real-time, proving especially beneficial in dynamic environments demanding flexibility.

2.5 Role of Instrumented Prototypes

Instrumented prototypes play a key role in DSD, especially during validation. They incorporate extensive sensors to validate concepts in realistic environments, where modelling may fall short. For instance, in automotive prototyping, additional sensors can provide critical data to inform final designs, transitioning to minimal sensors in production [14].

2.6 Example Case Studies

Traditional DFM [1] optimises for manufacturing, but DSD integrates sensors for vibration monitoring from conception. Optimised placement [3] ensures the observability of turbine states, enabling a digital twin [5] for predictive maintenance, which reduces downtime by 15%, as demonstrated for General Electric Renewable Energy's wind turbines. This approach achieves up to a 20% increase in efficiency [13]. Siemens Energy's Gas Turbine increased efficiency by 1.5%, while Australia's CS Energy's coal and gas-fired power stations increased their overall plant efficiency by 5% [13]. Industries such as aerospace, automotive, and energy are logical first adopters of DSD due to their emphasis on real-time data [4][13].

3. DSD: A CORE FRAMEWORK FOR MODERN ENGINEERING EDUCATION

DSD serves as both a design methodology and an educational paradigm, bridging the gap between traditional engineering disciplines and the demands of Industry 4.0. DSD provides a natural framework for integrating traditionally separate engineering disciplines and modern digital competencies into a cohesive curriculum. By centring design around sensing and digitalisation, DSD inherently connects fundamental engineering principles with essential digital age skills:

3.1 Integration of Core Competencies

DSD naturally incorporates key modern engineering skills through statistical analysis for sensor network design and data interpretation, programming and data science for digital twin implementation, machine learning for system optimisation and predictive analytics, computer-aided engineering (CAE) for system modelling and simulation, and optimisation techniques for sensor placement and signal path design. This framework promotes the development of modern engineering competencies by integrating statistical analysis, programming, and data science into the design process, thereby equipping students with the skills to effectively harness sensor-generated data through data analytics and AI proficiency. Engagement with sensor network design and digital twin implementation enables practical application of CAE, machine learning, and optimisation methods, ensuring digital skills are cultivated holistically in the context of engineering design rather than in isolation.

3.2 Bridging Traditional and Modern Engineering

The DSD framework seamlessly connects traditional design considerations [10] with digital capabilities by allowing manufacturing constraints to inform sensor placement and integration, maintenance requirements to drive monitoring strategy development, reliability engineering to guide sensor network redundancy, system dynamics to influence sensing frequency and placement, and material selection to consider both structural and sensing requirements.

3.3 Integrated Approach

A DSD-based curriculum emphasises an integrated approach that prepares engineers to address complex system design challenges while naturally accommodating traditional design considerations within a modern digital framework through project-based learning that combines physical design and digital implementation, integration of industry-standard tools and practices, real-world case studies demonstrating DSD principles, team-based projects reflecting modern engineering workflows, and continuous feedback between design and implementation phases.

4. CONCLUSION

The Design for Sensing and Digitalisation (DSD) paradigm shifts sensor integration and signal optimisation from secondary considerations to core design principles. DSD prioritises system observability through modelling and virtual inference to design optimised sensor networks that enhance signal integrity with strategic design choices, enabling digital twins for adaptive responses. This proactive approach enhances system performance by integrating maintenance, manufacturing, reliability, and design improvement considerations throughout the system's lifecycle. DSD also supports modern engineering education by integrating digital competencies with traditional principles. Hands-on experience with sensor networks, digital twins, and data-driven optimisation equips students with the skills necessary for Industry 4.0, reinforcing a cycle of improved design practices and capabilities. The result is a robust and adaptable methodology that meets the demands of digital engineering while preparing engineers to drive industry transformation. DSD is a natural and practical modern design methodology and educational framework that integrates technological advancement.

Acknowledgements

The authors acknowledge using artificial intelligence tools to assist in refining and expanding the language of initial manuscript drafts. All conceptual, analytical, and technical content remains the authors' original work, and AI tools were employed solely to support clarity and consistency in writing.

References

[1] Anderson, D.M. (2020). Design for Manufacturability: How to Use Concurrent Engineering to Develop Low-Cost, High-Quality

Products for Lean Production Rapidly. (2nd ed.). Productivity Press.

[2] Gullo, L. J., & Dixon, J. (Eds.). (2021). Design for Maintainability. (1st ed.). John Wiley & Sons, Ltd.

[3] Chae, Y., and D. N. Wilke. 2017. Heuristic Linear Algebraic Rank-Variance Formulation and Solution Approach for Efficient Sensor Placement. Engineering Structures, 153: 717–731.

[4] Madakam, S., Ramaswamy, R. and Tripathi, S. (2015) Internet of Things (IoT): A Literature Review. Journal of Computer and Communications, 3, 164-173.

[5] Wilke, D. N. (2022). Digital Twins for Physical Asset Lifecycle Management. In Digital Twins: Basics and Applications, 13–26.

[6] Long, D., Hong, X., & Dong, S. (2006). Signal-path driven partition and placement for analog circuit. Asia and South Pacific Conference on Design Automation, 2006, Yokohama, Japan, pp. 6.

[7] Wilke, D. N., Kok, S., & Groenwold, A. A. (2010). The application of gradient-only optimization methods for problems discretized using non-constant methods. Structural and Multidisciplinary Optimization, 40, 433-451.

[8] Ben Turkia, S., Wilke, D. N., Pizette, P., Govender, N., & Abriak, N. E. (2019). Benefits of virtual calibration for discrete element parameter estimation from bulk experiments. Granular Matter, 21, 1–16.

[9] Booyse, W., Wilke, D. N., & Heyns, S. (2020). Deep digital twins for detection, diagnostics, and prognostics. Mechanical Systems and Signal Processing, 140, 106612.

[10] Wilke, D. N. 2023. A Personal Perspective on the Engineering Design Process and Design Qualification. Medium, Month Day. https://medium.com/@wilkedn/a-personal-perspective-on-the-engineering-design-process-and-design-qualification-d6352a5cca55.

[11] Prasad, B. (1996). Concurrent Engineering Fundamentals: Integrated Product and Process Organization. (1st ed.). Prentice Hall.

[12] Kossiakoff, A., Sweet, W. N., Seymour, S. J., & Biemer, S. M. (2011). *Systems engineering: Principles and practice* (2nd ed.). John Wiley & Sons.

[13] Siemens. (2024). Digital Twins in Action: Enhancing Turbine Performance Monitoring and Optimization. Siemens Energy Publications.

[14] Ulrich, K. T., & Eppinger, S. D. (2016). Product Design and Development. (6th ed.). McGraw-Hill Education.

Beyond Silicon: Sensitivity and Propagation Investigation in Mechanical Feed-Forward Neural Networks

WJ van Vuuren[1*] and DN Wilke[1]

[1]*Department of Mechanical and Aeronautical Engineering,*

University of Pretoria, Pretoria, 0002, South Africa

u21458449@tuks.co.za

Abstract

This study addresses the practical challenges of error management in analogue computation, particularly within mechanical neural networks (MNNs). A Mechanical Feed-Forward Neural Network (MFFNN) testbed was developed alongside a virtual model to serve as a training environment. This setup enabled systematic investigation into the emergence and propagation of errors in mechanical computation. The study introduces a novel training strategy that allows parameter transfer from an equivalent artificial neural network (ANN) by injecting stochastic noise directly into the network's activation signals; a method that improves error resilience and facilitates transfer learning in analogue systems. Experimental results demonstrate enhanced robustness and repeatability of the MFFNN, supporting its potential for scalable and reliable analogue computation.

1. Introduction

Neural networks have experienced an exponential rise in attention over the past decade [1]. As they are applied to increasingly complex problems, the demand for larger and deeper architectures has grown accordingly, resulting in significantly higher computational requirements [2]. This surge in computational demand has prompted researchers to explore alternative paradigms, with analogue computation emerging as a promising candidate due to its potential for improved energy efficiency [1], [2], [3], [4], [5], [6], [7]. Most contemporary neural networks are implemented on von Neumann architectures, which are fundamentally limited in their ability to exploit the intrinsic parallelism of neural computation [1]. In contrast, analogue physical computing systems are inherently massively parallel, with execution speeds that are largely independent of system complexity [1]. This distinction has driven growing interest in alternative computing paradigms, particularly those inspired by biology. Neuromorphic hardware platforms, which emulate the brain's architecture and information processing strategies, are a prominent example [7]. The motivation behind such approaches lies in the remarkable energy efficiency of the brain: while it operates on roughly 20 watts, a state-of-the-art digital computer capable of matching its processing power would require approximately 0.5 gigawatts; a disparity of seven orders of magnitude [1]. Lee et al. [6] highlights the implications of error generation in analogue systems, stating that it is currently not feasible to train the network in a simulated environment and then transfer the obtained parameters to the physical system due to these inherent errors not present in the digital domain. Consequently, the networks must be trained directly on the physical system to learn and manage these errors effectively. This underscores the importance of understanding and mitigating the susceptibility of MNNs to these errors [8], [9], [10]. This study presents a novel approach to mitigating error propagation in MNNs by enabling virtual training and parameter transfer from an equivalent ANN. The method involves modelling Gaussian noise directly within the activation signals between layers of the ANN, rather than only at the input level. While input-level noise is a common regularisation technique, the deep injection of noise into the network's activations represents a new strategy for obtaining transferable parameters suitable for analogue neural networks during training. Notably, since the completion of this study, Manduca et al. [10] have independently proposed a related method of injecting noise into activation signals to improve robustness and generalisability in ANNs. However, their work does not address parameter transfer to physical systems, which remains a key focus of the present study.

2. MFFNN Design

Training MNNs often require solving implicit, and typically nonlinear, equilibrium equations of motion, which makes computation both challenging and resource-intensive [11]. To address this limitation, rotational encoding at steady state was selected as the information representation method for the MFFNN. This approach enables the development of an explicit mechanical neuron whose response can be directly determined, eliminating the need to solve equilibrium equations during inference.

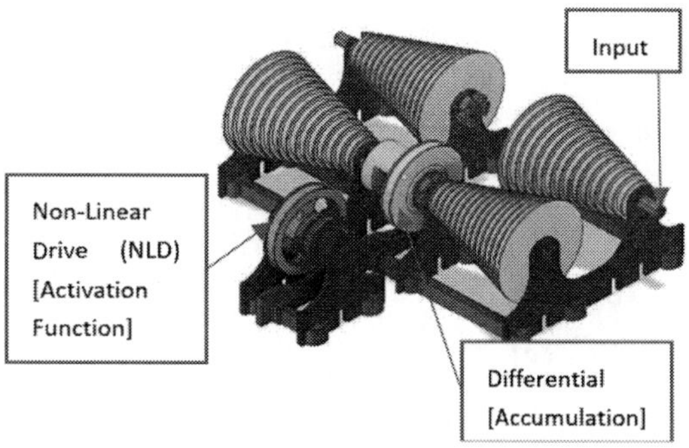

Figure 1: Single Neuron of the MFFNN Model

Figure 1 shows a single neuron assembly with the MFFNN consisting of a connected network of these neurons. Inputs to the neuron are given to the two initial pulleys as an angular encoding. The drive ratio between the pulleys then represents the weight or connection strength. The differential accumulates the inputs from the pulleys. The accumulated sum is then passed through the NLD acting as an activation function. The neuron output is then passed to subsequent neurons as inputs, arranged as shown in **Figure 2**.

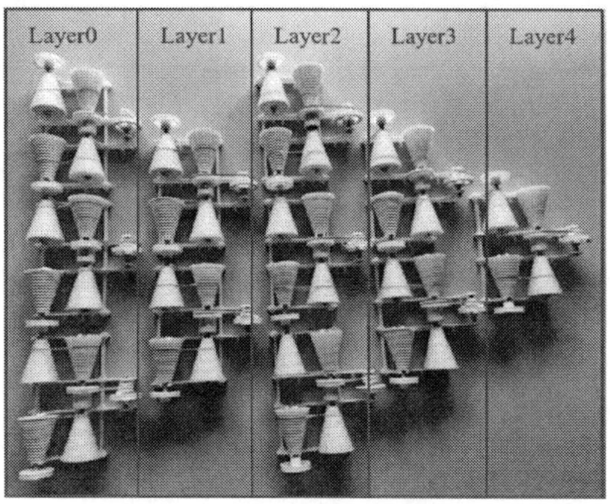

Figure 2: Photo of the MFFNN

As illustrated in **Figure 3**, each neuron performs three distinct operations. First, it receives two inputs and multiplies them by their respective weights using the pulley ratios. Next, the weighted inputs are accumulate using the differential. Finally, this weighted sum is transformed by the NLD.

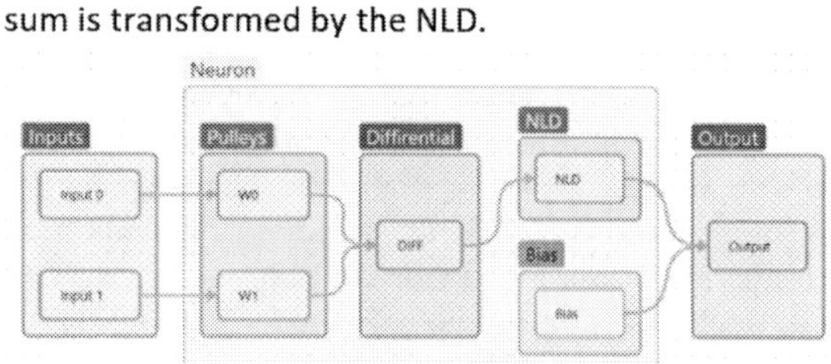

Figure 3: Schematic of a Single Neuron of the MFFNN

A key distinction between this neuron configuration and that of a standard artificial neuron lies in the treatment of the bias term: instead of being added before the activation function, the bias is incorporated as the initialised starting position of the subsequent connected neuron and is therefore applied after activation. **Figure 4** shows the MFFNN schematic, consisting of four inputs, 11 neurons, and one output. The network is five layers deep in a 3-2-3-2-1 height configuration over the depth.

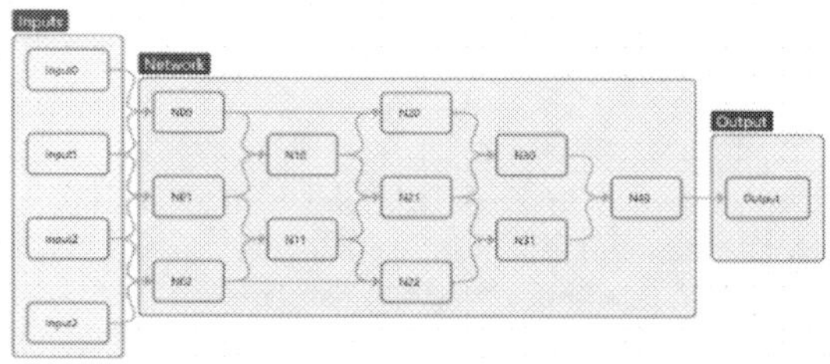

Figure 4: Schematic of the base MFFNN

Another key difference between this MFFNN and a fully connected FFNN is that, in this MFFNN design only neighbouring neurons are connected. Additionally, connections between neurons at the edges of layers 0 and 2 bypass layer one entirely. This configuration allows the MFFNN to conserve its dimensionality.

3. Artificial FFNN (AFFNN)

The following equations represent the mathematical equivalent of an idealised MFFNN and will be used to train the MFFNN digitally. Equation 1 replicates the structure of the NLD. The resulting matrices are notably sparse, a consequence of the local connectivity between neighbouring neurons, which mirrors the limited mechanical interactions in the physical implementation. Further a link to the python implementation can be found in Appendix A.

$$NLD(x) = \begin{cases} x+a, & x \leq -a \\ 0, & -a < x < a \\ x-a, & x \geq a \end{cases}, \quad where\ a = 0.16 \tag{1}$$

$$Layer\ 0: \begin{bmatrix} N_{00} \\ N_{01} \\ N_{02} \end{bmatrix} = NLD\left(\begin{bmatrix} w_{00}^{0} & w_{01}^{0} & 0 & 0 \\ 0 & w_{11}^{0} & w_{12}^{0} & 0 \\ 0 & 0 & w_{22}^{0} & w_{23}^{0} \end{bmatrix} \cdot \begin{bmatrix} I_0 \\ I_1 \\ I_2 \\ I_3 \end{bmatrix} \right) + \begin{bmatrix} b_0^0 \\ b_1^0 \\ b_2^0 \end{bmatrix} \tag{2}$$

$$Layer\ 1: \begin{bmatrix} N_{10} \\ N_{11} \end{bmatrix} = NLD\left(\begin{bmatrix} w_{00}^{1} & w_{01}^{1} & 0 \\ 0 & w_{11}^{1} & w_{12}^{1} \end{bmatrix} \cdot \begin{bmatrix} N_{00} \\ N_{01} \\ N_{02} \end{bmatrix} \right) + \begin{bmatrix} b_0^1 \\ b_1^1 \end{bmatrix} \tag{3}$$

$$Layer\ 2: \begin{bmatrix} N_{20} \\ N_{21} \\ N_{22} \end{bmatrix} = NLD\left(\begin{bmatrix} w_{00}^{20} & 0 \\ w_{10}^{20} & w_{11}^{20} \\ 0 & w_{21}^{20} \end{bmatrix} \cdot \begin{bmatrix} N_{10} \\ N_{11} \end{bmatrix} + \begin{bmatrix} w_{00}^{21} & 0 & 0 \\ 0 & 0 & 0 \\ 0 & 0 & w_{22}^{21} \end{bmatrix} \cdot \begin{bmatrix} N_{00} \\ N_{01} \\ N_{02} \end{bmatrix} \right) + \begin{bmatrix} b_0^2 \\ b_1^2 \\ b_2^2 \end{bmatrix} \tag{4}$$

$$Layer\ 3: \begin{bmatrix} N_{30} \\ N_{31} \end{bmatrix} = NLD\left(\begin{bmatrix} w_{00}^{3} & w_{01}^{3} & 0 \\ 0 & w_{11}^{3} & w_{12}^{3} \end{bmatrix} \cdot \begin{bmatrix} N_{20} \\ N_{21} \\ N_{22} \end{bmatrix} \right) + \begin{bmatrix} b_0^3 \\ b_1^3 \end{bmatrix} \tag{5}$$

$$Layer\ 4{:}\ Output = [N_{40}] = NLD\left([w_{00}^4 \quad w_1^4] \cdot \begin{bmatrix} N_{30} \\ N_{31} \end{bmatrix}\right) + [b_0^4] \quad (6)$$

4. Experimental Study

4.1. Baseline Training

In this study, a genetic algorithm was used (see Appendix A) as the model's weights are discrete, making gradient-based optimisation unsuitable. The AFFNN was trained to emulate the functions of AND, OR, and XOR logic gates as a synthetic task, with the binary inputs applied in analogue form. The network was optimised as a regression problem, allowing for direct control over the learnt function shape of the classifier, with mean squared error (MSE) loss to minimise the need for extensive output processing.

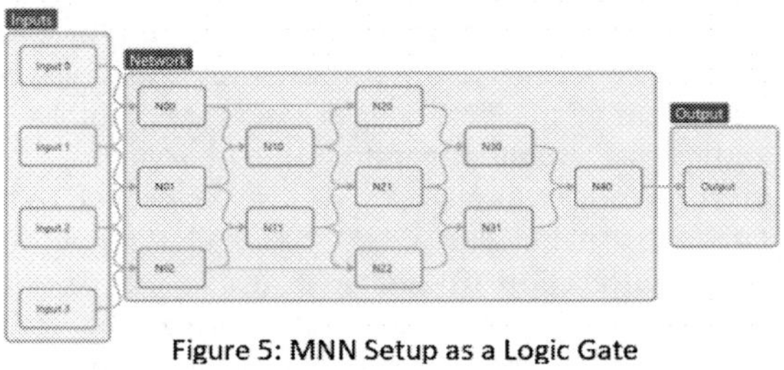

Figure 5: MNN Setup as a Logic Gate

The learned parameters, such as the weights and biases, from the AFFNN were transferred to the MFFNN by setting the pulleys to the drive ratios that correspond to the learned parameters. As expected, however, the MFFNN performed very poorly, sixty-three times worse than the AFFNN, with the baseline trained parameters, as seen in **Table 1**, as the digital training did not account for the error generation of the analogue system.

Table 1. MSE Loss of the trained AFFNN and MFFNN

MSE Loss	AND	OR	XOR
AFFNN	0.0001	0.0050	0.0034
MFFNN	0.1562	0.1065	0.2746

4.2. Error Propagation and Sensitivity Study

By comparing the activations of the MFFNN to the theoretical activations of the AFFNN, the error at each activation was calculated, and a distribution of these errors was created. A Gaussian was fitted to this distribution. Although the fitted Gaussian does not accurately represent the distribution, it was found to be adequate for simulating the error, **Figure 6**. Subsequently, the MFFNN's sensitivity to errors was analysed in relation to activation magnitude, layer depth, and subsequent weights. Negligible correlations were observed between error sensitivity, activation magnitude, and layer depth. However, a significant correlation of 0.67 was identified between the errors generated and the subsequent weights (**Figure 7**). This finding suggests that managing the weights within the MFFNN could effectively reduce error susceptibility.

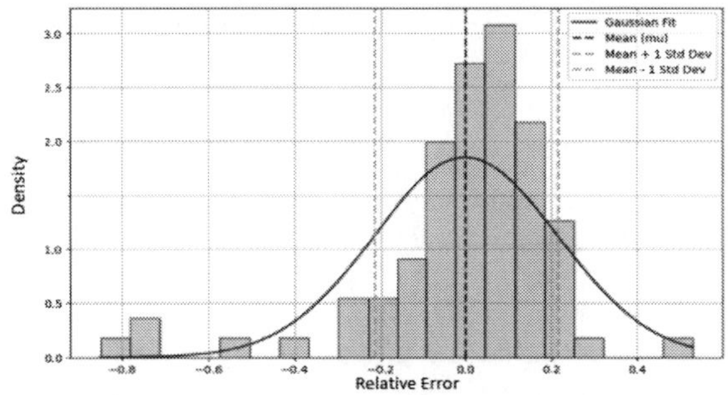

Figure 6: Distribution of the Relative Activation Error with a Gaussian Fit

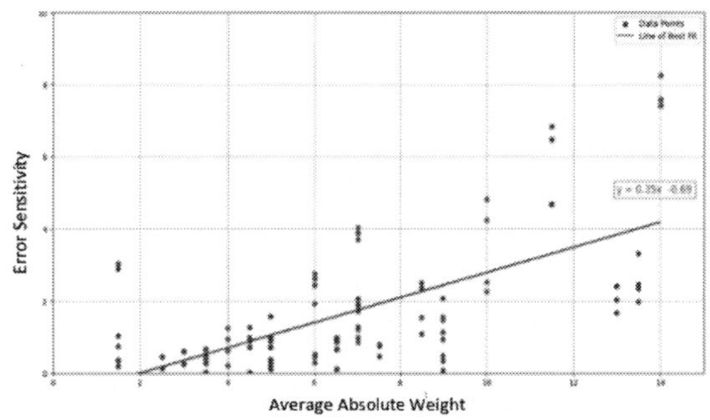

Figure 7: Relative Error Sensitivity vs Weights Plot

It was further determined that the decision boundaries learned by the AFFNN were sensitive to input errors. This sensitivity arose because the decision boundary was located very close to some of the main data points, left row of **Figure 8**, making the system prone to misclassification and amplifying its vulnerability to errors.

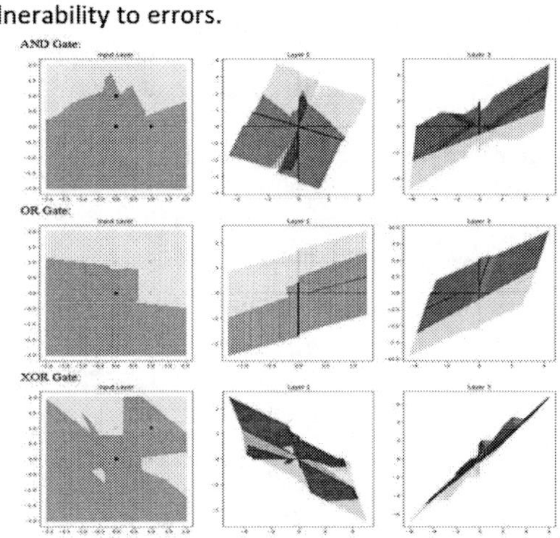

Figure 8: Latent Space Interception of the AFFNN

4.3. Simulated Analogue Training Method

Table 2: Performance of Standard Training

	AND	OR	XOR
Accuracy	86.55%	83.35%	68.87%
Precision	75.08%	100.0%	73.75%
Recall	69.16%	77.80%	58.61%
F1 Score	72.00%	87.51%	65.31%

The error sensitivity and generation depend on network parameters; thus, the network can manage these errors by adjusting its parameters. Since neural networks often settle in different local minima, training the network in the presence of errors enables it to find a solution that inherently manages these errors.

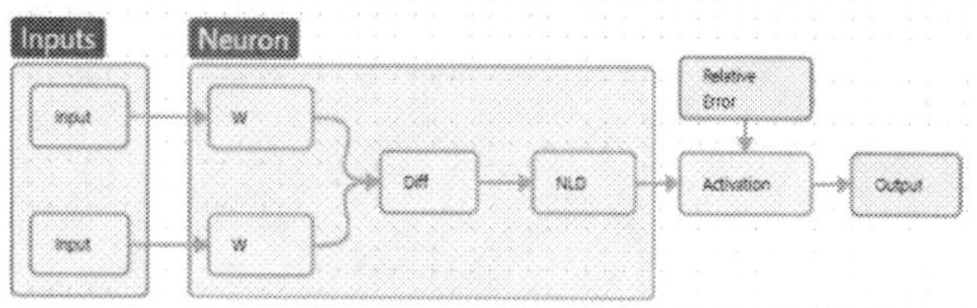

Figure 9: Error is Introduced in the virtual FFNN

For the simulated analogue training method, the network was trained using the same algorithm as the baseline. However, during the forward passes, we simulated the error generation characteristic of the MFFNN by injecting Gaussian noise into the activation signals between layers. The noise parameters were roughly estimated based on the data shown in **Figure 6**, and the noise injection process is illustrated in **Figure 9**. By adding Gaussian noise during training, we mimic the imperfect, stochastic computations that occur in a physical MFFNN [10]. This encourages the AFFNN to learn representations that are robust to such imperfections, ultimately improving the transferability of its parameters. Since the goal of the training algorithm was to increase robustness to computational errors, each training batch consisted of 5,000 repeated passes over the entire dataset. For each pass, new random internal errors were simulated. This strategy allowed the AFFNN to learn and adapt to the overall distribution of possible error patterns, rather than overfitting to any specific error instance.

Table 3: Performance of Simulated Analogue Training

	AND	OR	XOR
Accuracy	95.69%	88.58%	86.64%
Precision	99.20%	97.64%	91.82%
Recall	83.41%	86.87%	80.45%
F1 Score	90.62%	91.94%	85.76%

When examining the performance of the simulated analogue training strategy, **Table 3** shows significant improvements compared to the baseline model, **Table 2**. The average F1 score increased from 74.94 to 89.44, marking a substantial enhancement in performance. Furthermore, the system's overall accuracy also experienced a dramatic increase, with all three logic gates achieving higher accuracy than the baseline.

5. Conclusion

The MFFNN testbed developed in this study proved valuable for investigating error generation and accumulation within analogue neural networks. The accompanying AFFNN simulation environment was an essential tool for training the MFFNN, significantly reducing training time and enabling efficient experimentation of different training methods for better digital to analogue parameter transfer. The error analysis performed on the MFFNN, by comparing its activations to the ideal ones of the AFFNN's, provided key insights into the generation and propagation of the analogue errors. The error sensitivity was found to be largely associated with weight magnitudes, reinforcing the importance of parameter management for error reduction in MNN design. Additionally, the study revealed that the different mechanisms of error generation did not create bias in activation errors, validating the use of relative activation errors as an accurate model for error distribution modelling in the MFFNN. The insights gained from the error analysis informed the development of a targeted training strategy that reduces the impact of analogue errors. This approach involves injecting Gaussian noise into neuron activation signals between layers during AFFNN training, combined with repeated batching of the entire dataset. By exposing the network to simulated analogue errors during training, the method encourages the model to learn parameter configurations that are inherently robust to such imperfections. This simulated analogue training strategy effectively minimises sensitivity to errors through parameter management, resulting in more stable and reliable network behaviour. Although this approach increases computational training demands, it significantly improves the transferability of learned parameters from the digital AFFNN to the physical MFFNN. Thus, it enabled effective transfer learning in MFFNN by ensuring that information remains robust in the presence of the intrinsic variability and noise found in analogue mechanical implementations.

Acknowledgments

I am deeply grateful to Professor D.N. Wilke for his guidance, structured approach, and encouragement, which have been instrumental in shaping my research and learning experience. I would also like to thank Dr. Johannes Joubert, Mr. Carl-Henry Keyter, Mr. Peet Kruger, Mr. Myren Rajh, Mr. Ruben Grobler, Ms Eljuné Kemp. Additionally, OpenAI's ChatGPT and Google's Gemini were used for grammatical improvements in this document.

References

[1] M. Hermans, M. Burm, T. Van Vaerenbergh, J. Dambre, and P. Bienstman, "Trainable hardware for dynamical computing using error backpropagation through physical media," *Nat Commun*, vol. 6, 2015, doi: 10.1038/ncomms7729.

[2] R. Hamerly, L. Bernstein, A. Sludds, M. Soljačić, and D. Englund, "Large-Scale Optical Neural Networks Based on Photoelectric Multiplication," *Phys Rev X*, vol. 9, no. 2, May 2019, doi: 10.1103/PhysRevX.9.021032.

[3] J. B. Hopkins, R. H. Lee, and P. Sainaghi, "Using binary-stiffness beams within mechanical neural-network metamaterials to learn," *Smart Mater Struct*, vol. 32, no. 3, Mar. 2023, doi: 10.1088/1361-665X/acb519.

[4] S. Li and X. Mao, "Training all-mechanical neural networks for task learning through in situ backpropagation," *Nature Communications* , vol. 15, no. 1, Dec. 2024, doi: 10.1038/s41467-024-54849-z.

[5] L. G. Wright *et al.*, "Deep physical neural networks trained with backpropagation," *Nature*, vol. 601, no. 7894, pp. 549–555, Jan. 2022, doi: 10.1038/s41586-021-04223-6.

[6] R. H. Lee, E. A. B. Mulder, and J. B. Hopkins, "Mechanical neural networks: Architected materials that learn behaviors," 2022. [Online]. Available: https://www.science.org

[7] T. W. Hughes, I. A. D. Williamson, M. Minkov, and S. Fan, "Wave physics as an analog recurrent neural network," 2019. [Online]. Available: https://www.science.org

[8] T. Mei, Y. Zhou, and C. Q. Chen, "Mechanical Neural Networks with Explicit and Robust Neurons," *Advanced Science*, vol. 11, no. 33, Sep. 2024, doi: 10.1002/advs.202310241.

[9] T. Xu *et al.*, "Perfecting Imperfect Physical Neural Networks with Transferable Robustness using Sharpness-Aware Training," Nov. 2024, [Online]. Available: http://arxiv.org/abs/2411.12352

[10] A. F. & G. M. F. Giovanni Maria Manduca, "The Analog Layer: Simulating Imperfect Computations in Neural Networks to Improve Robustness and Generalization Ability," Dec. 2024.

[11] T. Mei, Y. Zhou, and C. Q. Chen, "Mechanical Neural Networks with Explicit and Robust Neurons," *Advanced Science*, vol. 11, no. 33, Sep. 2024, doi: 10.1002/advs.202310241.

Appendix A: AFFNN Code:

https://colab.research.google.com/drive/1HJhKiC5n-hybiHyV3euuqUtWJ2lP7v7n?usp=sharing

Analysis of blood flow for designing a prosthetic mitral valve using computational fluid dynamics method

Maral Taheri[a], Shahrokh Shojaei[a], Siamak Khorramymehr[b], Kamran Hassani[c,*]

[a]*Department of Biomedical Engineering, Islamic Azad University, Central Tehran Branch, Tehran, Iran*

[b]*Department of Biomedical Engineering, Science and Research Branch, Islamic Azad University, Tehran, Iran*

[c]*School of Mechanical, Industrial, and Aeronautical, University of Witwatersrand, South West Engineering Building, Braamfontein Campus East, Yale Road, Johannesburg, South Africa*

**corresponding author email: kamran.hassani@wits.ac.za*

Abstract

Valvular heart disease is described as a cardiovascular malfunction that affects people worldwide and requires surgery or complete valvular replacement in some cases. It affects the hemodynamics performance of the valve and often leads to serious cases that are known as flaws in cardiovascular fluid dynamics. The numerous cases of heart valve disease make the development of prosthetic valves a necessity. In this study, we extracted the real geometry of the mitral valve from MRI data, and then the dimensions of a prosthetic one were added to the geometry. The normal function of a prosthetic mitral valve has been investigated in the systolic phase for opening angles of 15,25,40,55,70, and 90. The inlet blood flow to the leaflets is pulsatile, and the hemodynamic parameters like velocity, pressure drops, and drag coefficient were studied. The commercial software Fluent was used for numerical simulations. The pressure drop decreased with an increasing rate of leaflet opening angle, where the drag coefficient was nearly constant after an opening angle of 40. At the end of the systolic phase, the back flows were detected more compared to the beginning of systole. The fluid-structure coupling modeling was not considered since we intend to investigate the valve's influence on the local hemodynamics rather than the effects on the valve and the surrounding physiological structure. Therefore, this model was sufficient since it has captured the physics associated with blood flow without being too complicated with a functional fluid-structure interaction.

Keywords: Fluid, Computational, Mitral, Modeling, Prosthetic

1. Background and Motivation

The mitral valve is located between the left atrium and the left ventricle of the heart, which has two tapered cusps. The mitral valve is one of four valves in the heart that regulate blood flow from the left atrium into the left ventricle. The left ventricle is the heart's main pumping chamber. Four decades ago, the first prosthetic valve was implanted in the heart successfully, and there are now more than 50 design types of designs for mechanical and bioprosthetic valves [1]. There are still two main problems for implanting the prosthetic valves in the mitral valve, including the initiation of thrombus due to induced stresses by blood flow and the stress concentration on the leaflets because of their motion during each heart cycle, which leads to the failure of prosthetic valves. Computational fluid dynamics is a useful tool that has been widely used, along with in vivo/in vitro experiments, to investigate the hemodynamics of blood flow in mitral prosthetic valves. The experimental methods play a crucial role in the analysis of the real valve function and numerical simulations. Moreover, they provide data for validation of the computations. Computational simulations are rapidly gaining importance in the research on the understanding of the complex physiology of the native and prosthetic valve dynamics. It was approved that prosthetic mitral valves affect the hemodynamic of the left ventricle [2,3]. The flow behaviors are complex, and they vary with time. The flow has many vortical structures. Therefore, we need advanced computational methods for better analysis of the flow [4]. Conducting a realistic numerical simulation for a prosthetic valve requires a complex three-dimensional geometry and real boundary conditions [3], and such modeling shall also be validated by clinical data. Several studies have simulated the mechanical and prosthetic mitral valve [5-9]. These investigations include the pressure, velocity, induced stress on the leaflets, shear stress on the leaflets, vortexes, and displacement of the leaves. This study aimed to investigate the effects of the blood flow on a prosthetic mitral valve hemodynamic using MRI data. A comprehensive computational fluid dynamics simulation was performed, and the results were compared to existing clinical and numerical data.

2. Methodology and Computational Approach

The valve geometry was built according to segmentations of the CT data at the end of diastole and systole. The CT scans were acquired using a Siemens Somatom Scope 16-slice scanner (Germany), operating in high-resolution mode. The scanning protocol included a 200-mm field of view, 130 kV, 300 mA, 0.5 seconds/rotation, and a slice thickness of 0.625 mm, yielding an in-plane resolution of approximately 0.39 mm × 0.39 mm. The prosthetic mitral valve was modelled as a generic leaflet mechanical valve. Figure 1 shows the geometry of the mechanical valve. The diameter of the mechanical mitral valve was like the annulus of the realistic valve. The geometry was built by ICEM software, and the meshing was generated by the software as per Figure 2. The fluid density and viscosity were assumed to be 1050 kg/m3 and 0.04 Pa · s, respectively. The tissue module of elasticity was assumed to be 1.5 MPa with a Poisson's ratio of 0.4[10,11]. We have used the K-Omega Wilcox method for our model in each cardiac cycle. The k-Omega Wilcox model is a turbulence model used in computational fluid dynamics (CFD) to simulate blood flow, including the flow through the mitral valve. It's a two-equation model, solving for turbulent kinetic energy (k) and specific dissipation rate (ω) to approximate turbulent behaviour in fluid flow. This model is particularly useful for simulating complex flows like those found in the heart, including the mitral valve, where accurate near-wall flow behaviour is crucial. The blood flow was assumed Newtonian [12]. Moreover, Turbulence in a prosthetic mitral valve, characterized by fluctuating blood flow, is a common finding and can indicate either normal valve function or the presence of valve dysfunction [13]. The models were introduced for the opening angles of 15,25,40, 55,70, and 90 degrees. The inlet was set based on a pulsatile velocity curve as shown in Figure 3 [14], and the outlet pressure was set to zero; all other surfaces were treated as no-slip walls. The leaflets were also considered fixed. The analysis of mesh independence was performed using three meshes for the full leaflet opening, including 697000,998000, and 1450000 elements, where the velocity of flow was computed, and we found that the model was independent of the mesh size for 998000. We have considered isotropic linear mechanical properties to present the mechanical properties of the mitral leaflets. Fluent software was used for our numerical simulations for time steps of 0.5 ms, and the hemodynamic parameters, including the velocity, pressure, and stresses, were analysed. Before starting the CFD simulation, it is important to look at the governing equations. Although there are some complex conditions, such as pulsatile flow and non-Newtonian fluid properties, the governing equations are the same as other fluid problems. The fundamental governing equations are the continuity equation (1) and the Navier-Stokes equations (2). Since blood is considered an incompressible fluid, the rate of change of density(ρ) will be zero:

$$\nabla \cdot v = 0 \tag{1}$$

$$\rho(\partial v/\partial t + v \cdot \nabla v) = -\nabla p + \mu \nabla^2 v + F \tag{2}$$

Where p is pressure, v is velocity, and μ is viscosity.

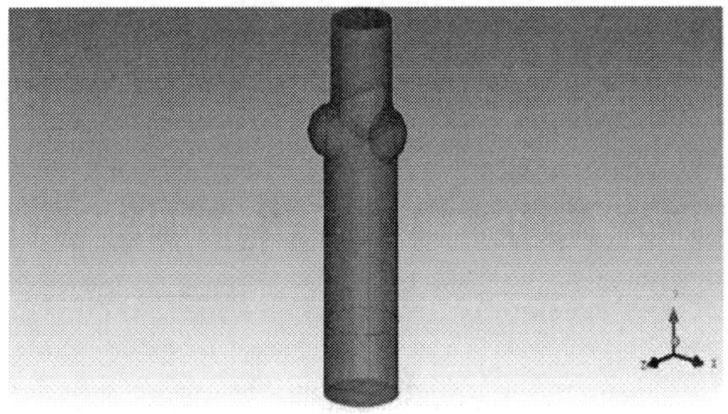

Fig. 1. The geometry of the model

Fig. 2. The meshed model of the leaflets.

3. Key Results and Discussions

We have incorporated some key results for each opening angle magnitude. Figure 4 shows the pressure variations across the leaflets when the leaflets are fully open, 90 degrees. Figure 5 presents the shear stress contour at the end of the systole for an opening degree of 70. The pressure distribution is shown for the opening degree of 55. Finally, the velocity vectors are presented in Figure 6 at the opening degree of 15. The pressure drop values versus the degree of opening, as well as the drag coefficient variations, were presented in Figures 7 and 8. The results show that the pressure drop decreased with an increase in leaflet opening degree. Furthermore, the pressure drop relative to the opening valve degree could be considered as a four-degree polynomial. The maximum drag coefficient was seen at 15 degrees, but the graph showed constant after a degree of 40 degrees. The variation of the drag versus opening degree could be assumed by a power curve. After the leaflets have reached the maximal opening, they begin to drift toward the head or anterior end of the leaflet. It happens before the blood flow reaches its maximum and the leaflet tips coapt in the neutral position. The leaflets get their neutral position because the pressure gradient of the left atrium and left ventricle is very low. The left atrium normally begins to contract at the end of the atrial diastole and increases the pressure gradient between the atrium and ventricle. Then, the mitral leaflets start to move toward the left ventricular wall. The opening degree of the mitral valve was found to be less during systole compared to diastole. The pressure gradient of the left ventricle and atrium affects the mitral valve flow at diastole, as it affects the degree of mitral leaflet opening. Therefore, it was confirmed that the mitral flow variations will not affect mitral leaflet opening. Furthermore, the leaflet opening was observed during each time step to see the asymmetry in the leaflet motion. As the leaflet moves through the cardiac cycle, the hemodynamics could affect the leaflets in different ways. It should also be noted that the opening of the mitral leaflets during diastole, along with their closure degree in systole, was influenced by left ventricular force. That is why the elevated magnitudes of blood pressure could lead to mitral regurgitation.

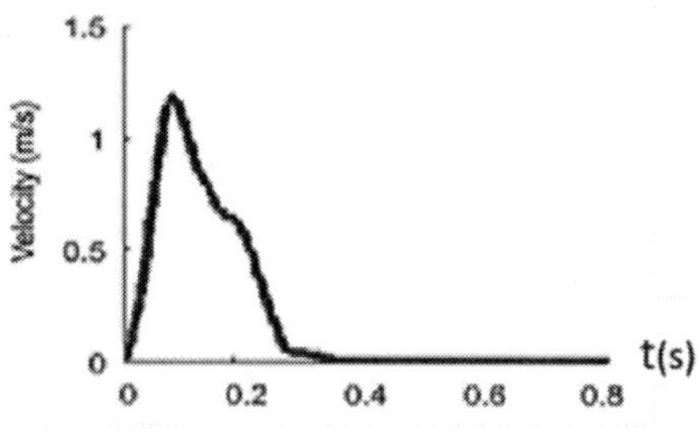

Fig. 3. The velocity graph for the model input

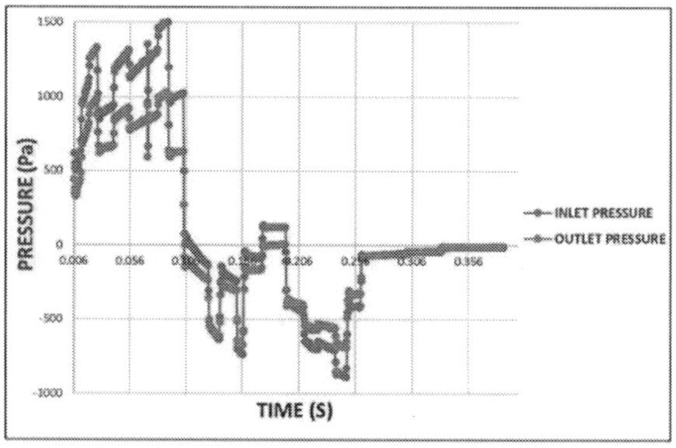

Fig. 4. The pressure variation for the opening angle of 90

Fig. 5. The shear stress distribution for an opening angle of 70 at the end of systole

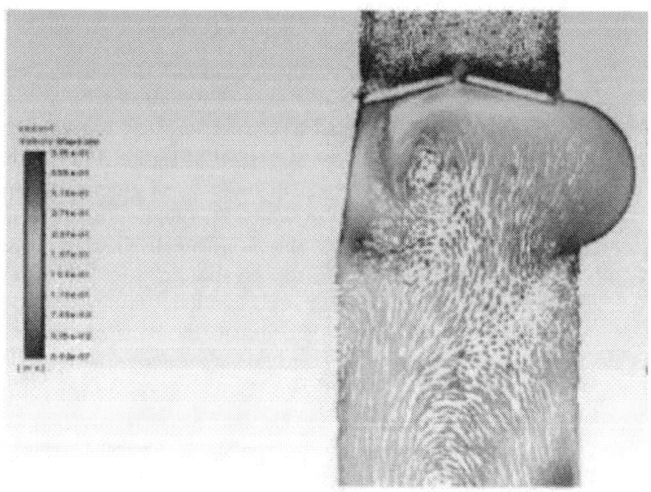

Fig. 6. The velocity vectors distribution for an opening angle of 15 at the end of systole

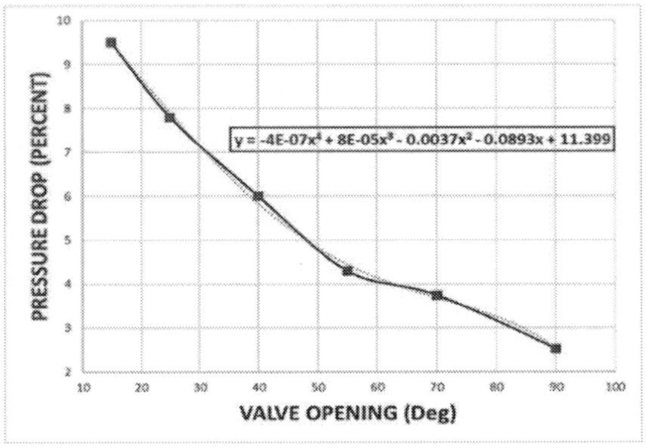

Fig. 7. The graph of pressure drops versus opening degree across the leaflets

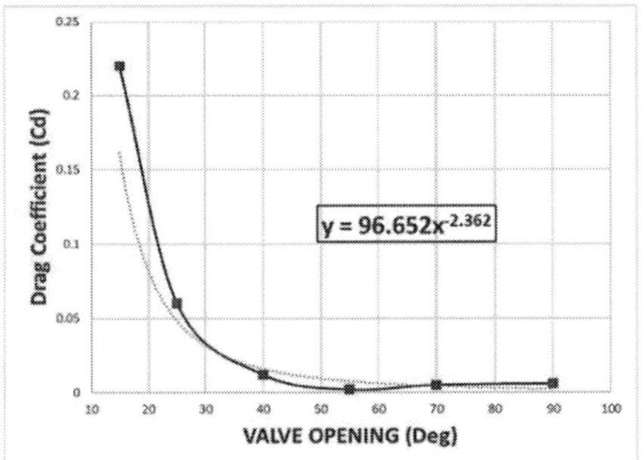

Fig. 8. The graph of drag coefficient versus opening degree across the leaflets.

4. Conclusion and Future Works

In this study, a computational model was developed to investigate the hemodynamic effects of blood flow on a prosthetic mitral valve. The results demonstrate that the blood flow induces significant stress concentrations on the valve leaflets, which are critical considerations for the future design and durability of prosthetic valves. While this study focused solely on the fluid domain, future work should incorporate fluid-structure interaction (FSI) modelling to more accurately capture the dynamic behaviour of the valve leaflets under physiological conditions. Additionally, patient-specific modelling and validation through experimental or clinical data will enhance the relevance and applicability of the findings to real-world scenarios.

References

[1] Baldwin J,Tarbell J, Deutsch S et al. (1991). Mean Velocities and Reynolds Stresses Within Regurgitant Jets Produced by Tilting Valves. American Society of Artificial Internal Organs Transactions, 37

[2] Bolger, A., Heiberg, E., Karlsson, M., Wigström, L., Engvall, J., Sigfridsson, A., Ebbers,T., Kvitting, J.-P.E., Carlhäll, C.J., Wranne, B. (2007). Transit of Blood Flow Through the Human Left Ventricle Mapped by Cardiovascular Magnetic Resonance. J.Cardiovasc. Magn. Reson. 9: 741–747.

[3] Chandran, K.B. (2010). Role of Computational Simulations in Heart Valve Dynamics and Design of Valvular Prostheses. Cardiovasc. Eng. Technol. 1: 18–38

[4] Sotiropoulos, F., Le, T.B., Gilmanov, A. (2016). Fluid Mechanics of Heart Valves and Their Replacements. Annu. Rev.. Fluid Mech. 48: 259–283.

[5] Saaid, H., Segers, P., Novara, M., Claessens, T., Verdonck, P. (2018). Single calibration multiplane stereo-PIV: the effect of mitral valve orientation on three-dimensional flow in a left ventricle model. Exp. Fluids 59, 49.

[6] Wu, W., Pott, D., Mazza, B., Sironi, T., Dordoni, E., Chiastra, C., Petrini, L., Pennati, G., Dubini, G.,Steinseifer, U., Sonntag, S., Kuetting, M., Magliavacca, F. (2016). Fluid–Structure Interaction Model of a Percutaneous Aortic Valve: Comparison with an In Vitro Test and Feasibility Study in a Patient-Specific Case. Ann. Biomed. Eng. 44: 590–603.

[7] Yoganathan, A.P., He, Z., Casey Jones, S. (2004). Fluid Mechanics of Heart Valves. Annu. Rev.. Biomed. Eng. 6: 331–362.

[8] Zajac, J., Eriksson, J., Dyverfeldt, P., Bolger, A.F., Ebbers, T., Carlhäll, C.-J. (2015). Turbulent kinetic energy in normal and myopathic left ventricles. J. Magn. Reson. Imaging 41: 1021–1029

[9] Westerdale, J.C., Adrian, R., Squires, K., Chaliki, H., Belohlavek, M. (2015). Effects of Bileaflet Mechanical Mitral Valve Rotational Orientation on Left Ventricular Flow Conditions. Open Cardiovasc. Med. J. 9: 62–68.

[10] Khalili F, Gamage PPT, Mansy HA. Hemodynamics of a bileaflet mechanical heart valve with different levels of dysfunction. J Appl Biotechnol Bioeng. 2017;2(5):187-191. DOI: 10.15406/jabb 2017.02.00044

[11] Khalili, F., Gamage, P.T., & Mansy, H.A. (2018). Prediction of Turbulent Shear Stresses through Dysfunctional Bileaflet Mechanical Heart Valves using Computational Fluid Dynamics. arXiv: Fluid Dynamics.

[12] Saha, S., Biswas, P., Das, A. N., Kumar, A., Kumar Singh, M. (2023). 'Analysis of Blood Flow Bifurcation Phenomena in Mitral Valve: A Numerical Approach to Predict Cardiac Arrest', Journal of Applied Fluid Mechanics, 16(3), pp. 491-504. doi: 10.47176/jafm.16.03.1351

[13] Kvitting JP, Dyverfeldt P, Sigfridsson A, Franzén S, Wigström L, Bolger AF, Ebbers T. In vitro assessment of

flow patterns and turbulence intensity in prosthetic heart valves using generalized phase-contrast MRI. J Magn Reson Imaging. 2010 May;31(5):1075-80.

[14] Choi, Y.J., Vedula, V., Mittal, R. (2014). Computational study of the dynamics of a bileaflet mechanical heart valve in the mitral position. Ann. Biomed. Eng. 42:1668 1680

14th South African Conference on Computational and Applied Mechanics (SACAM 2025)
21-23 January 2025, Johannesburg, South Africa

Fluid structural interaction simulation of a carotid artery with flexible arterial walls

Hanneke Strübing[1*], Joshua Lowe[2], Patrick Geoghegan[2], Wei Hua Ho[3]

[1]University of Witwatersrand
Johannesburg
Private Bag 3
Wits 2050
South Africa

[2]College of Engineering and Physical Sciences
Aston University
Birmingham
B4 7ET
United Kingdom

[3]University of Cape Town
CERECAM (https://cerecam.uct.ac.za)
Private Bag X3
Rondebosch 7701
South Africa

*2779930@students.wits.ac.za

Abstract

Vascular haemodynamics (study of blood flow through the vascular system) is an emerging field in biomedical engineering as medical practitioners and engineers find greater coherence between their fields. Arterial walls are highly flexible structures and are a crucial consideration for accurate prediction of blood flow features and arterial wall shear stress (WSS). This is exacerbated by the pulsatile nature of blood flow and the expansion (and contraction) of the arterial walls causes a dampening effect, which influences haemodynamics. This consideration is often simplified in models due to computational complexity and difficulty in conducting in-vivo measurements. Computational fluid dynamic (CFD) studies of blood flow often assume rigid walls. This study looks at using fluid-structure interaction (FSI) simulations to incorporate the dampening effect and investigate its influence on hemodynamic characterisation specifically focusing on WSS as it is one of the key parameters influencing cardiovascular disease progression.

Keywords: FSI, Fluid structural interaction, carotid, wall shear stress, WSS, elastic models

1. Background and Motivation

Cardiovascular disease (CVD) is the leading cause of death worldwide (World Health Organization, 2020). CVD is a broad term used for a variety of diseases that can affect the heart or blood vessels. Typically, these diseases lead to a restriction of blood flow to the heart, brain, organs, or body. These restrictions can manifest as atherosclerosis which is the narrowing of the artery due to thickening of the arterial wall or thrombosis which is a blood clot that formed in the artery.

Wall shear stress (WSS) is an indication of the frictional force on the area of the internal wall of the artery (Lopes, et al., 2019). Endothelial cells align with the flow and are more elongated in areas with higher shear stress and rounder in areas with low shear stress (Kornet, et al., 1998). Arterial studies agree that atherosclerotic plaques tend to occur near arterial bifurcations and bends.

The arterial wall is viscoelastic and expands and contracts as the heart beats. This movement of the wall has an impact on the forces within the blood and on the arterial wall itself. With the development of more fluid structural interaction (FSI) studies, it is now known that CFD-only (rigid wall) studies typically overpredict both the wall shear stress and velocity because the dampening effect of the arterial walls, due to their viscoelastic nature, is not considered (Lopes, et al., 2019). The knowledge of this overprediction caused a shift towards FSI studies and important boundary conditions include the elasticity of the arterial wall, wall thickness and the structural support of the artery being modelled to allow for correct expansion. All of these will have an impact on the dampening and thus the predicted WSS.

The areas of the carotid artery prone to high oscillatory shear index numbers (OSI) and low wall shear stress (WSS) is the outer wall of the internal carotid artery near the carotid sinus and the base of the external carotid artery (Kumar, et al., 2022). Figure 1 shows the layout of the carotid arteries with the indication of the sinus.

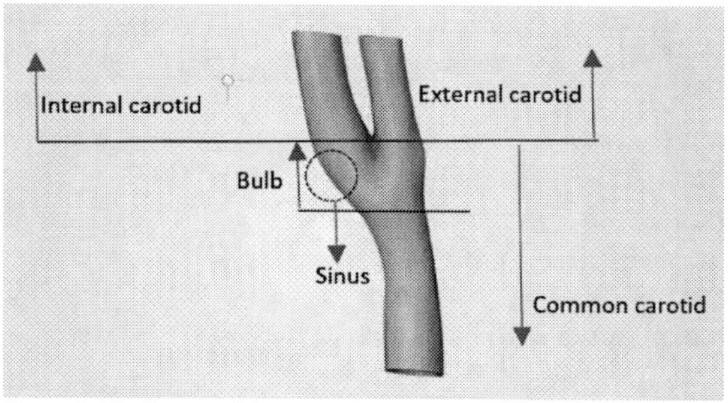

Figure 1: The layout of the carotid artery

Different elastic material models also have different responses to stress and strain. In a linear elastic material, the stress-strain relationship is linear, while with a hyperelastic material, it is non-linear. Viscoelastic materials exhibit hysteresis where there is a delay returning to their original shape after a load is applied.

The WSS is not only impacted by the viscoelasticity of the arterial walls, but also the non-Newtonian nature of blood (Geoghegan, et al., 2017). The shear-thinning properties of blood dominate the impact of its viscoelasticity (Gijsen, et al., 1999).

The objective of this paper is to give an overview of setting up a two-way coupled arterial FSI simulation.

2. Methodology and Computational Approach

This study uses an existing carotid geometry used in a collaborative study by Aston university.

The software packages used in the simulations are ANSYS Fluent, Ansys Mechanical APDL, and ANSYS System Coupling.

Firstly, the meshing and mechanical setup will be addressed. This will be followed by the setup on the fluid side, and lastly the coupling setup.

2.1. *Meshing*

Both on the structural and the fluid side, the mesh needs to move. This means the cells are constantly deformed by the forces generated by the flow and heartbeat. Following explicit meshing guidelines that consider the cell size and deformation required per time step is the first step in the simulation process.

2.2. *Mechanical model*

The fluid geometry was suppressed as a first step. Linear elastic properties were populated for the material of the artery. The contact zone for system coupling consists of all the inner faces on the arterial wall. For this work, the boundaries were set up as fixed supports on the ends of the geometry. Getting to a hexahedral mesh for a biological structure is very time-consuming, and a tetrahedral mesh was used for the carotid geometry. The thickness of the arterial wall is 0.6 mm in this setup.

2.3. *Fluid model*

Blood is a non-Newtonian fluid, and the Carreau model was used to represent the shear-thinning behaviour.

$$\mu = (\mu_0 - \mu_\infty)[1 + (\lambda\gamma)^2]^{(n-1)/2} \qquad (1)$$

Where μ is the viscosity, $\mu 0$ is zero shear viscosity, γ is the strain rate, λ is the time constant, and n is the power law index. Relevant values for these parameters can be found in literature and is dependent on the hematocrit levels of blood.

Blood was also assumed to be incompressible. The equations associated with solving the mass and momentum for the fluid flow are the continuity and Navier stokes equations represented by (2) and (3).

$$\nabla \cdot v = 0 \qquad (2)$$

$$\rho(\partial v/\partial t + v \cdot \nabla v) = -\nabla p + \mu\nabla^2 v + f \qquad (3)$$

Setting up the fluid model with data from a heart model results in both the flow and pressure varying with time. Transient tables containing the mass flow versus time for the inlet and pressure versus time for the outlet were used. The data from the tables were obtained from Lopes D. , Puga, Teixeira, & Teixeira (2019).

To effectively converge this on the fluid side, small time steps with a large number of iterations are required. The final time step and iterations will require some trial and error to optimise them for the geometry being used. It is critical to ensure that the fluid time step converges with each iteration. This is due to the error building up resulting in larger changes required in the mesh which in turn result in cell collapse and failure of the simulation.

2.4. *Fluid Structure Interaction*

Ansys System Coupling used the Arbitrary-Lagrangian-Eulerian (ALE) method.

$$\rho(\partial u/\partial t + ((u - ug) \cdot \nabla)u) = -\nabla p + \mu\nabla^2 u \qquad (4)$$

Where ρf is the fluid density, p the pressure, u the fluid velocity and ug is the moving coordinate velocity. To account for the movement of the mesh the term (u-ug) is added to the conventional Navier-Stokes equtions.

The fluid and structural coupled simulation is a 2 way closely coupled simulation due to the small differences in density between the arterial wall and blood.

Transient tables containing the mass flow versus time for the inlet and pressure versus time for the outlet were used. The data from the tables were obtained from Lopes et al (2019).

2.5. *Stabilisation and Coupling setup*

In Ansys, there is the option to stabilize both the structural and fluid simulations in Mechanical APDL and FLUENT, respectively. This should be turned on together with "large deflections" in Mechanical APDL. The artery wall is either modelled as linear elastic, hyper-elastic or viscous-elastic. In the elastic artery the initial deformation due to the transfer of force from the fluid model is large. This is the reason most models fail at the first time step. Pre-stressing the geometry is one possibility. There is a tutorial on the ANSYS Learning Hub for a Flexible Valve FSI which covers methodology for both linear elastic and hyper-elastic materials.

Mostly picking sufficiently small time steps so that the fluid side converges, but not too small so that the number of iterations is too large, will address the initial and subsequent deformations. Relaxation settings in System Coupling should also be low. In the various setups simulated, it was as low as 0.1.

In this simulation of the Carotid artery, the Coupling time step was 0.001 s, and the relaxation factors on both Fluent and Mechanical were set to 0.2. On the Fluent side, 500 iterations per time step were set. Coupling iterations were set to a minimum of 30 and a maximum of 150.

3. Key Results and Discussion

3.1. *CFD*

The CFD results show low WSS on the carotid sinus and below the external carotid artery as predicted by Kumar et al (2022). The typical WSS range in an artery is between 1-7 Pa, with vasoregulatory mechanisms attempting to keep it around 1.5 Pa (Papaioannou, et al., 2006). WSS values below 0.4 Pa in zones indicates areas at risk for atherosclerosis (Lopes, et al., 2019).

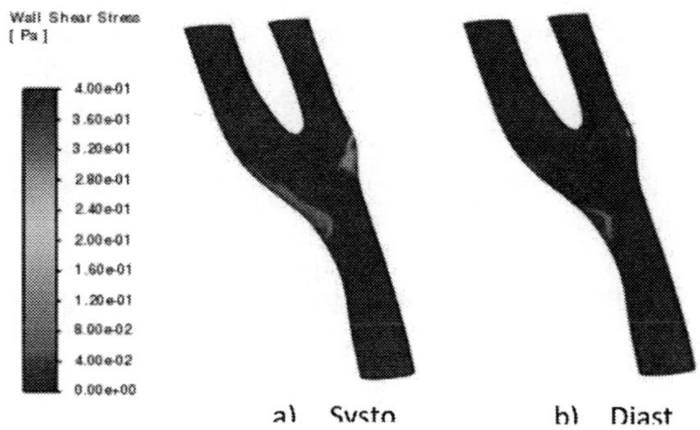

Figure CFD results showing WSS below 0.4 Pa on the carotid sinus and below the external carotid artery.

3.2. *FSI*

An initial FSI similation was set up as described in the methodology. Kruizinga et al, 2013 measured the distension of the carotid artery over a typical heartbeat cycle. The average literature value for the distension varies between 6-12% and is subject to the age and condition of the artery. During the test simulation the maximum distension at the bifurcation was 11.7%, with the average distension around 7%.

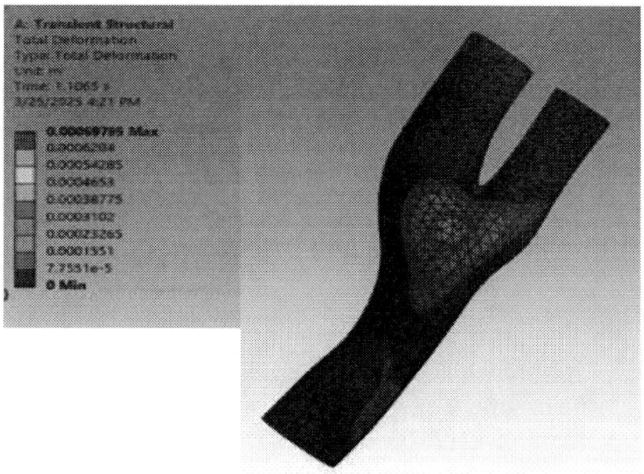

Figure 2: Results of FSI simulation using Carreau model for blood and a linear elastic property model.

4. Conclusion and Future Work

The CFD model predicts similar areas of concern for high OSI and WSS than was found by Kumar et al (2022) when shear stress is capped at a maximum of 0.4 kPa.

The preliminary FSI results correspond to distention ranges found in literature.

Four items play an important role in FSI simulation of an artery:

1. Time step size
2. Low relaxation values
3. Large number of coupling iterations
4. The most important lesson learnt in FSI of an elastic artery is that the fluid time step should converge at every coupled iteration. This means a large number of iterations per time step on the fluid side.

Future work will focus on verification of the FSI simulation methodology by duplicating experimental results obtained by Joshau Lowe at Aston University with FSI simulation.

References

1. ANSYS, 2024. *Flexible Valve FSI - Application Best Practice,* s.l.: s.n.
2. Geoghegan, P. H., Jermy, M. C. & Nobes, D. S., 2017. A PIV comparison of the flow field and wall shear stress in rigid and compliant models of healthy carotid arteries. *Jouranl of Mechanics in Medicine and Biology,* 17(3).
3. Gijsen, F. J. H., N, V. d. V. F. & D, J. J., 1999. The influence of the non-Newtonian properties of blood on the flow in large arteries: Steady flow in a carotid bifurcation model. *Journal of Biomechanics,* Volume 32, pp. 601-608.
4. Grøttum, P., Svindland, A. & Walløe, L., 1983. Localization of early athreosclerotic lesions in the righ carotid bifurcation in humans. *Acta Pathologica et Microbiologica Scandinavica,* Volume 91, pp. 65-70.
5. Kornet, L., Lambregts, J., Hoeks, A. P. G. & Reneman, R. S., 1998. Differences in Near-Wall Shear Rate in the Carotid Artery Within Subjects Are Associated With Different Intima-Media Thicknesses. *Arteriosclerosis, Thrombosis, and Vascular Biology,* Volume 18, pp. 1877-1884.
6. Kruizinga, P. et al., 2014. High-Definition Imaging of Carotid Artery Wall Dynamics. *Ultrasound in Medicine & Biology,* 40(10), pp. 2392-2403.
7. Kumar, N., Pai, R., Abdul Khader, S. M. & Kyriacou, P. A., 2022. Influence of blood pressure and rheology on oscillatory shear index. *Journal of the Brazilian Society of Mechanical Sciences and Engineering,* Issue 44, p. 510.
8. Lopes, D., Puga, H., Teixeira, J. C. & Teixeira, S. F., 2019. Influence of arterial mechanical properties on carotid blood flow: Comparison of CFD and FSI studies. *International Journal of Mechanical Sciences,* Volume 160, pp. 209-218.
9. Lopes, D., Puga, H., Teixeira, J. & Lima, R., 2020. Blood flow simulations in patient-specific geometries of the carotid artery: A systematic review. *Journal of Biomechanics,* Volume 111.
10. Oyejide, A. J., Awonusi, A. A. & Ige, E. O., 2023. Fluid-structure interaction study of hemodynamics and its biomechanical influence on carotid artery artherosclerotic plaque deposits. *Medical Engineering and Physics,* Volume 117.
11. Papaioannou, T. G. et al., 2006. Assessment of vascular wall shear stress and implications for atherosclerotic disease. *International Journal of Cardiology,* Volume 113, pp. 12-18.
12. Tzeng, Y. C., 2012. *Applied Aspects of Ultrasonography in Humans - Chapter:The Role of Ultrasonography in the Assessment of Arterial Baroreflex Function.* s.l.:s.n.
13. World Health Organization, 2020. *The top 10 causes of death.* [Online] Available at: https:// www.who.int/news-room/fact-sheets/detail/the-top-10-causes-of-death

Computational Fluid Dynamics of Aqueous Humor and Gas Bubble Interactions in Endothelial Keratoplasty: Implications for Graft Adherence

E. Cheng[a*], N. Basson[a], R. Hollhumer[a] and W.H. Ho[b]

[a]University of the Witwatersrand, Johannesburg, Private Bag 3, Wits 2050, South Africa

[b]University of Cape Town, Private Bag X3, Rondebosch 7701, South Africa

*2095188@students.wits.ac.za

Abstract

Graft detachment remains a critical complication following endothelial keratoplasty (EK), often requiring additional interventions. This study investigates the interaction between aqueous humour (AH) flow and intraoperative gas bubbles within the anterior chamber (AC) using computational fluid dynamics (CFD) simulations. Employing the Volume of Fluid (VOF) model in ANSYS Fluent, we modelled multiphase dynamics and patient motion through time-dependent gravitational effects.

Simulations revealed a 23% velocity increase and peak wall shear stress (WSS) of 0.6 Pa, approximately six times higher than the reported endothelial damage threshold, during supine-to-upright motion.

These results indicate that abrupt patient movements can significantly increase mechanical stress on the graft. This framework could guide surgical planning and patient positioning protocols to reduce detachment risk.

Keywords: Endothelial Keratoplasty, Graft Adherence, Aqueous Humor Flow, Air Bubble Dynamics, Computational Fluid Dynamics

1. Background and Motivation

Endothelial keratoplasty (EK) is a transformative corneal transplant technique that selectively replaces diseased layers of the cornea, such as Descemet's membrane and the endothelium. While EK offers improved visual outcomes and reduced rejection rates compared to full-thickness transplants, postoperative graft detachment remains a critical complication. The introduction of an intraoperative air bubble to secure the graft is pivotal; however, its interaction with the anterior chamber (AC) fluid—the aqueous humour (AH)—is not fully understood [1].

The anterior chamber, bordered by the cornea and iris, is essential for maintaining corneal health through AH circulation, see Fig 1. The interaction of AH with the air bubble and its influence on graft adherence during postoperative patient motion is underexplored. Existing literature highlights the significance of wall shear stress (WSS) and velocity changes within the AC, especially under transitional movements such as supine-to-upright and upright-to-supine positioning [2-4]. This study seeks to address these gaps using computational fluid dynamics (CFD) simulations to evaluate the mechanical and fluidic forces affecting graft stability. Specifically, it seeks to quantify the impact of postoperative patient movement on associated WSS and velocity profiles within the AC thereby providing insights into optimising surgical and postoperative protocols to mitigate graft detachment risks.

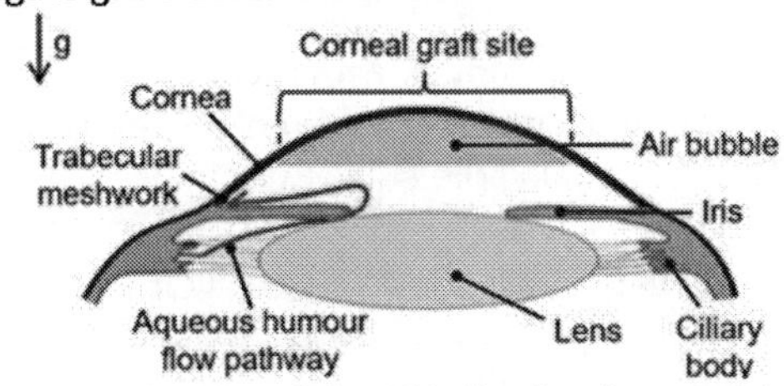

Fig. 1. Cross-section of the AC indicating the corneal graft site and red arrow indicating AH flow pathway

2. Methodology

A pressure-based, transient CFD model was developed using ANSYS Fluent. An idealized 2D cross-sectional geometry of the human anterior chamber was constructed, see Fig 2. AH flow, possessing properties similar to water, was simulated using the Boussinesq model for buoyancy-driven circulation, given the temperature difference between the 34°C and 37°C of the cornea and iris, respectively [5]. Air Bubble Interaction was modeled using the Volume of Fluid (VOF) approach to account for multiphase dynamics. Fluid and material properties are summarized in Table 1. Time-dependent, sinusoidal gravitational effects simulated a 4-second transition between supine and upright positions to model typical patient motion based on planar rotation.

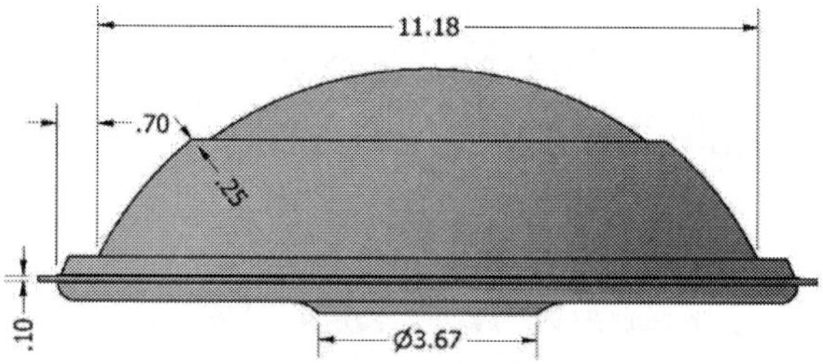

Fig. 2. Idealized, 2D AC geometry, showing graft cutout in mm (Autodesk Inventor® 2024)

Boundary conditions included an inlet velocity, for an AH inflow rate of 2-3 µl/min, on the AC's right boundary [6]. Pressure outlets facilitated the drainage of AH based on an IOP of 15 mmHg which is within the healthy range of 10-20 mmHg [2]. Under normal, healthy conditions, the drainage of AH is approximately the same rate as the inflow rate. Mesh independence tests ensured numerical accuracy. Local mesh sizing was applied to the four fluid bodies to accurately resolve the geometrical features, see Fig 3. A global maximum cell length of 0.08 mm was used. The average mesh skewness and minimum orthogonal quality gave acceptable values of 0.18 and 0.12, respectively. The fluid properties of AH is given in Table 1.

The simulation model, without graft and bubble, was validated against literature before the graft was added before finally, the bubble using the volume-of-fluid model.

Table 1. Material and fluid properties [6]

Property	AH (37°C)	Air (35°C)	Cornea
Density (kg/m^3)	1000	1.1455	1200
Viscosity (Pa·s)	0.001	1.81E-5	-
Thermal Conductivity (W/m·K)	0.6	0.025	0.58
Specific Heat Capacity (J/kg·K)	4180	1006	3300

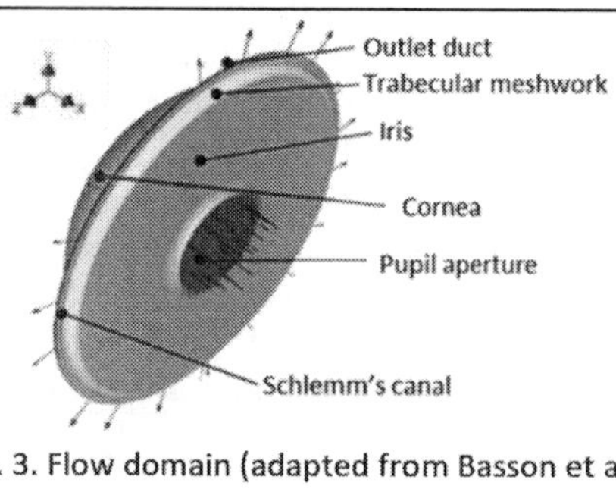

Fig. 3. Flow domain (adapted from Basson et al. [6])

3. Key Results and Discussion

3.1. Aqueous humour flow validation

Initial simulations revealed increased corneal endothelium stress due to elevated flow velocities in the presence of the graft. Maximum velocity increased by 23%, compared to a normal AC [2], to 0.79 mm.s^{-1}, even without the air bubble, simulating scenarios post-bubble dissipation. Maximum WSS along the corneal endothelium increased, with notable peaks at locations adjacent to the graft.

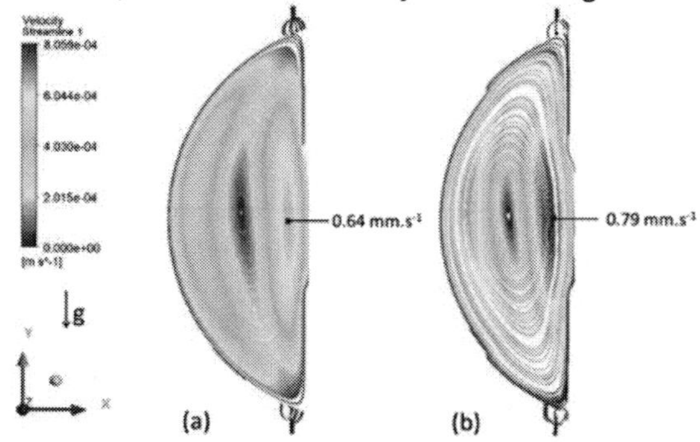

Fig. 4. Velocity streamlines for (a) normal and (b) graft inclusion conditions

3.2. Air bubble effects and graft adherence

The inclusion of an air bubble in the computational model revealed distinct changes in aqueous humour (AH) circulation patterns, which significantly influenced wall shear stress (WSS) and graft stability.

Higher velocities and wall shear stresses near the graft and corneal curvature were observed in the thicker grafts, potentially leading to endothelial cell damage and graft detachment, see Fig 5. Changes in bubble shape during transitional motion led to fluctuations in shear stress values, underscoring the importance of accurate bubble modelling in predicting postoperative outcomes.

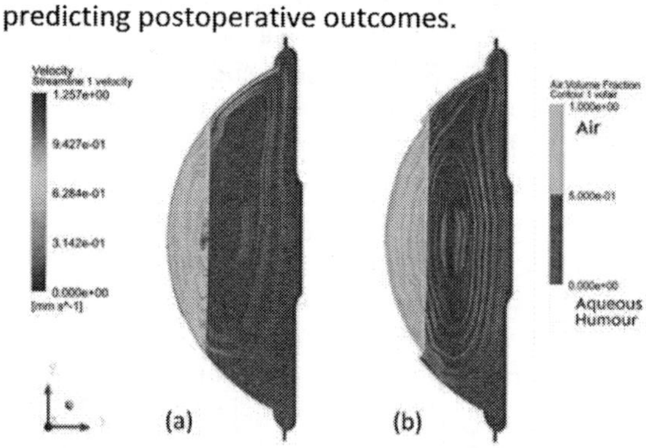

Fig. 5. Velocity streamlines and VOF phase contour immediately after upright-to-supine motion for (a) 15 µm and (b) 250 µm graft thicknesses

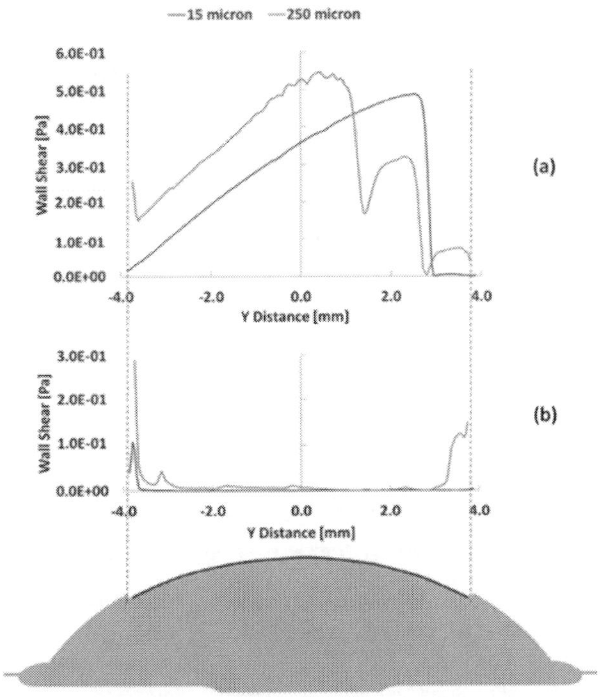

Fig. 6. WSS along the graft-AC interface for (a) supine-to-upright and (b) upright-to-supine transitions

During the supine-to-upright transition, the model showed a sharp increase in maximum WSS, particularly along the central axis of the graft, see Fig 6. Maximum WSS reached close to 0.6 Pa, which is higher than Kaji et al.'s critical point of endothelial cell damage of 0.1 Pa when tested on pigs [4]. While species differences mean the 0.1 Pa threshold may not directly apply to human corneal endothelium, the relative increase suggests a heightened mechanical risk during abrupt postoperative motion. In contrast, the upright-to-supine transition exhibited a more gradual redistribution of AH flow, with lower peak WSS values of about 0.3 Pa. Based on simulation results, thinner grafts with more gradual patient motions will lessen the likelihood of endothelial cell damage.

The sharp density difference between air and AH resulted in localized high-shear zones near the bubble boundary, affecting the nearby graft regions. These effects were amplified during transitional motion.

Elevated WSS near the central graft curve and bubble interface suggests increased detachment risk. These findings underscore the need for tailored patient positioning protocols to minimize abrupt movements during early recovery. Furthermore, educating patients on postoperative motion restrictions to enhance graft stability.

4. Conclusion and Future Work

This study provides a CFD framework for analyzing AH-bubble interactions in EK procedures. Bubble dynamics alters the AH circulation, causing an elevation of WSS, particularly during supine-to-upright transition, which may cause graft detachment. Limitations include the absence of direct measurements of graft attachment forces and the use of an idealised 2D geometry, which may not fully capture complex anterior chamber morphologies. Corneal geometry varies among patients, where steeper corneas amplify curvature-driven flow and peak WSS, increasing detachment risk, whereas flatter corneas produce more uniform flow with milder shear variations.

Future work will refine bubble modeling, incorporate patient-specific geometries, and include boundary conditions for AH generation and drainage. Developing a patient-specific surgical planning tool could optimize graft outcomes and improve postoperative care.

References

[1] S. Liu, Y. L. Wong, and A. Walkden (2022). *Current Perspectives on Corneal Transplantation*. Dove Medical Press Ltd. doi: 10.2147/OPTH.S289359.

[2] Z. Qin, L. Meng, F. Yang, C. Zhang, and B. Wen (2021). *Aqueous humor dynamics in human eyes: a lattice Boltzmann study.*

[3] J. I. Fernández-Vigo et al., (2018). Computational simulation of aqueous humour dynamics in the presence of a posterior-chamber versus iris-fixed phakic intraocular lens. PLoS One, vol. 13, no. 8, pp. e0202128-.

[4] Y. Kaji, T. Oshika, T. Usui, and J. Sakakibara (2005). Effect of Shear Stress on Attachment of Corneal Endothelial Cells in Association With Corneal Endothelial Cell Loss After Laser Iridotomy. Cornea, vol. 24, pp. S55–S58.

[5] C. R. Canning, M. J. Greaney, J. N. Dewynne, and A. D. Fitt (2002). Fluid flow in the anterior chamber of a human eye. Math Med Biol, vol. 19, no. 1, pp. 31–60.

[6] N. Basson, F. Alimahomed, et. al. (2022). An aqueous humour fluid dynamic study for normal and glaucomatous eye conditions. 44th Annual International Conference of the IEEE Engineering in Medicine & Biology Society, pp. 3963–3966.

Comparative Analysis of Shear-Driven and Coupled Biochemical-Haemodynamic Models for Thrombus Formation

Jordan Bougardt[a,*], Randall Paton[a], Malebogo Ngoepe[b], Cecil Pule[c], Karabo Moisane[d], Jayandiran Pillai[c], Wei Hua Ho[b]

[a]*School of Mechanical, Industrial and Aeronautical Engineering, University of the Witwatersrand, 1 Jan Smuts Avenue, Johannesburg 2000, South Africa*

[b]*Department of Mechanical Engineering, University of Cape Town, Rondebosch, Cape Town 7701, South Africa*

[c]*Department of Surgery, Donald Gordon Medical Center, University of the Witwatersrand, 21 Eton Road, Parktown, Johannesburg 2193, South Africa*

[d]*Division of Orthopaedic Surgery, School of Clinical Medicine, University of the Witwatersrand, 7 York Road, Parktown, Johannesburg 2193, Johannesburg*

**jordan.bougardt@gmail.com*

Abstract

Thrombus (clot) formation is a critical factor in the study of vascular flow and related pathologies, as thrombi act as porous media that inhibit blood flow. This resistance may have adverse health effects (e.g., strokes, seizures), while being beneficial to some medical treatments such as embolisation (deliberate vessel blockage). Clot structures are complex, and their effects on health and surgical efficacy are difficult to determine; thus, computational fluid dynamics (CFD) techniques have become more prevalent in investigating their dynamics and impact.

This study presents the first direct comparison between a shear-driven and coupled biochemical-haemodynamic model of thrombosis, evaluated against an experimentally grown clot within an idealised aneurysm. The shear-driven model, developed in this work, parameterises clotting as a function of shear and offers computational simplicity (lower cost) but cannot capture temporal or biochemical clot evolution. Conversely, the coupled biochemical model, initially developed by Ngoepe and Ventikos [1] and expanded upon, by Jimoh-Taiwo et al [2], with results presented here, integrates flow and the biochemistry of the coagulation cascade, such as thrombin generation and fibrin polymerization.

The shear-driven model closely matches the end-state experimental thrombus, differing in percentage occlusion by less than 0.5%, while the biochemical model captures both the final distribution and thrombus dynamics. The results suggest that the shear-driven model is well suited for rapid screening or flow obstruction prediction, whereas the biochemical model is better positioned for investigating thrombus progression and composition in detail-oriented studies.

Keywords: Thrombus, aneurysmal thrombosis, shear, biochemical-haemodynamic, clot modelling

1. Background and Motivation

Thrombus formation is a fundamental process in vascular haemostasis, wherein a damaged blood vessel wall (e.g., due to device placement or ballooning) triggers the formation of a network of blood, platelets, and plasma proteins [3] that serve to prevent bleeding. This process has the potential to become pathological/harmful under certain conditions, such as aneurysms and thromboembolic events. Aneurysms, characterised by localised vessel dilation [4], alter flow dynamics, leading to regions of low shear rate and flow recirculation that promote thrombus formation. The thrombi, which serve as the focus of this investigation, are referred to as fibrin clots, the formation of which is illustrated in Figure 1.

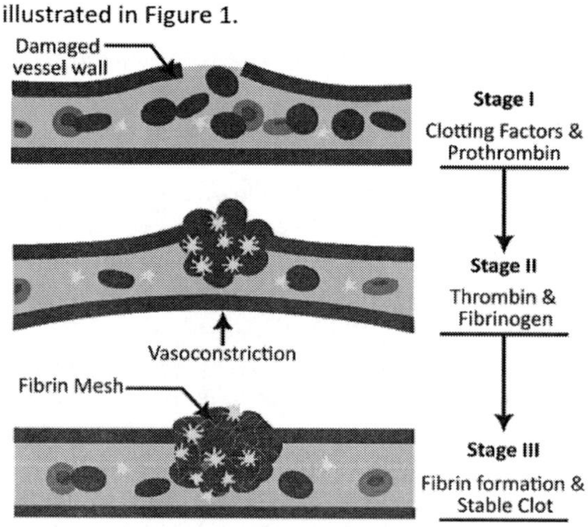

Fig. 1. Fibrin clot formation process

As shown, an injured blood vessel triggers vasoconstriction so as to reduce blood flow. Blood platelets adhere to exposed collagen fibres at the injury site, whereafter a series of biochemical reactants (clotting factors), most importantly, prothrombin, are exposed [5]. Coagulation (reaction cascade) follows wherein, fibrin (insoluble protein) is generated from the reaction of thrombin and the soluble plasma protein fibrinogen, which solidifies into a fibrous mesh surrounded by red blood cells [6] and activated platelets, thus forming the stable clot structure.

To capture this complex multiphysics phenomenon, advanced CFD has been increasingly employed to model the interacting haemodynamic and biochemical systems involved in thrombus formation [7-9].

2. Objectives and Scope

This study aims to develop and evaluate a shear-driven numerical model of thrombus formation and compare it to a published biochemical-haemodynamic model. The objectives are thus:

1. To implement a shear-driven model that captures thrombus formation based solely on haemodynamic parameters.
2. To benchmark this model against a coupled biochemical-haemodynamic model, to assess the trade-offs between computational cost, temporal resolution, and physiological fidelity.

Both models are assessed using a common experimental reference, thrombus growth data obtained from a flow phantom aneurysm study [10], where thrombin was introduced into a saline-fibrinogen mixture to generate a fibrin clot.

3. Methodology and Computational Approach

3.1 Aneurysmal geometry

The computational domain for this study is an idealised saccular aneurysm, as shown in Figure 2, resembling a spherical out-pouching (berry) attached to a parent vessel. The clot origin face, denoted in Figure 2, refers to the site of injury as this is the location at which thrombin is introduced into the domain for the experimental and biochemical model cases.

Steady boundary conditions were utilised as the variables of interest are the growth and stability of clots and these conditions align with those of the experimental study conducted by Ngoepe *et al* [10].

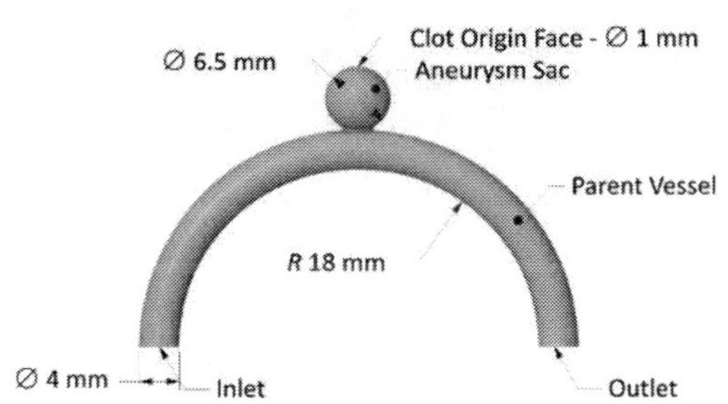

Fig. 2. Computational domain representing an idealised saccular aneurysm

3.2 Shear-Driven Model

The shear-driven model developed using the framework of Diamond *et al* [11-12], models clot formation based on haemodynamic parameters. It neglects biochemistry and thus cannot produce reliable data concerning transient clot growth, rather replicating end-state clot formation through the use of shear mediated computational cell-based functions.

The haemodynamic parameters and thresholds, including fibrinolysis (breaking down of blood clots), are defined according to empirically derived values shown in Table 1. The model operates in computational time, leveraging numerical controls such as cell-adjacency criteria and mesh refinement to replicate the final clot distribution and flow blockage.

Table 1. Shear rate conditions and clot properties

Property	Value	Reference
Shear rate limit (s^{-1})	50	[13]
Clot porosity	0.75	[11]
Clot permeability (m^2)	1.0e-12	[12]
Fibrinolysis threshold (s^{-1})	5300	[12]

Clot development relies on connectivity macros and indices, enforcing that clots grow from clotted cells (computational), as shown in Figure 3. This mirrors the in vivo process where a thrombus originates from injury sites and propagates outward, akin to the mechanisms of fibrin deposition and platelet aggregation.

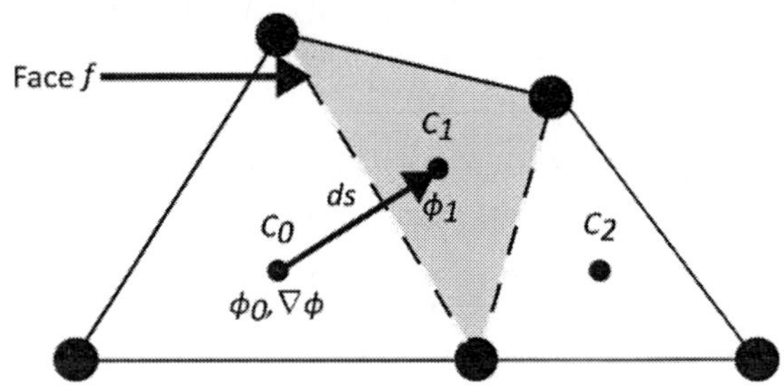

Fig. 3. Adjacency controls and indexing

Using Figure 3 as an example, if cell C_0 is clotted and C_1 meets the shear rate criteria, C_1 can be marked as clotted. However, C_2, adjacent to C_1 but not sharing a boundary with C_0, cannot clot in the same step, ensuring logical clot propagation based on proximity.

The mesh adaption used during the implementation of the shear-driven model, is isolated to the clot front to ensure that computational cells adjacent to the clot , i.e., the boundary between clotted cells and fluid cells are refined sufficiently.

The refinement was scoped to the clot front through cell registers modified to target computational cells with a high porosity gradient. Cells at the clot front exhibit the largest gradients, as shown in Figure 4, as the clot boundary forms an interface between the unobstructed flow and newly formed clot structures, resulting in a sharp change in local porosity, while cells internal to the clot have lower porosity gradients.

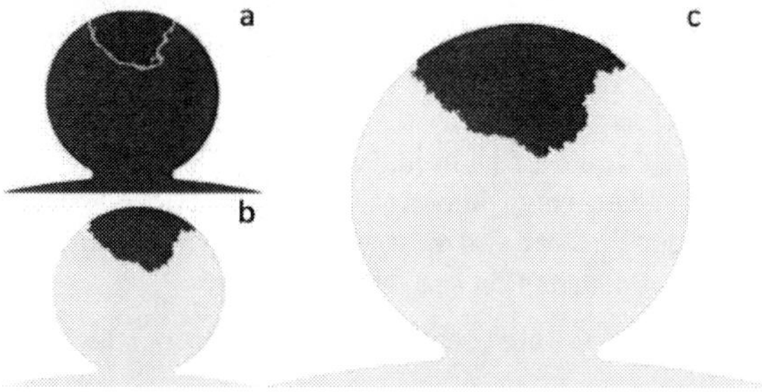

Fig. 4. Isolation of the clot front for adaptive refinement (a) Clot formation (blue), (b) clot structure by cell and (c) adaption applied to clot front (blue adaption cells)

3.3 Coupled Biochemical-Haemodynamic Model

The coupled model, as developed by Ngoepe and Ventikos [1] and Jimoh-Taiwo *et al* [2], integrates haemodynamic properties with the biochemical processes involved in clot formation. This includes thrombin generation, fibrinogen consumption, fibrin polymerisation and shear rate tracking. Specifically, this model monitors changes in thrombin and fibrin concentrations over time.

Thrombin is inserted at the site of injury, Clot Origin Face in Figure 2, at concentrations specified by parameterised functions known as thrombin generation curves, detailed in Wagenvoord *et al* [14]. This concentration is tracked through implementation of User Defined Scalars (UDS). Additionally, fibrinogen is initialised and allowed to flow through the domain at specified concentrations from the inlet.

The rate of the enzymatic action of thrombin on fibrinogen to form fibrin is given by the Michaelis-Menten equation, shown in Equation 1, wherein fibrinogen consumption (sink) and fibrin generation (source) source terms are equivalent, with the reaction rate being dependent on the concentration of fibrinogen and thrombin within the flow field.

$$S_{FN} = -S_{FG} = \frac{k_{cat} \cdot [FG][TH]}{k_m + [FG]} \tag{1}$$

, where S denotes source terms, FN fibrin production, FG is fibrinogen consumption, k_{cat} is the turnover number and k_m is the Michaelis-Menten constant.

3.4 Experimental clot growth

The experimental clot growth data, collected from Ngoepe *et al* [6], forms the basis for comparison for both the shear-driven and coupled biochemical-haemodynamic models of clot formation.

Thrombin was injected into the apex of the aneurysm sac after the aneurysmal flow phantom was filled with a saline mixture which included free fibrinogen. The subsequent formation of the pure-fibrin clot was monitored over the course of four minutes (240 s), at which time the clot was taken to be stable (having reached end-state).

4. Results and Discussion

4.1 Aneurysmal flow and shear-driven clotting

The flow in the parent vessel is directed from left to rate at the base of the aneurysm sac, hence, it is expected that an anti-clockwise flow would arise, as shown in Figure 5a, with the highest shear flow entering from the parent vessel and moving up the right wall of the aneurysm.

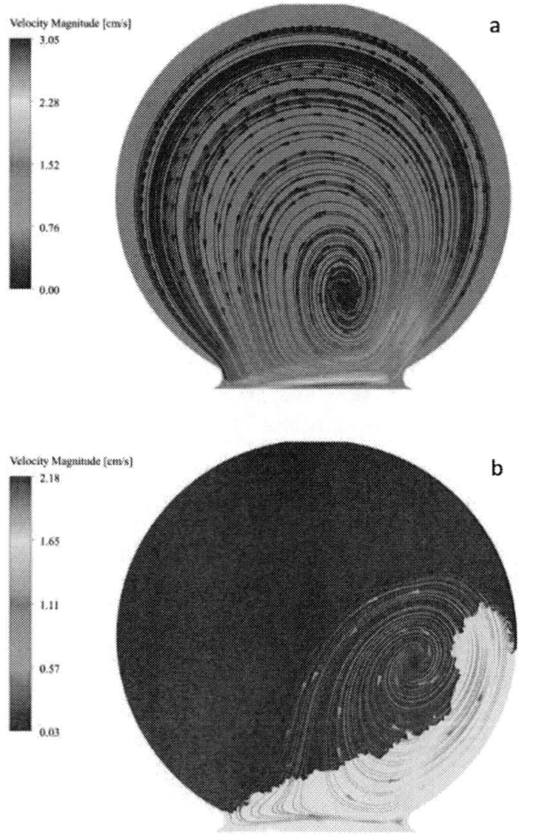

Fig. 5. Flow through the idealised aneurysm sac pre-clotting (a) and post- shear-driven clotting (b).

Figure 5b illustrates the persistence of the anti-clockwise vortex within the aneurysm after thrombus formation, which overlaps the clot as it is a 3D structure that does not fill the aneurysm uniformly, with higher velocity zones on the aneurysm's right side. This localised higher shear inhibits clot development in that region. The lower shear rates on the aneurysm's left side foster a favourable environment for thrombosis. The resulting differential shapes the final clot as it is constrained by elevated shear zones. This interaction between flow dynamics and clotting establishes a self-regulating mechanism that dictates the thrombus's spatial distribution and total size.

This inconsistency may be explained by considering that biochemical model was computed in 2D to reduce computational cost, which exacerbates the blockage effect for transported scalars, as reactants, such as the free-flowing fibrinogen are restricted from entering the domain from the out-of-plane dimension. In contrast, although the shear-driven model runs as a transient simulation, the time taken to reach a steady state clot does not correlate to the experimental duration of 240 s or 4 minutes, making time-dependent comparisons inconclusive. Nonetheless, for applications focused on final occlusion rather than clotting dynamics, the model remains effective.

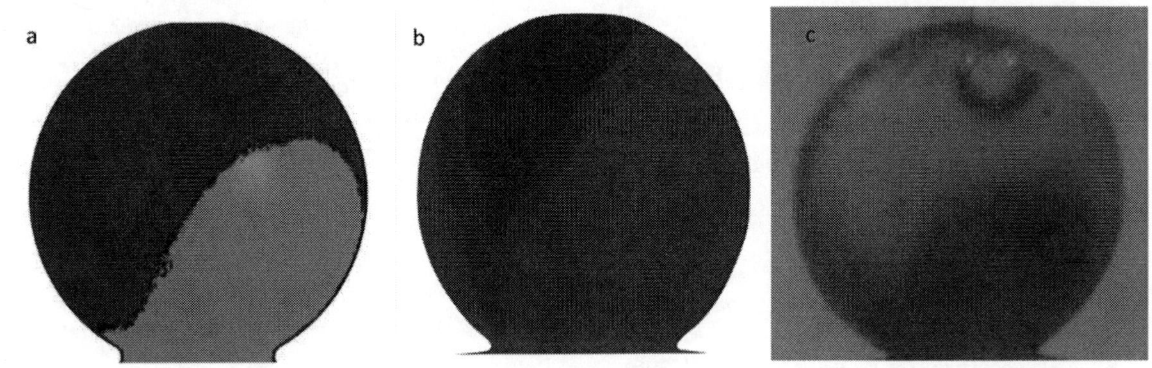

57.18	55.27	57.60

Fig. 6. Clot morphologies and percentage occlusion [%]. Shear-driven (a), biochemical (b) and experimental clot morphologies (c).

4.2 Comparison of computational and experimental clot growth

Figure 6 depicts the final clot morphologies achieved through the shear-driven (a) and biochemical models (b), redrawn from Ngwenya & Ngoepe [15], as well as the experimental fibrin clot growth (c). The results for all modelling regimes exhibit similarity in their stable clot distribution, in that there is preferential clot formation on the left side of the aneurysm in all three. This similarity is particularly clear when comparing the shear-driven (a) and experimental (c) thrombi.

Notably, both clots terminate and have comparable thickness on the left side of the aneurysm while also showing a minor growth on the right-hand side, which is inhibited from further expansion along the wall due to the high shear that forms at the clot boundary as the clot nears the vortex center, as illustrated in Figure 5b.

This similarity is reinforced by the total percentage occlusion accomplished in each case, wherein the shear-driven and experimental model clots differ in occlusion percentage by $\approx 0.5\%$, whereas the coupled biochemical haemodynamic model clot is slightly smaller at 55.27%, particularly because of the void between the clot and the left wall of the aneurysm.

5. Conclusion and Future Work

Both the shear-driven and coupled biochemical haemodynamic models have significant merit and limitations. The shear-driven model, while accurate in determining the final clot morphology and distribution, cannot provide reliable transient data as it neglects biochemical reaction and transport. Conversely, the biochemical model solves additional equations (three scalar) to capture coagulation biochemistry; hence, its computational expense may be inhibitive

The choice between these models depends on the specific application. The shear-driven model is best-suited for fast, large-scale simulations aimed at evaluating end-state occlusion, or screening of device designs. The biochemical model is more appropriate for research applications that require insight into clot progression dynamics, reactant depletion, etc.

Future work will explore the development of hybrid strategies, combining shear-based controls with simplified scalar transport to approximate biochemical effects. Additionally, optimisation techniques such as model order reduction or GPU acceleration could reduce simulation time for the biochemical model.

References

1. Ngoepe, M. N., & Ventikos, Y. (2016). Computational modelling of clot development in patient-specific cerebral aneurysm cases. *Journal of Thrombosis and Haemostasis*, *14*(2), 262–272. https://doi.org/10.1111/jth.13220
2. Jimoh-Taiwo, Q., Haffejee, R., & Ngoepe, M. (2022). A Mechano-Chemical Computational Model of Deep Vein Thrombosis. *Frontiers in Physics*, *10*. https://doi.org/10.3389/fphy.2022.886193
3. Tomaiuolo, M., Brass, L. F., & Stalker, T. J. (2017). Regulation of Platelet Activation and Coagulation and Its Role in Vascular Injury and Arterial Thrombosis. In *Interventional Cardiology Clinics* (Vol. 6, Issue 1, pp. 1–12). Elsevier Inc. https://doi.org/10.1016/j.iccl.2016.08.001
4. Teixeira, F. S., Neufeld, E., Kuster, N., & Watton, P. N. (2020). Modeling intracranial aneurysm stability and growth: an integrative mechanobiological framework for clinical cases. *Biomechanics and Modeling in Mechanobiology*, *19*(6), 2413–2431. https://doi.org/10.1007/s10237-020-01351-2
5. Savage, B., Saldívar, E., & Ruggeri, Z. M. (1996). Initiation of Platelet Adhesion by Arrest onto Fibrinogen or Translocation on von Willebrand Factor. In *Cell* (Vol. 84).
6. Chapin, J. C., & Hajjar, K. A. (2015). Fibrinolysis and the control of blood coagulation. *Blood Reviews*, *29*(1), 17–24. https://doi.org/10.1016/j.blre.2014.09.003
7. Grande Gutiérrez, N., Mukherjee, D., & Bark, D. (2024). Decoding thrombosis through code: a review of computational models. *Journal of Thrombosis and Haemostasis : JTH*, *22*(1), 35–47. https://doi.org/10.1016/j.jtha.2023.08.021
8. Anand, M., & Rajagopal, K. (2017). A Short Review of Advances in the Modelling of Blood Rheology and Clot Formation. *Fluids*, *2*(3), 35. https://doi.org/10.3390/fluids2030035
9. Yesudasan, S., & Averett, R. D. (2019). Recent advances in computational modeling of fibrin clot formation: A review. In *Computational Biology and Chemistry* (Vol. 83). Elsevier Ltd. https://doi.org/10.1016/j.compbiolchem.2019.107148
10. Ngoepe, M. N., Pretorius, E., Tshimanga, I. J., Shaikh, Z., Ventikos, Y., & Ho, W. H. (2021). Thrombin–Fibrinogen In Vitro Flow Model of Thrombus Growth in Cerebral Aneurysms. *TH Open*, *05*(02), e155–e162. https://doi.org/10.1055/s-0041-1728790
11. Diamond, S. L. (1999). *Engineering Design of Optimal Strategies for Blood Clot Dissolution.*
12. Brass, L. F., & Diamond, S. L. (2016). Transport physics and biorheology in the setting of hemostasis and thrombosis. In *Journal of Thrombosis and Haemostasis* (Vol. 14, Issue 5, pp. 906–917). Blackwell Publishing Ltd. https://doi.org/10.1111/jth.13280
13. Casa, L. D. C., Ku, D. N., & Woodruff, G. W. (2017). *Thrombus Formation at High Shear Rates.* https://doi.org/10.1146/annurev-bioeng-071516
14. Wagenvoord, R., Hemker, P. W., & Hemker, H. C. (2006). The limits of simulation of the clotting system. *Journal of Thrombosis and Haemostasis : JTH*, *4*(6), 1331–1338. https://doi.org/10.1111/j.1538-7836.2006.01967.x
15. Ngwenya T. & Ngoepe, M. (2023). *The Development of a Reduced Order Model for Prediction of Haemodynamic and Biochemical Changes in a Computational Cerebral Aneurysm Thrombosis Model*

Biomechanical Characterisation of Rat Myocardia through Hyperelastic Model Fitting

Thanyani Pandelani[a]*, Harry Ngwangwa[a], Fulufhelo Nemavhola[b], Alex G. Kuchumov[c],

[a]*College of Science Engineering and Technology, University of South Africa, Corner of Christiaan de Wet Road & Pioneer Avenue, Florida, 1709, South Africa*

[b]*Durban University of Technology, 41/43 M L Sultan Rd, Greyville, Durban, 4001, South Africa*

[c]*Biofluids Laboratory, Perm National Research Polytechnic University, Perm, Komsomol'skiy Prospekt, 29, Perm, Perm Krai, 614990, Russia,*

**corresponding author email: epandet@unisa.ac.za*

Abstract

This study focuses on developing an understanding of regional differences in wall mechanics of the heart, specifically targeting the left ventricle (LV), septal wall (STW), and right ventricle (RV) of rat hearts. To accomplish this, biaxial tests were performed on these three myocardial regions, using a selection of hyperelastic models including Fung, Choi-Vito, Polynomial (Anisotropic), Four-Fiber Family, and two versions of the Holzapfel model (2000 and 2005). These models were evaluated to determine which could best be utilised in developing computational heart models, using the Coefficient of Determination (R2).

Results indicated that the Polynomial (Anisotropic) model provided the most accurate fit across all myocardial tissues, achieving EIs over 90%. The Four-Fiber Family and the two Holzapfel models showed moderate performance in the LV and STW, with EIs ranging from 50% to 70%. In contrast, the Fung and Choi-Vito models demonstrated poor goodness of fit in these tissues.

The findings of this research are significant for the advancement of computational models that aim to more precisely simulate the mechanics of cardiovascular diseases, contributing essential data that enhances our understanding of heart disease mechanisms through accurate modeling of myocardial mechanics.

Keywords: Cardiac mechanics; experimental mechanics; rat heart mechanics; hyperelastic constitutive model fitting; soft tissue mechanics; biaxial testing.

1. Introduction

Understanding cardiac wall mechanics is crucial for unraveling the mysteries of heart failure and for crafting precise computational models that predict cardiac performance. These computational models are indispensable for deciphering both local and global heart function post-medical intervention, especially in the context of cardiovascular diseases, which rank as the leading cause of death worldwide. Heart diseases such as myocardial infarction and myocardial ischemia often culminate in heart failure and cardiac arrest. Traditionally, research has predominantly concentrated on the left ventricle (LV), with the importance of the right ventricle (RV) only gaining recognition in the late 1950s [1]. This LV-centric approach has led to an overemphasis on LV mechanics in the development of computational models, potentially compromising their accuracy. Recent insights indicate that the mechanics of the left ventricle, septal wall, and right ventricle may vary significantly. Therefore, a comprehensive exploration of the mechanics of all heart walls, including the left ventricle, septal wall, and right ventricle, is imperative to enhance the precision of heart models during their development.

The study of soft tissue mechanical behaviour has garnered significant interest for its potential to shed light on disease progression [2-5]. Previous investigations have utilized pig hearts to probe mechanical properties across various regions [4, 5]. Given the substantial deformation of the cardiac wall, researchers have increasingly turned to computational models like Finite Element Models to investigate the mechanisms of heart diseases, including myocardial infarction and myocardial ischemia. Consequently, comprehending the mechanical properties of heart myocardia is of paramount importance to both basic and clinical researchers. Numerical analyses have played a pivotal role in understanding both global and regional cardiac function [6-14]. Previous work employed Finite Element analysis to explore the impact of myocardial infarction using a subject-specific 3D model of the rat heart [12-14].

While computational models have made significant strides in understanding disease progression, the assumption that the LV shares identical mechanical properties with the RV and septal wall (STW) may be unfounded. Most developed computational models have erroneously assumed uniform mechanical properties across the entire 3D heart model [15].

Although biomechanical analysis of the LV has illuminated aspects of computational cardiac analysis [16], it is imperative to discern the mechanical distinctions between the LV, STW, and RV.

Algorithms have been employed globally to ascertain the material properties of various biological tissues. Material parameters for healthy myocardium have primarily been derived from a single region, the left ventricle (LV). This study presents biaxial data from three pivotal regions within the rat heart: LV, RV, and STW. Primarily most researchers use biaxial tensile testing to determine the mechanical properties of myocardial tissue. According to Harrison John P, biaxial test is a scientific method used to determine the elastic constants of a material such as stress by subjecting it to pressure in two perpendicular directions. In this paper we are using biaxial tensile testing to determine viscoelasticity data which is then used in Hyperfit [ref]. The mechanical properties of these regions were meticulously characterized through biaxial testing. While most studies have focused on the mechanical characterization of the LV under various conditions, there is a pressing need to determine the passive mechanical material parameters for an anisotropic constitutive model of myocardial infarcts in rats. In this study, the mean biaxial stress-strain data from the three rat heart regions (RV, LV, and STW) were harnessed to determine material properties for the Fung [17], Choi-Vito model [18], Holzapfel (2000) [19], Holzapfel (2005) [20], Four-fiber family model [21, 22], and the Polynomial (Anisotropy) model [23]. The passive material parameters for these regions were determined using an algorithm that minimizes the error between experimental and predicted data.

The primary objective of this study was to examinine regional variations in cardiac mechanics within healthy rat hearts by fitting six hyperelastic models. While prior research has predominantly concentrated on studying the mechanical properties of single heart chambers, particularly the LV free wall, limited attention has been devoted to the septal wall. We hypothesized that heart walls, including the left ventricle, septal wall, and right ventricle, may exhibit distinct mechanical properties. To validate this hypothesis, we subjected heart myocardium to biaxial testing to obtain experimental data. Subsequently, we converted force and displacement data into stress and strain for curve fitting using various hyperelastic models.

2. Materials and Methods

The mechanical testing of the rat myocardia in the LV, RV and STW regions were conducted using the customed biaxial tensile material apparatus (BioTester 5000 CellScale, Waterloo, Canada) [17, 18]. The clamps of BioRake design were used for piercing the tissue samples, see Figure 1. A 5x5 mm of the sample size was clamped on the biaxial BioRake allowing the overlap of approximately 2.5 mm as shown in Figure 2, the rat myocardia were clamped and subjected to eqi-biaxial testing. Vernier calliper was utilised in measuring the thickness of the samples. This measurement was done carefully in trying to minimise the error to poor or inaccurate thickness measurement. To reduce these errors, four to five measurement were taken from each sample and the average thickness was then used for further process of the results. The precondition was performed before testing and collecting data by subjecting each sample to a 10 % strain at a rate of 0.08 per second. During testing 0.005 N of force was applied for 0.53 seconds as a preload. Also, all samples were preconditioned by applying 40 N/m of eqi-biaxial tension as previously described [16]. The body temperature was mimicked by immersing all harvested samples in the saline 0.91 % w/v of NaCl solution and heated and maintained at 37 oC. in collecting the biaxial testing data, in each sample, 40 % of strain was applied in both the longitudinal and circumferential direction at the strain rate of 0.08 per seconds. Most studies have reported between 25 % [16], 35 % [19] and 60 % [20] of strain in the rat heart; however, in this study, 40 % strain was then selected.

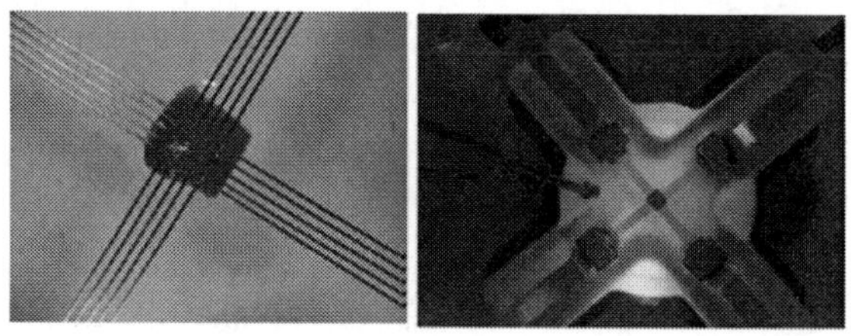

Fig. 1. Experimental set up of equi-biaxial mechanical testing of rat heart looking at different regions in left ventricle (LV), septal wall (STW) and right ventricle (RV). (Left) depicts the 9 × 9 mm left ventricle (LV) myocardium sample along with the rake assembly designed for clamping, while (right) illustrates the utilization of the biotester system employed for performing biaxial testing on rat LV myocardium [21]

2.1. Numerical Analysis

A Powell method implemented as 3rd Party library algorithm in HYPERFIT software (v8.1.0.604) was used to fit the Fung, Choi-Vito, Polynomial (anisotropic), Holzapfel (2000), Holzapfel (2005) and Four-fibre-amily strain energy functions to the biaxial experimental data, the details of these strain functions or constitutive models are described and presented in our previous work [21]. All curves were then averaged to obtain the material parameters of each hyperelastic model from the average stress - strain curves [22]. Therefore, to obtain average curves of each specimen, engineering stresses and strains data was obtained using the same frequency and stresses and strains were calculated using Eqs. (1) and (3).

The recorded force and displacement from both longitudinal and circumferential directions were converted to the engineering stress and strain, respectively.

$$\sigma_{c,l} = \frac{F_{c,l}}{A_{c,l}} \tag{1}$$

Where $A_{c,l}$ is the cross-sectional area in the circumferential and longitudinal direction and calculated as follows:

$$A_{c,l} = l_{c,l} x b_{c,l} \tag{2}$$

The engineering strain is calculated as follows:

$$\varepsilon_{c,l} = \frac{l_o - l_i}{l_0} \tag{3}$$

Where;

$l_{c,l}$ is the length of the prepared sample in the circumferential and longitudinal directions,

$b_{c,l}$ is the breadth of the prepared sample (see Figure 1) in the circumferential and longitudinal directions,

l_i is the initial or original length of the sample before loading and l_0 is the final length of the sample after loading.

3. Results

Figure 2 a-c shows the non-linear force-displacement curves for the LV, RV and STW for selected tests in fibre and cross-fibre direction. The areas between the loading and unloading curves show the amount of energy dissipated due to internal material friction in both directions. This is attributed to the viscoelastic property of the cardiac tissue. This implies that cardiac tissue does not really return all the absorbed tension energy on contraction. The dissipated energy was found to be greater in myocardial tissues extracted from the left ventricular free wall and the mid-wall layer of the myocardium, compared to those from the right ventricular free wall.

The coefficients of determination or correlation coefficients (R2) for each of the six constitutive models are plotted in Figure 3 on fourteen test samples. The three plots represent results from (a) the left ventricle – LV, (b) the mid-wall – STW, and (c) the right ventricle walls.

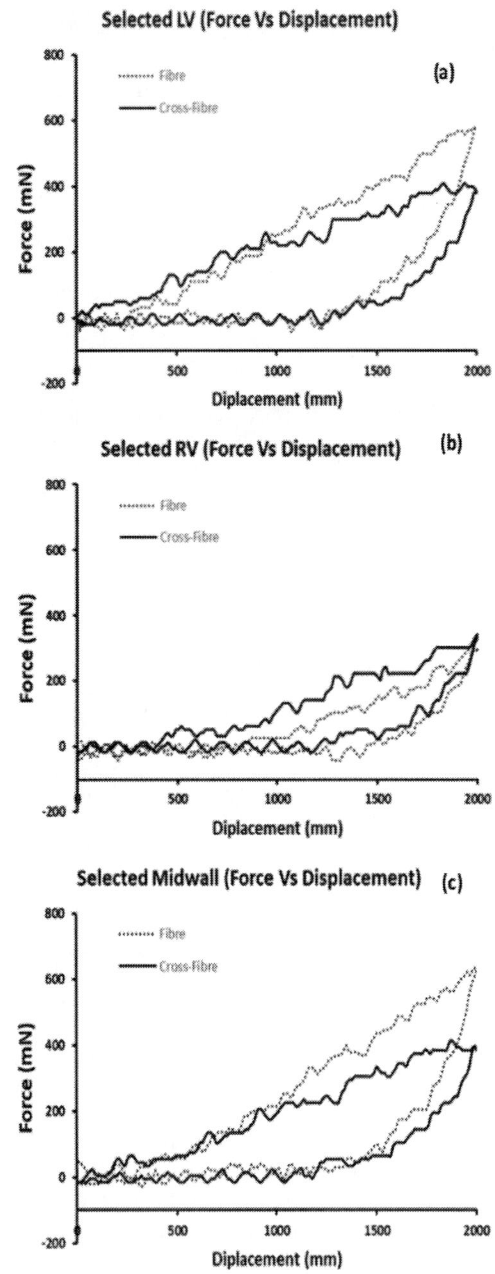

Fig. 2. Experiential data force- displacement sample of (a) LV, (b) RV and (c) STW-wall

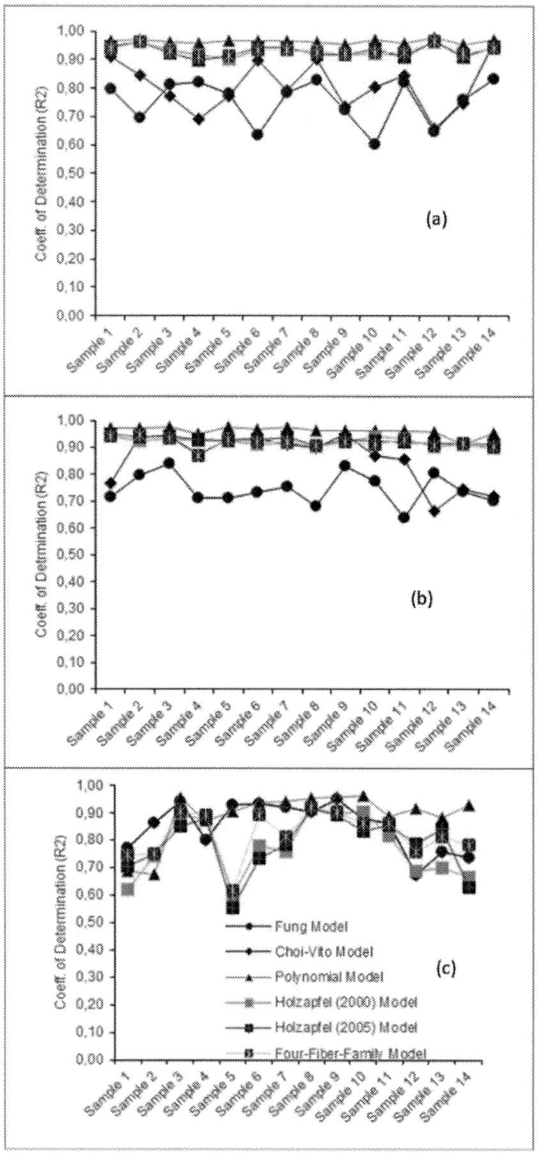

Fig. 3. Coefficient of determination (R2) plotted again all samples comparing fitting of Fung, Choi-Vito, Polynomial (anisotropic), Holzapfel (2000), Holzapfel (2005) and Four-fibre-family hyperelastic models in the (A) LV, (B) STW and (C) RV.

4. Discussion

Six constitutive models were applied to biaxial tensile results obtained from the three different myocardial walls of the rat heart. Fourteen (14) specimen for each myocardial wall were tested. The results show that the Polynomial model outperforms the other five models. However, this Polynomial model is twelve-parameter model which incorporates both the fiber-orientation and anisotropy effects. It was found that the performance of this model was closely matched by the Holzapfel and Four-fiber family models. It is quite remarkable that the Holzapfel with only five parameters was able to match the performance of the Polynomial model on the LV and STW tissue. Thus, the Holzapfel has much more computational benefits than the other models.

The RV myocardial tissue behaviour is the hardest to predict apparently due to its higher degree of anisotropy. Although the Polynomial model yields around 90 % correlation, it was observed that this was relatively lower than its correlation on the LV and STW tissue behaviour. There was also largest variability in the EI's c-values for the RV tissues as compared to the LV and STW. At this stage, it therefore seems that anisotropy is the one of the contributing factors to the difficulties in predicting RV tissue behaviour. There might be other factors which require further investigation.

Thehe poor performance of the Fung and Choi-Vito models is further evidence to the fact that myocardial tissue behaviour cannot be accurately represented by models that do not incorporate fiber orientation and anisotropy. Notice that all models that yield better performance have at least one of the two factors. On the other hand, the Fung and Choi-Vito models have neither of these factors. Despite the large number of material parameters in the Fung model, it still yielded poor results.

The area enclosed between the loading and unloading curves represents the energy dissipated due to internal friction within the material, indicative of its viscoelastic behavior. In cardiac tissue, this energy dissipation appears substantially greater than that observed in other connective soft tissues, such as tendons (15).

5. Conclusion

We successfully employed the biaxial mechanical data to demonstrate the relationship between the mechanical l behaviour of various regions in the rat heart. The established anisotropic materials properties could be utilised in the understanding of various disease mechanisms by using the rat model.

Inverse finite element model and genetic algorithm were employed to identify the parameters of Fung orthotropic hyperelastic strain energy function in three regions (RV, LV and STW) in the rat heart. The identified material parameters will provide new platform for FE investigations of mechanical aspects of various therapies when using rat models.

Acknowledgements

This study was approved by the Faculty of Health Sciences Animal Ethics Committee of the University of Cape Town on 6 May 2019 under reference number FHS AEC REF 019-019. The study was funded by NRF Grant CSUR240511218301. Alex G. Kuchumov thank for funding Ministry of Science and Higher Education of the Russian Federation (Project № FSNM-2024-0009).

References

1. Murphy, E. and B. Shelley, The right ventricle—structural and functional importance for anaesthesia and intensive care. BJA Education, 2018. 18(8): p. 239-245.
2. Ngwangwa, H.M., T. Pandelani, and F. Nemavhola, The Application of Standard Nonlinear Solid Material Models in Modelling the Tensile Behaviour of the Supraspinatus Tendon. Preprints, 2021. 2021080298.
3. Ngwangwa, H.M. and F. Nemavhola, Evaluating computational performances of hyperelastic models on supraspinatus tendon uniaxial tensile test data. Journal of Computational Applied Mechanics, 2021. 52(1): p. 27-43.
4. Pandelani, T., Semakane, L., Msibi, M., Kuchumov, A. G., & Nemavhola, F. (2025). Passive biaxial mechanical properties of sheep myocardium. Frontiers in bioengineering and biotechnology, 13, 1549829.
5. Nemavhola, F., Study of biaxial mechanical properties of the passive pig heart: material characterisation and categorisation of regional differences. International Journal of Mechanical and Materials Engineering, 2021. 16(1): p. 1-14.
6. Guccione, J.M., K.D. Costa, and A.D. McCulloch, Finite element stress analysis of left ventricular mechanics in the beating dog heart. Journal of biomechanics, 1995. 28(10): p. 1167-1177.
7. Masithulela, F.J., Computational biomechanics in the remodelling rat heart post myocardial infarction. 2016.
8. Ngwangwa H., Nemavhola F., Pandelani T., Msibi M., Mabuda I., Davies N., et al. (2022). Determination of cross-directional and cross-wall variations of passive biaxial mechanical properties of rat myocardia. Processes 10 (4), 629. 10.3390/pr10040629.
9. Masithulela, F. The effect of over-loaded right ventricle during passive filling in rat heart: A biventricular finite element model. in ASME International Mechanical Engineering Congress and Exposition. 2015. American Society of Mechanical Engineers.
10. Masithulela, F. Analysis of passive filling with fibrotic myocardial infarction. in ASME international mechanical engineering congress and exposition. 2015. American Society of Mechanical Engineers.
11. Masithulela, F. The Effect of Over-Loaded Right Ventricle During Passive Filling in Rat Heart: A Biventricular Finite Element Model. in ASME 2015 International Mechanical Engineering Congress and Exposition. 2015.
12. Nemavhola, F., Detailed structural assessment of healthy interventricular septum in the presence of remodeling infarct in the free wall–A finite element model. Heliyon, 2019. 5(6): p. e01841.
13. Nemavhola, F., Fibrotic infarction on the LV free wall may alter the mechanics of healthy septal wall during passive filling. Bio-medical materials and engineering, 2017. 28(6): p. 579-599.
14. Nemavhola, F., Mechanics of the septal wall may be affected by the presence of fibrotic infarct in the free wall at end-systole. International Journal of Medical Engineering and Informatics, 2019. 11(3): p. 205-225.
15. Yin, Z., et al., Cardiac wall mechanics analysis in hypertension-induced heart failure rats with preserved ejection fraction. Journal of Biomechanics, 2020. 98: p. 109428.
16. Sirry, M.S., et al., Characterisation of the mechanical properties of infarcted myocardium in the rat under biaxial tension and uniaxial compression. journal of the mechanical behavior of biomedical materials, 2016. 63: p. 252-264.
17. Anssari-Benam, A., et al., Rate-dependency of the mechanical behaviour of semilunar heart valves under biaxial deformation. Acta biomaterialia, 2019. 88: p. 120-130.
18. van Disseldorp, E.M., et al., Reproducibility assessment of ultrasound-based aortic stiffness quantification and verification using Bi-axial tensile testing. Journal of the mechanical behavior of biomedical materials, 2020. 103: p. 103571.
19. D'Amore, A., et al., Bi-layered polyurethane–extracellular matrix cardiac patch improves ischemic ventricular wall remodeling in a rat model. Biomaterials, 2016. 107: p. 1-14.
20. Caulk, A.W., et al., Quantification of the passive and active biaxial mechanical behaviour and microstructural organization of rat thoracic ducts. Journal of The Royal Society Interface, 2015. 12(108): p. 20150280.
21. Ngwangwa, H., et al., Biomechanical analysis of sheep oesophagus subjected to biaxial testing including hyperelastic constitutive model fitting. Heliyon, 2022. 8(5): p. e09312.
22. Robertson, D. and D. Cook, Unrealistic statistics: how average constitutive coefficients can produce non-physical results. Journal of the mechanical behavior of biomedical materials, 2014. 40: p. 234-239.

14th South African Conference on Computational and Applied Mechanics (SACAM 2025)
21-23 January 2025, Johannesburg, South Africa

A Future of Computational Mechanics: Opportunities with Quantum Computing

Daniel N. Wilke[a]

[a]Department of Mechanical and Aeronautical Engineering, University of Pretoria, Pretoria, 0086, South Africa
*corresponding author email: nico.wilke@up.ac.za

Abstract

Quantum computing promises to transform computational mechanics by addressing challenges that exceed the limits of classical computing methods. Quantum computing's impact is particularly significant in complex areas, such as structural analysis, fluid dynamics, and materials science, where traditional methods struggle due to high dimensionality and nonlinear behaviour. Leveraging quantum superposition and entanglement, quantum algorithms offer the potential for exponential speedups in certain tasks. For instance, the Harrow-Hassidim-Lloyd (HHL) algorithm accelerates the solution of linear systems, which are vital for finite element analysis. At the same time, the Quantum Approximate Optimisation Algorithm (QAOA) improves optimisation in structural design and material selection. Furthermore, quantum computing enables atomic-scale simulations of materials, driving progress in material engineering. Although current hardware limitations and algorithmic challenges remain, integrating quantum computing into computational mechanics holds promise for tackling previously intractable problems, advancing design optimisation, and unlocking deeper understanding of complex mechanical systems.

***Keywords*:** Quantum Computing, Quantum Algorithms, Computational Mechanics, Lagrangian Methods, Particle-Based Solvers, Finite Element Analysis, Materials Simulation, Optimisation

1. INTRODUCTION

Computational mechanics plays a critical role in engineering and science, enabling the modelling, analysis, and prediction of complex physical systems across structural mechanics, fluid dynamics, materials science, and multiphysics domains. Its development has aligned with computational advancements, evolving from manual computations to digital simulations utilising Central Processing Units (CPUs), Field Programmable Gate Arrays (FPGAs), Graphical Processing Units (GPUs), and edge computing. Quantum computing represents the forthcoming paradigm shift, offering the potential for exponential speedups on problems that are intractable on classical hardware. Given the dependence on numerical techniques, such as finite element, finite volume, and material point methods to tackle nonlinear, high-dimensional, and fine-scale issues, quantum computing holds potential to revolutionise these simulations. This paper serves as an invitation to the mechanics community to actively participate in the development of quantum solutions rather than passively adopting them. Herein, we outline the fundamentals of quantum computing, explore relevant quantum algorithms for computational mechanics, and assess the integration challenges and opportunities.

2. QUANTUM COMPUTE FUNDAMENTALS

Quantum computing employs qubits rather than classical bits. In contrast to bits, which represent either 0 or 1, qubits can occupy a superposition of both states concurrently. When paired with quantum entanglement, this facilitates computations that may outperform classical counterparts in efficiency for certain problems.

2.1. Key Concepts

- **Superposition:** A qubit resides in a probabilistic blend of 0 and 1 until measured, permitting simultaneous exploration of multiple states.
- **Entanglement:** Linked qubits exhibit correlated states, irrespective of separation, yielding computational power exceeding that of isolated qubits.
- **Quantum Gates:** These operations manipulate qubits, akin to classical logic gates, leveraging superposition and entanglement for intricate calculations.

2.2. Quantum Algorithms

Numerous quantum algorithms exhibit potential for theoretical exponential speedups over classical methods for targeted problems [1][2]. Some relevant to computational mechanics include:

- **Harrow-Hassidim-Lloyd (HHL) Algorithm:** This algorithm resolves systems of linear equations exponentially faster than classical methods under certain specific conditions [3][4][44].
  - **Mechanism:** HHL employs Quantum Phase Estimation to ascertain eigenvalues of the N×N matrix **A** in **Ax=b**, followed by controlled rotations to invert eigenvalues and an inverse quantum Fourier transform to yield a quantum state $|\mathbf{x}\rangle$ proportional to the solution. The HHL algorithm has a fundamental limitation in that it stores the solution **x** in the amplitudes of a

quantum state |**x**⟩, but these amplitudes are not directly measurable. The algorithm only provides access to expectation values ⟨**x**|**M**|**x**⟩ rather than the full solution vector.

- **Comparison to classical:** Classical methods, such as Gaussian elimination, exhibit polynomial runtimes (e.g., $O(N^3)$). For sparse, well-conditioned matrices, HHL provides a theoretical speedup, with a complexity of $O(\text{poly}(\log N, \kappa, 1/\varepsilon))$, where κ is the condition number and ε is the precision [3]. Recent enhancements include Enhanced Hybrid HHL, reducing errors by 57% on ideal processors and 13-20% on hardware like IBM Torino and IonQ Aria-1 for 2x2 systems via higher-precision eigenvalue estimates [29]; Post-selection-improved HHL (Psi-HHL) framework, achieving optimal linear O(κ) scaling for large κ (up to ~10^6 for 64x64 matrices and ~393 for 256x256 in quantum chemistry), with fewer shots and minimal circuit depth increase compared to original HHL's $O(\kappa^2)$ [30]; post-selection-free variants for symmetric Toeplitz and Hankel matrices, eliminating amplitude amplification needs [31]; and variable-time amplitude amplification refining to $O(\kappa \log^3 \kappa \log N)$ [32].

  However, the quantum state output necessitates strategies for classical readout, such as amplitude amplification or tomography, which can negate speedups for full vector extraction; solutions include exploiting sparsity for efficient downloading in machine learning contexts [33], designing observables for partial information [34], or hybrid predictor-corrector frameworks [4]. Quantum Singular Value Estimation (QSVE) enhancements address this with advances like analytic polynomials for optimal QSVT-based inversion (reducing circuit depth for large $\kappa \log(\kappa/\varepsilon)$) [35] and numerical techniques for angle estimation in ill-conditioned matrices [36].

- **Relevance:** Directly applicable to finite element methods (FEM) and other techniques reliant on large linear systems, with recent hardware demonstrations via hybrid HHL++ solving up to 4x4 matrices on trapped-ion systems [37].

- **Quantum Approximate Optimisation Algorithm (QAOA):** A hybrid quantum-classical algorithm for approximating solutions to combinatorial optimisation problems [5].
  - **Mechanism:** QAOA iterates via a parameterised quantum circuit (ansatz) on quantum hardware, evaluates the solution quality, and refines parameters using classical optimisation between iterations.
  - **Comparison to Classical:** Classical optimisers traverse solution spaces sequentially, often getting trapped in local minima; QAOA exploits superposition for parallel exploration, potentially evading getting trapped in local minima.
  - **Relevance:** Applicable to combinatorial issues in structural design, manufacturing processes [6], and machine learning [7][8][9].
- **Variational Quantum Eigensolver (VQE):** A hybrid algorithm for determining the ground-state energy (lowest eigenvalue) of quantum systems [10].
  - **Mechanism:** Analogous to QAOA, VQE uses a classical optimiser to tune a parameterised quantum circuit. The computer prepares a trial wavefunction (ansatz) and measures its energy. The classical computer suggests new parameters to lower the measured energy, iterating until the system's minimum energy is found.
  - **Comparison to Classical:** Classical eigensolvers scale exponentially with system size. VQE delegates state preparation to quantum hardware to improve the computational efficiency.
  - **Relevance:** Pertinent to atomic-scale material simulations and eigenvalue problems in solid mechanics and structural engineering [11], bridging quantum interactions with mechanics analyses. Recent integrations with commercial FEM software, such as ABAQUS, on 5-7 qubit processors enable structural analyses of hexagonal trusses, Timoshenko beams, and plane-strain continua, with AI-driven extensions for multi-scale modelling and structural health monitoring.

3. CHALLENGES AND FUTURE OUTLOOK

Although quantum computing offers considerable advantages, substantial obstacles must be surmounted for broad adoption, particularly in the Noisy Intermediate-Scale Quantum (NISQ) era [12].

3.1. Hardware Limitations

Current quantum computers are still in their early stages of development. They suffer from:

- **Limited Qubit Coherence Times:** Qubits lose their quantum properties (coherence) quickly, limiting the duration of computations.
- **High Error Rates:** Quantum gates are prone to errors, which can accumulate and affect the accuracy of calculations.

- **Scalability Issues:** Building more stable quantum computers is a major technological hurdle.
- **Cooling and Environmental Controls:** Quantum computers must be cooled to very low temperatures and isolated from external interferences.

3.2. Algorithm Development

Advancing practical quantum algorithms that surpass classical ones in computational mechanics remains an active area of research. Challenges include:

- **Mapping Problems to Quantum Computers:** Translating complex engineering problems to quantum formats, which may not be intuitive for engineering contexts.
- **Algorithm Optimisation:** Optimisation tailored to hardware and problem specifics, including readout issues in HHL where full vector extraction requires exponential measurements; recent shifts emphasise application-specific strategies like sparsity exploitation, observable design, and hybrid subroutines [33][34].
- **Hybrid Quantum-Classical Approaches:** Hybrid quantum-classical paradigms to harness respective strengths, with AI integration accelerating ecosystem maturation toward NISQ advantage.

3.3. Software and Integration

Integrating quantum computing into existing computational mechanics workflows demands:

- **New Software Tools:** Quantum programming languages, compilers, and simulators are needed to develop and test quantum algorithms. Qiskit is an open-source quantum computing framework that enables users to design, simulate, and execute quantum circuits on real quantum hardware [24].
- **Hybrid Software Frameworks:** Frameworks for seamless quantum-classical integration are essential for large-scale adoption and practical applications.
- **Training and Education:** Educational initiatives to equip engineers with quantum principles.

3.4. Future Outlook

Notwithstanding the challenges, the potential of quantum computing in computational mechanics is significant [1].

Near-Term (5-10 years): Hybrid algorithms will expedite targeted tasks, such as linear solves and optimisations, with an initial quantum advantage in niches like materials simulations by 2026-2030, potentially incorporating fault-tolerant elements for error mitigation and HHL variants that handle large κ in engineering domains.

Mid-Term (10-20 years): Fault-tolerant systems will manage complex issues, including detailed fluid dynamics and material simulations, extending advantages to numerous engineering domains by 2035-2045.

Long-term (20+ years): Unprecedented simulations at quantum-engineering scales are likely to drive breakthroughs in multiphysics applications, material science research, and material design after 2045.

4. APPLICATIONS IN COMPUTATIONAL MECHANICS

Mesh-based and particle-based strategies are core paradigms for discretising partial differential equations (PDEs) in analyses like Computational Structural Analysis (CSA) and Computational Fluid Dynamics (CFD). These convert continuous PDEs into large algebraic systems, differing in domain handling, physics, flexibility, and suitability for complex or dynamic problems.

4.1. Mesh-Based Solvers and Quantum Acceleration

Mesh-based methods, including Finite Difference Method (FDM), Finite Volume Method (FVM), and Finite Element Method (FEM), underpin PDE solutions in CFD and CSA. They discretise domains into grids or meshes, generating large linear and nonlinear systems that dominate costs, while providing accuracy for defined geometries at the expense of assembly and refinement overheads. Quantum algorithms can accelerate these via faster linear solves or mesh optimisations, though NISQ-era noise and scalability persist as hurdles [12].

4.1.1 Finite Element Method (FEM)

FEM partitions domains into elements for structural, thermal, or electromagnetic analyses, yielding sparse matrices. Quantum enhancements that target bottlenecks include:

- **Linear System Solution**: Harrow-Hassidim-Lloyd (HHL) or variational solvers accelerate elliptic PDE resolutions, offering potential exponential speedups for well-conditioned matrices, as applied to electromagnetic problems [25] and hybrid classical-quantum implementations [26].
- Adaptive Mesh Refinement: Quantum-assisted techniques, such as hλ-adaptive methods [13] or quantum algorithms for Maxwell's equations, enable dynamic grid optimisation, reducing classical iteration counts while maintaining precision in singularly perturbed problems [22].

4.1.2 Finite Volume and Finite Difference Methods

Finite Volume Methods (FVM) and Finite Difference Methods (FDM) approximate conservation laws and derivatives for Navier-Stokes-governed flows, often necessitating the use of iterative nonlinear solvers. Quantum improvements include:

- Nonlinear PDE Resolution: Variational algorithms and hybrid frameworks enable precise incompressible flow simulations, scaling logarithmically with mesh size [14][15][16].
- Turbulence and Multiscale Modelling: Quantum-inspired tensor networks accelerate 2D turbulence by an order of magnitude [17], extendable to 3D CFD via Hydrodynamic Schrödinger Equations (HSEs) [18].

4.2. Particle-Based Solvers and Quantum Acceleration

Particle-based methods, such as Smoothed Particle Hydrodynamics (SPH), the Material Point Method (MPM), the Lattice Boltzmann Method (LBM), and the Discrete Element Method (DEM), simulate large deformations, fractures, and multiphase flows. These excel over meshes in flexibility but incur costs from particle management. Quantum subroutines can enhance neighbour searches and linear solves, with broader acceleration via Monte Carlo techniques [27].

4.2.1 Smoothed Particle Hydrodynamics (SPH)

SPH models fluids or solids via kernel-interpolated particles. Quantum accelerations target:

- **Nearest Neighbour Search:** Grover's algorithm [19] reduces complexity from O(N) to O(√N) for large simulations with fixed-radius neighbour searches or general nearest-neighbour problems [20].
- **Pressure Poisson Equation:** HHL expedites solutions in incompressible formulations via hybrid predictor-corrector methods [4].

4.2.2 Material Point Method (MPM)

The Material Point Method (MPM) integrates particle representations with Eulerian grids. Quantum enhancements may include:

- **Interpolation Between Particles and Grids:** Quantum algorithms may offer speed-ups in mapping information between particles and the grid [4].
- **Solving Governing Equations on the Grid:** Quantum linear system solvers could improve implicit formulations [21].

4.2.3 Discrete Element Method (DEM)

DEM simulates the discrete motions and interactions of particles, incurring the costs associated with contact detection and forces. Quantum opportunities:

- **Contact Detection:** Grover variants accelerate identifications, paralleling nearest-neighbour efficiencies [1][19].
- **Force Calculations:** Quantum machine learning can enhance model accuracy [9].

4.2.4 Lattice Boltzmann Method (LBM)

LBM tracks mesoscopic particle distributions on lattices for fluid flows. Potential quantum impacts include:

- **Boltzmann Equation Solves:** Linear solvers or quantum walks can provide speedups [23] for solving the discrete Boltzmann equation.
- **Complex Collision Operators:** The speed and accuracy of simulations of non-ideal or multiphase flows can be enhanced, with recent turbulent flow advancements reducing complexity [28].
- **Data Parallelism:** Quantum parallelism can improve the handling of large lattices and geometries.

5. CONCLUSION

Quantum computing possesses the potential to fundamentally reshape computational mechanics by enabling theoretical exponential speedups in simulation and optimisation tasks. Domains such as structural analysis, fluid dynamics, and material modelling stand to benefit from accelerated linear solvers, improved optimisation algorithms, and quantum-scale simulations that address limitations of classical methods.

However, this promise is moderated by ongoing hardware limitations, the need for algorithmic maturity, and the complexity of integration into existing workflows. In the near term, hybrid quantum-classical methods may yield incremental improvements. Fully unlocking these capabilities will require sustained interdisciplinary collaboration. For the computational mechanics community, active involvement in quantum research and development is essential to ensure that emerging tools are both revolutionary and practically applicable.

Acknowledgements

The author acknowledges using AI-powered tools to refine language, including paraphrasing and grammatical corrections. All ideas, analyses, and insights presented in this paper are entirely the author's original work.

References

[1] Au-Yeung, R., Camino, B., Rathore, O., & Kendon, V. (2024). Quantum algorithms for scientific computing. Reports on Progress in Physics, 87(11), 116001.

[2] Montanaro, A. (2016). Quantum algorithms: an overview. npj Quantum Information, 2(1), 15023.

[3] Dervovic, D., Herbster, M., Mountney, P., Severini, S., Usher, N., & Wossnig, L. (2018). Quantum linear systems algorithms: a primer. arXiv:1802.08227.

[4] Rathore, O., Basden, A., Chancellor, N., & Kusumaatmaja, H. (2025). Integrating quantum algorithms into classical frameworks: A predictor–corrector approach using HHL. Quantum Science and Technology, 10(2), 025041.

[5] Farhi, E., Goldstone, J., & Gutmann, S. (2014). A Quantum Approximate Optimization Algorithm. arXiv:1411.4028.

[6] Kurowski, K., Pecyna, T., Slysz, M., Różycki, R., Waligóra, G., & Węglarz, J. (2023). Application of quantum approximate optimization algorithm to job shop scheduling problem. European Journal of Operational Research, 310(2), 518–528.

[7] Mcclean, J. R., Boixo, S., Smelyanskiy, V. N., Babbush, R., & Neven, H. (2018). Barren plateaus in quantum neural network training landscapes. Nature communications, 9, 4812.

[8] Aïmeur, E., Brassard, G., & Gambs, S. (2013). Quantum speed-up for unsupervised learning. Machine Learning, 90, 261-287.

[9] Biamonte, J., Wittek, P., Pancotti, N., Rebentrost, P., Wiebe, N., & Lloyd, S. (2017). Quantum machine learning. Nature, 549, 195-202.

[10] Peruzzo, A., McClean, J., Shadbolt, P., Yung, M. H., Zhou, X. Q., Love, P. J., Aspuru-Guzik, A. & O'Brien, J. L. (2014). A variational eigenvalue solver on a photonic quantum processor. Nature communications, 5(1), 4213.

[11] Liu, Y., Liu, J., Raney, J. R., & Wang, P. (2024). Quantum computing for solid mechanics and structural engineering – A demonstration with Variational Quantum Eigensolver. Extreme Mechanics Letters, 67, 102117.

[12] Angara, P. P., Stege, U., Müller, H. A., & Bozzo-Rey, M. (2020). Hybrid quantum-classical problem solving in the NISQ era. Proceedings of the 30th Annual International Conference on Computer Science and Software Engineering (CASCON '20), 247–252.

[13] Drebotiy, R. H., & Shynkarenko, H. A. (2025). Quantum-assisted hλ-adaptive finite element method. Partial Differential Equations in Applied Mathematics, 13, 101120.

[14] Lubasch, M., Joo, J., Moinier, P., Kiffner, M., & Jaksch, D. (2020). Variational quantum algorithms for nonlinear problems. Physical Review A, 101(1), 010301.

[15] Peddinti, R. D., Pisoni, S., Marini, A., Lott, P., Argentieri, H., Tiunov, E., & Aolita, L. (2024). Quantum-inspired framework for computational fluid dynamics. Communications Physics, 7, 135.

[16] Chen, Z.-Y., Ma, T.-Y., Ye, C.-C., Xu, L., Bai, W., Zhou, L., Tan, M.-Y., Zhuang, X.-N., Xu, X.-F., Wang, Y.-J., Sun, T.-P., Chen, Y., Du, L., Guo, L.-L., Zhang, H.-F., Tao, H.-R., Wang, T.-L., Yang, X.-Y., Zhao, Z.-A., Wang, P., Zhang, S., Zhao, R.-Z., Zhang, C., Jia, Z.-L., Kong, W.-C., Dou, M.-H., Wang, J.-C., Liu, H.-Y., Xue, C., Zhang, P.-J.-Y., Huang, S.-H., Duan, P., Wu, Y.-C., & Guo, G.-P. (2024). Enabling large-scale and high-precision fluid simulations on near-term quantum computers. Computer Methods in Applied Mechanics and Engineering, 432(Part B), 117428.

[17] Hölscher, L., Rao, P., Müller, L., Klepsch, J., Luckow, A., Stollenwerk, T & Wilhelm, F.K. (2025). Quantum-inspired fluid simulation of two-dimensional turbulence with GPU acceleration. Physical Review Research, 7(1), 013112.

[18] Meng, Z., & Yang, Y. (2023). Quantum computing of fluid dynamics using the hydrodynamic Schrödinger equation. Physical Review Research, 5(3), 033182.

[19] Grover, L. K. (1996). A fast quantum mechanical algorithm for database search. Proceedings of the twenty-eighth annual ACM symposium on Theory of computing, 212-219.

[20] Au-Yeung, R., Williams, A. J., Kendon, V. M., & Lind, S. J. (2024). Quantum algorithm for smoothed particle hydrodynamics. Computer Physics Communications, 294, 108909.

[21] Sulsky, D., Zhou, S.-J., & Schreyer, H. L. (1995). Application of a particle-in-cell method to solid mechanics. Computer Physics Communications, 87(1–2), 236–252.

[22] Guilkey, J. E., & Weiss, J. A. (2003). Implicit time integration for the material point method: Quantitative and algorithmic comparisons with the finite element method. International Journal for Numerical Methods in Engineering, 57(9), 1323–1338.

[23] Childs, A. M. (2010). On the relationship between continuous-and discrete-time quantum walk. Communications in Mathematical Physics, 294(2), 581-603.

[24] Qiskit. (2019). Qiskit: An Open-Source Framework for Quantum Computing. Retrieved from https://www.ibm.com/quantum/qiskit.

[25] Deiml, M., & Peterseim, D. (2025). Quantum realization of the finite element method. arXiv:2403.19512.

[26] Arora, A., Ward, B.M., & Oskay, C. (2025). An implementation of the finite element method in hybrid classical/quantum computers. Finite Elements in Analysis and Design, 248, 104354.

[27] Olivier, N., & Nowak, M. (2024). Monte Carlo particle transport on quantum computers. arXiv:2410.19489.

[28] Malinverno, G., & Blasco Alberto, J. (2025). A Review of the Current State-of-the-Art of Quantum Computing for CFD Approaches, Advantages, and Limitations. Aerotecnica Missili & Spazio.

[29] Morgan, J., Ghysels, E. & Mohammadbagherpoor, H. (2024). An enhanced hybrid HHL algorithm. arXiv:2404.10103.

[30] Tsemo, P.B., Jayashankar, A. Sugisaki K., Baskaran, N. Chakraborty, S., & Prasannaa, V. S. (2024). Enhancing the Harrow-Hassidim-Lloyd (HHL) algorithm in systems with large condition numbers. arXiv:2407.21641.

[31] Li, Y.-L., Peng, B., Wang, Y.-H., Zhang, Z.-C., & Wang, Y.-K. (2025). HHL algorithm without ancilla postselection for Toeplitz and Hankel systems. *Physics Letters A*, *531*, 130154.

[32] Ambainis, A. (2012). Variable time amplitude amplification and quantum algorithms for linear algebra problems. *In 29th International Symposium on Theoretical Aspects of Computer Science (STACS 2012), Leibniz International Proceedings in Informatics (LIPIcs), 14*, 636–647.

[33] Liu, J., Liu, M., Liu, J.P., Ye, Z., Wang, Y., Alexeev, Y., Eisert, J., & Jiang, L. (2024). Towards provably efficient quantum algorithms for large-scale machine-learning models. *Nature Communications,* **15**, 434.

[34] Archpaul, J., VijayaKumar, E. N., Rajendran, M., & et al. (2024). Enhancing quantum state tomography: Utilizing advanced statistical techniques for optimized quantum state reconstructions. *Journal of the Korean Physical Society*, *85*, 677–690.

[35] Sünderhauf, C., Németh, Z., Walayat, A., Patterson, A., & Berntson, B. K. (2025). Matrix inversion polynomials for the quantum singular value transformation. *arXiv:2507.15537.*

[36] Novikau, I., & Joseph, I. (2024). Estimating QSVT angles for matrix inversion with large condition numbers. arXiv preprint arXiv:2408.15453.

[37] Yalovetzky, R., Minssen, P., Herman, D., & Pistoia, M. (2024). Solving linear systems on quantum hardware with hybrid HHL++. *Scientific Reports*, *14*(1), 20610.

[38] Li, P., & Liu, X. (2018). Bilinear interpolation method for quantum images based on quantum Fourier transform. *International Journal of Quantum Information*, *16*(04), 1850031.

[39] Ramos-Calderer, S. (2022). Efficient quantum interpolation of natural data. *Physical Review A*, *106*(6), 062427.

[40] Pang, C.-Y., Zhou, R.-G., Hu, B.-Q., Hu, W., & El-Rafei, A. (2019). Signal and image compression using quantum discrete cosine transform. *Information Sciences*, *473*, 121–141.

[41] Tumbiolo, E., Roncallo, S., Macchiavello, C., & Maccone, L. (2025). Quantum frequency resampling. *npj Quantum Information*, *11*(1), 123.

[42] Chen, J., Childs, A. M., & Hung, S.-H. (2018). Quantum algorithm for multivariate polynomial interpolation. Proceedings of the Royal Society A: Mathematical, Physical and Engineering Sciences, 474(2209), 20170480.

[43] Rosenkranz, M., Brunner, E., Marin-Sanchez, G., Fitzpatrick, N., Dilkes, S., Tang, Y., Kikuchi, Y., & Benedetti, M. (2025). Quantum state preparation for multivariate functions. Quantum, 9, 1703.

[44] Harrow, A. W., Hassidim, A., & Lloyd, S. (2009). Quantum algorithm for linear systems of equations. Physical Review Letters, 103(15), 150502.

Vehicle dynamics parameter estimation through automated covariance scheduling

Silvia L. Eldring[a], Herman A. Hamersma[a,*], Andries J. Peenze[a], P. Schalk Els[a]

[a]*Department of Mechanical and Aeronautical Engineering, University of Pretoria*

c/o Lynnwood Road and Roper Street, Hatfield, Pretoria, 0083

**corresponding author email: herman.hamersma@up.ac.za*

Abstract

Parameter estimation makes use of measured system states alongside some estimation algorithm to determine the parameters of a system. However, to estimate, these states used in estimation must be sufficiently excited, or else the algorithm won't converge.

In this paper, a Modified Load Transfer Model is used to estimate the sprung mass, centre of mass location and the roll and pitch moment arms of a four-wheeled road vehicle as developed by Sandu et al. [1] and modified to make use of an Unscented Kalman Filter by Botha & Els [2]. Through implementation and testing, it was found that the moment arms, unlike the other estimated parameters, require additional excitement of lateral and longitudinal accelerations to estimate. These accelerations are not always excited under normal driving conditions.

During these periods of insufficient excitation, the process noise covariance relating to the moment arms experience a linear increase, indicating the lack of filter convergence. These parameters don't remain constant, thus providing unreliable estimates.

Through the implementation of Process Noise Covariance, this unreliability was addressed. Here, the process noise covariance value relating to the moment arm was scaled, restricting changes in the estimate during periods of insufficient excitation, and allowing the estimates to otherwise move freely, allowing the filter to perform accurately over changing use cases without the need for multiple models or significant increases in computation time.

1. Background and Motivation

System control relies on accurate knowledge of system states and parameters. Whilst some control is designed to be robust and rely on few, measurable states, other control using prediction and complex physical models rely on accurate estimates of unmeasurable states and parameters.

For this purpose, Kalman Filters are often used. These filters, through the combination of physical models and measurements, allow for the estimation of unmeasured states and parameters. [3,4]

However, a single physical model does not always accurately define a system through all use cases, as the estimation of parameters relies on the excitation of related states. Whilst multiple models can be set up to represent the system during various excitations, as done in a Switching Kalman Filter [5], this method is computationally expensive.

In this paper, an Unscented Kalman Filter is used to estimate the mass parameters of a road vehicle. This method, developed originally by Sandu et al. [1] and modified to use an Unscented Kalman Filter (UKF) by Botha & Els [2], estimates mass parameters including the moment arms between the centre of mass and the roll and pitch axes.

To accurately estimate these moment arms, the vehicle must experience sufficient roll and pitch excitation induced through lateral and longitudinal acceleration. Since these accelerations are not always present, accurate estimation of the moment arms is not always possible. To rectify this issue in a way that keeps computation time low, Process Noise Covariance Scheduling (PNCS) is used.

2. Objectives and Scope

The objective of the work presented in this paper is to implement and improve the performance of a previously developed mass parameter estimation algorithm for on-line use. The final algorithm should aid in the implementation of other control; thus, it is crucial to keep the computational cost low.

Since a lack of convergence in the moment arms was found by the previous studies, the source of the behaviour leading to a lack of convergence is first investigated.

From this information gathered, three solution strategies are proposed and outlined in Section 3.4. Whilst the first two methods were attempted, they were deemed unsuitable for this application. The third method explored results in the implementation of PNCS.

Lastly, the performance of the algorithm before and after the implementation of PNCS is provided.

3. Methodology

The problems in filter performance originally observed consisted of two parts. Firstly, moment arm estimates were seen to occasionally drift to unrealistic values. Secondly, periods of extended linear growth in state covariance values indicating a lack of convergence were observed.

It was found that this behaviour was dictated by three parts of the filter. The subsequent sections describe how this behaviour develops, and how each component of the filter can be used to rectify the issue.

The equations of the UKF in use are unchanged from the algorithm implemented by Botha & Els [2]. The only change made is to simplify the moment arms into a single value for roll and a single value for pitch.

3.1. Problem Root: MLTM

The cause of the behaviour stems from the model used in the state transition function. This model, named the Modified Load Transfer Model (MLTM) contains five equations defining the forces generated in each suspension strut, as well as the bounce force of the vehicle body. Equation 1, provided as an example, shows the equation relating to the front left strut.

$$F_{fl} = m_{fl}\ddot{z}_{fl} - \frac{m_{fl} + m_{fr}}{W}\ddot{y}_{CG}h_r - \frac{m_{fl} + m_{rl}}{L}\ddot{x}_{CG}h_p \tag{1}$$

The front left strut force (F_{fl}) is calculated using three terms. The first term determines the inertial force of the corner mass, calculated using the front left corner mass (m_{fl}) and its vertical acceleration ($\ddot{z}_{fl}$). The second and third terms determine the effects of lateral and longitudinal load transfer on the suspension strut, respectively.

The MLTM only defines the first five values in the state vector, provided as Equation 2. The remaining values in the state vector, representing the corner masses and moment arms, which are the parameters this algorithm aim at determining, are defined as random walk variables.

$$\boldsymbol{x} = \{F_{fl}\ F_{fr}\ F_{rl}\ F_{rr}\ F_b\ m_{fl}\ m_{fr}\ m_{rl}\ m_{rr}\ h_r\ h_p\}' \tag{2}$$

When the vehicle experiences no longitudinal or lateral accelerations ($\ddot{x}_{CG}$ or $\ddot{y}_{CG}$) the corresponding load transfer components of the MLTM equations fall away. The strut forces now are only defined as the inertial force on the corner masses. Whilst all corner masses still feature in the MLTM equations, the moment arms (h_r and h_p) do not.

As a result, during periods of insufficient lateral or longitudinal acceleration, the moment arms are only defined as random walk variables. Thus, the moment arms cannot be estimated as there is insufficient information available, and any estimate made during this period is random.

3.2. Predict Step Behaviour

In the predict step, sigma points ($\boldsymbol{s}_{k-1}$) are calculated using the states ($\hat{\boldsymbol{x}}_{k-1}$) and state covariance ($\boldsymbol{P}_{k-1}$) from the previous time step. These sigma points are passed through the state transition function, resulting in the predicted sigma points ($\boldsymbol{s}^*_{i,k|k-1}$) which are then combined into the predicted state vector ($\hat{\boldsymbol{x}}_{k|k-1}$). Using these values alongside the process noise covariance matrix ($\boldsymbol{Q}$) the predicted state covariance ($\boldsymbol{P}_{k|k-1}$) can be found. The relevant equations in this process are provided in Equations 3 and 4.

$$\boldsymbol{s}_{k-1} = \left[\hat{\boldsymbol{x}}_{k-1} \quad \hat{\boldsymbol{x}}_{k-1} + \gamma\sqrt{\boldsymbol{P}_{k-1}} \quad \hat{\boldsymbol{x}}_{k-1} - \gamma\sqrt{\boldsymbol{P}_{k-1}}\right] \tag{3}$$

$$\boldsymbol{P}_{k|k-1} = \boldsymbol{P}_{k-1} + \boldsymbol{Q} + \sum_{i=0}^{2l_x} \boldsymbol{W}_i^c\left(\boldsymbol{s}^*_{i,k|k-1} - \hat{\boldsymbol{x}}_{k|k-1}\right)\left(\boldsymbol{s}^*_{i,k|k-1} - \hat{\boldsymbol{x}}_{k|k-1}\right)^T \tag{4}$$

In Equation 3, the sigma points are calculated using both the previous states as well as the state covariance. This is done so that the resulting sigma points provide a representation of the mean and distribution of each value in the state vector.

When the distribution of a state has no effect on the rest of the system, the summation in the last term of Equation 4 will return a value of 0 for these states. As a result, the predicted state covariance for this state will equal the value from the previous time step plus the relevant process noise covariance value ($\boldsymbol{Q}$).

This behaviour is by design, as it will cause the state covariance, and thus the spread of the sigma points, to increase until they affect the system, after which the system can converge to an accurate estimate.

However, since the parameters discussed in this case periodically have no effect on the system no matter their distribution, this increase will continue at a linear rate of the process noise covariance multiplied by the filter operating frequency until sufficient excitement of the system occurs.

3.3. Update Step Behaviour

Lack of convergence in the predict step can often be counteracted using the update step. Here, sigma points are once again calculated, this time using the predicted state vector and predicted state covariance. These sigma points are passed through the measurement function, and combined to determine the predicted measurements. Using all sigma points, predicted measurements, predicted states and the measurement noise covariance, the Kalman gain ($\boldsymbol{K}_k$) and covariance of the transformed points ($\boldsymbol{S}$) can be found. These are then used alongside the predicted state covariance to determine the corrected state covariance at this time step ($\boldsymbol{P}_k$) as shown in Equation 5.

$$\boldsymbol{P}_k = \boldsymbol{P}_{k|k-1} - \boldsymbol{K}_k \boldsymbol{S} \boldsymbol{K}_k^{\ T} \tag{5}$$

This process allows for the correction of predicted states and parameters using real measurements, and can be used to correct values otherwise not represented in the system. Since the moment arms are not represented in the measurement function, they are not corrected in this step.

3.4. Implemented Solution: PNCS

From the above understanding of the filter behaviour, three solutions appear possible.

The first option is to make use of the Update step. Here, a false measurement with large measurement noise is used to rectify drift. Whilst this method can be applied successfully in other use cases, it is not applicable here, as it counteracts any progress in estimation, instead forcing the estimates to move towards the inaccurate measurement supplied.

The second option is to modify the MLTM in some way. This resulted in multiple versions of the MLTM, the difference between them being what moment arms are or are not estimated. This results in a simplified version of a Switching Kalman Filter [5]. Whilst this method works, smooth switching requires multiple filters to operate simultaneously, which is computationally expensive, making this solution undesirable.

The final option is to make use of the process noise covariance matrix, thus rectifying the issue in the predict step. Here, periodically setting the process noise covariance to 0 during periods of insufficient excitation should eliminate the linear growth as well as latching the most recent, reliable moment arm estimates.

Since the lateral and longitudinal vehicle accelerations directly correlate to estimation behaviour, these two measurements will be used to perform the scheduling. A 0.5 s window moving root mean square (RMS) is first performed on the measurements to ensure switching is not made erratic by measurement noise. This RMS value is then turned into a gain of 0 below 0.5 m/s^2 to pause estimation or 1 above 1 m/s^2 for full estimation with a smooth transition between the two using a lookup table. This gain is then applied to the process noise covariance values related to the roll and pitch moment arms.

This method is referred to as Process Noise Covariance Scheduling (PNCS). Since it makes use of simple processes, the computational cost should be low. A similar method was applied by Wittmer et. Al. [7], although scaling was applied to states used in estimation, not the estimated parameters themselves.

4. Results and Discussion

The following improvement in behaviour was found when using PNCS. Figure 2 shows the periods of estimation, indicated as the shaded area. The manoeuvre used here is an ISO 3888 [8] Double Lane Change (DLC) performed using a Land Rover 110 as the test vehicle at 40 km/h at the Gerotek Test Facilities [9], shown in Figure 1.

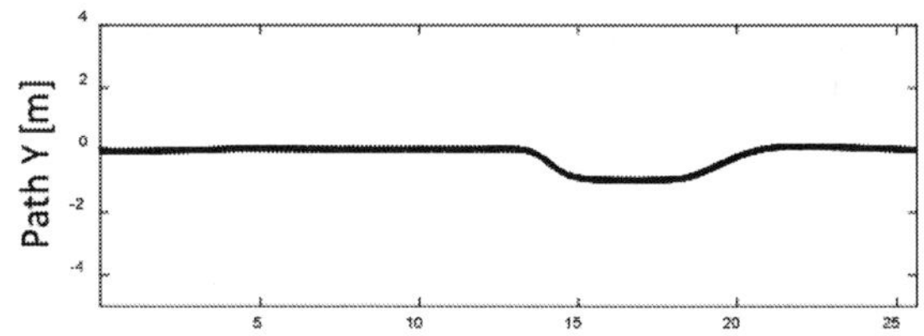

Fig. 1. Path used for DLC

The longitudinal acceleration (blue) shows the vehicle accelerating up to speed and later returning to a full stop, and the lateral acceleration (red) shows the vehicle performing the DLC. The results clearly show the periods where significant excitation is present and thus should result in the algorithm only estimating moment arms when sufficient information is available.

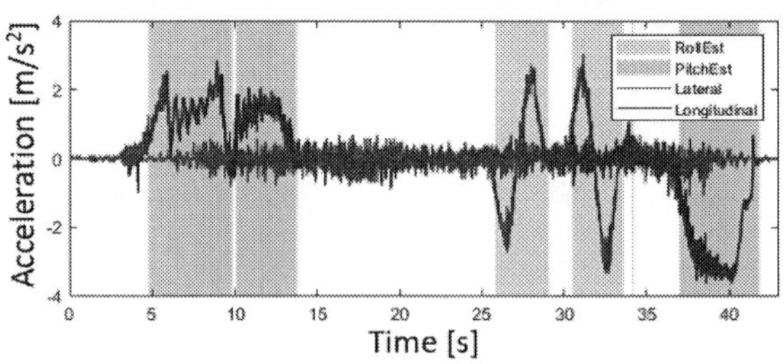

Fig. 2. Estimation behaviour compared to accelerations

Figure 3 shows the state covariances. Before the implementation of PNCS, large periods of linear increase are present (cf. the left plot in Figure 3), with a gradient of 100 caused by the filter running at 10 Hz and using a process noise covariance of 10 for the moment arms. This behaviour is restricted using PNCS (cf. the right plot in Figure 3).

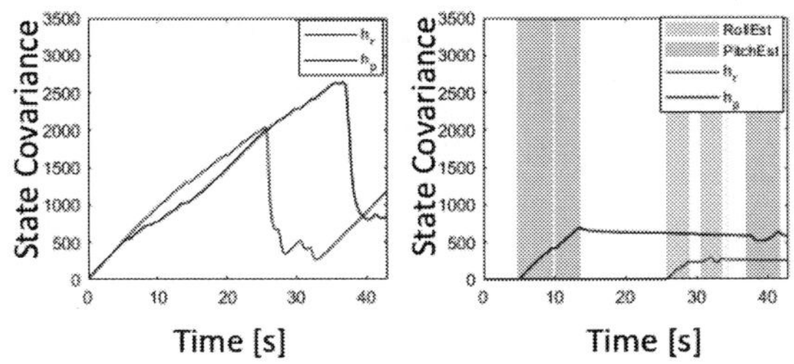

Fig. 3. Moment Arm State Covariance (Left) Before (Right) After PNCS

Lastly, the estimates are investigated. As can be seen in Figure 4, most of the motion in the moment arm estimates is halted outside of periods of estimation. Some motion is still possible due to the state covariance during these periods being non-zero, but it is minimal and cannot be avoided without forcing the state covariances to 0 and increasing convergence time.

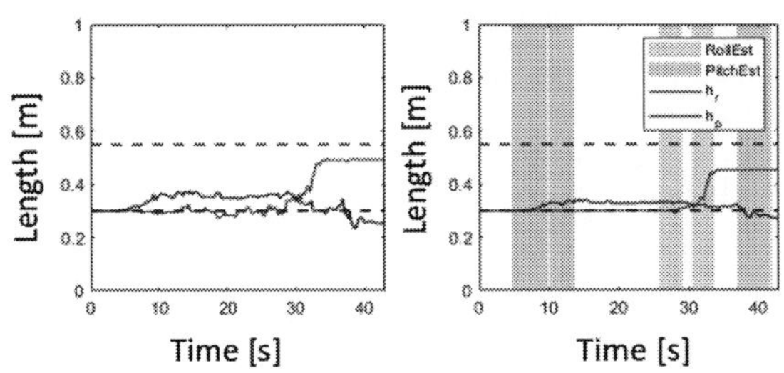

Fig. 4. DLC estimates (Left) Before (Right) After PNCS

Figure 4 does not show significant improvements with PNCS. Thus, Figure 5 provides estimates for an extended urban driving test, and Figure 6 provides those for a stationary test.

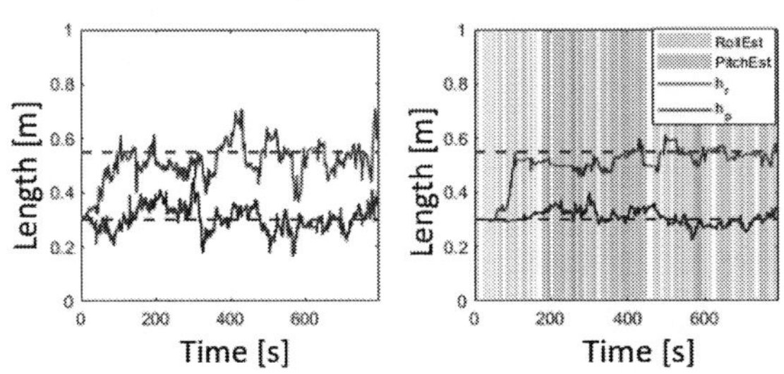

Fig. 5. Urban Drive estimates (Left) Before (Right) After PNCS

The benefits of PNCS are seen more clearly in Figures 5 and 6. Figure 5 shows that, using PNCS, estimates over long periods of time remain closer to the expected value, and no longer deviate based on unreliable information. Figure 6 (left) shows that, during extended periods of insufficient excitation, drift in estimates to unrealistic values seen before the implementation of PNCS is now eliminated, as the lack of shaded areas indicate no estimation of moment arms is performed. Thus, the implementation of PNCS is deemed successful.

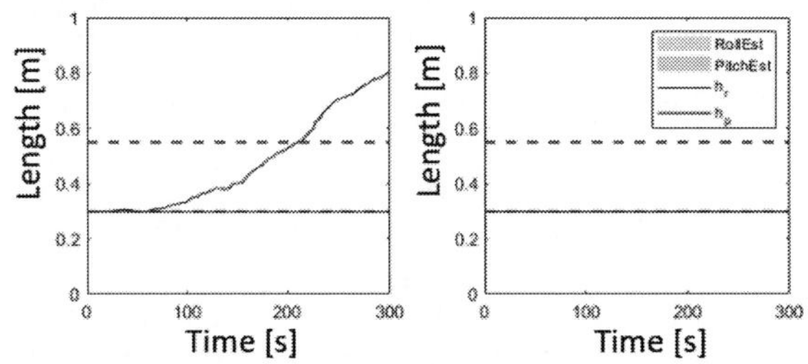

Fig. 6. Stationary Test estimates (Left) Before (Right) After PNCS

Lastly, it should be noted that, on a dSpace MABX II [8], the system on the test vehicle which will be used to run this estimation algorithm, the UKF executes in 0.13ms, and the process noise covariance scheduling executes in approximately 0.06ms, less than half the execution time of the UKF.

Since, in a switching filter, multiple UKF's would need to run simultaneously to avoid discontinuities, and the UKF is the most computationally expensive process in the algorithm, this execution time would be significantly increased. The use of PNCS avoids this, along with being computationally inexpensive.

5. Conclusion and Future Work

Through the application of PNCS, challenges seen in the convergence and accuracy of moment arm estimation of this mass parameter estimation algorithm were successfully improved without using multiple filters or processes drastically increasing the computational cost. Both the infinite linear growth in state covariance as well as the development in unrealistic estimates was eliminated, with final estimates converging to the expected values.

The full mass parameter estimation algorithm can thus be developed further for the purpose of on-line implementation, expanded to determine mass moments of inertia, and potentially implemented for the improvement of control and prediction algorithms reliant on accurate information of these parameters.

References

[1] Sandu, C., Kolansky, J., Botha, T., & Els, S. (2015). *6.3. Multibody Dynamics Techniques for Real-Time Parameter Estimation*. Advanced Autonomous Vehicle Design for Severe Environments, **44**, 221-241.

[2] Botha, T. R., & Els, P. S. (2019). *Vehicle centre of mass, roll-centre and pitch-centre height estimation*. International Journal of Vehicle Systems Modelling and Testing, **13**(4), 319-339.

[3] Kalman, R. E. (1960). *A new approach to linear filtering and prediction problems.*

[4] Julier, S. J., & Uhlmann, J. K. (2004). *Unscented filtering and nonlinear estimation.* Proceedings of the IEEE, **92**(3), 401-422.

[5] Murphy, K. P. (1998). *Switching Kalman Filters.* Technical Report. Department of Computer Science. UC Berkley.

[6] International Organization for Standardization. (2018). *ISO 3888 Passenger Cars - Test Track for a Severe Lane-Change Manoeuvre*. In. Geneva, Switzerland.

[7] Wittmer, K., Sawodny, O., & Henning, K.-U. (2023). *Model-Based Estimation of Vehicle Center of Gravity Height and Load.* Journal of Dynamic Systems, Measurement, and Control, **145**(5), 1-11.

[8] Armscor. (2024). *Gerotek Test Facilities.* https://www.armscor.co.za/gerotek/

[9] dSPACE. (2024). *dSPACE MicroAutoBox 2.* https://www.dspace.com/en/pub/home/products/hw/micautob/microautobox2.cfm

Improving brake performance with changeable suspension settings on rough terrain

F. Achamer[a,*], H.A. Hamersma[a], and P.S. Els[a]

[a]*University of Pretoria, Lynwood Rd, Pretoria 0002, South Africa*

**corresponding author email: herman.hamersma@up.ac.za*

Abstract

Anti-Lock Braking Systems (ABS) improve braking performance during emergency braking events. It is well-known that that the braking performance of a vehicle fitted with ABS is degraded by road roughness which can be compounded by fluctuating wheel speeds, tyre loads, and longitudinal load transfer. Initial studies suggest that optimised suspension settings can improve ABS performance on such terrains. This study, conducted with a 1997 Land Rover Defender that has been modified with a semi-active suspension system, investigates optimal suspension configurations for braking on rough surfaces. Co-simulations with MATLAB/Simulink and ADAMS/View revealed that reducing pitch rate and load transfer variations enhance brake performance. A reduced-order pitch-and-bounce model, paired with Bayesian inference, was developed to efficiently determine ideal suspension settings for various rough roads. Future work involves applying these settings to the test vehicle for experimental validation.

Keywords: Brake performance; rough terrain; ABS; semi-active suspension; simulation model

1. Introduction

The brake performance of a vehicle is considered an active safety feature [1]. It is of utmost importance to decelerate a vehicle as quickly as possible while allowing the driver to maintain directional full control over the vehicle. The development of Anti-Lock Braking Systems (ABS) with integrated Electronic Stability Control (ESC) has improved the deceleration of a vehicle until safely coming to a complete stop on smooth terrains [2]. The ABS is primarily used to prevent any of the four wheels of a vehicle from locking, such that lateral forces within the tyre contact patch are maintained, allowing the driver to obtain directional control over the vehicle.

1.1. Problem statement

Although ABS has improved the safety of vehicles on smooth terrains, it was determined that their efficacy deteriorates on rough terrains [3]. It was found that the system is negatively affected by the fluctuations in the angular wheel speeds and the vertical conditions of the tyre as a result of the excitations caused by the rough terrain [4]. Also, vehicle body motions such as roll, pitch and yaw are excited when braking on rough terrains, disturbing the slip control of the wheel [5]. Initial studies show that the vehicle's suspension system can mitigate the effects of the rough terrain; thus, the following research question is proposed [6]:

"Which suspension settings improve brake performance of a vehicle equipped with conventional ABS and a semi-active suspension system on rough terrains?"

1.2. Opportunity

The University of Pretoria has a 1997 Land Rover Defender test vehicle, whose OEM suspension system was replaced with a hydro-pneumatic semi-active suspension called the four-state semiactive suspension system ($4S_4$). The suspension system installed on the test vehicle is an improvement of the suspension system developed by Els, et al. [7], allowing spring and damper characteristics to be altered at each corner of the vehicle. This test vehicle provides a platform for conducting experimental research in vehicle dynamics.

2. High-fidelity Land Rover model

A high-fidelity simulation model of the available 1997 Land Rover Defender test vehicle was developed by Thoresson, et al. [8] in an ADAMS/View [9] and MATLAB/Simulink [10] co-simulation environment, housing a semi-active suspension, an ABS control algorithm, and a validated FTire tyre model. The model, depicted in Fig. 1, was used to conduct straight-line brake tests from 80 km/h on various terrains, each with increasing roughness.

A mathematical brake model proposed by Budynas, et al. [11] was used to capture the working principle of disc brakes that are fitted to the actual test vehicle. The brakes are modelled as a torque (T_{Brake}) that is applied in the opposing direction of the rotating tyre. Also, the model assumes uniform wear (f) and is highlighted in eq. 1 as follows:

$$T_{Brake} = \frac{1}{2} \cdot \left[\frac{\theta_{pos} \cdot \pi}{180}\right] \cdot f \cdot P_{Brake} \cdot r_i \cdot \left(r_o^2 - r_i^2\right) \tag{1}$$

Fig. 1. Representation of high-fidelity Land Rover model in ADAMS/View

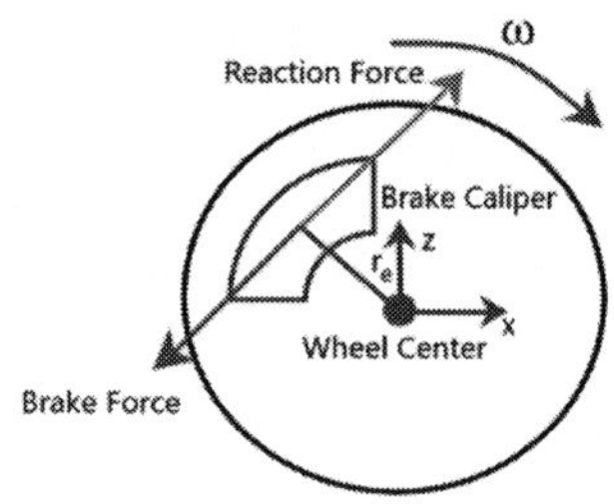

Fig. 2. Representation of brake forces on rotating tyre

Fig. 2 represents the forces on the wheel as it is rotating in the clockwise direction. The brake calliper's position causes a downward brake force, whose reaction force acts in the opposite direction. The brake forces are derived from brake line pressure inputs in the simulation model.

2.1. High-fidelity model validation

The model was validated against experimental brake tests at 80 km/h with the test vehicle, where an actuator-controlled brake pedal targeting 500 kPa brake pressure was used. During the tests the vehicle was set to its softest suspension configuration (soft springs and low damping at all four corners). The experimental inputs were applied to the simulation environment to assess model accuracy. Fig. 3 shows that key vehicle states—velocity, deceleration, and brake distance—match closely between simulation and experiment, while pitch angle and rate show minor discrepancies but follow similar trends. These results show that the simulation model is an accurate representation of the real test vehicle; the model is thus deemed experimentally validated.

2.2. High-fidelity simulation results

For the analyses, a Class-C Road, described by ISO 8608:2016 [12], and the Belgian Paving, a very rough

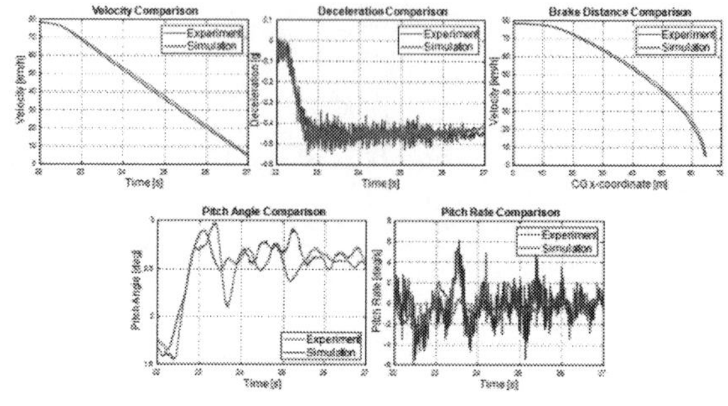

Fig. 3. Soft suspension: Experimental vs simulated results

road located at the Gerotek vehicle test facility, were used. The tests involved varying the suspension's damper and spring characteristics, with a 10 x 10 grid determining damping coefficients and spring stiffness values. The objective of the simulations is to gain greater insight in vehicle states that contribute to the improvement of brake performance due to the vehicle's suspension characteristics.

2.2.1 Class-C Road braking

The study analyses stopping distance, pitch angle, and deceleration for best and worst suspension configurations during braking from 80 km/h to 5 km/h. The results indicate that lower front-axle spring stiffness and higher rear-axle spring stiffness reduce stopping distance. The optimal configuration maintains steadier pitch angles and higher deceleration rates, yielding improved braking, as seen in Fig. 4. Load transfer and pitch rate stability were also crucial, as lower pitch rate variation minimises vehicle rocking, leading to a more consistent load distribution between the front and rear axles. This is shown in Fig. 5. Additionally, stiffer rear-axle suspension reduces the likelihood of wheel lock-up by sustaining tyre-road contact. This configuration supports a larger tyre-road contact patch, facilitating greater longitudinal braking forces and enhancing braking stability and effectiveness.

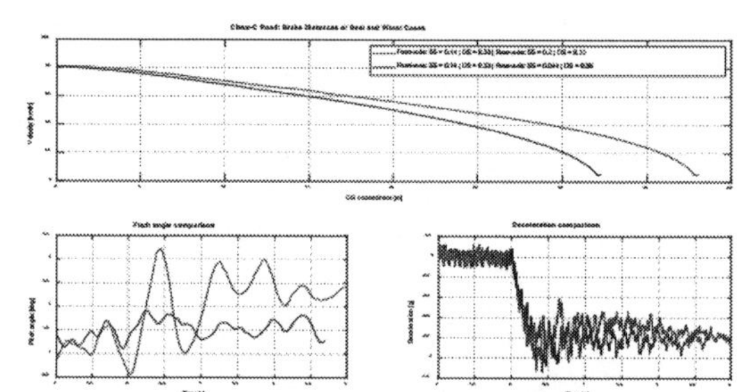

Fig. 4. Brake distance, pitch angle and deceleration of best and worst suspension configurations on a Class-C Road

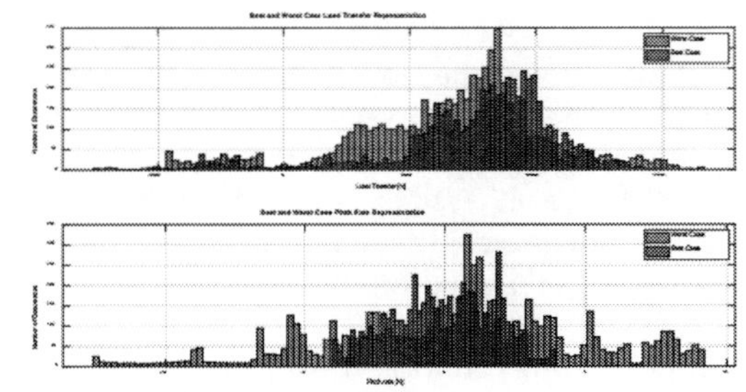

Fig. 5. Load transfer and pitch rate distributions of best and worst suspension configurations on a Class-C Road

2.2.2 Belgian Paving braking

Brake test simulations on the Belgian paving in Fig. 6 showed less improvement in braking performance compared to the Class-C Road analysis, though similar trends were observed. Softer front springs and stiffer rear springs still yield better performance, but the damper setting is the same for both axles. The vehicle's pitch angle and deceleration results were similar for both the best and worst suspension configurations, making it challenging to discern clear performance differences.

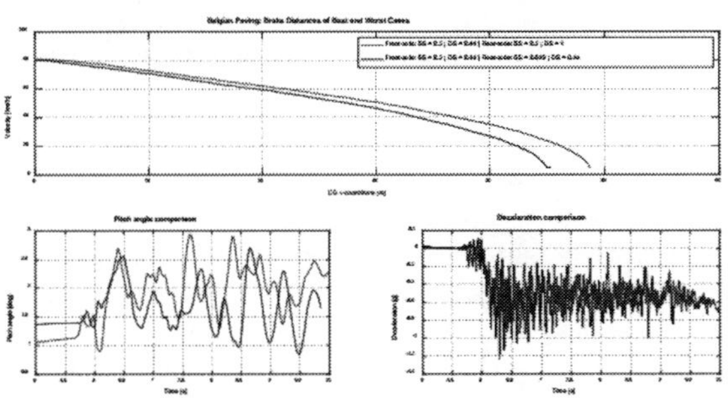

Fig. 6. Brake distance, pitch angle and deceleration of best and worst suspension configurations on the Belgian Paving

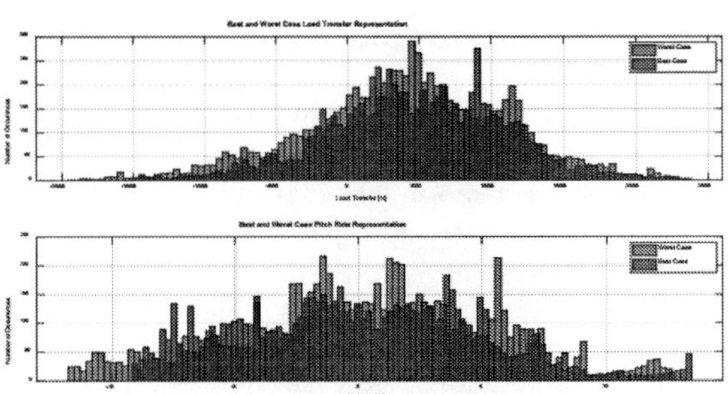

Fig. 7. Load transfer and pitch rate distributions of best and worst suspension configurations on the Belgian Paving

Analysis of load transfer and pitch rate distributions, shown in Fig. 7, revealed little variation between suspension configurations. Additionally, rough terrain caused wheel hop, with the rear-right tyre experiencing inconsistent loading, leading to reduced braking performance.

2.2.3 Summary of results

Table 1 summarises the findings of the simulations. Rougher terrain, such as the Belgian paving, increases stopping distances due to wheel oscillations that reduce tyre-road contact, impairing longitudinal braking force generation. Lower standard deviations (SD) in load transfer and pitch rate correlate with shorter stopping distances. However, on the Belgian Paving the significance of these observed vehicle states is reduced. Nevertheless, an improvement of 5 m in stopping distance is achieved.

Table 1. Summary of observed vehicle states

Parameter	Best Case		Worst Case	
	Class-C	Belgian Paving	Class-C	Belgian Paving
Load transfer SD [N]	3208	6163	3898	7033
Pitch rate SD [deg/s]	1.76	5.24	4.45	5.05
Brake distance [m]	32.02	44.96	38.43	50.39

However, the use of the high-fidelity model was computationally expensive. Therefore, it was worthwhile to investigate if a reduced-order vehicle model can be created to determine suspension configurations that minimise pitch rate, since this vehicle state indicated an improvement of brake performance.

3. Reduced-order model: Pitch-plane Land Rover model

The pitch-plane model, lacking a tyre model, does not account for longitudinal forces generated in the tyres due to its single-point contact assumption. Instead, it serves to optimise pitch rate efficiently, avoiding the high computational demands of a high-fidelity model. This model, shown in Fig. 8, enables the identification of the best spring and damper characteristics for the front and rear axles to minimise pitch rate variations.

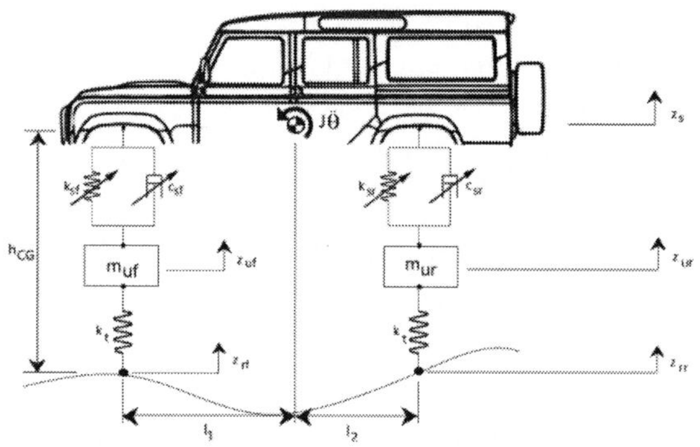

Fig. 8. Reduced-order Land Rover pitch-plane model

The reduced-order model makes use of the same non-linear springs and dampers that were mathematically modelled in the high-fidelity model.

The optimisation technique was founded on Bayesian inference, a statistical method that updates the probability of a hypothesis as new evidence or information becomes available. This approach was chosen for its sample efficiency, allowing exploration without investigating the entire 10 x 10 design space, thereby achieving computational efficiency. Also, Euler's method, along with kinematic relationships, were used to simulate the model over a given road profile.

3.1. Reduced-order model validation

The model was validated against the same experimental brake tests shown in Section 2.1. The results are summarised in Fig. 9, where a strong correlation between the pitch-plane model and the actual test vehicle is achieved. All analysed vehicle states follow the same trend as that of the actual test vehicle, giving great confidence in the developed model.

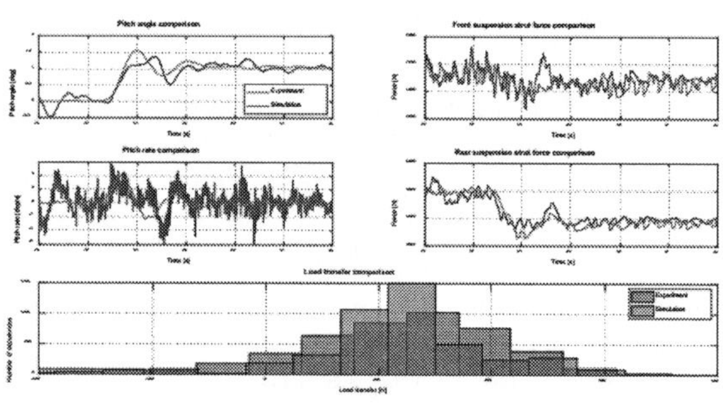

Fig. 9. Soft suspension: Pitch-plane model validation

3.2. Reintroduction of dynamic complexity

The recommended suspension settings from the pitch-plane model were applied to the validated ADAMS Land Rover model, reintroducing dynamic complexity through the model's full-scale FTire and ABS systems. The next section compares brake performance between these recommended settings, the initial ADAMS results, and the comfort setting (softest suspension setting) as a baseline, commonly used for rough terrain driving.

3.2.1. Comparing comfort, ADAMS model, and pitch-plane model suspension settings on a Class-C Road

The suspension configurations obtained from the pitch-plane model optimisation were implemented on the high-fidelity model where brake tests from 80 km/h on a Class-C Road were simulated. The load transfer and pitch rate distributions for all three cases are shown in Fig. 10.

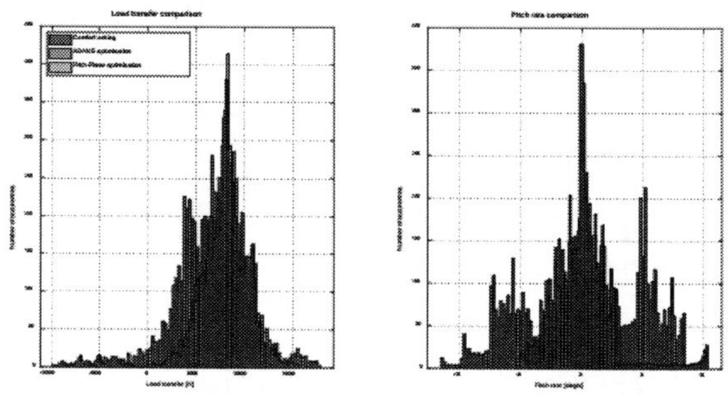

Fig. 10. Load transfer and pitch rate distributions

The results indicate that the pitch-plane model effectively identified suspension configurations that closely matched the trends of the high-fidelity model. Additionally, the pitch-plane model successfully reduced pitch rate and load transfer variations, leading to a shorter stopping distance. Both models achieved a 5 m reduction in stopping distance compared to the comfort setting.

4. Conclusion and Future Work

Using a high-fidelity ADAMS Land Rover model determined that a softer front and stiffer rear suspension reduced pitch rate and load transfer distributions, shortening stopping distances. However, rougher terrains limited these benefits due to wheel hop and reduced tyre contact. A reduced-order pitch-plane model confirmed similar results, validating that simple vehicle models can be used to determine suspension configurations for a vehicle travelling on various terrains in a more computational efficient manner. For future work, it is worthwhile to transfer the obtained suspension configurations that lead to greater brake performance to the actual test vehicle. Thereafter, conducting experimental brake tests to verify the findings in simulation are recommended.

References

[1] Š. Liščák, L. Moravčík, and M. Jaśkiewicz, "Safety requirements for road vehicles," (in Eng), *Zeszyty Naukowe Akademii Morskiej w Szczecinie,* no. 39 (111), pp. 94-99, // 2014.

[2] S. Jitesh, "Antilock braking system (ABS)," *International Journal of Mechanical Engineering and Robotics Research,* vol. 3, no. 4, p. 253, 2014.

[3] N. A. van der Merwe, P. S. Els, and V. Žuraulis, "ABS braking on rough terrain," *Journal of Terramechanics,* vol. 80, pp. 49-57, 2018.

[4] W. C. W. Penny and P. S. Els, "The test and simulation of ABS on rough, non-deformable terrains," *Journal of Terramechanics,* vol. 67, pp. 1-10, 2016.

[5] M. Reul and H. Winner, "Enhanced braking performance by integrated ABS and semi-active damping control," in *Proc. 21st Enhanced Safety Vehicles (ESV) Conf,* 2009, pp. 1-14.

[6] H. A. Hamersma and P. S. Els, "Improving the braking performance of a vehicle with ABS and a semi-active suspension system on a rough road," *Journal of Terramechanics,* vol. 56, pp. 91-101, 2014.

[7] P. S. Els, N. J. Theron, P. E. Uys, and M. J. Thoresson, "The ride comfort vs. handling compromise for off-road vehicles," *Journal of Terramechanics,* vol. 44, no. 4, pp. 303-317, 2007.

[8] M. J. Thoresson, P. E. Uys, P. S. Els, and J. A. Snyman, "Efficient optimisation of a vehicle suspension system, using a gradient-based approximation method, Part 1: Mathematical modelling," *Mathematical and Computer Modelling,* vol. 50, no. 9-10, pp. 1421-1436, 2009.

[9] Hexagon. "MSC Software ADAMS." MSC Software Corporation. https://hexagon.com/products/product-groups/computer-aided-engineering-software/adams (accessed 30 October, 2024).

[10] Mathworks. "MATLAB and Simulink." https://www.mathworks.com/ (accessed 30 October, 2024).

[11] R. G. Budynas, J. K. Nisbett, and J. E. Shigley, *Shigley's mechanical engineering design*, 9th ed. (McGraw-Hill series in mechanical engineering). New York: McGraw-Hill (in English), 2011.

[12] INTERNATIONAL ORGANISATION OF STANDARDISATION *ISO 8608:2016 Mechanical vibration - Road surface profiles - Reporting of measured data*. Geneva, Switzerland, 2016.

Method for characterisation of wall surface roughness in bulk solids handling using spectral analysis

Terrance J. Frangakis[1,*]

[1] School of Mechanical, Industrial and Aeronautical Engineering, University of the Witwatersrand, Johannesburg, South Africa
*Terrance.Frangakis@wits.ac.za

Abstract

Wall friction plays a significant role in bulk materials handling. The frictional characteristics of a particulate sliding on a wall surface are measured in flow properties tests but there is no method as yet to predict wall friction. An algorithm for characterisation of a wall surface roughness profile is proposed in which dominant wavelengths in the profile are determined. The algorithm includes pre-processing to remove misalignment and waviness, followed by spectral analysis which yields signal power in relation to wavelengths present in the waveform. Post-processing of the results allows dominant wavelengths and amplitudes to be extracted. This may be used to generate surfaces for use in discrete element simulations, and may lead to a better understanding and prediction of wall friction.

Keywords: particulates, wall surface roughness, spectral analysis, dominant wavelengths

1 Background and Motivation

Bulk materials such as ores consist of individual particles with sizes ranging from micro-millimetres (clays and powders) to tens or even hundreds of millimetres (coarse particles to boulders). When such particulates are handled, relative motion occurs both between particles and between particles and wall surfaces of mechanical handling equipment. The flowability of particulates is dictated by factors relating to the particles (such as shape, size distribution, mineralogy, affinity for moisture), wall surfaces (roughness, surface free energy, hardness, resistance to wear), and environmental conditions (moisture, temperature etc.).

Standardised laboratory tests are used to assess particulate flowability but rely on the sample material being representative of the material and conditions in the application (including moisture, wall material type and condition, etc.). Laboratory test results indicate how the particulate will behave in the specific application but cannot be extrapolated to other materials and conditions. There is currently limited scope for prediction of particulate behaviour without laboratory tests.

Simulation of particulates (e.g. Discrete Element Modelling, DEM) is increasingly being used to better understand and predict flowability, provided various physical and model parameters are first measured or calibrated. Since particle–wall interaction plays a critical role in the design of handling equipment, and is dominated by friction [1], there is an opportunity to improve the understanding of the mechanisms of friction in relation to the wall surface. Friction causes a resistance to sliding of particles at a wall surface, which in turn is communicated into the bulk of particulate as particles move relative to one another. Improved characterisation of wall surfaces may lead to improvements in simulation and prediction of real particle motion.

Previous researchers [2], [3], [4] investigated various surface roughness parameters (e.g. arithmetic mean roughness, R_a) and their relationship to wall friction, but correlation was limited as surfaces with similar R_a values had different wall friction angles. Roberts et al. [3] explored spectral analysis of surface roughness profiles to investigate wavelengths present in the profiles but no further progress has been made. In this current research, the aim is to develop a practical algorithm to assess dominant wavelengths in a surface profile through spectral analysis. The dominant wavelengths arising from the spectral analysis may indicate the potential for suitably sized particles to interlock with surface features, and may be used to create realistic surface profiles for the purposes of simulation.

2 Objectives and Scope

The aim of the research is to develop the surface characterisation algorithm which has the following steps: de-trending and pre-filtering of the input profile, spectral analysis to identify dominant wavelengths and post-processing of the data to reconstruct the profile with the dominant wavelengths. The algorithm is tested on three roughness profiles of a grit-blasted surface, where the particle size distribution of the grit is known. The dominant wavelengths are extracted and used to reconstructed a surface profile which is compared with the original profile to assess the performance of the algorithm.

3 Algorithm Development

Before exploring the algorithm methodology, there are some relevant considerations from bulk solids handling, metrology and signal processing.

3.1 Bulk solids handling

The wall yield locus defines the relationship between shear stress and normal stress for a particulate sliding on a wall surface. It measures the combined effects of friction and adhesion (when moisture is present), but friction is dominant [1]. Wall friction depends on the characteristics of the wall surface.

Figure 1 shows the interaction of different sized particles with a surface profile. Particles A and B may have restricted motion due to interlocking, while larger particles such as C and D may slide over the top of local protrusions, or be assisted by the wedging action of smaller particles such as E. Small particles can move into and out of long wavelength surface features without significant resistance.

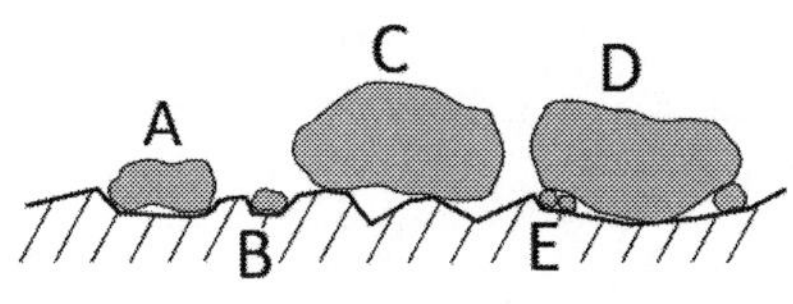

Figure 1: Particles interacting with surface features

Of particular relevance is the following:

- Particles smaller than 4 mm dominate the cohesive behaviour of particulates, particularly when moisture is present [5]. Wavelengths larger than 4 mm will therefore be ignored.
- Stylus-based profilers have tip sizes in the range 2.5 – 5 μm and will not resolve features smaller in size. The size of the stylus tip used will dictate the lower wavelength limit.

3.2 Metrology

Surface roughness measurement is detailed in standards such as ISO 3277, 4287, 4288, 16610. Of particular relevance is the following:

- The evaluation length is the total length of the profile over which the roughness measurements are made. As the arithmetic roughness increases, the sampling length should ideally also increase (see Table 1).
- Cut-off wavelengths distinguish roughness from long wavelength waviness which may have significant spectral power but is unlikely to affect particle interlocking. Cut-off wavelengths are used to attenuate larger wavelengths in the signal. The cut-off wavelength values for spline filtering are indicated in Table 1 and will typically be λ_c = 0.8 mm or λ_c = 2.5 mm, depending on R_a (see Equation 7 in Appendix A.2).

Table 1: Cut-off wavelengths and evaluation lengths as a function of R_a [6]

Roughness R_a (μm)	Sampling length/cut-off $l_r = \lambda_c$ (mm)	Evaluation length l_n (mm)
0.02 - 0.1	0.25	1.25
0.1 - 2	0.8	4
2 - 10	2.5	12.5
10 - 80	8	40

3.3 Digital signal processing

Signal processing of spatial (displacement-based) signals is analogous to temporal (time-based) signals. Spatial signals have sampling frequency f_{sx} (in samples per mm) and spectral frequency bins with units 1/mm (see Appendix A.1). Of particular relevance is the following:

- In spectral analysis it is assumed that the signal repeats itself. Start and endpoints of the waveform should be attenuated by a windowing function to avoid spectral leakage.

- The total power contained in the signal may be estimated in the spatial or temporal domains using Parseval's theorem (see Appendix A.1).
- The reciprocal of the frequency bin values yields the wavelengths contained in the spatial profile.
- If FFT sample size, N, is equal to the signal length, n, then bin 1 of the FFT will correspond to a base wavelength equal to the stylus displacement x_{max}. Bins 2 onwards will be ratios of the base wavelength. If $\lambda_c < x_{max}$ it will attenuate bin 1.
- Dominant wavelengths (bin frequencies) may be extracted from the frequency spectrum and used to reconstruct a representative spatial waveform.

3.4 Analysis algorithm

The surface profile analysis described below is for a one dimensional profile measured in the direction of shearing motion of the particulate. The proposed algorithm is shown schematically in Figure 2 and the steps described below.

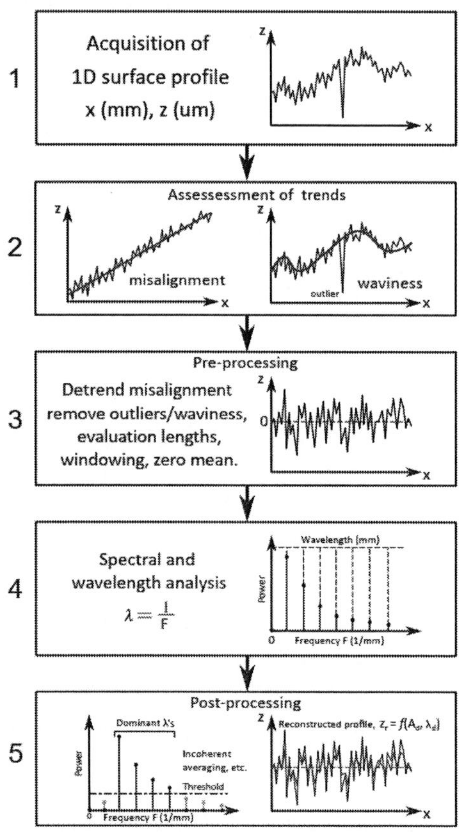

Figure 2: Proposed algorithm flowchart for surface roughness characterisation

1. The x-z profile data is obtained using a contact or optical profiler, with x in mm and z in μm.
2. Trends in the raw data are assessed to determine if there is sensor/test piece misalignment, outliers and waviness due to manufacturing.
3. The profile is pre-processed as follows:
   (a) Misalignment is removed with a linear least-squares curve-fit subtracted from the data.
   (b) A robust spline method [7] is used to remove outliers (using a Tukey estimator) and remove waviness using a cut-off wavelength λ_c (Table 1). See Appendix A.3 for details of the mathematical equations.
   (c) Evaluation lengths are generated, if needed.
   (d) Windowing is applied if necessary to reduce spectral leakage, and the evaluation lengths zero-mean adjusted.
4. The adjusted profile is processed using a single-sided FFT. The Power Spectrum is generated with scaling based on the percentage contribution of a particular bin to the total power.
5. In post-processing, incoherent averaging may be applied to evaluation lengths to reduce noise. Dominant wavelengths and amplitudes are then identified and used to reconstruct an equivalent spatial profile.

4 Results and Discussion

To test the algorithm a steel surface was roughened by means of grit-blasting. The material was Hardox 600 (hardness approx. 55 HRC). The size range of the corundum grit was 0.25 – 1 mm, with the majority of the particles between 0.5 – 0.7 mm. The surface roughness profile was measured on a Hommel Etamic T8000 profiler, with 2.5 μm TKV600 stylus probe. The sample test length was 4.8 mm with 2088 data points. The results of Test 12 are presented below (results from Tests 13 and 14 show similar trends, see Appendix B).

Figure 3 shows the original profile of Test 12 with two spline fitted curves with cut-off wavelengths of λ_c = 0.8 and 2.5 mm. The mean roughness was R_a = 6.43 μm, so cut-off wavelength λ_c = 2.5 mm was used (see Table 1). No outliers were present. The lower plot shows the final profile after removing the waviness.

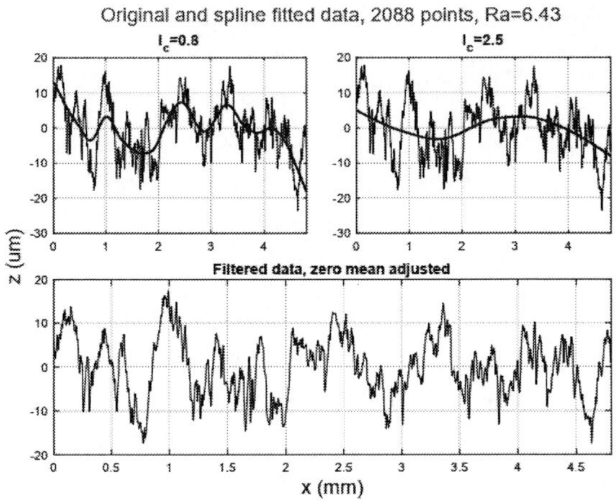

Figure 3: Spline curve fits to Test12 profile and final profile

The profile was zero mean adjusted and a windowing function applied, after which the FFT and power spectra were determined as a function of wavelength. Figure 4 shows the power spectra with wavelength indicated at the top of the plot, and a cumulative power indicated with a dashed line.

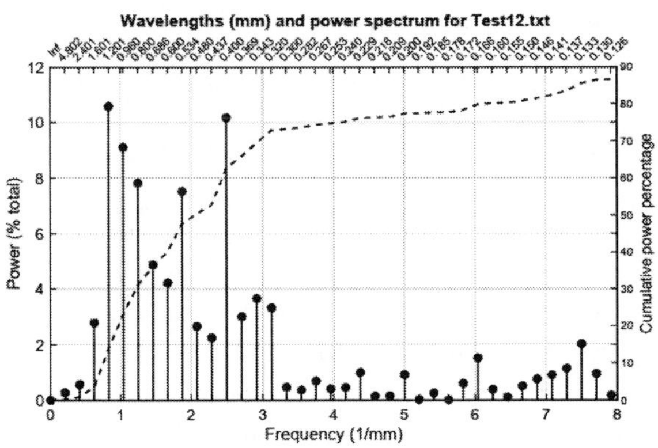

Figure 4: Wavelength spectral analysis for Test 12

Visually, the dominant wavelengths were in the range 0.32 – 1.2 mm with the peak at 0.8 mm. There was reasonable agreement between the dominant wavelengths and the grit particle size. Most of the power in the signal (approximately 80%) was in the range of wavelengths 0.17 – 1.6 mm. For Test 13 and 14 in Appendix B the dominant wavelengths are also in the range 0.3 – 1.2 mm. Their reconstructed profiles are also in good agreement.

Figure 5 shows the reconstructed profile for Test 12 based on the dominant wavelengths and amplitudes whose power was above 30% of the power of the dominant wavelength (0.8 mm). The reconstructed profile captures the main features of the original profile.

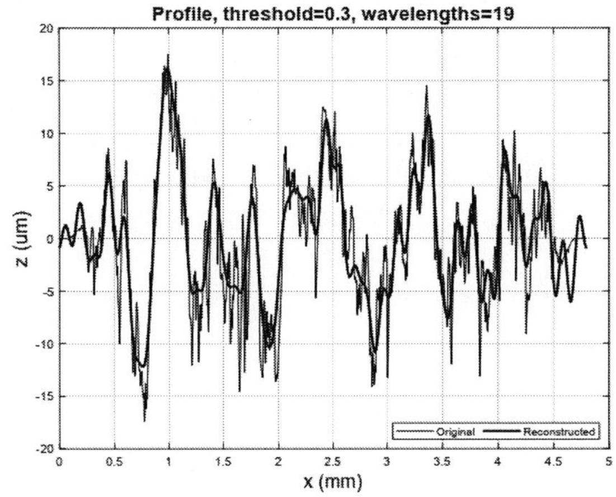

Figure 5: Reconstructed profile for Test 12

5 Conclusions and Future Work

An algorithm for surface roughness characterisation was developed. It incorporated components to assess and remove misalignment and waviness in a profile, and carry out a spectral analysis to determine the power per wavelength contained in that surface roughness profile. Dominant wavelengths and amplitudes in the spectra could be identified and extracted, and used to reconstruct a waveform in the spatial domain with similar characteristics.

A grit-blasted steel surface was prepared and its roughness profile measured using a stylus-based profiler. The profile was processed using the algorithm which indicated dominant wavelengths that were reasonably consistent with the size of the blasting media. The dominant wavelengths and associated amplitudes were used to reconstruct a profile in the spatial domain, which had good agreement with the original profile.

Recommendations for future work on the algorithm include assessing the sensitivity of the output to variations of the input parameters (e.g. number of data points and sampling frequency, FFT sample size, effect of longer sample lengths), and potential benefit of incoherent frequency averaging to see whether noise is reduced.

Experimentation should include additional surface roughness profile measurements of surfaces prepared using different methods (e.g. grinding, polishing), and laboratory friction measurements of particulates on surfaces that have been characterised to see if there is correlation between dominant wavelengths and wall friction angles.

References

[1] Frangakis TJ and Sheer TJ, *A review of the wall yield locus in the context of wet and sticky ores, International Conference on Bulk Materials Storage, Handling and Transportation (13th: 2019: Surfers Paradise, QLD.)* Engineers Australia Barton, ACT, 2019, pages 488–499.

[2] Ooms M and Roberts A, *Significant influence on wall friction in the gravity flow of bulk solids,* Bulk Solids Handling **5.6** (1985).

[3] Roberts A, Sollie L, and De Silva S, *The interaction of bulk solid characteristics and surface parameters in surface or boundary friction measurements,* Tribology International **26.5** (1993), pages 335–343.

[4] Hastie D and Halford W, *Preliminary observations of the effect of plate roughness and particle size on the determination of wall friction* (2016).

[5] Roberts AW, *Basic principles of bulk solids storage, flow and handling,* Institute for Bulk Materials Handling Research, 1993.

[6] Mitutoyo, *Quick guide to surface roughness measurement. Reference guide for laboratory and workshop. Bulletin No. 2229.* (2016).

[7] Lou S, Zeng WH, Jiang XQ, and Scott PJ, *Robust filtration techniques in geometrical metrology and their comparison,* International Journal of Automation and Computing **10** (2013), pages 1–8.

[8] Zeng W, Jiang X, and Scott P, *A generalised linear and nonlinear spline filter,* Wear **271.3-4** (2011), pages 544–547.

[9] Tong M, Zhang H, Ott D, Zhao X, and Song J, *Analysis of the boundary conditions of the spline filter,* Measurement Science and Technology **26.9** (2015), page 095001.

[10] *ISO 16610-32, Geometrical product specifications (GPS) – Filtration: Robust profile filters: Spline filters,* Technical Report, ISO, 2023.

Acknowledgements

The author would like to thank Prof. Rudolph Laubscher (Department of Mechanical Engineering Science at the University of Johannesburg) who provided access to the Hommel Etamic surface roughness profiler where the test profiles were measured.

Appendix A. Analysis

A.1 Spectral analysis

The spatial spectral analysis uses a frequency that is defined as the number of samples per mm of displacement, as compared with, say, a vibration signal that has a frequency of samples per second.

The following equations are used to define the sampling frequency, FFT bin resolution and frequency values, and the equivalent wavelengths (which are the reciprocals of the bin frequencies). Where relevant, a subscript 'x' is included to indicate that the parameter is referring to the spatial domain. 'n' refers to the number of samples in the roughness profile measurement, 'N' refers to the size of the FFT. In most instances $n = N$. The analysis assumes that $x_{\text{min}} = 0$.

$$\text{Sampling freq.} \quad f_{sx} = \frac{\text{samples}}{\text{mm}} = \frac{n}{x_{\text{max}}} = \frac{1}{\Delta x} \tag{1}$$

$$\text{Bin resolution} \quad \Delta f = \frac{f_{sx}}{N} = \frac{n}{N x_{\text{max}}} = \frac{1}{N \Delta x} \tag{2}$$

$$\text{Bin freq's} \quad [0,\ 1\Delta f,\ 2\Delta f,\ 3\Delta f, \cdots, (N/2)\Delta f] \tag{3}$$

$$\text{Wavelengths } \lambda = [\infty,\ \frac{1}{1\Delta f},\ \frac{1}{2\Delta f},\ \frac{1}{3\Delta f}, \cdots, \frac{1}{(N/2)\Delta f}] \tag{4}$$

$$\text{or } \lambda = [\infty,\ \frac{N x_{max}}{1n},\ \frac{N x_{max}}{2n},\ \frac{N x_{max}}{3n}, \cdots, \frac{N x_{max}}{(N/2)n}] \tag{5}$$

Parseval's theorem may be used to relate the signal power in the spatial domain to that in the frequency domain. A useful feature is that coefficients that have small magnitudes in the frequency domain may be ignored and not substantially affect the signal in the spatial domain. The total signal power is calculated using Equation 6 and individual bin power magnitudes scaled as a percentage of the total power.

$$\text{Signal power} \quad \sum_{i=0}^{N-1} |z[i]|^2 = \frac{1}{N} \sum_{k=0}^{N-1} |Z[k]|^2 \tag{6}$$

where: $z[i]$ is the deviation of the profile from a zero mean in the spatial domain and $Z[k]$ is the unscaled Fourier transform of the spatial data in the frequency domain.

A.2 Surface roughness

$$\text{Arithmetic mean roughness:} \quad R_a = \frac{1}{n} \sum_{i=1}^{n} |z[i]| \tag{7}$$

A.3 Robust spline

The following description is based on [7], [8], [9] and [10]. Filtration of a non-periodic surface profile allows the separation of waviness from roughness by means of a filtering equation. Gaussian filtering is widely used but suffers from end effects, while linear spline filtering does not. However, spline filtering does not have the same transmission characteristics but this may be improved with a tension parameter β; however, end effects still persist. End-effects may be reduced by modifying the boundary conditions in an approximating spline filter [9].

In addition, outliers may affect the resulting waveform and may be eliminated by means of a weighting function such as the Tukey biweight influence function [10]. Equation 8 is the spline filter equation:

$$w = \left[\delta + \beta\alpha^2 P + (1-\beta)\alpha^4 Q\right]^{-1}\delta z \tag{8}$$

where

- w is the underlying waviness in the surface profile,
- δ is a diagonal matrix of vertical weights which is either the identity matrix for the case of no outliers, or an iteratively determined weight if outliers are present:

$$\delta_i = \begin{cases} \left[1 - \left(\frac{z_i - w_i^{(m)}}{c^{(m)}}\right)^2\right]^2 & \text{for } \left|z_i - w_i^{(m)}\right| \le c^{(m)} \\ 0 & \text{for } \left|z_i - w_i^{(m)}\right| > c^{(m)} \end{cases} \tag{9}$$

- $i = 1, \ldots, n$ where n is number of z profile data points ,
- $\alpha = \left(2\sin\left(\frac{\pi\Delta x}{\lambda_c}\right)\right)^{-1}$ for cut-off wavelength λ_c and sampling interval Δx,
- β is a tension parameter and set to either 0 or 0.625 242 (see notes below),
- $w_i^{(m)} = (w_1^{(m)}, \ldots, w_n^{(m)})^T$ where m is the iteration number,
- $c = 4.4478 \times \text{median}|z - w|$ (for z and w arrays), iteration terminates when $\left|c^{(m+1)} - c^{(m)}\right| / c^{(m)} \le 10^{-5}$ or $m \ge 12$.
- P and Q matrices as per ISO [10] (or modified boundary conditions as per [9] in brackets).

$$P = \begin{bmatrix} 1(0) & -1(0) & 0 & \cdots & \cdots & 0 \\ -1 & 2 & -1 & \ddots & & \vdots \\ 0 & -1 & 2 & -1 & \ddots & \vdots \\ \vdots & \ddots & -1 & 2 & -1 & 0 \\ \vdots & & \ddots & -1 & 2 & -1 \\ 0 & \cdots & \cdots & 0 & -1(0) & 1(0) \end{bmatrix} \tag{10}$$

$$Q = \begin{bmatrix} 1(-1) & -2(2) & 1(-1) & 0 & \cdots & \cdots & 0 \\ -2(-1) & 5(3) & -4(-3) & 1 & \ddots & & \vdots \\ 1 & -4 & 6 & -4 & 1 & \ddots & \vdots \\ 0 & \ddots & \ddots & \ddots & \ddots & \ddots & 0 \\ \vdots & \ddots & 1 & -4 & 6 & -4 & 1 \\ \vdots & & \ddots & 1 & -4(-3) & 5(3) & -2(-1) \\ 0 & \cdots & \cdots & 0 & 1(-1) & -2(2) & 1(-1) \end{bmatrix} \tag{11}$$

Notes:

1. If no outliers are present then use $\delta = I$ (identity matrix) and solve Eq. 8 using a least squares minimum norm approach.
2. If $\beta = 0$ there are no end-effects, but the transmission characteristic is not the same as the Gaussian filter.
3. To reduce end-effects and have similar transmission characteristics as the Gaussian filter, set $\beta = 0.625\,242$ and use the P and Q matrices with the values in brackets. Iteration is then required to solve for w, the underlying waviness in the profile.

Appendix B. Comparative results

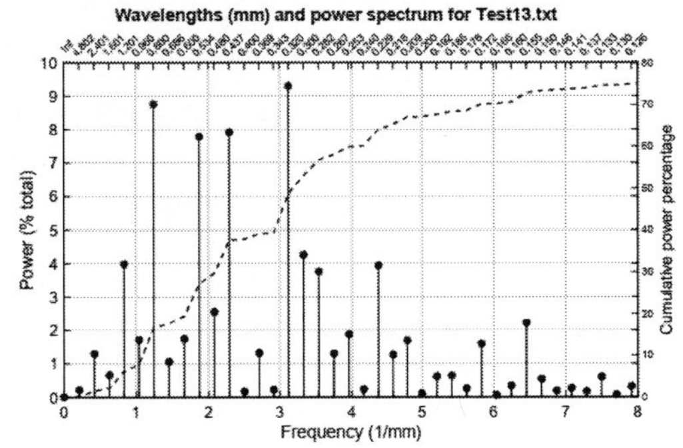

Figure 6: Wavelength spectral analysis for Test 13

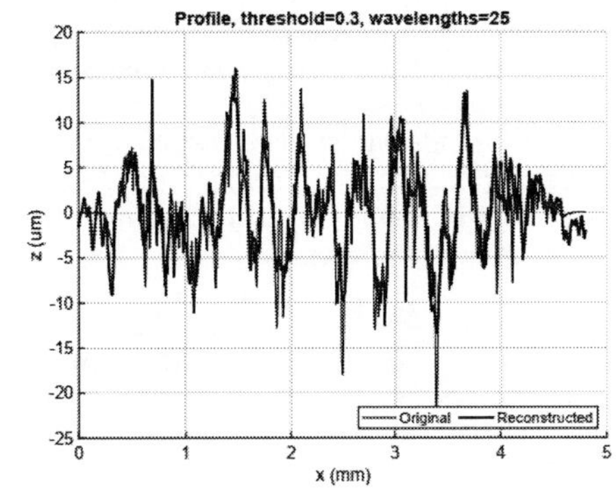

Figure 7: Reconstructed profile for Test 13

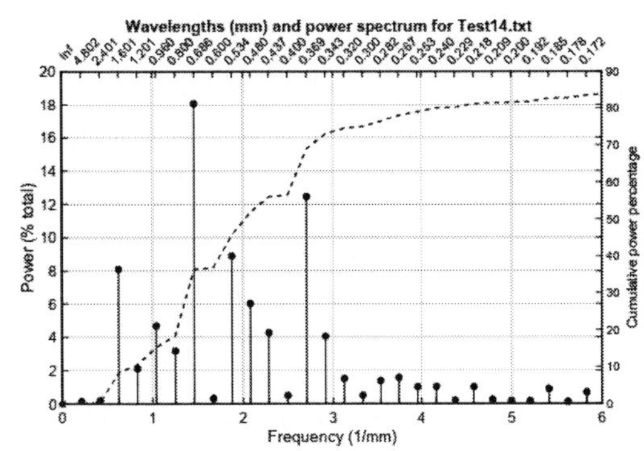

Figure 8: Wavelength spectral analysis for Test 14

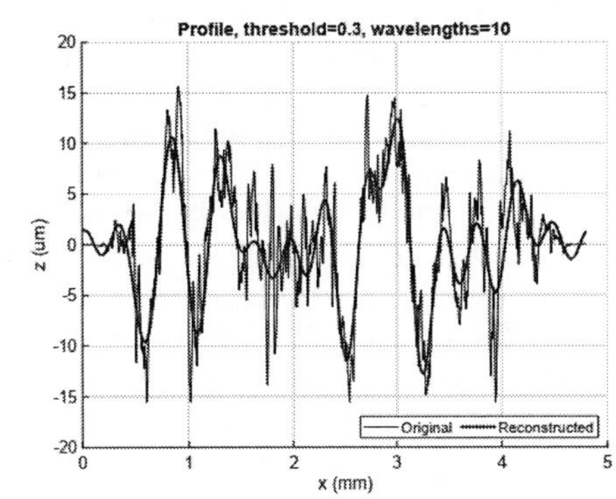

Figure 9: Reconstructed profile for Test 14

Dynamics of a nutating mill with vibration isolation mounts

HC (Christo) Janse van Vuuren[a,*], Johann Bredell[a], Corné Coetzee[a]

[a]*Stellenbosch University, Department of Mechanical and Mechatronic Engineering, Stellenbosch 7600, South Africa*

**21642478@sun.ac.za*

Abstract

This study investigates the structural dynamics of a laboratory-configured nutating test mill (NuMill) equipped with vibration isolation mounts, intended to emulate the compliance observed in industrial systems such as the Hicom mill. The mounts introduce a rigid body resonance during startup, which was captured through experimental measurement and subsequently replicated in a rigid-body multi-body dynamics (MBD) simulation using Ansys Motion.

Dynamic properties of the rubber mounts and tyre coupling were experimentally determined and implemented in the simulation. The comparison between measured and simulated frame displacements showed good agreement in resonant behaviour and transient response envelope shape, though displacement amplitudes differed. The experimentally obtained damping coefficients were retained to preserve accurate overall behaviour. The results provide a validated foundation for continued two-way coupled simulation development involving charge-structure interaction.

Keywords: MBD, Experimental Validation, Rubber Vibration Isolation, Structural Dynamics, Rigid Body Resonance.

1. Background and motivation

The De Beers Group mining company makes use of the Hicom, shown in Figure 1, a high intensity nutating mill for some of their operations. A nutating mill utilises a vertical cylindrical chamber that has its top end constrained so it can only rotate about the horizontal axes, as would be the case if one were to fix one end of a universal joint. The bottom end is then gyrated by cranking it with a pin eccentrically connected to a shaft. A nutating mill has an advantage over conventional horizontal ball mills in that its effectiveness is not limited by gravity [2]. The De Beers Group has encountered mechanical reliability issues during their specific operations and therefore a research study was undertaken to understand the dynamics of the mill better [1]. Analysis of a nutating mill is complicated by the unpredictable behaviour of the particle charge.

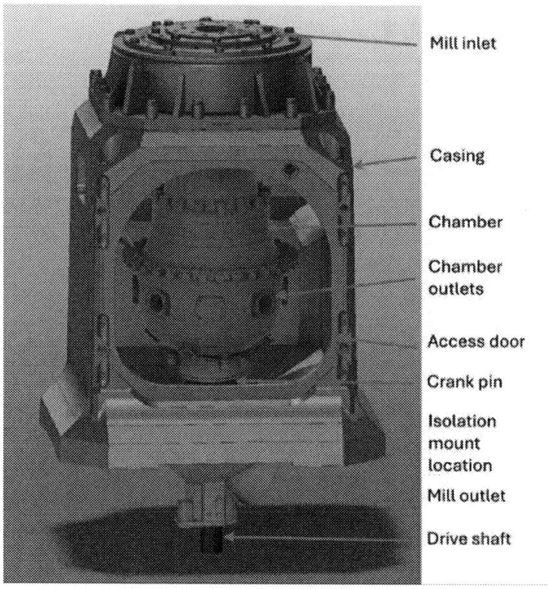

Figure 1: Hicom nutating mill

The previous work had the objective to develop an analysis method to determine the interface forces between components. For this, an experimental laboratory nutating mill with rigid mounts and a smooth chamber interior was developed for experimental measurements, dubbed the NuMill. A rigid multi-body dynamics (MBD) simulation of the NuMill was then developed. The two types of test cases that were investigated were the case with rigid mass charge representations and the case with particle charges. For the particle case, forces from a discrete element method (DEM) simulation were measured and included in the MBD simulation.

After characterisation and comparison of component interface forces of the experimental and simulation tests, two relevant conclusions were reached. The largest interface force is the radial force experienced by the crank pin connected at the bottom of the chamber, due to centrifugal forces. Secondly, it was concluded that the difference between the radial forces generated for the rigid mass charge and the particle charge is significant [1].

2. Objectives and scope

The current study has been continued with the aim to develop a more comprehensive analysis method for nutating mills that includes more aspects of the Hicom. For this reason, the objectives of the study that will be discussed in this paper includes:

1. Experimental characterisation of flexible component dynamic parameters – namely translational and rotational stiffness and damping.
2. Development of a rigid MBD simulation of the NuMill with vibration isolation mounts.
3. Experimental measurement of the NuMill's structural dynamics.
4. Experimental validation of the MBD simulation.

The scope of the study is limited to external interface forces, global frame responses, and the characterisation of compliant components. The mill structure is assumed to be rigid, and internal deformation is neglected. Particle breakage and grinding phenomena are excluded. The simulation focuses on rigid-body motion and does not incorporate DEM modelling in this paper. Validation of the MBD simulation is based on experimental data from tests involving an empty chamber and rigid mass charge analogues.

3. Methodology and analysis approach

The methodologies utilised in this project include analytical, experimental and simulation methodologies. Analytical computations were mostly performed on simplified models as a first step to inform the expected results of certain experiments and simulations.

3.1 Experimental

Figure 2 shows the modified NuMill design. All labelled components – excluding the crank pin and rigid mass charge – reflect alterations introduced for this study. The local constraints of joints interior to the NuMill that facilitate the nutating movement of the chamber are as described in Section 1. The top frame is connected to the base frame legs with five connections, four rubber bobbin mounts and one flexible tyre coupling on the shaft that is free to rotate, thus all connections between the top frame and bottom legs are through rubber material. The bottom legs are bolted to the floor through the foot spacers, creating a fixed connection.

The axial stiffness of the mounts and coupling were experimentally determined by performing limited linear displacement tension and compression tests in a universal testing machine (UTM). The mounts were bolted onto a mounting adaptor that was pinned into the UTM's crosshead. For the tyre coupling's linear stiffness tests, it was connected to the same mount adaptor with its taper bushing. The testing procedure consisted of separate rounds of limited tensile and compressive displacement, starting from an unloaded state.

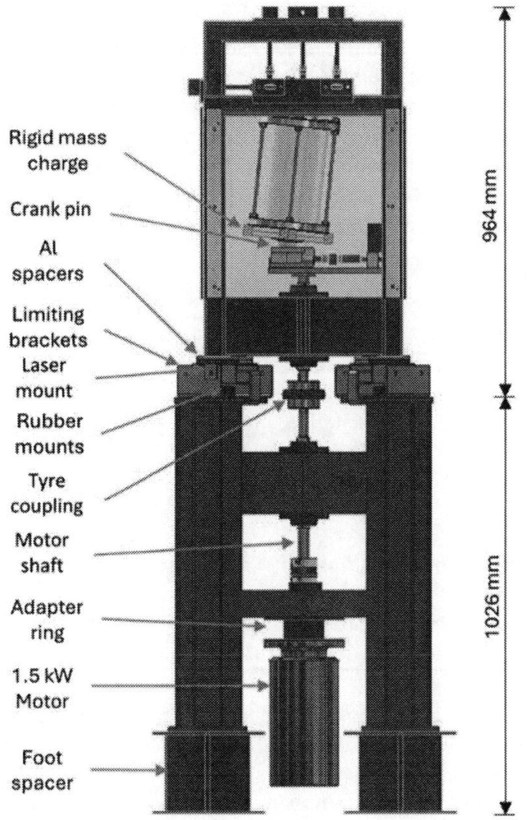

Figure 2: Modified NuMill design

Subsequently the axial damping coefficient of the mounts was determined experimentally, informed by the ASTM standard for '*Dynamic Testing of Vulcanized Rubber and Rubber-Like Materials Using Vibratory Methods*'. The experiment consisted of the generation of a discrete acceleration transfer function for a simple linear mass, spring and damper system with base excitation on a shaker. In this case the rubber mount is both the spring and the damper. The physical setup consisted of a shaker bolted to the floor, the rubber mount bolted into the top of the shaker and a rigid mass bolted onto the mount. An accelerometer was installed onto the shaker mounting plate to measure the acceleration input of the system. An accelerometer was installed on the rigid mass to measure the response. The shaker was then used to excite the system at specified constant frequencies to generate an acceleration transfer function. The transmissibility curve indicated by the transfer function could then be used to determine the equivalent viscous damping coefficient by fitting the base excitation transfer function [3].

$$H(r) = \sqrt{\frac{1 + (2\zeta r)^2}{(1 - r^2)^2 + (2\zeta r)^2}} \quad (1)$$

Finally, the NuMill behaviour was measured during startup and during its steady state. Considering the goal of validating the MBD simulation among the measured parameters, two categories were established: independent input parameters and dependent reaction parameters. Furthermore, among the reaction parameters there are some that are local to the internal joints of the NuMill, as shown in Table 1.

Table 1: Measured parameters

Parameter	Category	Sensor	Location
Motor speed	Input	Tachometer	Motor shaft
Motor torque	Reaction, local	Torque transducer	Motor shaft
Radial force	Reaction, local	Force transducer	Crank pin
Top frame displacement	Reaction, global	Laser displacement	Bottom corner
Frame acceleration	Reaction, global	Piezoelectric accelerometer	Top corner

Specific attention was given to the frame global displacement for comparison between the MBD simulation and the physical experimental NuMill machine, since simulation validation is the current focus. Displacement was measured at the bottom corner of the top frame at the laser mount indicated on Figure 2. Acceleration was not prioritised since it is the second derivative of displacement with respect to time.

3.2 Simulation

The MBD simulation of the NuMill was conducted using Ansys Motion. Only the top frame is included in the simulation since the base frame is fixed to the floor with no possible rigid body motion. All MBD simulation geometry was declared as rigid. Both the rubber mount and tyre coupling flexible connection components, which connect the top frame to the bottom feet, were defined as zero dimensional bushing elements with six degrees of freedom (DOF) that are defined by both stiffness and viscous damping coefficients. A description of each kinematic joint as defined in Ansys Motion is shown in Table 2.

Table 2: Simulation kinematic joints

Components	Joint type	DOF
Ground – frame	Four bushings	6 DOF
Ground – motor shaft	Revolute	1 RDOF: Global Z-axis
Motor shaft – input shaft	Bushing	6 DOF
Input shaft – frame	Revolute	1 RDOF: Frame Z-axis
Crank pin – chamber	Sphere and cylinder	3 RDOF, 1 TDOF: Crankpin axis
Chamber – bearing cross	Revolute	1 RDOF: Frame X-axis
Bearing cross – frame	Revolute	1 RDOF: Frame Y-axis

4. Key results and discussion

4.1 Component characterisation

The results of the characterisation of the dynamic parameters of the flexible connections, which were then used in the simulation, are shown in Table 3. The flexible components are cylindrical. Thus, the Z direction is axial, and X and Y are radial, when using a local coordinate system.

Table 3: Rubber component dynamic parameters

	Translational		
Stiffness, k (kN/m)	**X radial**	**Y radial**	**Z axial**
Mount	72.25	72.25	425
Coupling	234	234	343.8
Damping, c (Ns/m)	**X radial**	**Y radial**	**Z axial**
Mount	343	343	343
Coupling	0	0	0
	Rotational		
Stiffness, k (Nm/rad)	**X rotation**	**Y rotation**	**Z rotation**
Mount	0	0	0
Coupling	300	300	286.5
Damping, c (Nms/rad)	**X rotation**	**Y rotation**	**Z rotation**
Mount	0	0	0
Coupling	0	0	0

4.2 Structural dynamics – MBD validation

To validate the MBD simulation, the frame displacement reaction of the NuMill during the startup ramp of an empty chamber, as obtained from the simulation, is compared with experimental results. A graph of the transient displacement response of the simulated and the experimental frame, can be seen in Figure 3. Unfortunately, no reliable data for the experimental rotational velocity is available for the current set of experiments due to equipment availability. The simulation rotation speed shown follows the speed measured in an earlier set of experiments with similar motor settings where a tachometer was available.

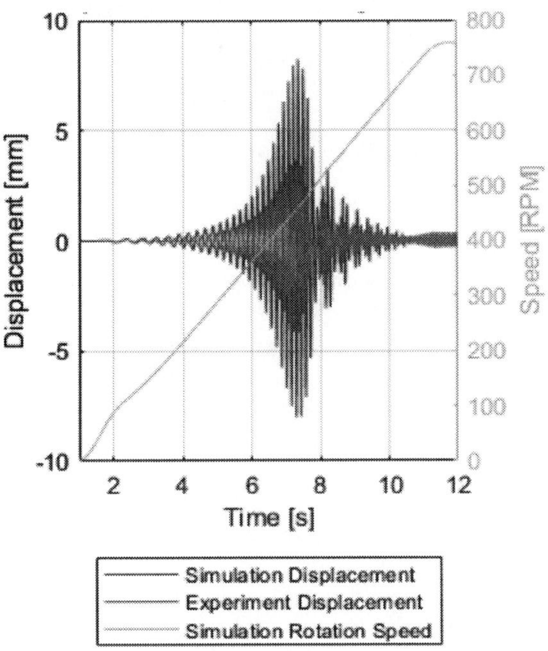

Figure 3: Frame Displacement with Velocity for c = 343 Ns/m

From Figure 3 it can be seen that the maximum displacement at the rigid body mode is 3.64 mm for the experiment and that of the simulation is 226% larger at 8.23 mm, for the experimentally determined axial mount damping coefficient of 243 Ns/m. At steady state the experimental displacement is 177% larger than the simulated steady-state. Since the simulation rotation speed is based on experimental data from a case similar to the one shown, the overlapping experimental and simulation peaks implicate a similar resonant frequency. Conducting a fast Fourier transform (FFT) on the displacement data confirms this, indicating a clear resonance frequency at 6.758 Hz (405.5 RPM) and 6.933 Hz (416.0 RPM) from the experimental and simulation's respective datasets.

Simulation iteration showed that to decrease the peak simulated acceleration with 10.6% the damping coefficient needed to be increased with 23.1% from 343 Ns/m. When extrapolating the linear relationship shown, for the peak simulation acceleration to match the peak experimental acceleration, the damping coefficient needs to be increased with 54.8% to 759 Ns/m, decreasing the acceleration by 69.3%. However, when making such a large change to the damping coefficient the overall shape of the simulated acceleration and displacement time series envelope differs significantly from the experimental one, causing vibrations to die out too quickly. Considering the future use of the MBD simulation in a coupled MBD-DEM simulation, the matching of the overall dynamic structural behaviour was prioritised, thus the experimentally determined damping coefficient will be used.

The validation process was improved by measuring acceleration in six directions. Three points in the horizontal X-axis and three points in the vertical Z-axis, thus capturing more DOF. Since the flexible components are flexible in six DOF, the global reaction needs to be measured in enough directions to capture that behaviour for validation. The Y-axis can be omitted since the frame has a 90-degree symmetry about the Z-axis.

5. Conclusion and future work

The dynamic parameters of the vibration isolation mounts and tyre coupling were experimentally characterised and applied in the multi-body dynamics (MBD) simulation of the NuMill. The experimental testing revealed a rigid body resonance during startup, which was then replicated in the simulation. While the simulated displacement magnitudes differ from the experimental measurements, the resonance frequency and overall dynamic behaviour match well. As such, the experimentally determined damping coefficient will be retained, prioritising accurate qualitative behaviour over amplitude calibration. These validated parameters will serve as the basis for further simulation development.

Future work includes the remaining research objectives of the study:

- To develop an experimentally validated coupled rigid MBD-DEM simulation.
- To investigate the effect of changing from rigid mounts to rubber vibration isolation mounts.
- To investigate the effect of changing the chamber interior surface from smooth to vertically ribbed.

Acknowledgements

Special thanks to my family for their support and to Mr Bredell, Prof Coetzee and Dr Nickerson for their input. This work is based on the research supported in part by the National Research Foundation of South Africa (Grant Number: 137952).

References

[1] Van Tonder, J.J. (2023). *Experimental Testing and Simulation of a Nutating Grinding Mill*. Master's Thesis, Department of Mechanical and Mechatronic Engineering, Stellenbosch University, Stellenbosch.

[2] Boyes, J. (1988). *High-intensity centrifugal milling — A practical solution*. International Journal of Mineral Processing, Vol. 22, no. 1-4, pp. 413-430. ISSN 03017516.

[3] Inman, D.J. (2014). *Engineering vibration. Always Learning*, 4th edn. Pearson, Boston, Mass. München. ISBN 978-0-273-76844-9.

Appendices

This section contains images of scenes and results discussed in the body of the paper. Figure 4 shows one of the bobbin mounts during a tension test on the UTM.

Figure 4: Mount tension test

Figure 5 shows cross sections of the mounting configurations used on the UTM as described.

Figure 5: UTM mounting cross-sections

Figure 6 shows a force-displacement curve generated by the UTM.

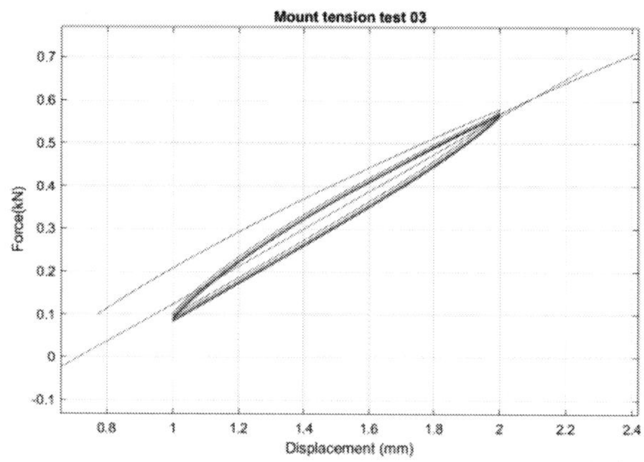

Figure 6: Mount tension displacement test curve

Figure 7 shows a lateral stiffness test where the top frame is pulled laterally, in line with the mounts, by winding a turnbuckle and then measuring the force on the cable with a force transducer and the displacement with a linear variable displacement transducer (LVDT). This test was done in two configurations, one where all flexible components were connected and one where only the mounts were connected. The difference was used to determine the lateral stiffness of the tyre coupling.

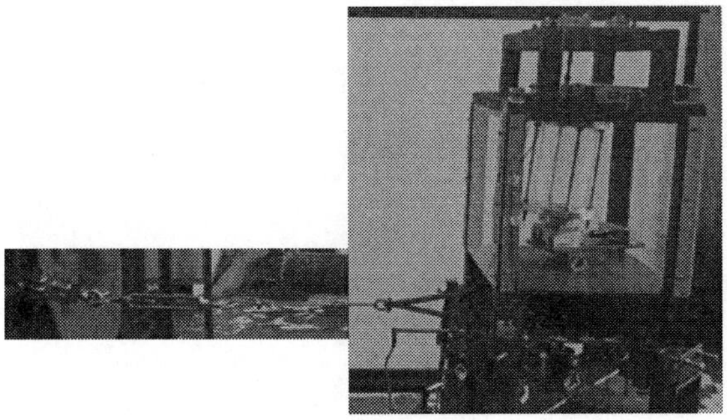

Figure 7: Lateral stiffness test

Figure 8 shows the experiment conducted to determine the bending stiffness of the tyre coupling, similar equipment was used with the lateral stiffness experiments.

Figure 8: Coupling bending stiffness test

Figure 9 shows the linear base excitation experiment used to determine the axial damping coefficient of the bobbin mounts.

Figure 9: Mount shaker experimental setup

Figure 10 shows the transfer function generated from the mount-shaker tests and the theoretical curve that was fitted to it.

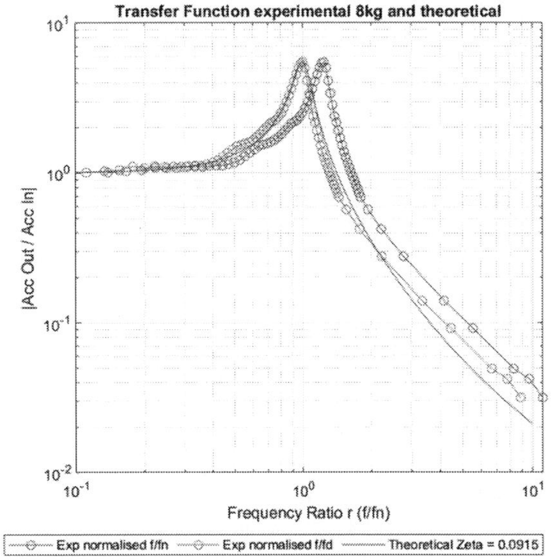

Figure 10: Mount axial acceleration transfer function

Figure 11 shows a damping sensitivity study done with the MBD simulation for steady states. The baseline damping coefficient is the one that was determined experimentally.

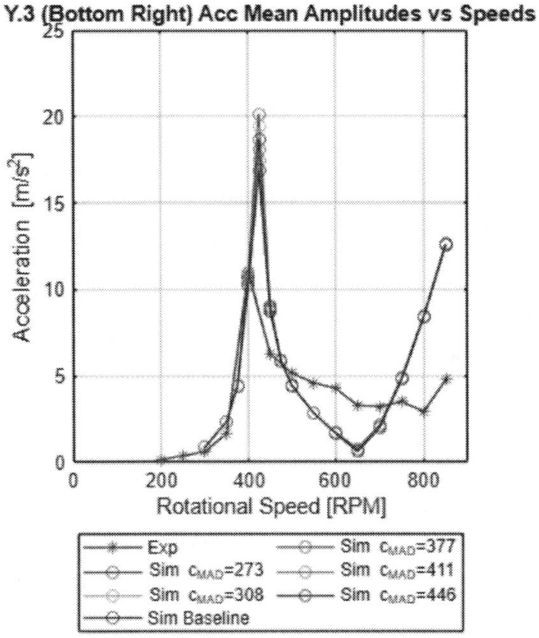

Figure 11: Simulation steady-state damping sensitivity study

Articulating Drop Foot Orthosis Testing Device

K. Zapke[a,*], M.P. Venter[a]

[a] *Mechanical and Mechatronic Engineering, Stellenbosch University, Private Bag X1, Matieland, 7602, South Africa*

**23671866@sun.ac.za*

Abstract

Drop foot weakens the upward tilt of the foot, leading to an altered gait often treated with ankle-foot orthoses (AFOs). This study evaluates a digital prototype testing device designed to test a new AFO. The objectives are to identify component properties that meet the deformation requirement and to assess whether the prototype satisfies the dynamic requirements. Finite element analyses (MSC Apex) indicated total deformation of 1.83 mm with suitable component properties, within the ISO 22675 limit of 2 mm. Multibody dynamic simulations (Hexagon Adams Modeler) yielded root mean square errors of 5.36° and 3.61°, with a maximum deviation of 4.5° in vertical foot tilt during normal and drop foot gait cycles, compared to literature. Simulated load profiles showed root mean square errors of 18.2 N (cyclic) and 22.7 N (static) for the loads transferred between components, with a maximum deviation of 46.8 N relative to ISO 22675. These results demonstrate the prototype's capability to replicate gaits and meet relevant mechanical requirements. Future work will involve experimental validation and the development of the control system.

Keywords: Drop Foot, Digital Prototype, Finite Element Analysis, Multibody Dynamic Simulation

1. Background and Motivation

Drop foot is a condition that weakens the upward tilt of the foot. This leads to an altered walking style (gait). Brockett and Chapman (2016) determined that a normal gait had a vertical foot tilt profile ranging from 17.4° to -17.9° [1], and Shin et al. (2023) determined a vertical tilt profile ranging from 15° to -18° of the foot for a drop foot patient [2]. A treatment for drop foot involves the use of Ankle-Foot orthoses (AFOs), assisting the movement of the foot [3]. Current evaluation of ankle-foot devices largely relies on the BS EN ISO 22675:2016 standard, which specifies test methods and requirements for ankle-foot prostheses and defines a testing device as a vertically actuated unit stepping on a rotating platform, aimed at assessing non-articulating foot units [4]. However, ISO 22675 does not adequately address the testing of articulating AFOs, presenting a significant gap in standardized evaluation methods for articulating AFOs. To address this, a new testing device capable of evaluating articulating AFOs is needed. In this study, a digital prototype of such a device was developed using a "Design by Simulation" approach. The prototype design adhered to relevant requirements from ISO 22675, enabling virtual testing and iterative refinement before manufacturing. The design requirements were defined based on vertical foot tilt profiles observed in literature [1,2], as well as maximum deformation and dynamic requirements specified in ISO 22675 [4].

2. Objectives and Scope

This study aims to analyse a digital prototype designed as a testing device for articulating AFOs. The objectives are to identify component properties that satisfy the deformation requirement and to evaluate whether the prototype meets the dynamic requirements. Specifically, this involves assessing the mechanical properties of the four main structural components and examining the prototype's ability to replicate normal and drop foot gaits. Additionally, the dynamic evaluation addresses the transfer of cyclic and static load profiles specified by ISO 22675 within the prototype. This study is limited to virtual analyses.

3. Methodology and Computational Approach

A simulation approach was employed, consisting of three main phases: formulating design specifications, computational analysis, and iterative refinement, adapted from Budynas and Nisbett [5]. A foot length of 320 mm, being the maximum in ISO 22675, was selected, assuming it would cause the greatest deformation [4]. Brockett and Chapman (2016)'s and Shin et al. (2023)'s normal and drop foot gait vertical foot tilt profiles were used during some of the multibody dynamic (MBD) simulations [1,2]. ISO 22675 establishes the cyclic and static loading and deformation requirements [4]. The measurements included the total maximum deformation of the four structural components, the angle between the prosthetic foot's sole and the horizontal plane during replication of the two vertical tilt gait profiles, and the comparison between the input load profiles (cyclic and static) and the load at the ankle joint. Figure 1 shows a digital prototype concept of the testing device.

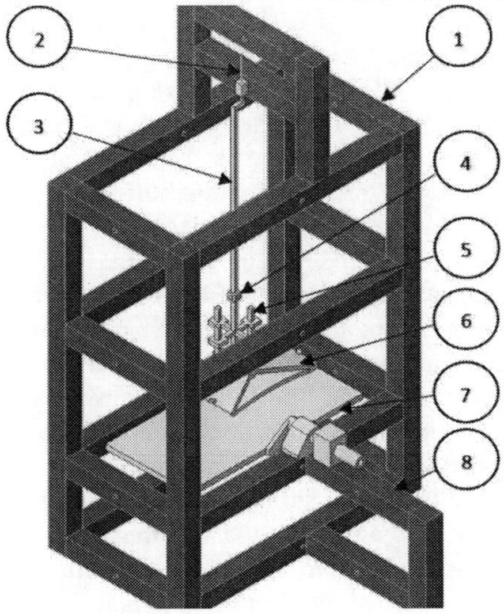

Fig. 1 Digital prototype concept of the testing device, comprising the frame (1), linear actuator 1 (2), prosthetic leg (3), guiding bracket (4), linear actuators 2 (5), prosthetic foot (6), rotating platform (7), and rotational mechanism connection point (8). The overall dimensions of the prototype were approximately 560 mm × 650 mm × 1200 mm.

MSC Apex 2024.1.1 was used for the finite element (FE) analyses, using isotropic material models, for evaluating the structural components' (frame, leg, foot, and rotating platform) deformations. The models employed 1D elements for the frame and leg, and 2D elements for the foot and platform. Various beam elements, surface thicknesses, and material properties were used to identify the combination that minimises deformation. The node with the maximum deformation was used in each case. Rahman and Alam used a similar approach in their research [6]. Three FE simulations were conducted for each component: static heel loading, static forefoot loading, and cyclic loading. This resulted in a total of twelve FE simulations. After determining the above-mentioned properties, mesh convergence studies were performed for each component. Convergence was considered achieved when the displacement of the node adjacent to the one exhibiting the maximum displacement varied by less than 1% upon mesh refinement. Figures 2 through 5 illustrate the FE models for the frame, leg, foot, and rotating platform, respectively. The converged mesh sizes were 40 mm (391 elements) for the frame, 10 mm (66 elements) for the leg, 9 mm (912 elements) for the foot, and 12 mm (635 elements) for the rotating platform.

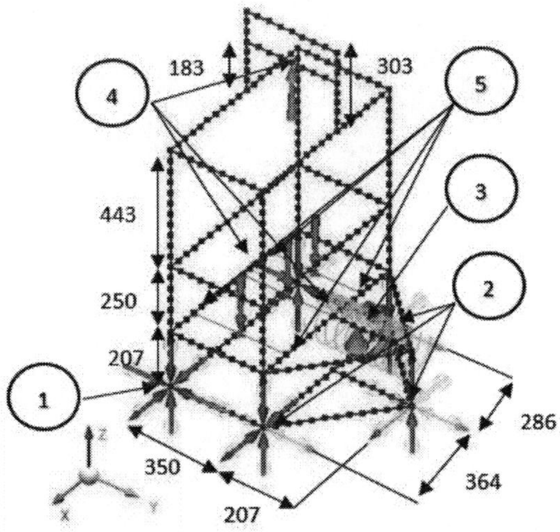

Fig. 2 Finite element model of the frame (dimensions in mm) showing boundary conditions: support (1) restricting translation in all directions; supports (2) restricting x- and y-translation to prevent rigid body motion. A 213 Nm torque was applied at (3) to simulate the rotational mechanism's reaction torque. Maximum cyclic loads were applied at actuator 1's connection point and the platform connections (4). Static vertical loads were applied at (4), being 4.5 kN (linear actuator 1), 1.5 kN (heel) and 2.2 kN (forefoot). The front and rear supports of the rotating platform exerted 3.0 kN and 2.3 kN vertically, positioned 120 mm and 508 mm from the front, respectively.

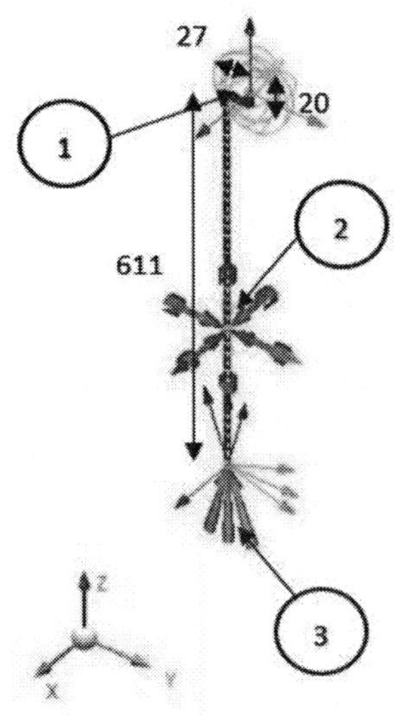

Fig. 3 Finite element model of the prosthetic leg (dimensions in mm). Node (1) was constrained to allow rotation but prevent translation, simulating a ball-and-socket joint, required by ISO 22675. Position (2), 120 mm from the base, permits translation only along the z-axis to model corrective loads. Loads applied at (3) represent maximum static and cyclic conditions.

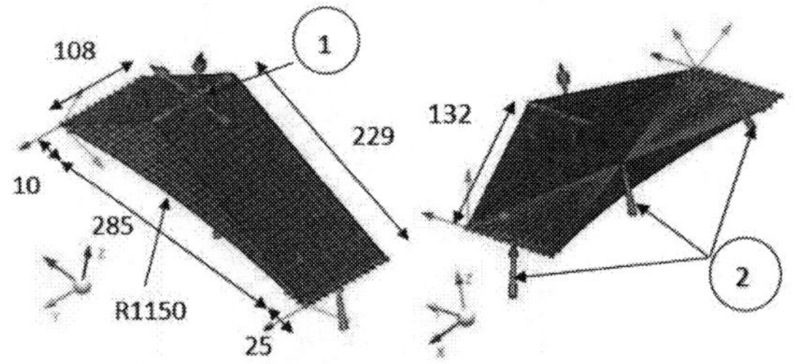

Fig. 4 Finite element model of the prosthetic foot (dimensions in mm) with a width of 108 mm, based on Jurca et al. (2019) [7]. A fully fixed constraint at (1) represents the rigid leg-foot joint, and static and cyclic loads were applied at the sole (2).

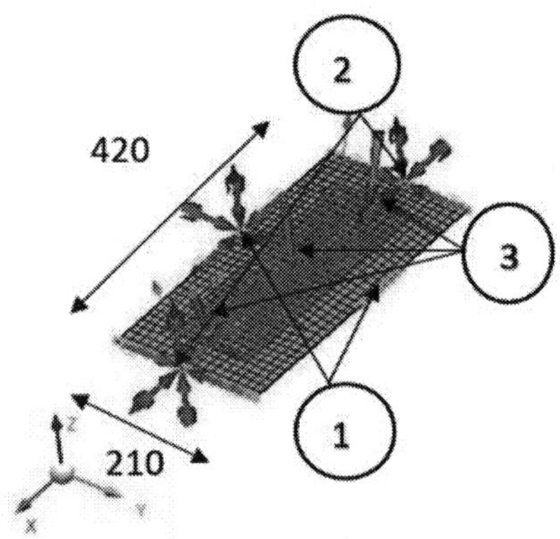

Fig. 5 Finite element model of the rotating platform (dimensions in mm) showing translation constraints at points 46 mm from the centre on the long sides (1) and front/rear supports during static loading (2). The prosthetic foot centre aligns with the platform centre. Cyclic and static loads were applied across the platform surface (3).

Hexagon Adams Modeler 2023.2 was used for MBD simulations as it allows for evaluating the dynamics of a system using simulations. The MBD model was constructed to match the dimensional requirements specified by ISO 22675 [4], without incorporating the component dimensions from the FE analyses and utilising rigid body models exclusively. Thus, focusing on dynamic behaviour rather than structural strength analysis. The joint selections were made to best resemble the physical interaction between components. The frame, ground, rotating platform, linear actuator 1's bracket, and platform rotation support brackets were interconnected using fixed joints. Linear actuator 1 attached to its bracket via a cylindrical translation joint and to the leg with a spherical joint. The leg guiding bracket was connected to the leg using a rectangular translational joint and to the frame with springs. The rotating platform was connected to its brackets through rotational joints, and a contact force was defined between the foot and the rotating platform. A fixed joint linked the leg and foot for analyses of ISO 22675's cyclic and static loading [4]. This configuration resulted in the model having 14 degrees of freedom. The springs were assigned a stiffness of 100 N/mm and a damping coefficient of 1 Ns/mm. Contact properties consisted of the Impact method for normal force, with a stiffness of 100 N/m, a force exponent of 2, a damping of 30 Ns/mm, and a penetration depth of 0.1 mm. The loads were recorded at linear actuator 1 and the fixed ankle joint. Figure 6 illustrates a modification to the fixed ankle joint, enabling articulation and the replication of the normal and drop foot gait. The gait profiles were assigned to the rotational platform and linear actuators 2. This modified ankle joint increased the model's degrees of freedom to 32. The springs connecting linear actuators 2 to the foot were assigned a stiffness of 5 N/mm and damping of 1 Ns/mm, except for the spring connecting the front part of the foot to its actuator, which had a stiffness of 1 N/mm and damping of 1 Ns/mm. The articulation of the analytical universal joint was measured and used to analyse the simulations.

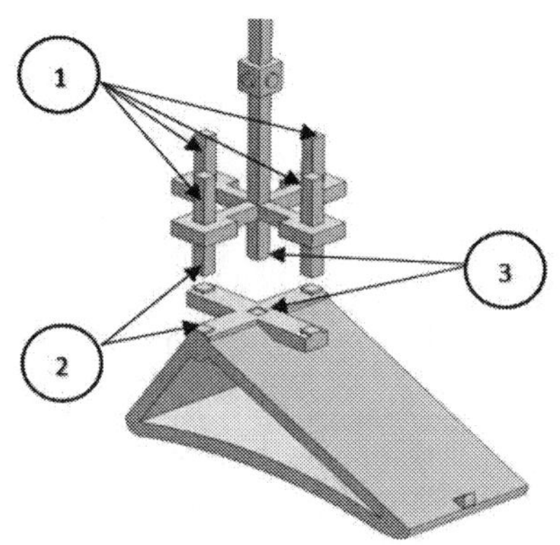

Fig. 6 Modified ankle joint model for multibody dynamic simulations: linear actuators 2 (1) connected to the leg via rectangular translation joints and to the foot through springs (2) for limited movement. The foot and leg were linked by an analytical universal joint (3) during motion testing.

4. Results and Discussion

FE simulations modelled the frame with 50×50×2 mm square tubing, the prosthetic leg with 25×38×2 mm rectangular tubing, the foot as 15 mm thick plates, and the rotating platform as a 6 mm thick plate. All components were assumed to be mild steel ($E = 200$ GPa, $\nu = 0.3$, $\rho = 7850$ kg/m^3) [5]. Static forefoot loading caused a maximum deformation of 1.83 mm, below the 2 mm limit [4]. The Root Mean Square Error (RMSE) was used to compare simulated and literature gait data, quantifying average deviation in degrees. Figure 7 shows that the simulated vertical foot tilt for a normal gait closely follows experimental profiles [1], with an RMSE of 5.36° (~13% of data range) and a maximum deviation of 4.5°, likely due to input parameter overestimation. Despite this, the prototype replicated normal gait vertical foot tilt.

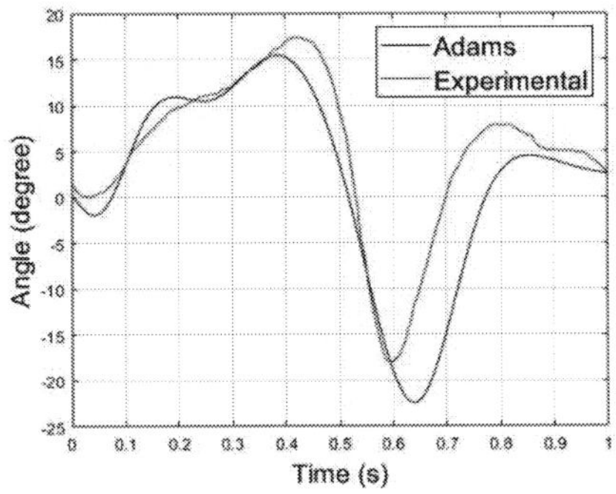

Fig. 7 Simulated vertical foot tilt profile for a normal gait compared to an experimental profile from literature [1].

Figure 8 compares the simulated vertical foot tilt for a drop foot gait with literature data [2], yielding an RMSE of 3.61° and a maximum deviation of 2.4°, demonstrating the prototype's ability to replicate a drop foot gait vertical foot tilt profile.

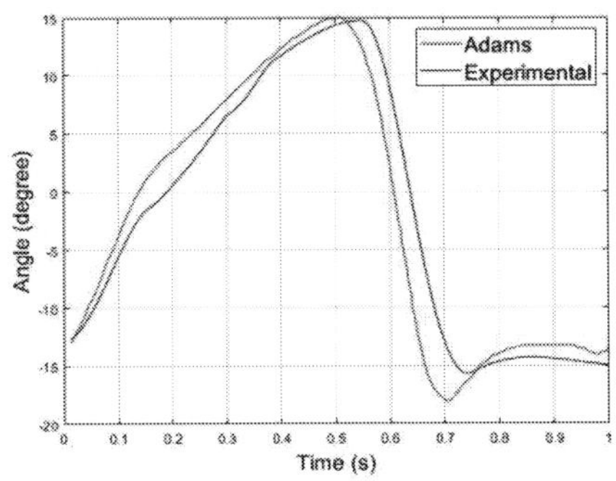

Fig. 8 Simulated vertical foot tilt profile for a drop foot gait compared to an experimental profile from literature [2].

MBD simulations using ISO 22675's cyclic and static load profiles as linear actuator 1's inputs showed RMSEs of 18.3 N and 32.7 N compared to the loads measured at the ankle joint, with a maximum deviation of 46.8 N, indicating load transfer between components with minimum deviation [4]. Figures 9 and 10 display these comparisons.

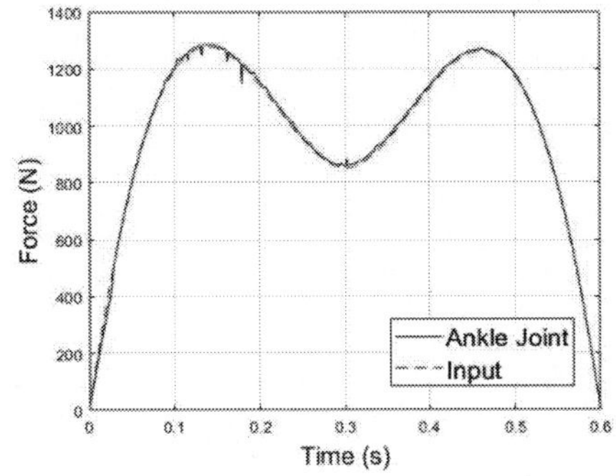

Fig. 9 ISO 22675's cyclic load input profile applied to linear actuator 1 compared to the load measured at the ankle joint [4].

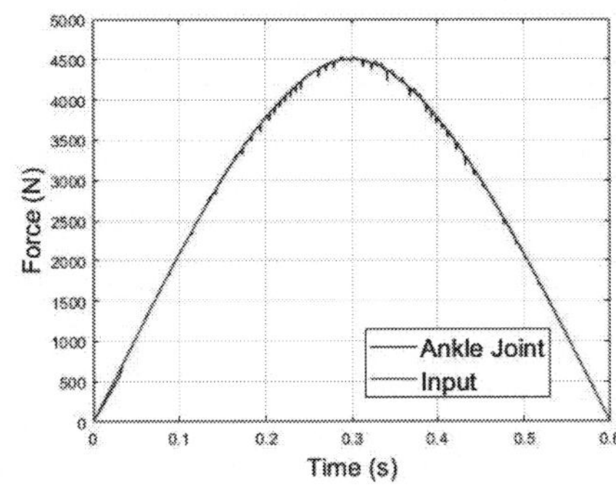

Fig. 10 ISO 22675's Static load input profile applied to linear actuator 1 compared to the load measured at the ankle joint [4].

5. Conclusion and Future Work

This study evaluated a digital prototype for AFO testing via simulations, showing a maximum structural deformation of 1.83 mm. The prototype replicated normal and drop foot vertical tilt profiles with root mean square errors of 5.36° and 3.61°, and maximum deviation of 4.5°. Load transfer accuracy was confirmed with root mean square errors of 18.3 N and 32.2 N, and a maximum deviation of 46.8 N. These results demonstrate the prototype's capability to replicate gaits and meet relevant mechanical requirements. Future work will include experimental validation and the development of a control system to replicate simulated inputs.

References

[1] Brockett, C.L. & Chapman, G.J. 2016. Biomechanics of the ankle. Orthop Trauma, 30(3):232-238. DOI: 10.1016/j.mporth.2016.04.015

[2] Shin, W., Nam, D., Ahn, B., Kim, S.J., Lee, D.Y., Kwon, S. & Kim, J. 2023. Ankle dorsiflexion assistance of patients with foot drop using a powered ankle-foot orthosis to improve the gait asymmetry. J Neuroeng Rehabil, 20(1):140. DOI: 10.1186/s12984-023-01261-1

[3] Alam, M., Choudhury, I.A. & Mamat, A.B. 2014. Mechanism and Design Analysis of Articulated Ankle Foot Orthoses for Drop-Foot. The Scientific World Journal, 2014:1-14. DOI: 10.1155/2014/867869

[4] BS EN ISO 22675, (2016) Prosthetics – Testing of ankle-foot devices and foot units – Requirements and test methods, London: British Standards Institution.

[5] Budynas RG, Nisbett JK (2020). Shigley's Mechanical Engineering Design. 11th ed. New York: McGraw-Hill Education; p. 5-8

[6] Rahman, N.u. & Alam, M.N. 2018. Finite element modelling to assess the effect of position and size of the piezoelectric layer of a hybrid beam. Latin American Journal of Solids and Structures, 15, Available: https://www.scielo.br/j/lajss/a/nLKJDkP7ggVYK7cVf55Y9cv/?lang=en

[7] Jurca, A., Žabkar, J. & Džeroski, S. 2019. Analysis of 1.2 million foot scans from North America, Europe and Asia. *Scientific Reports*, 9(1):19155. DOI: 10.1038/s41598-019-55432-z

14th South African Conference on Computational and Applied Mechanics (SACAM 2025)
21-23 January 2025, Johannesburg, South Africa

Propulsion Motor Sizing for a Battery-Powered New Energy Vehicle: A Retrofitting Approach using Vehicular Dynamics

Nicholas Xavier[a,*], Antoine F. Mulaba-Bafubiandi[b], Pathmanathan Naidoo[a]

[a]University of Johannesburg, PO Box 524, Auckland Park, Johannesburg, 2006, South Africa

[b]University of Johannesburg, PO Box 2028, Doornfontein, Johannesburg, 2094, South Africa

**corresponding author email: nicholasixavier317@gmail.com*

Abstract

The journey to carbon-neutrality through the adoption of new energy vehicles in the global automobile markets has never been this clear. While developed countries smoothly transition to new energy vehicles, developing countries are delayed. The leading contributing factors include demographic economics, the ongoing energy crisis, and unstructured public transportation systems. Coupled with the challenge of what happens to the already manufactured and existing fossil fuel-powered vehicles, a vehicle retrofitting model is a necessity. Although the vehicle retrofitting model is viewed as a model that can be used to design and develop a common base platform for converting internal combustion engine vehicles, homologation requirements, especially those relating to the powertrain design, are still mandatory.

This work employs vehicle dynamics to size a propulsion unit for a retrofitted battery-powered new energy vehicle. To achieve this, a moderate fidelity drive cycle is developed based on a test scenario case then used to compute the speed and torque of the propulsion unit, as well as its power. Findings have shown that the propulsion unit required must have its power ranging between 13-80 kW, a speed between 850-1200 rpm, and its torque ranging between 130-200 Nm. The selection sizing effort applied a safety factor of 1.5 to select a commercially available propulsion motor. Further work looks at sizing other aspects of the vehicle powertrain, like the battery pack.

***Keywords*:** New Energy Vehicles, Zero Emission Vehicles, Electrification Retrofitting, Electric Drivetrain, Electric Powertrain, Electric Propulsion Motor

1. Introduction

1.1. Background, Motivation, and Scope

The nascent of the engineering marvels of new energy vehicles, catalyzed by the global environmental qualms, has heightened hopes for a sustainable future. While a smooth transition is observed in developed countries, because of demographic economics, the ongoing energy crisis, and overreliance on paratransit transportation systems, developing countries are delayed.

Coupled with the challenge of what happens to the already manufactured and existing fossil fuel-powered vehicles, a vehicular electrification retrofitting model is a necessity. This work therefore employs vehicular dynamics to size the propulsion motor of a gasoline-powered Toyota Avanza – to – a retrofit battery-powered new energy vehicle (shown in Fig. 1).

Fig. 1. The Retrofitting ambition of the Project

Electrification retrofitting insights based on literature have been explored, addressing battery and powertrain sizing issues [1-3], business models and case studies [4, 5], techno-economic studies [6], and homologation-centered approaches associated with the retrofit model, specifically for Sub-Saharan Africa [1, 8]. However, few have delved into the nitty-gritty of sizing propulsion motors considering their characteristic torque-power performances based on the retrofit drive cycle propulsion requirements.

1.2. Contribution

This work contributes to vehicle electrification by making use of driving cycle data to efficiently mimic the performance characteristics of a conventional vehicle to a retrofit vehicle to size its propulsion motor.

2. Methodology

The methodology adopted in this work is based on the performance analysis of battery-powered vehicles

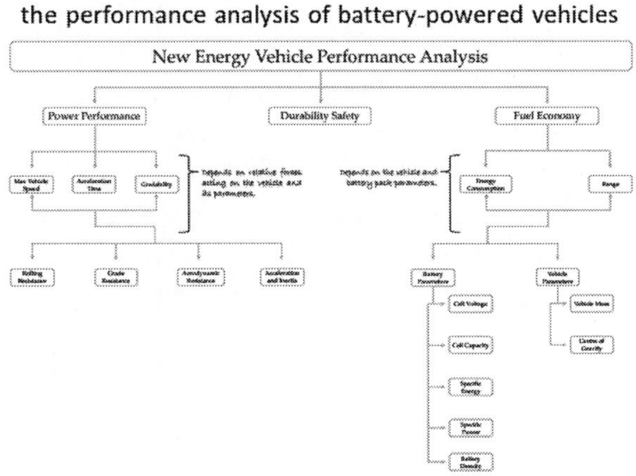

Fig. 2. The Performance Analysis Chart for New Energy Vehicles

shown in Fig. 2. To replicate the performance of the conventional energy vehicle to the new energy vehicle, driving cycle data is collected and applied vehicle dynamics to size the propulsion motor.

2.1. Method Overview

The propulsion motor sizing method is summarized in the follow figure:

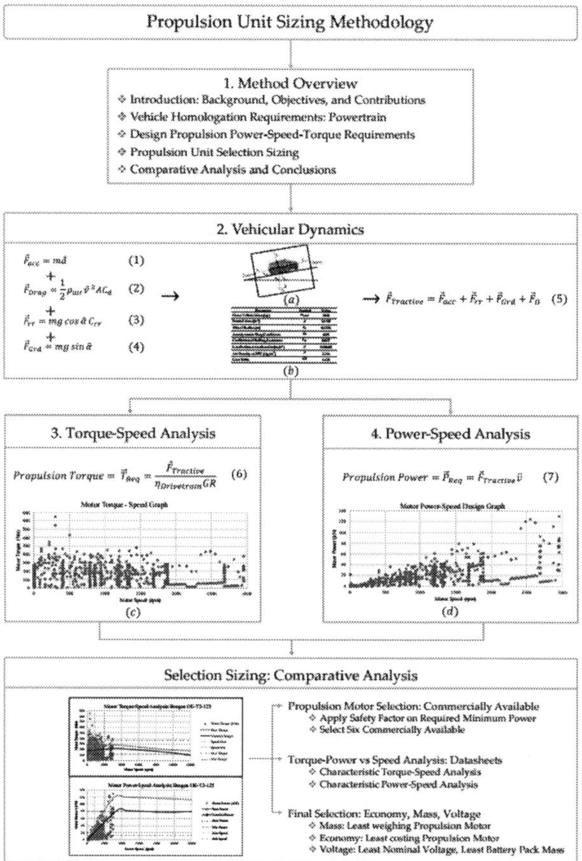

Fig. 3. The Method Summary adopted for sizing the Propulsion Unit

2.2. Field Data: Drive Cycle Data

Drive Cycle data was collected as field data based on an intercampus shuttling test case, combining speed data in Fig. 4(a), acceleration data (+ve) in Fig. 4(b), and slope variations of the route travelled in Fig. 4(c).

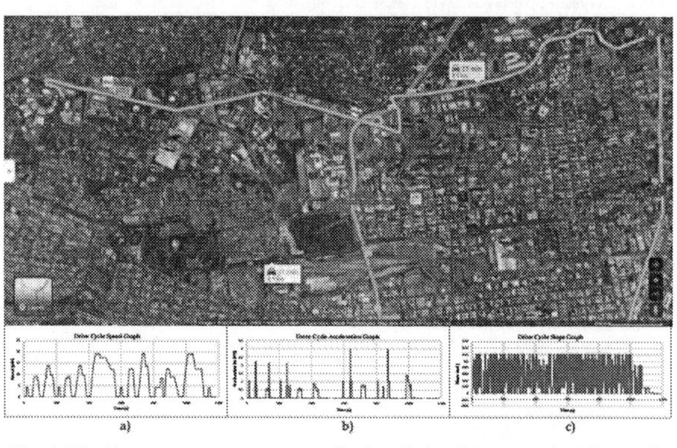

Fig. 4. The intercampus route travelled and the driving cycle (a) speed data, (b) acceleration data, and the (c) slope variations of the route

3. Homologation Requirements: Vehicle Modifications and Powertrain

Homologation requirements based on the relevant rules and regulations dictate that the retrofitting exercise pays cognizance to two things to pass for road-use:

- The gross mass (GVM) of the retrofit vehicle must not exceed the original gross mass of the original, this will mean, the mass of the retrofit powertrain must relate to the mass of the removed powertrain based on the following formula:

$$m_{Powertrain_{Max}} = m_{Veh_{Taremass}} - m_{Veh_{Post-Strp}} \quad (1)$$

- Modifications on the vehicle chassis are not allowed, this includes any drilling, welding, and cutting techniques.

These two homologation requirements are adhered to during the design and sizing effort to ensure compliance for the vehicle post-retrofitting.

4. Propulsion Power-Torque Requirements: Vehicle Dynamics

The field data and vehicle parameters (Fig. 3(a) & Fig. 3(b)) are used as inputs for vehicular dynamics. Speed data in Fig. 4(a) is used to compute the drag using equation 2 of Fig. 3, acceleration data in Fig. 4(b) is used to compute the vehicle's acceleration force using equation 1 of Fig. 3, and the slope data in Fig. 4(c) is used to calculate the grade and rolling resistance of the vehicle using equations 3 and 4 of Fig. 3.

With the use of equations 6 and 7 of Fig. 3, the propulsion requirements of the vehicles are obtained using the computed traction force calculated using equation 5. All the propulsion requirements are summarized using the following figure, where the minimum and maximum requirements, obtained quantitatively, are shown along with their graphical presentations:

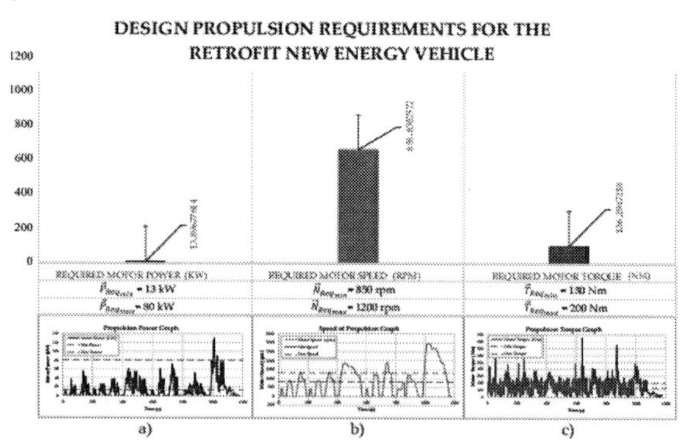

Fig. 5. Minimum and Maximum Propulsion a) Power, b) Speed, and c) Torque of the retrofit vehicle

5. Propulsion Unit Selection Sizing

Propulsion Unit sizing is conducted using the minimum and maximum propulsion requirements of the retrofit vehicle in Fig. 5. Applying a safety of 1.5 ($n_s = 1.5$), the selection of commercially available propulsion motors is based on the following propulsion requirements; 120 kW, 1800 rpm, and 300 Nm.

Fig. 6 shows the selected motors and Table 1 summarizes their design specifications.

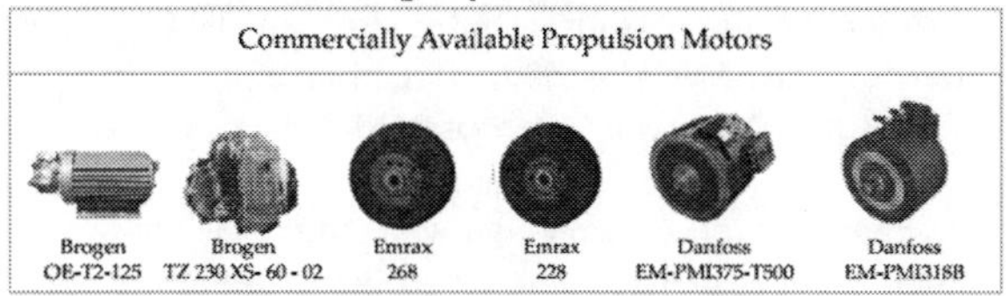

Fig. 6. Commercially available Propulsion Motors considered

Table 1. The Design Specifications of the Considered Propulsion Motors

Motor		Motor Parameter							
Type	Name	Max Power (kW)	Max Torque (Nm)	Max Speed (rpm)	Rated Speed (rpm)	Voltage (V)	Energy Efficiency (%)	Mass (kg)	Cost (Rands)
Permanent Magnet Synchronous Motors	Brogen OE-T2-125 [9]	120	280	12000	1317	240~460	92	85	55 852.80
	Brogen TZ 230 XS- 60 – 02 [10, 11]	120	400	8500	3000	270~540	85	68	40 286.83
	Emrax 228 [12]	124	220	6500	5500	50~830	92-98	13.5	81 905.34
	Emrax 268 [13]	210	500	5400	2600	100~830	92-98	22.3	84 022.00
	Danfoss EM-PMI375-T500 [14]	158	400	4000	1100	270~500	96	172	96 391.17
	Danfoss EM-PMI318B [15]	250	700	5500	3500	250~440	95	81	89 827.57

The selected propulsion motors are scrutinized based on their characteristic Torque and Power curves vs the retrofit vehicle requirements translated from the driving cycle data in Fig. 3 (c) and (d).

5.1. Torque-Speed Analysis and Brief Discussions

An analysis based on the characteristic Torque-Speed curves of the selected propulsion motors compared to the driving cycle Torque-Speed curve in Fig. 3(c) is adopted in this work. The goal is to ensure that the selected propulsion motors can meet the torque requirements of the retrofit vehicle based on the actual drive experience on the road, that they do not over- or under-deliver the torque demands.

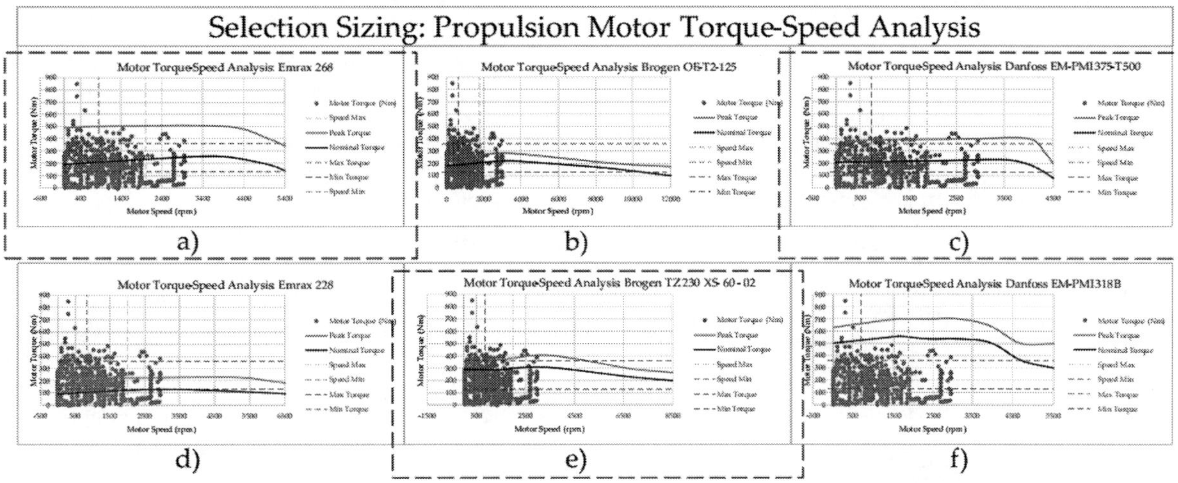

Fig. 7. Propulsion Motor Selection Sizing based on their characteristic Torque-Speed behaviour compared with the actual driving Torque-Speed demands of the retrofit vehicle (scarlet scatterplot)

Fig. 7 combines the Torque-Speed performance graphs of all the selected propulsion motors in the following order: a) Emrax 268, b) Brogen OE-T2-125, c) Danfoss EM-PMI375-T500, d) Emrax 228, e) Brogen TZ230-XS-60-02, and f) Danfoss EM-PMI318B.

Each graph vividly shows the Torque-Speed characteristic scatter plot according to the drive cycle data in scarlet, the minimum required torque in a blue dotted line plot, the maximum required torque in a green dotted line plot, the minimum required speed in a light blue dotted line plot, and the maximum speed in a yellow dotted line plot. The peak performance characteristic torque curve is displayed in a solid grey plot while the nominal characteristic torque performance curve is displayed in solid black, these performance curves were found in datasheets referenced to in Table 1.

Visually, technically performing propulsion motors will have their nominal and peak torque curves above (or within the range of) the minimum and maximum required torque dotted line plots respectively. This pattern is observed in the highlighted graphs (dotted scarlet) of Fig. 7(a), Fig. 7(c), and Fig. 7(e). Fig. 7(b) and Fig. 7(d) are clearly showing a Torque undersize while a pronounced speed gap is observed, especially for (b), this is not bad in terms of performance, but it is an oversize for this application. Fig. 7(f) is clearly an oversize based on both torque and speed.

5.2. Power-Speed Analysis and Brief Discussions

The following figure combines the Power-Speed performance graphs of all the selected propulsion motors in the same order and styling legend as adopted in Fig. 7. Based on the selection sizing analysis in Fig. 8, figures (a), (b), (c), and (e) are observed delivering based on the power requirements.

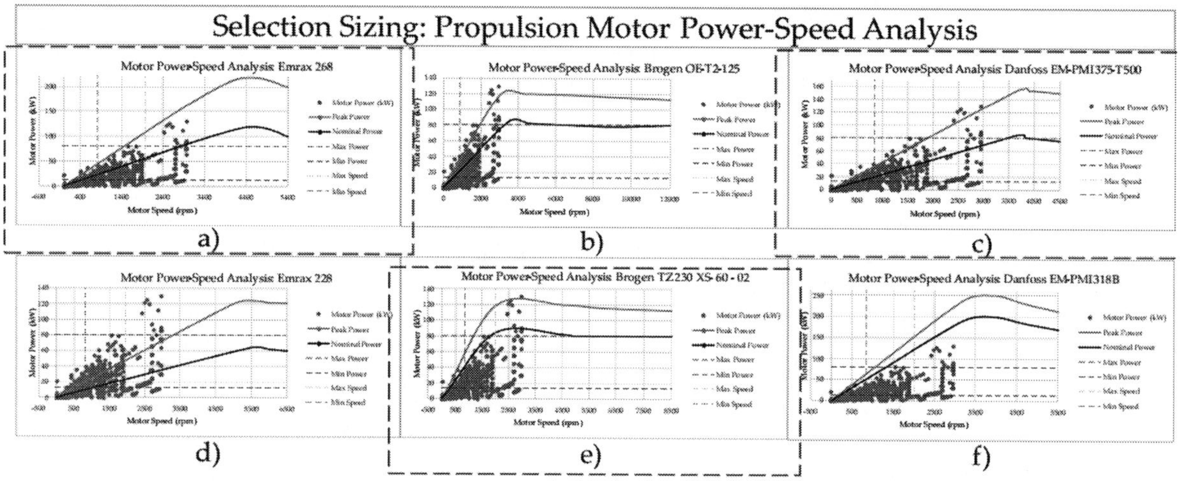

Fig. 8. Propulsion Motor Selection Sizing based on their characteristic Power-Speed behaviour compared with the actual driving Power-Speed demands of the retrofit vehicle (scarlet scatterplot)

Fig. 8(d) demonstrates an undersize while Fig. 8(f) demonstrates an oversize on power. However, based on both torque and power analyses, motor (a), (c), and (e) are chosen for final selection.

The final selection is based on three things, voltage requirements, mass, and price. A heavy and high voltage motor will mean more added weight to the vehicle, and a pricey one will strain the project budget.

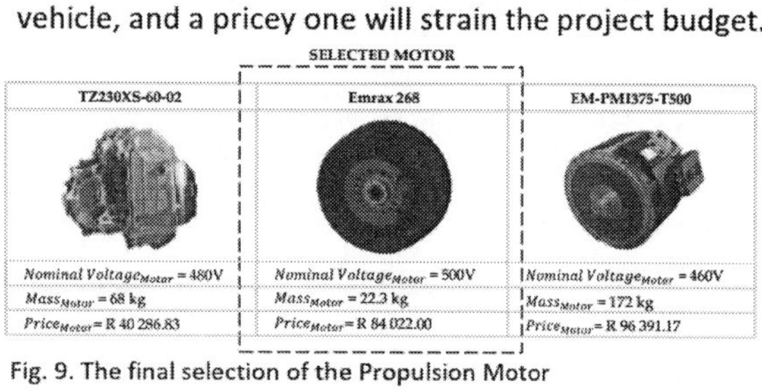

Fig. 9. The final selection of the Propulsion Motor

Based on this analysis then, Fig. 9 shows how the Emrax 268 motor was selected as the most suited motor for this application.

Conclusions

An engineering propulsion unit sizing method using drive cycle data to translate the performance characteristics of a conventional energy vehicle to a new energy vehicle using vehicular dynamics for an electrification retrofitting application is proposed in this work. The sizing effort was a success, meeting the objectives of the work. The Emrax 268 propulsion motor was selected for this application, having not only met the technical performance requirements, but it also met the homologation mass constraints for this vehicle model [16].

References

[1] S. Lacock, A. A. du Plessis and M. J. Booysen, "Using Driving-Cycle Data to Retrofit and Electrify Sub-Saharan Africa's Existing Minibus Taxis for a Circular Economy," *World Electric Vehicle Journal,* vol. 296, no. 14, pp. 1-24, 2023.

[2] E. Tara, S. Shahidinejad, S. Filizadeh and E. Bibeau, "Battery Storage Sizing in a Retrofitted Plug-in Hybrid Electric Vehicle," *IEEE Transportation Vehicle Technology,* no. 59, pp. 2786-2794, 2010.

[3] S. Kaleg, A. Hapid and M. R. Kumia, "Electric Vehicle conversion based on Distance, Speed and Cost Requirements," *Energy Procedia,* no. 68, pp. 446-454, 2014.

[4] F. Hoeft, ". Internal combustion engine to electric vehicle retrofitting: Potential customer's needs, public perception and business model implications," *Transportation Research Interdisciplinary Perspectives,* vol. 9, no. 100330, 2021.

[5] N. Xavier, "Calculating the Aerodynamic Drag Coefficient of a Toyota Avanza Car CAD Model using CFD Analysis," in *2023 31st Southern African Universities Power Engineering Conference (SAUPEC)*, Johannesburg, 2023.

[6] A. Anosike, H. Loomes, C. K. Udokporo and A. Garza-Reyes, "Exploring the challenges of electric vehicle adoption in final mile," *International Journal of Logistics Research and Applications parcel delivery,* vol. 26, pp. 683-707, 2021.

[7] Mpumalanga Green Cluster Agency, "Electric vehicle retrofitting: A case study of Electric Powered Vehicles Africa," 2023. [Online]. Available: https://mpumalangagreencluster.co.za/wp-content/uploads/2022/11/MPGA-Case-Study-EV-Retrofitting.pdf. [Accessed 29 September 2024].

[8] N. Xavier, A.F. Mulaba-Bafubiandi, P. Naidoo, and Rampersad, "Powertrain Sizing for a Retrofit Zero Emission Vehicle using Vehicular Dynamics: An Intercampus Test Scenario Case". In Proceedings of the 33rd Southern African Universities Power Engineering Conference, Institute of Electrical and Electronics Engineers Inc., 33rd Southern African Universities Power Engineering Conference, SAUPEC 2025, Pretoria, South Africa.

[9] Brogen Motors, "Passenger Cars Powertrain Solutions: OE-T2-125," Brogen Motors, [Online]. Available: https://brogenmotors.com/application/passenger-car/#VCGX5C. [Accessed 08 November 2024].

[10] Brogen Motors, "Passenger Cars Powertrain Solutions: TZ 230 XS- 60 - 02," Brogen Motor, [Online]. Available: https://brogenmotors.com/application/passenger-car/#VCGX5C. [Accessed 08 November 2024].

[11] everythingPE, "Everything Power Electronics: EV Traction Motor by Brogen - TZ 230 XS- 60 - 02," everthingPE, [Online]. Available: https://www.everythingpe.com/products/ev-traction-motors/brogen/1069-2098-tz-230-xs-60-02. [Accessed 08 November 2024].

[12] EMRAX INNOVATIVE E-MOTORS, "EMRAX E-MOTORS: 228 (124kW | 230Nm)," EMRAX Innovative E-Motors, [Online]. Available: https://emrax.com/wp-content/uploads/2024/02/EMRAX_228_datasheet_v1.5.pdf. [Accessed 08 November 2024].

[13] EMRAX INNOVATIVE E-MOTORS, "EMRAX E-MOTORS: 268 (210kW | 500Nm)," EMRAX Innovative E-Motors, [Online]. Available: https://emrax.com/wp-content/uploads/2024/02/EMRAX_268_datasheet_v1.5.pdf. [Accessed 08 November 2024].

[14] Danfoss, "High Voltage Multipurpose Motors - Electric Motors: EM-PMI375-T500," Danfoss, [Online]. Available: https://assets.danfoss.com/documents/430739/AI269157472701en-000202.pdf. [Accessed 08 November 2024].

[15] Danfoss, "High Voltage Multipurpose Motors - Electric Motors: EM-PMI318B," Danfoss, [Online]. Available: https://assets.danfoss.com/documents/157056/AI362410058222en-000102.pdf. [Accessed 08 November 2024].

[16] N. Xavier and A. F. Mulaba-Bafubiandi, "Techno-economic Selection Sizing of a Propulsion Motor for a Retrofit Zero Emission Vehicle," November 2024. [Online]. Available: http://dx.doi.org/10.13140/RG.2.2.12549.15847.

Modelling of bubble population statistics in slag foaming using the dynamic multiple marker method

Quinn G. Reynolds[a,b,*], Oliver F. Oxtoby[c], and Rooi Mamburu[a,d]

[a]Mintek, Private Bag X3015, Randburg 2125, South Africa

[b]Stellenbosch University, Private Bag X1, Matieland 7602, South Africa

[c]ENGYS, John Archer Way, London, SW18 3SX, UK

[d]University of Johannesburg, PO Box 524, Auckland Park 2006, South Africa

*corresponding author email: quinnr@mintek.co.za

Abstract

Dispersed multiphase flows, in which one or more immiscible fluids are intermixed with each other, occur in many natural phenomena as well as chemical and metallurgical engineering processes. Dispersed flows are particularly challenging to simulate numerically as the effects of interest cover a wide range of scales in length, time, and the physical properties of the fluids involved. In the present paper, the novel dynamic multiple marker (DMM) computational fluid dynamics method is used to investigate bubble population dynamics in gas-liquid systems such as metallurgical slag foaming. Simulations of single bubbles are used to assess the effect of capillary Courant number on model stability, and determine appropriate modifications of gas-phase properties to mitigate parasitic current effects. The model is then applied to the dynamics of phase separation in systems containing large numbers of bubbles, comparing the evolution over time of different bubble size classes and the dependence of the behaviour on physical properties of the fluids present.

Keywords: VOF, multiphase flow, dispersed phase flow, thin film phenomena, metallurgy

1. Introduction

Metallurgical smelting furnaces are used to produce iron and steel, chromium, manganese, nickel, titanium dioxide, and a wide range of other industrial commodities. These furnaces typically operate at elevated temperatures (1500°C or higher) with at least two liquid phases inside the vessel – a light and viscous slag, consisting of metal oxides, and a dense alloy, consisting of a mixture of metallic products [1].

During normal operation the reduction reactions in the furnace convert metallic oxides to metals while continuously evolving gaseous products, and these form as small bubbles either dispersed through the slag phase or concentrated at the slag-alloy interface. In order to escape into the furnace freeboard to be extracted from the vessel, the gas bubbles must rise through the bath of molten process material and may accumulate in a low-density foam layer near the slag surface (see Fig. 1). If the growth rate of the foam is not controlled it can rapidly fill the freeboard spaces in the furnace and overflow from the vessel. This can cause injuries to personnel, equipment damage, and costly plant shutdowns.

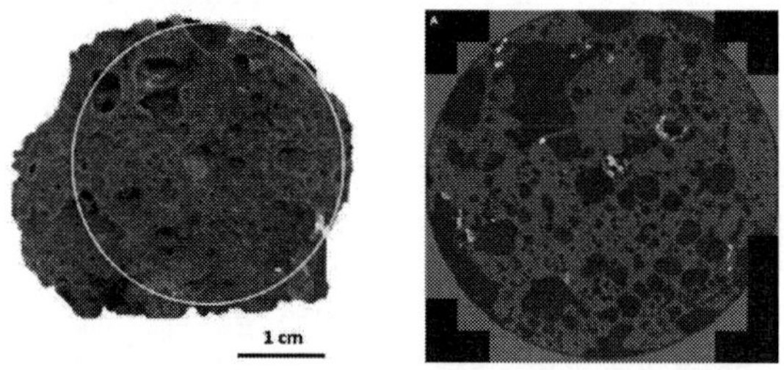

Fig. 1: (l) sectioned sample of solidified slag foam from an electric arc furnace smelting chromite ore, (r) electron backscatter image of circular area showing voids (dark grey)

In industrial metal production processes, slag foaming is often quantified in terms of its steady state behaviour. This is generally estimated using the foam index Σ, which has units of time and represents a ratio between the rate of gas generation by the process (represented by the superficial gas velocity) and the resulting equilibrium thickness of the foam layer. Σ is a useful indicator of a slag's propensity to foam and has been correlated to physical properties such as slag density, viscosity, surface tension, bubble diameter, and others by many authors [2], [3], [4]. Such empirical correlations do however have limitations – their applicability is usually constrained to a particular process and slag type, and the foam index also fails to capture the transient growth and decay dynamics of slag foam layers. Transient behaviour may be particularly significant during the initial phases of uncontrolled foaming events.

Slag foaming is part of the field of dispersed-phase multiphase fluid flow, in which one of the phases is present as small bubbles or droplets within the other.

Numerical modelling of such flows for industrial applications is challenging, and is traditionally performed using unresolved methods such as Eulerian (in which the dispersed phase is approximated as a continuum) or mixed Eulerian-Lagrangian (where the dispersed phase is represented by a collection of solid particles which exchange forces with the continuum fluid) [5], [6].

In the present work a meso-scale modelling approach with the resolved Eulerian dynamic multiple marker (DMM) method [7] is used, which permits detailed study of the dynamics of bubble motion and coalescence in slag foam layers. Numerical calculation of bubble population statistics and their evolution in time is made possible with this approach, and this can in turn be used in developing new empirical closures for the unresolved methods described above.

2. Modelling Methodology

In the DMM method, standard multiple marker algorithms are combined with a contact detection and dynamic marker phase reassignment step inside a volume-of-fluids (VOF) framework to facilitate the direct, resolved modelling of fluid flow in both the dispersed and continuum phases. This approach eliminates the problem of premature coalescence when adjacent bubbles of the dispersed phase approach each other closer than the computational mesh resolution, while still retaining acceptable performance for problems with large numbers of bubbles. It also implicitly allows natural breakup mechanics for large bubbles. When bubbles are in contact at the VOF mesh level, the thin film of continuum fluid between them (μm or less) evolves according to a sub-grid model. In the present work the standard Reynolds equation from lubrication theory is used, and is transformed from film coordinates to VOF mesh coordinates to permit solution of the film component on the same computational mesh as the bulk fluid dynamics. Once the film thickness between adjacent bubbles drops below a certain threshold, coalescence of the two bubbles is triggered by assigning them to the same marker phase.

The primary novelty and advantage of the DMM method is the computationally efficient extension of the multiple marker algorithm to problems with very large numbers of bubbles or droplets of the dispersed phase, and the ability to model natural bubble break-up and coalescence dynamics without explicit closures. Full details of the DMM method including a description of the contact-detection and marker phase reassignment algorithms, as well as the sub-grid film drainage model, are available in earlier publications [7], [8]. Validation of the DMM method against well-established empirical slag foaming correlations and laboratory experiment data presented in [2], [3], and [4] was conducted in [8], with acceptable agreement found between modelled and calculated slag foaming indices for a range of cases.

One of the prevalent problems with VOF and related methods is the presence of parasitic currents, which are unphysical flow artifacts arising from the discretisation of surface tension forces at the phase interfaces [8]. These artifacts are greatly exaggerated in cases where one or more of the phases has low density, or the surface tension between the phases is high. Unfortunately slag foaming falls into both of these categories, and any means of mitigating the parasitic current noise is therefore desirable.

One approach to which has been explored in earlier work is to artificially increase the density of the gas phase [9]. This can however only be done as long as the gas density remains negligible in comparison to that of the liquid, otherwise the bulk flow behaviour will be impacted. In the present study the effect of increasing both the density and viscosity of the dispersed phase is examined.

In many problems involving multiphase flow with thin films, parasitic current noise can also be reduced by considering the capillary Courant number Co_c, which is determined by the velocity of capillary waves at the phase interface [10]. This poses an additional stability constraint on the numerical time step size Δt as shown in equation (1).

$$\Delta t = \left(\frac{\rho_a + \rho_b}{2\pi\sigma}\right)^{\frac{1}{2}} \Delta x^{\frac{3}{2}} \tag{1}$$

Here, ρ_i are the fluid densities, σ is the interfacial surface tension, and Δx is the computational mesh resolution. It is interesting to note that this time step is *not* linearly proportional to Δx, as is the case for other Courant-Friedrichs-Lewy limits. As pointed out in [10], this implies that Co_c increasingly dominates the time step size (and hence reduces computational performance and efficiency) as the spatial scale of the problem decreases and mesh resolutions get smaller. This is particularly relevant in slag foaming problems, where bubble sizes may cover a wide range from < 1 mm up to several cm.

3. Results and Discussion

The parameters used for modelling of problems related to slag foaming are shown in Table 1. The parameters chosen here are broadly representative of those encountered during production of ferrochromium alloy from chromite ores [11].

Table 1. Parameters used in slag foaming models

Parameter	Slag	Gas[a]
Density	2700 kg/m^3	27 kg/m^3
Viscosity	0.1 Pa·s	0.001 Pa·s
Interfacial tension	0.45 N/m	
Initial bubble diameter	0.005 m	
Capillary Co number	5	
Critical film thickness[b]	10^{-7} m	
Mesh resolution[c]	2.5 x 10^{-4} m	

[a]Gas phase properties are chosen empirically to mitigate parasitic currents – see section 3.1
[b]Film rupture thickness estimated from [12]
[c]Mesh dependence study available in [9]

3.1. Single-bubble tests

The first set of simulations involves a single symmetrical gas bubble rising to the surface of a quiescent slag layer and its subsequent rupturing into the bulk gas phase, as shown in Fig. 2. This case is used to investigate the effect of different measures to mitigate parasitic currents.

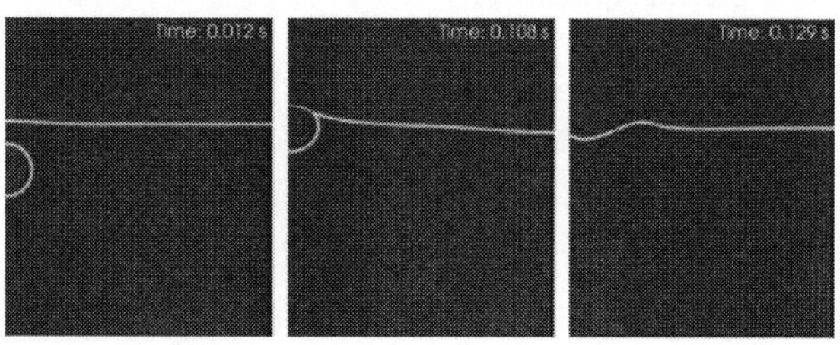

Fig. 2: Phase fields (blue=gas, red=slag) in single-bubble model at different times, model parameters per Table 1

Under realistic slag foaming conditions the density and viscosity of the gas phase would be very low (< 1 kg/m^3 and < 10^{-4} Pa·s respectively). Here, the gas properties were instead treated as an empirical numerical stability factor and adjusted in proportion to the slag properties – so for example, a property factor p_f of 0.01 would indicate a gas density of 27 kg/m^3 and a gas viscosity of 10^{-3} Pa·s. The results of varying p_f are shown in Fig. 3 and 4.

From these results it is seen that gas properties that are close to the realistic values (p_f = 0.001) result in a large amount of parasitic current noise at the surface of the bubble after it reaches the slag-gas interface. However, using values that are excessively high (p_f = 0.1) affects the bulk flow patterns in the region during the initial rising of the bubble to the surface, and also unphysically delays the rupture of the bubble film. The default values shown in Table 1, representing p_f = 0.01 between both extremes, would appear to be a reasonable choice.

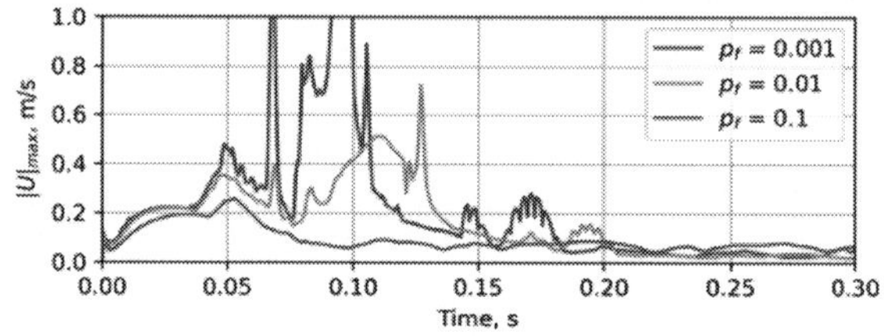

Fig. 3: Evolution of maximum velocity magnitude in simulation region for different values of p_f

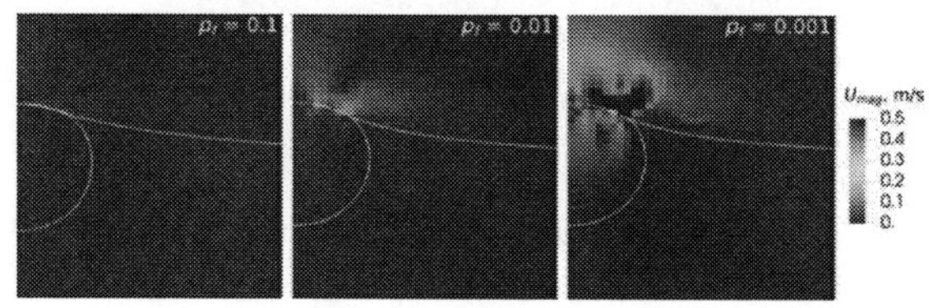

Fig. 4: Velocity magnitudes in vicinity of bubble at surface of slag, at 0.099 s into the simulation for different values of p_f

The effect of varying the capillary Courant number is shown in Fig. 5 and 6. It is desirable to increase Co_c as much as possible, as this directly influences the time step size and hence the performance of larger simulations.

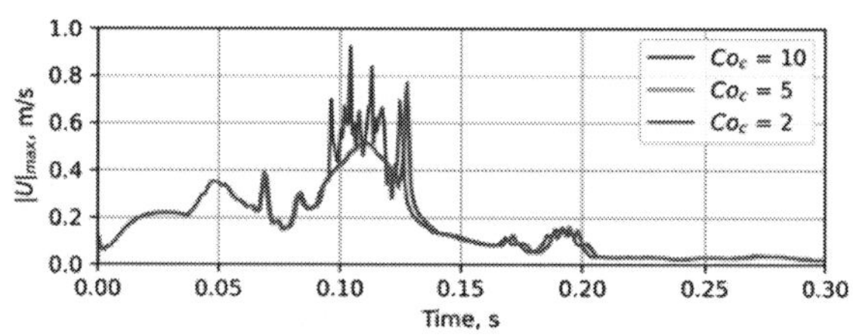

Fig. 5: Evolution of maximum velocity magnitude in simulation region for different values of Co_c

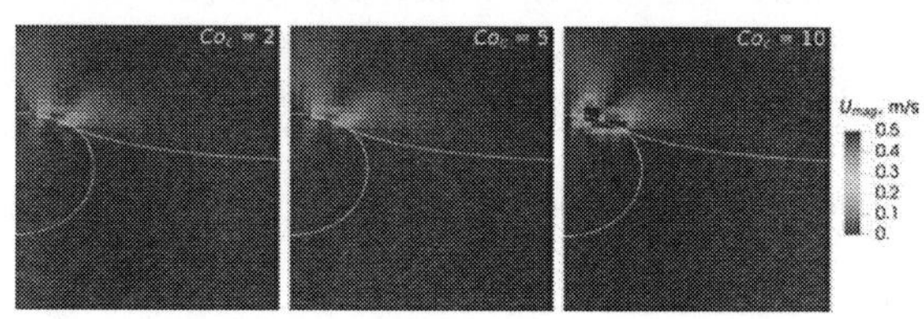

Fig. 6: Velocity magnitudes in vicinity of bubble at surface of slag, at 0.102 s into the simulation for different values of Co_c

These results show that for the model parameters given in Table 1 the critical value lies between 5 and 10, with significant increase in parasitic current noise visible in the Co_c = 10 case. A conservative value of 5 was therefore chosen for all subsequent simulations.

3.2. Population statistics in slag foam layers

In the second set of simulations, over 2000 bubbles were dispersed evenly through a slag layer in a two-dimensional rectangular region 0.2 x 0.4 m in size. The bubbles rise through the slag and form a temporary foam layer at the surface, which begins to decay as the bubbles burst and merge into the bulk gas phase. The left and right boundaries are treated as cyclic to emulate a broader region.

The behaviour is compared between two cases – one using the parameters in Table 1, and one with an increased slag viscosity μ_s of 0.5 Pa·s. It is possible for such changes to occur in chromite smelting as a result of a drop in operating temperature, or variations in the slag chemistry. Visualisations of the phase fields at different times in the simulation for both cases are shown in Fig. 7 and 8.

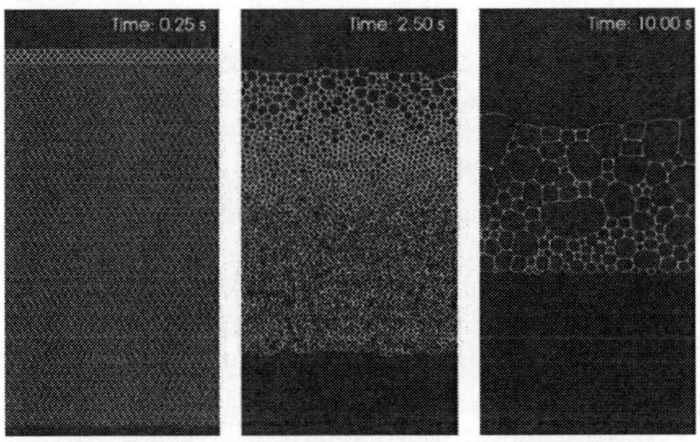

Fig. 7: Evolution of slag foam layer, μ_s = 0.1 Pa·s

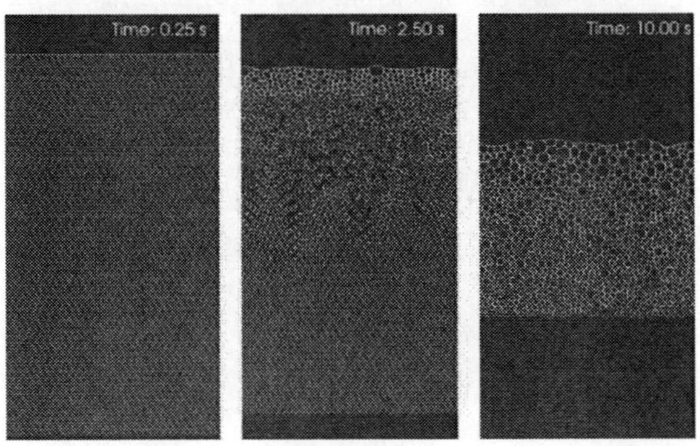

Fig. 8: Evolution of slag foam layer, μ_s = 0.5 Pa·s

It is interesting to observe that although the initial rise rates of the bubbles through more viscous slags are slower as expected, the final foam layer decays at approximately the same rate in both cases. This can be seen in the phase field visualisations, with the foam layers in the later stages of the simulations (e.g at 10 s) being of similar heights. This is a somewhat counter-intuitive result considering the empirical correlations for slag foaming index (e.g. [2], [3]) which are typically strongly proportional to μ_s. It is however possible that these correlations sometimes describe a related condition called an *expanded slag*, in which the gas bubbles are more evenly dispersed through the bulk continuum phase than in a true low-density foam. In the current simulations the system starts as an expanded slag and then evolves toward a true foam as time progresses. It can be seen from the middle images in Fig. 7 and 8 that the expanded slag state persists for longer when μ_s is high.

Although the overall dynamics of foam decay are somewhat similar, it is clear from the phase fields that the distribution of bubble sizes is substantially different in the two cases. The equivalent spherical diameters of each individual bubble at every time step in the simulation were calculated from bubble volume information extracted from the DMM field data, and cumulative distribution functions (CDF) were generated. The evolution of the CDFs over time for the two cases are shown in Fig. 9.

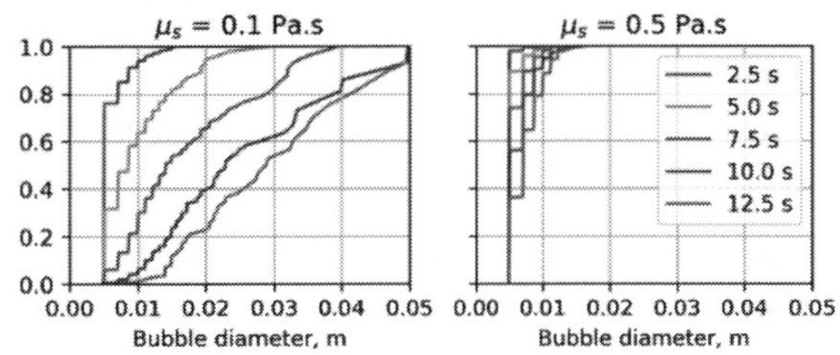

Fig. 9: CDF curves at different times for low and high slag viscosity cases

As seen qualitatively in Fig. 7 and 8, the system with low slag density produces a slag foam containing relatively few larger bubbles, while the high-density case contains many small bubbles close to the initial diameter (0.005 m).

4. Conclusions and Recommendations

A preliminary investigation of transient behaviour in slag foaming problems using the dynamic multiple marker method was conducted. Parasitic current errors were mitigated by artificially increasing gas density and viscosity combined with time-step limits imposed by the capillary Courant number. Application of the model to large-scale foam-layer formation problems showed that although the overall decay rate was comparable between cases with low and high slag viscosity, the bubble population statistics were very different with smaller bubbles retained for longer in the high-viscosity case. One potential implication of this is that standard empirical slag foaming correlations may not give a complete picture of the problem when assessing the risk of foaming in a given pyrometallurgical process. Foam stability may be overestimated in more viscous slags due to expanded-slag effects, and, in addition to steady-state foam thickness, it may be important to account for the transient evolution of foam layers in response to fluctuating gas production rates in the process.

Work is in progress to extend the present study with an experimental investigation on gas injection into different fluids. This will use room-temperature analog liquids to simulate slags with different properties, and will apply photogrammetric methods to capture quantitative information about the motion and coalescence behaviour of individual bubbles in the foam for additional validation studies. Future developments on the DMM method itself will include exploration of variations on the sub-grid model for film drainage, as well as more robust methods to manage parasitic current noise.

Acknowledgements

This paper is published by permission of Mintek, and was supported by internal funding on Science Vote Project PDR-24003. The authors acknowledge the Centre for High Performance Computing (CHPC), South Africa, for providing computational resources to this research project. Fig. 1 and input on the estimation of slag properties were provided by Mr Markus Erwee from Samancor Chrome Ltd.

References

[1] S. Seetharaman, A. McLean, R. I. L. Guthrie, and S. Sridhar, *Treatise on process metallurgy - Volume 3 - Industrial Processes*. Amsterdam Waltham Paris: Elsevier, 2014.

[2] R. Jiang and R. J. Fruehan, "Slag foaming in bath smelting," *Metall. Trans. B*, vol. 22, no. 4, pp. 481–489, Aug. 1991, doi: 10.1007/BF02654286.

[3] Y. Zhang and R. J. Fruehan, "Effect of the bubble size and chemical reactions on slag foaming," *Metall. Mater. Trans. B*, vol. 26, no. 4, pp. 803–812, Aug. 1995, doi: 10.1007/BF02651727.

[4] D. Lotun and L. Pilon, "Physical Modeling of Slag Foaming for Various Operating Conditions and Slag Compositions," *ISIJ Int.*, vol. 45, no. 6, pp. 835–840, 2005, doi: 10.2355/isijinternational.45.835.

[5] D. C. Guo, L. Gu, and G. A. Irons, "Developments in modelling of gas injection and slag foaming," *Appl. Math. Model.*, vol. 26, no. 2, pp. 263–280, Feb. 2002, doi: 10.1016/S0307-904X(01)00060-9.

[6] Q. Li and P. C. Pistorius, "Toward Multiscale Model Development for Multiphase Flow: Direct Numerical Simulation of Dispersed Phases and Multiscale Interfaces in a Gas-Stirred Ladle," *JOM*, vol. 73, no. 10, pp. 2888–2899, Oct. 2021, doi: 10.1007/s11837-021-04806-8.

[7] Q. G. Reynolds, O. F. Oxtoby, M. W. Erwee, and P. J. A. Bezuidenhout, "An extension of the multiple marker algorithm for study of phase separation problems at the mesoscale," *MATEC Web Conf.*, vol. 347, p. 00025, 2021, doi: 10.1051/matecconf/202134700025.

[8] D. J. E. Harvie, M. R. Davidson, and M. Rudman, "An analysis of parasitic current generation in Volume of Fluid simulations," *Appl. Math. Model.*, vol. 30, no. 10, pp. 1056–1066, Oct. 2006, doi: 10.1016/j.apm.2005.08.015.

[9] Q. G. Reynolds and O. F. Oxtoby, "Toward Meso-scale Modelling of Slag Foaming Phenomena in Pyrometallurgy," in *Materials Processing Fundamentals 2023*, S. Wagstaff, A. Anderson, and A. S. Sabau, Eds., in The Minerals, Metals & Materials Series. , Cham: Springer Nature Switzerland, 2023, pp. 113–123. doi: 10.1007/978-3-031-22657-1_10.

[10] F. Denner, F. Evrard, and B. Van Wachem, "Breaching the capillary time-step constraint using a coupled VOF method with implicit surface tension," *J. Comput. Phys.*, vol. 459, p. 111128, Jun. 2022, doi: 10.1016/j.jcp.2022.111128.

[11] C. T. Nguyen *et al.*, "Film drainage and the lifetime of bubbles," *Geochem. Geophys. Geosystems*, vol. 14, no. 9, pp. 3616–3631, Sep. 2013, doi: 10.1002/ggge.20198.

[12] I. J. Geldenhuys, "Aspects of DC chromite smelting at Mintek - an overview," in *Efficient Technologies in Ferroalloy Industry: Proceedings of the Thirteenth International Ferroalloys Congress*, in International Ferroalloys Congress. Almaty, Kazakhstan: Public Association INFACON XIII, Jun. 2013, pp. 31–47.

CFD analysis of heat transfer in a shaft furnace using the immersed boundary method

Mopeli Khama[a,*], Hendrik Johannes Vermaak[a], Quinn Gareth Reynolds[a,b], Buhle Sinaye Xakalashe[a]

[a]*Mintek, 200 Malibongwe Drive, Randburg 2194, South Africa*

[b] *Stellenbosch University, Private Bag X1, Matieland 7602, South Africa*

**Mopelik@mintek.co.za*

Abstract

Heat transfer is an important phenomenon in the operation of heterogeneous reaction vessels such as shaft furnaces. In metallurgical applications, shaft furnaces are typically used for the solid-state pre-reduction of various ore types. The kinetics of these heterogeneous gas-solid reactions are influenced by temperature, diffusion and advection, among other factors. It is therefore crucial to firstly understand the underlying heat transfer mechanisms in these systems.

The current study focuses on the heat transfer induced by passing a pre-heated inert gas upwards through the bottom part of a fixed bed of equally-sized spherical particles to simulate transient pre-heating conditions in a shaft furnace. The heat transfer simulations were conducted using the OpenFOAM® computational fluid dynamics (CFD) code. The convective, radiative and conductive heat transfer were considered by employing a special type of immersed boundary method (IBM), termed the fictitious domain method.

The findings reveal that the heat intensity initially concentrates at the bottom section of the shaft furnace and then gradually spreads to the upper section of the bed. Important transient phenomena such as the duration required to heat up a cold shaft furnace were quantified. The understanding of the associated fundamental heat transfer phenomena obtained from studies such as this will help inform the design and operation of pre-reduction shaft furnaces at industrial scale.

The current CFD approach is a computationally efficient and easy implementation that allows for rapid numerical computation of heat transfer phenomena in a heterogeneous gas-solid system. This is especially important for future studies wherein the fictitious domain method has to be incorporated into larger and more complex, coupled multi-physics reaction systems involving porous media.

Keywords: Shaft furnace, packed beds, heat transfer, IBM, OpenFOAM®

1. Introduction

Heat transfer in porous media has various applications in areas such as heat exchangers, packed bed reactors, filtration, CO_2 storage, etc. Several researchers [1-3] have carried out research to gain an improved understanding of the underlying heat transfer mechanisms in porous media. Porous media systems are known for their large surface areas and erratic fluid flow through the porous regions resulting from continuous mixing. The large surface areas coupled with the chaotic fluid flow behaviour can significantly improve heat transfer [4].

Packed bed reactors such as shaft furnaces can be used in the metallurgical industry as heterogeneous gas-solid reaction vessels for the solid-state pre-reduction of various ore types. A typical example is hydrogen pre-reduction of manganese (Mn) ores where the kinetics of these systems are strongly dependent on temperature [5]. This necessitates either the introduction of a pre-heated gas or external heating of the reaction vessel.

In the former case, the pre-heated gas exchanges thermal energy with the porous bed of solid particles until thermal equilibrium is reached. In the latter case, the vessel wall is typically heated externally using electrical heating elements. An inert gas such as nitrogen can be used to facilitate the cold heat-up stage to reduce operational costs and to comply with air emission regulations related to hydrogen use.

To help inform the design and operation of industrial pre-reduction shaft furnaces, test work is required to determine the time needed to reach the required reaction temperature and to determine operating conditions such as the optimum inert gas flow rate. CFD can be used as an inexpensive tool to safely investigate multiple sets of shaft furnace test conditions in a relatively short timeframe. From the CFD results, an optimal set of operating conditions can be produced that can subsequently be validated through pilot-scale test work, thereby eliminating much of the trial and error associated with physical test work.

2. Background

2.1. Heat transfer in porous media

Heat transfer in porous media is a well-studied and developed subject. However, the modelling approach that employs the representation of the porous medium as a field inside the main computational domain as developed by [6] is a new approach. The current study analyses the heat transfer characteristics of Mn ore pellets when hot nitrogen gas is distributed upwards through a cold bed of Mn pellets. The associated heat transfer mechanisms include particle-particle and fluid-particle heat transfer.

Fluid flow has a pronounced effect on heat transfer as the film thickness between the pellets changes with flow, resulting in a change of the apparent thermal conductivity of the fixed bed.

2.2. Convective heat transfer in porous media

The convective heat transfer in porous media is influenced by factors such as the physical properties of both the porous medium and the fluid as well as the fluid flow rate. The difference in temperature between the fluid and the solid matrix results in convective, conductive and radiative heat transfer within the porous medium.

The role of hydrodynamic effects on the temperature field is captured accurately by taking both the macroscopic and microscopic velocity fields into account. The inclusion of only macroscopic velocity fields in the energy equation does not account for all the hydrodynamic effects since the intra-particle dispersion effects can also have an impact on the temperature distribution in porous media [7].

The fluid flow and convective heat transfer in a porous medium is described by Darcy's law as well as the Brinkman and Forchheimer equations [4,8]. Darcy's law relates fluid velocity to pressure drop in a porous medium, while Forchheimer's equation is an extension of Darcy's model that accounts for non-linear flow resistance effects at higher Reynolds numbers.

In certain gas-solid reaction systems, the kinetics are favoured by high temperatures [5]. This requires external heating through either the introduction of a pre-heated gas into the solid matrix or through electrical heating of the reaction vessel walls. The introduction of the pre-heated gas is an example of the influence of entrance effects on the heat distribution.

The current study investigated the influence of entrance effects by introducing hot nitrogen gas through the bottom part of a shaft furnace and its distribution through a packed bed of cold Mn ore pellets to study transient pre-heating conditions. Upon introducing the hot gas, a local disturbance is created which results in lateral and axial diffusion-convection. Heat transfer coefficients were calculated from the correlations shown in Equations (1) and (2).

$$Sh = 2 + 0.6Re^{1/2}Sc^{1/3} \quad (1)$$

$$Nu = 2 + 0.6Re^{1/2}Pr^{1/3} \quad (2)$$

Where Sh, Re, Sc, Nu and Pr are the dimensionless Sherwood, Reynolds, Schmidt, Nusselt and Prandtl numbers, respectively.

2.3. Radiative heat transfer in porous media

The porous medium is treated as a collection of fixed bed of particles and the interaction of radiation with the particles is considered. It is often the case that the scattering, absorption and emission by the solid matrix is significant compared to the same phenomenon by the gas.

The Mn ore pellets are modelled as a grey body and nitrogen as a grey gas. The assumption of a grey body simplifies the implementation of thermal radiation since the net radiative heat transfer between surfaces is calculated by considering the net outgoing radiation.

$$\nabla \cdot (\Gamma \nabla G) + 4\pi\left(\frac{a\sigma T_g^4}{\pi} + E_S\right) - (a + a_s)G = 0 \quad (3)$$

Where G refers to the incident radiation, σ is Stefan-Boltzmann constant, E_s is emission of the solid matrix, a and a_s are the respective absorption coefficients of the gas mixture and solid matrix, while T_g is the gas temperature. Γ is defined in Equation (4).

$$\Gamma = \frac{1}{3(a + a_s + \sigma_{scatter}) + C\sigma_{scatter}} \quad (4)$$

Where $\sigma_{scatter}$ is the scattering coefficient and C is the coefficient that represents either backward or forward scattering [6].

In the current study, the Radiative Transfer Equation (RTE) was used to describe the radiative intensity. The conservation of the radiative energy is achieved by the summation of the contributions of radiative energy from all wavenumbers and all possible directions. The assumption of a grey, diffuse surface was made and the P1-approach was used to derive the RTE as described by Equation (3).

In this study, CFD was used to investigate gas-solid heat transfer in a shaft furnace when hot nitrogen gas is used to pre-heat a packed bed of cold Mn ore pellets. The geometry of the porous Mn ore pellets was defined by the fictitious domain method which is a computationally efficient method that can form the basis of more complex, coupled multi-physics reaction systems studies.

3. Model Development

3.1. Fictitious domain method

The Mn ore pellets were included into the computational domain using a special form of the immersed boundary method (IBM) named the fictitious domain method. This special type of IBM is used to describe stationary or moving bodies into the spatial discretisation without having to use a body conforming mesh [9].The strength of this approach lies in the ability to deploy fast solvers for auxiliary problems on complicated geometries. For discretization, a single rectangular grid is projected over the fictitious domain that is independent of the shape of the porous material. In essence, the geometry of Mn ore pellets is described by their physical properties such as density, thermal conductivity and porosity.

The governing equations for the gas and solid phase are shown in Equations (5) to (11) as derived from [6]. In the modelling of the convective heat transfer between the gas and the solid phase, each phase was considered to be a homogeneous continuum with a non-equilibrium temperature distribution.

$$\frac{\partial}{\partial t}\rho^G u + \nabla\cdot(\rho^G uu) + \varepsilon\nabla P - \nabla\cdot(\mu_{eff}\nabla u) - \rho^G g = -\mu_{eff}\, D\cdot u - F\cdot u \tag{5}$$

$$\frac{\partial}{\partial t}\varepsilon\rho^G + \nabla\cdot(\rho^G u) = 0 \tag{6}$$

$$\frac{\partial}{\partial t}\varepsilon\rho^G Y_i^G + \nabla\cdot(\rho^G u Y_i^G) - \nabla\cdot(\varepsilon\rho^G D_{eff}\nabla Y_i^G) = 0 \tag{7}$$

$$\frac{\partial(\varepsilon\rho^G C_p^G T^G)}{\partial t} + \nabla\cdot(\rho^G C_p^G T^G u) - \nabla\cdot\nabla(\varepsilon k_{eff}^G T^G) = -h_{conv}S_{Av}(T^G - T^S) + S^{G,Radiation} \tag{8}$$

$$\frac{\partial}{\partial t}(1-\varepsilon)\rho^S = 0 \tag{9}$$

$$\frac{\partial}{\partial t}(1-\varepsilon)Y_k^S\rho^S = 0 \tag{10}$$

$$\frac{\partial}{\partial t}(1-\varepsilon)(\rho^S C_p^S T^S) - \nabla\cdot\left((1-\varepsilon)\mathbb{K}k_{eff}^S\cdot\nabla T^S\right) = h_{conv}S_{Av}(T^G - T^S) + S^{S,Radiation} \tag{11}$$

Where the G and S superscripts refer to the gas and solid phase, respectively. Moreover, ρ, u, ε, P, μ_{eff}, g, D, F, Y_i, D_{eff}, Cp, T, t, k_{eff}, h_{conv}, S_{Av}, $S^{Radiation}$ and $\mathbb{K}$ refer to the density, fluid velocity, porosity, pressure, effective fluid dynamic viscosity, gravitational force, viscous resistance tensor, Forchheimer term, species mass fraction, effective diffusion coefficient, heat capacity, temperature, time, effective thermal conductivity, convective heat transfer coefficient, surface area to volume ratio, thermal radiation source term and the heat conductivity anisotropy tensor in the porous medium.

3.2. Geometry and meshing

The study considered a 3-dimensional shaft furnace with a height of 0.3 m and a diameter of 0.11 m. The shaft was packed with spherical Mn ore pellets in a hexagonal packing configuration with a bed height of 0.18 m as shown in Fig.1. The spherical particles had a diameter of 0.01 m and the bulk porosity was 0.3.

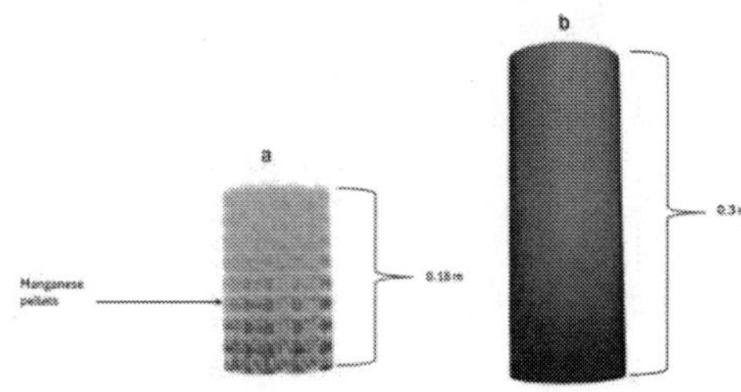

Fig. 1 Computational domain (a) bed of Mn ore particles; (b) entire domain

3.3. Results and discussions

Pre-heated nitrogen gas at a temperature of 1073 K was introduced upwards through the bottom of a shaft furnace at a volumetric flow rate of 10 L/min to a bed of Mn pellets with an initial temperature of 300 K. The distribution of a hot gas through a bed of particles creates a local disturbance that results in lateral and axial diffusion-convection. Upon contact with the cold bed, heat transfer ensued, and the lower section of the bed became hotter than the upper section as it was the first point of contact with the pre-heated gas. This is shown in Fig. 2 (a) and (c).

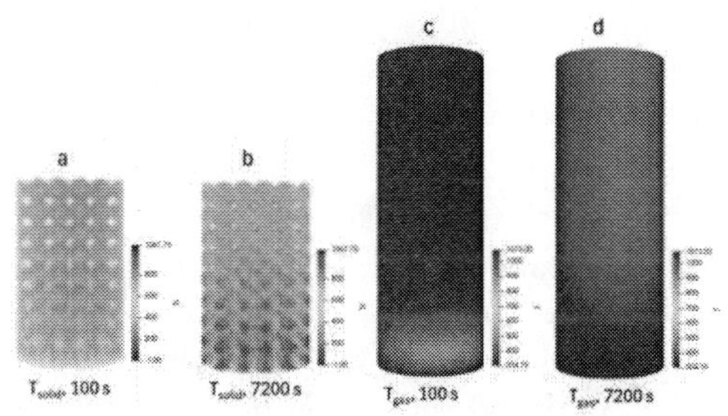

Fig. 2 Solid and gas phase temperatures at various times: (a) solid at 100 s; (b) solid at 7200 s; (c) gas at 100 s; (d) gas at 7200 s

Although the bottom section of the shaft furnace initially exhibited the most intensive heat transfer, the temperature of the upper section of the packing gradually increased. The results from the current study follow the same trend as the work of [1] who studied transient gas-solid heat transfer in a packed bed. Steady-state Mn pellet temperatures were reached before 3600 s (1 hour) with a maximum Mn pellet temperature of 1067 K – see Fig. 2 (a) and (b).

3.4. Influence of intra-particle porosity

Changes in intra-particle porosity result in changes in the gas flow rate through the Mn pellets and affects the rate of convective heat transfer. Effective thermal conductivity decreases with an increase in intra-particle porosity and, if conductive heat transfer dominates, the rate of heat transfer decreases.

However, in the current study, the Biot number was determined to be less than 0.5 indicating that convective heat transfer dominates. An increase in the intra-particle porosity increases the flow of gas through Mn pellets resulting in an increased rate of convective heat transfer. This is evident in Fig. 3, which shows that an increase in the intra-particle porosity from 0.05 to 0.40 results in improved convective heat transfer.

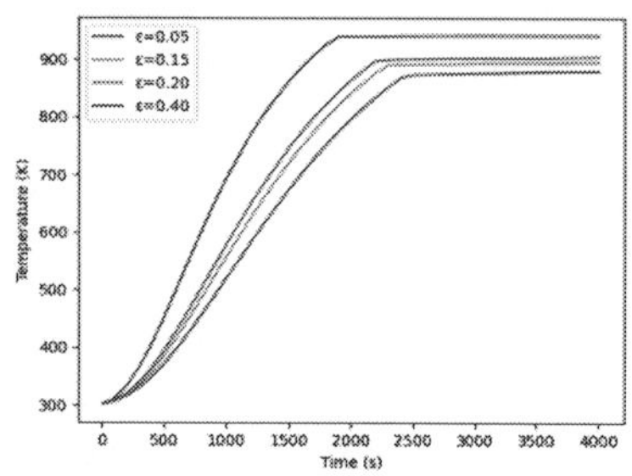

Fig. 3 Temperature of pellets at varying porosity

3.5. Influence of gas flow rate on heat transfer

The pre-heated gas flowrate was varied from 10 L/min to 25 L/min to assess the impact of gas flow rate on heat transfer. An increase in the gas flow rate decreases the thermal boundary layer thickness and increases the convective heat transfer coefficient. This results in an increased heat transfer rate.

This is illustrated in Fig. 4, where volume averaging was performed radially at an axial position of 0.1 m to determine temporal variations in temperature. As observed, cases with higher gas flow rates showed an increased heat transfer rate as evidenced by rapid increase in temperature. The model results were validated against the heat transfer experimental work by [10] and as observed in Fig.5, initially, there was a discrepancy between model and experimental results. However, the steady state temperatures are in close agreement.

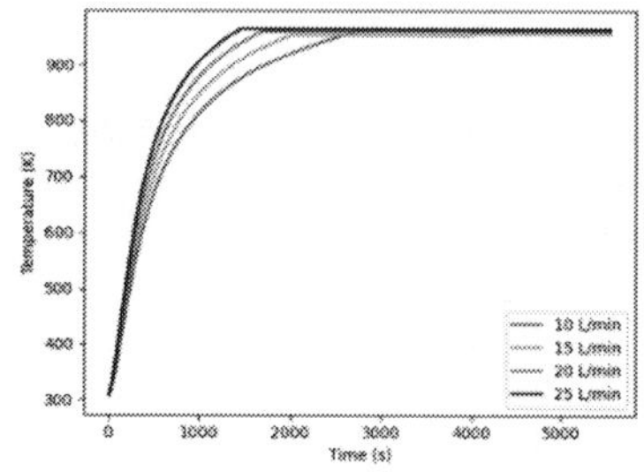

Fig. 4 Temperature of Mn ore pellets at varying gas flow rates

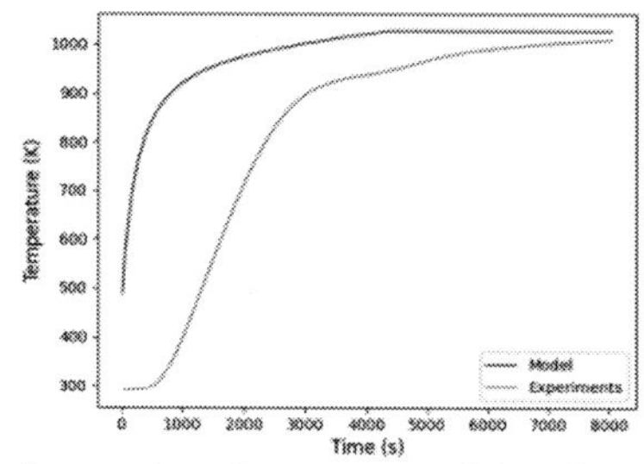

Fig. 5: Comparison between model and experimental results

3.6. Conclusions

The CFD model was able to capture the thermal behaviour of pre-heating a packed bed of solid particles with a hot gas. As some authors reported in the literature [1], at the initial stage of pre-heating, the bottom region of the shaft furnace exhibit the highest intensity of gas-solid heat transfer and heat intensity is distributed upwards through the bed over time.

The Biot number was determined to be less than 0.5 which indicates that convective heat transfer dominates. Higher intra-particle porosities resulted in increased convective heat transfer due to increased gas flow through the particles. It was also found that higher gas velocities resulted in improved heat transfer rates. The quantified heating time and Biot number could inform the design of shaft furnaces by guiding the choice of optimum particle size, furnace dimensions and gas flow rate.

The current CFD approach is a computationally efficient and easy implementation that allows for rapid numerical computation of heat transfer phenomena in a heterogeneous gas-solid system. This is especially important for prospective studies wherein the fictitious domain method has to be incorporated into larger and more complex, coupled multi-physics reaction systems involving porous media.

Acknowledgements

This paper is published by the permission of Mintek and was supported by internal funding on Science Vote Project PDR-24023. The authors acknowledge the Centre for High Performance Computing (CHPC), South Africa, for providing computational resources to execute this research project.

References

[1] Che H, Yue Y, Jiang Z, Li D (2024). *Numerical investigation of transient gas-solid heat transfer in a packed bed: Impact of intra-particle thermal diffusion*. Particuology, **90**: 404-411.

[2] Hassan OH, Sultan GI, Sabry MN, Hegazi AA (2022). *Investigation of heat transfer and pressure drop in a porous media with internal heat generation*. Case Studies in Thermal Engineering, **32**: 101849.

[3] Vafai K, Tien CL (1981). *Boundary and inertia effects on flow and heat transfer in porous media*. International Journal of Heat and Mass Transfer, **24**: 195-203.

[4] Esfe MH, Bahiraei M, Hajbarati H, Valandkhani M (2020). *A comprehensive review on convective heat transfer of nanofluids in porous media: Energy-related and thermohydraulic characteristics*. Applied Thermal Engineering, **178**: 115487.

[5] Sarkar A, Schanche TL, Wallin M, Safarian J (2024). *Evaluating the reaction kinetics of the H_2 reduction of a manganese ore at elevated temperatures*. Journal of Sustainable Metallurgy, **10**: 2085-2103.

[6] Źuk PJ, Tuźnik B, Rymarz T, Kwiatkowski K, Dudyński M, Galeazzo FCC, Filho GCK (2022). *OpenFOAM solver for thermal and chemical conversion in porous media*. Computer Physics Communications, **278**: 108407.

[7] Kaviany M (1995). *Principles of Heat Transfer in Porous Media*. New York, Springer.

[8] Avramenko AA, Shevchuk IV, Kovetskaya MM, Kovetska YY (2020). *Darcy-Brinkman-Forchheimer Model for Film Boiling in Porous Media*. Transport in Porous Media, **134**: 503-536.

[9] Isoz M, Šourek MK, Studeník O, Kočí P (2022). *Hybrid fictitious domain-immersed boundary solver coupled with discrete element method for simulations of flows laden with arbitrarily-shaped particles*. Computers and Fluids, **244**: 105538.

[10] S. N. Sambo, L. Hockaday, T. Seodigeng, and Q. G. Reynolds, *Numerical and Experimental Study of Packed Bed Heat Transfer on the Preheating of Manganese Ore with Air up to 600 °C*. Metals (Basel), vol. 15, no. 3, Mar. **2025**, doi: 10.3390/met15030269.

Wake Redirection Prediction for Wind Farms

K. Surujhlal[a*], A. Biobaku[a], B. Zang[b], W. Ho[c]

[a]*University of the Witwatersrand, 1 Jan Smuts Avenue, Braamfontein, Johannesburg, 2000, South Africa*

[b]*University of Bristol, Beacon House, Queens Road, Bristol, BS8 1QU, United Kingdom*

[c]*University of Cape Town, Rondebosch, 7700, South Africa*

**corresponding author email: kayla36252@gmail.com*

Abstract

An increase in the demand for renewable energy has seen an increasing need for optimization of wind farms. In a wind farm, the upstream turbine has several negative effects on the downstream turbine which leads to uneven loads being exerted and reduced power outputs of the wind farm. Wake redirection techniques could reduce these negative effects and improve the overall power output of a wind farm. The aim of this study is to investigate the ability of a surrogate model to predict the flow field when the upstream turbine is at different yaw angles.

A Reduced Order Model (ROM) was used to analyze the dominant dynamic behavior of the wake while using less computational resources. The modes generated were analyzed and selectively used to reconstruct the most accurate flow fields. A Machine Learning (ML) model used the ROM outputs to predict the wake at yaw angles other than those it was trained on. Three models were compared – and LES-trained model (used a baseline), a DMD-trained model and a DMD-trained (reconstruct before) model. The DMD-trained model had an average percentage difference of 6.93%. This is attributed to the inability of accurately tracking of a mode through different yaw angles and the difficulty in generating accurate state-space matrices at predicted yaw angles.

Keywords: Wind turbines, Wake-redirection, Surrogate Modelling, Fluid Dynamics

1. Introduction and Motivation

As the world aims to produce more energy from renewable sources the need for wind farm optimization increases [1]. In most wind farms the turbines are aligned behind one another to maximize the number of the turbines that can be placed in a specific area. As a result, the upstream turbine wake has several negative effects on the downstream turbine. The upstream turbine extracts a lot of the energy, leaving little energy to be extracted by the downstream turbine and the wake exerts uneven loads on the downstream turbine [2]. Wake redirection techniques (such as yaw misalignment) could reduce these negative effects and possibly increase the overall power output of the wind farm. Although there is a contradiction in research as to whether tilt or yaw misalignment leads to a greater percentage gain in power output [3], [4], [5], [6], it was chosen to investigate yaw misalignment because there is less risk of the tip of the blade hitting the tower [7] and more turbines already have the ability to yaw – potentially making this method simpler to implement. The aim of this study is to investigate the ability of a surrogate model to predict the flow field when the upstream turbine is at different yaw angles. This could assist with real-time wake analysis and optimization at a wind farm, without the need for running high-fidelity simulations to predict the wake when the yaw angle of the turbine is changed.

2. Description of the Dataset

The dataset identified to create the surrogate model is an open-source dataset containing Large-Eddy Simulations (LES) of a two-turbine array and it was generated using the Simulator for Wind Farm Applications (SOWFA). The model contains an upstream and downstream turbine that are placed five diameters (5D) apart [8]. The wind turbines are horizontal axis, 3-blade, DTU 10MW wind turbines. Additional characteristics of the turbine modelled can be found in Bak et. al [9]. The model is set up such that there is a uniform inflow condition, with no boundary layer or ground effects. The upstream turbine is yawed between 10° and 30° periodically (remaining at intermediate angles for two seconds), whereas the downstream turbine remains constant at 0°. Therefore, there is only information about the fully developed flow field at the two angles.

3. Method

3.1 Surrogate Model

A surrogate model is a combination of a ROM and a ML algorithm. In this study, a ROM will be used to simplify the complex flow model and analyze the physical properties of the wake while using less computational resources, and the ML algorithm will be used to predict the wake at different yaw angles. Various types of ROMs exist, such as proper orthogonal decomposition (POD) and dynamic mode decomposition (DMD) [8]. DMD, and specifically input-output dynamic mode decomposition (IODMD), was chosen because it can capture more physical information of the flow field [10], which is important for understanding the factors that influence the wake dynamics and thus the loading on the downstream turbine in the wake.

To create the ROM, the original 4D LES dataset (x, y, z and time) was first rearranged into 2D matrices that require less computational resources to work with. A snapshot matrix (X) was created by arranging single snapshot matrices containing spatial-velocity data in sequential columns so that each column contained velocity field data at a single time step. Illustration of this process is given by Surujhlal et. al [11]. Singular Value Decomposition (SVD) was then performed on matrix X to reduce its size and obtain a low-rank approximation that still captured the required dynamics. SVD decomposes matrix X it into left (U), right (V) and diagonal (Σ) singular value matrices [10], as seen in Equation 1 below. Truncation, which is directly related to the number of modes generated, was then performed which results in the least important columns being discarded first.

$$X = U\Sigma V = [\tilde{U} \quad \tilde{U}_{rem}] \begin{bmatrix} \tilde{\Sigma} & 0 \\ 0 & \tilde{\Sigma}_{rem} \end{bmatrix} \begin{bmatrix} \tilde{V} \\ \tilde{V}_{rem} \end{bmatrix} = \tilde{U}\tilde{\Sigma}\tilde{V} \quad (1)$$

Validation was then performed by comparing the validation dataset to the identification dataset. This was done using Variance Accounted For (VAF) (see Equation 2). VAF is a quantitative measure of how much the model differs from the original data.

$$VAF_i\big(y_i(k), \hat{y}_i(k)\big) = \left(1 - \frac{\frac{1}{N}\sum_{K=1}^{N}|y_i(k) - \hat{y}_i(k)|^2}{\frac{1}{N}\sum_{K=1}^{N}|y_i(k)|^2}\right) \quad (2)$$

To apply the traditional method of DMD, the low-dimensional linear model of the dynamical system ($\tilde{A}$) is calculated by projecting the states onto the subspace [8].

$$\tilde{A} = U^{*}X'V\Sigma^{-1} \quad (3)$$

where * represents a conjugate transpose, X' is the time-shifted snapshot matrix and Σ^{-1} is the inverse of the sigma matrix from the SVD process. The reduced order model is given by the following equation:

$$\tilde{x}_{k+1} = \tilde{A}\tilde{x}_k \quad (4)$$

where $\tilde{x}_{k+1}$ represents the projected future state and $\tilde{x}_k$ represents the projected current state.

IODMD is two steps further from the traditional DMD method, with Dynamic Mode Decomposition Control (DMDc) between the two. DMDc differs from DMD in that the future state (x_{k+1}) is not only dependent on the current state (x_k), but also on a control variable (u_k) [8]. IODMD is an extension of DMDc in that not only are the inputs mapped, but the full-order model is projected onto the output subspace which is also specified by a function (y_k) [8]. The following equations represent the reduced order model:

$$\tilde{x}_{k+1} = \tilde{A}\tilde{x}_k + \tilde{B}u_k \quad (5)$$

$$y_k = \tilde{C}\tilde{x}_k + \tilde{D}u_k \quad (6)$$

where $\tilde{x}_{k+1}$ is the reduced order future states, $\tilde{x}_k$ is the reduced order state-space matrix, u_k is the input vector and A, B, C and D are the state-space matrices.

3.2 Identification and grouping of modes

The modal characteristics were generated using mode matrices A-D. DMD modes are characterized by their frequency, damping ratio, energy contribution and growth rate. By analysing this information, it can be understood how each mode contributes to the wake shape and dynamics and thus which combination of modes would be best for prediction of the flow field. 100 modes were analysed since it is most common [8], [12]. Upon analysis it was found that multiple modes correspond to the same physical properties of the wake and were thus grouped as such. These five properties were the mean flow, tip vortices, hub vortices, wake deflection and noise. Further descriptions and illustrations of these mode groups can be found in Surujhlal et. al [11]. Since the flow field changes from 10° to 30°, the mode values also change. Hence, the modes needed to be generated and grouped at both angles (see Figure 1).

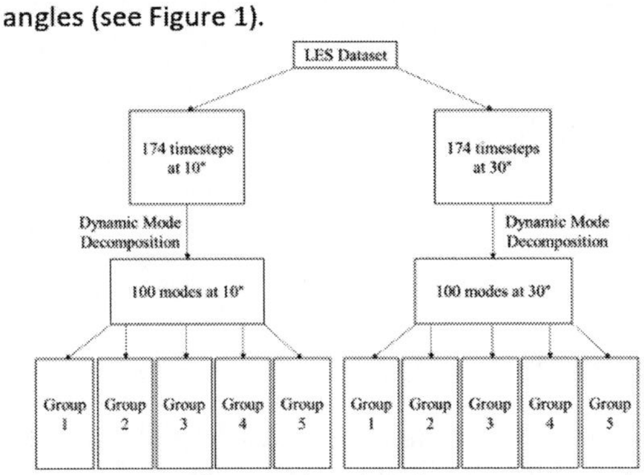

Figure 1. Process for grouping of modes at two yaw angles

3.3 Machine Learning Algorithm for Prediction

As there are only two yaw angles at which the flow field is fully developed, only these two datasets can be used for training the ML algorithm. Four ML algorithms were tested to predict the wake – a Convolutional Neural Network (CNN), a Deep Learning Network, Time Series Forecasting and a Neural Network Fitting (NNF) algorithm. The NNF model worked the best. This model requires two corresponding input variables, namely a

Predictor (independent) variable that contains the yaw angle, and a Response (dependent) variable that contains the velocity field at the corresponding yaw angle in the Predictor variable. After training, a yaw angle can be inputted, and the model will return the predicted flow field at that yaw angle.

The first attempt was to train the algorithm with the entire 3D flow field at a sample of 20 yaw angles, resulting in a matrix of size 51543 x 20. This proved to be not feasible as the memory required exceeded that which was available at that time (84 GB). Therefore, the flow field was split and reshaped into 2D "slices" (along the x, y and z axes), each of size 1998 x 20 for 20 yaw angles, that were fed into the NNF model separately. The design of the NNF model has, by default, an input layer, an output layer and a minimum of one hidden layer. The example of the network diagram shown below in Figure 2 is a model with 1 hidden layer. Through experimentation it was found that the Levenberg-Marquardt training algorithm resulted in the lowest overall percentage error and the optimal number of layers for this dataset was one.

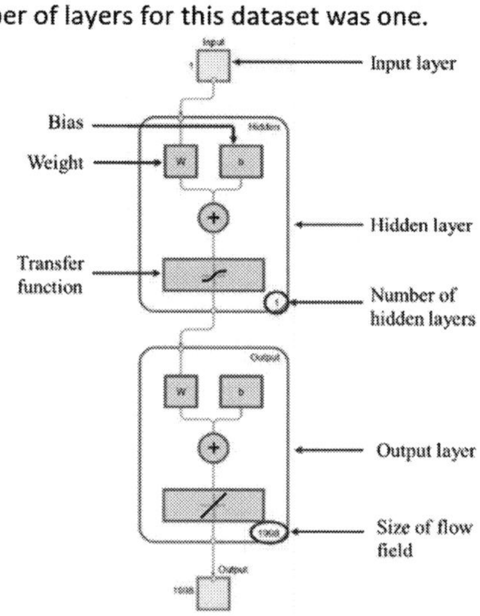

Figure 2. Network diagram for NNF model

4. Results

Three methods of prediction were tested – an LES-trained model, a DMD-trained model and a DMD-trained (reconstructed before) model. The LES-trained model was trained on the LES data, the DMD model trained on DMD data in the form of modes and the DMD-trained (reconstruct before) also on DMD data, but reconstruction of the flow field using the modes was performed before feeding these reconstructions into the ML model. The key difference between the second and third approach is in the second approach, the modes themselves were the training data but in the third approach, the modes were used to reconstruct the flow field and this reconstructed flow field was the training data. The accuracy of the model predictions was compared in two ways – comparing the accuracy of individual velocity field points and comparing the angle of wake deflection. The individual velocity points were compared by calculating the percentage difference in normalized velocity deficit between the prediction and the baseline. The angle of wake deflection was measured by setting a threshold RGB value that marks the edge of the wake, finding where this threshold lies at different downstream positions and then calculating the gradient (and thus angle) of the wake deflection (see Figure 3).

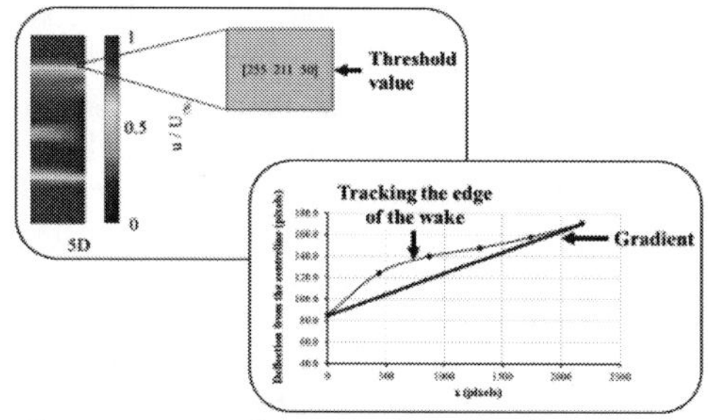

Figure 3. Process for measuring wake deflections

4.1 Comparison of Velocity Fields

The three models were compared to each other and compared to the original LES flow field. An example of the predictions produced by each model at a yaw angle of 30° is shown in Figure 4 as well as the average percentage difference of each to the original LES.

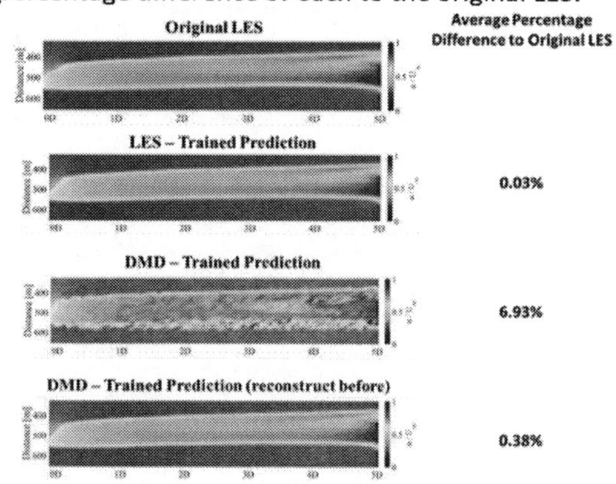

Figure 4. Model predictions and percentage differences

The top view of the flow field is pictured in Figure 4. The shape of the wake is a result of the anti-clockwise deflection experienced due to the clockwise yaw angle of the upstream turbine [13]. The LES-trained model compares well to the Original LES, which was expected. This indicates good accuracy in the ML model. The DMD-trained model predictions are comparatively worse, but the deflection of the wake and normalized velocity measurements compare well to the Original LES. The limitations of this method will be discussed later in section 4.2 below.

The DMD-trained (reconstruct before) model predicts better than the DMD-trained model, but a marginally higher percentage difference is noted compared to the LES-trained. This indicates a good use of mode selection without a significant loss in accuracy.

To further investigate the deviations, the percentage differences observed at different parts of the flow field was mapped for each yaw angle. Figure 5 below shows an example of the predicted flow field using the DMD-trained model, the LES-trained prediction, and the percentage difference at a yaw angle of 10°. Since there is no original LES data at this yaw angle, the DMD-trained model is compared to the LES-trained model.

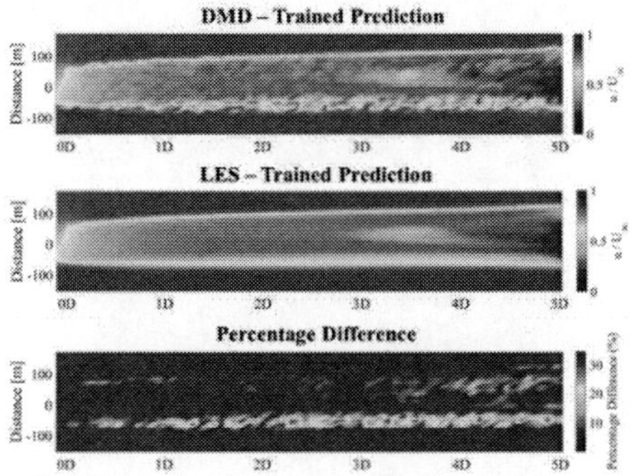

Figure 5. DMD-trained model prediction and percentage difference at 10°

It was observed that the largest errors occur at the edge of the wake where it should deflect away from. This is where there is entrainment of clean air from the freestream into the wake path and where it interacts with the wake. Due to the turbulent dynamics present here, the behaviour of this section of the wake is more difficult to predict and thus leads to higher percentage differences. This indicates that this model does not predict the deflection of the wake particularly well. However, this is investigated in the following section.

4.2 Comparison of Angle of Wake Deflection

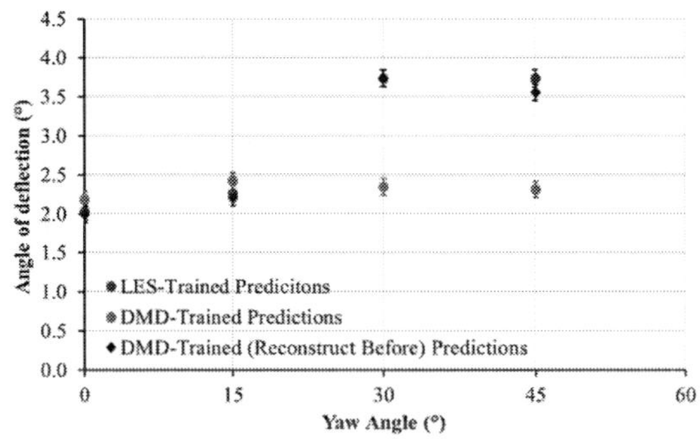

Figure 6. Graph of angle of deflection using different models

Similar trends are observed for the LES-trained model and DMD-trained (reconstruct before) models. The larger the yaw angle, the larger the angle of deflection. Since the training data fed into the ML model was at 10° and 30°, it is hypothesized that the minimal change in angle of deflection before 10° and after 30° is attributed to the limits of the ML model. This could indicate that the ML model is good at interpolation but not at extrapolation. It is expected that a limit of deflection should exist because at a certain yaw angle, the turbine will reach a stall angle thus the yaw angle would not have a significant effect on the wake deflection. However, previous experimental results indicated this limit is beyond 45°.

The DMD-trained model does not predict a significant change in the deflection of the wake at different yaw angles, indicating poor predictive capability. The sources of error can be attributed to the quality of data that it is trained on and the error in reconstruction. Firstly, there is inaccuracy induced in the tracking of a mode through the different yaw angles. Secondly, there is error in the generation of the state-space matrices at intermediate angles.

The surrogate model method (DMD-trained) significantly reduces the computational resources required since a single mode can be used instead of an entire LES flow field. However, due to the challenges with reconstruction, the end predictions are not good. To combine achieving both an accurate prediction and a reduction in computational resources, complications such as the mode tracking and the prediction of accurate corresponding state-space matrices for different yaw angles would need to be overcome.

5. Conclusion

The surrogate model tested can predict the wake of a two-turbine array when the upstream turbine is at non-zero yaw angles. Although computational resources are significantly reduced, the prediction accuracy is not very good. This is attributed to the inaccuracy in tracking of a mode through different yaw angles and the difficulty in generating accurate state-space matrices at predicted yaw angles, which would be useful for future work to investigate. The DMD-trained (reconstruct before) model predicts the wake very similarly to the LES-trained model (average difference of 0.38%). This model does not directly reduce the computational resources required when compared to the LES-trained model, but it could have an advantage in that only select modes were used instead of the entire flow field – meaning there could be a reduction in computational resources in that the model does not have to account for the complexities of the dynamics associated with the modes that were excluded, which could increase efficiency. To test this, a second dataset is required so that the predictions can be compared to LES data that it is not trained on.

References

[1] F. Porté-Agel, M. Bastankhah, and S. Shamsoddin, "Wind-Turbine and Wind-Farm Flows: A Review," *Boundary Layer Meteorol*, vol. 174, no. 1, pp. 1–59, Jan. 2020.

[2] K. W. Van Treuren and T. A. Burdett, "Experimental Testing of Wind Turbines Using Wind Tunnels With an Emphasis on Small-Scale Wind Turbines Under Low Reynolds Numbers," 2014, pp. 67–110.

[3] C. Cossu, "Evaluation of tilt control for wind-turbine arrays in the atmospheric boundary layer," *Wind Energy Science*, vol. 6, no. 3, pp. 663–675, May 2021.

[4] J. Bartl *et al.*, "Wind tunnel experiments on wind turbine wakes in yaw: Effects of inflow turbulence and shear," *Wind Energy Science*, vol. 3, no. 1, pp. 329–343, Jan. 2018.

[5] F. Trigaux, M. Moens, P. Chatelain, and G. Winckelmans, "Tilted wind turbines in farm configuration for improved global efficiency," in *Journal of Physics: Conference Series*, IOP Publishing Ltd, Sep. 2020.

[6] J. Cutler, A. P. J. Stanley, J. J. Thomas, and A. Ning, "Optimization of turbine tilt in a wind farm," in *AIAA Scitech 2021 Forum*, American Institute of Aeronautics and Astronautics Inc, AIAA, 2021, pp. 1–10.

[7] R. Scott, "Characterizing Tilt Effects on Wind Plants," Portland, OR, Jan. 2000.

[8] N. Rodrigues Cassamo, J.-W. van Wingerden João Miguel da Costa Sousa, C. Frederico Neves Bettencourt da Silva Supervisor, J.-W. van Wingerden, and M. Afonso Dias de Ayala Botto Paulo Jorge Coelho Ramalho Oliveira, "Model Predictive Control for Wake Steering: A Koopman Dynamic Mode Decomposition Approach Mechanical Engineering Examination Committee," 2020.

[9] C. Bak *et al.*, "General rights The DTU 10-MW Reference Wind Turbine). The DTU 10-MW Reference Wind Turbine. Sound/Visual production (digital)," 2013.

[10] K. Taira *et al.*, "Modal analysis of fluid flows: An overview," *AIAA Journal*, vol. 55, no. 12, pp. 4013–4041, 2017.

[11] K. Surujhlal, A. Biobaku, N. Zang, and W. H. Ho, "Wake Redirection Techniques for Wind Farms", 13th South African Conference on Computational and Applied Mechanics, 2024.

[12] Z. Wu, D. Laurence, S. Utyuzhnikov, and I. Afgan, "Proper orthogonal decomposition and dynamic mode decomposition of jet in channel crossflow," *Nuclear Engineering and Design*, vol. 344, pp. 54–68, Apr. 2019.

[13] K. Surujhlal, A. Biobaku, X. Liu, B. Zang, and W. H. Ho, "Some observations on wake redirection techniques of tandem wind turbines," *Physics of Fluids*, vol. 37, no. 3, Mar. 2025.

An unsupervised machine learning algorithm-based approach to Dynamic Mode Decomposition mode selection

A. Biobaku[a], B. Zang[b,c], W. Ho[a,d,e]*

[a] *University of the Witwatersrand, Johannesburg, South Africa*

[b] *University of Bristol, Bristol, United Kingdom*

[c] *Queen Mary University, London, United Kingdom*

[d] *University of Cape Town, Cape Town, South Africa*

[e] *Centre for Research in Applied Mechanics (CERECAM), Cape Town, South Africa*

**Corresponding author email: ayodimejilab@gmail.com*

Abstract

Reduced Order Models (ROM) are used to identify and reduce the number of variables that can represent the characteristics of a system. This reduces the complexity of the scenario while providing reasonably accurate results. High-resolution models like Computational Fluid Dynamics (CFD) are used to train the ROMs. Dynamic Mode Decomposition (DMD) is one of the most popular types of ROM used in fluid dynamics problems. It identifies important coherent structures by decomposing the flow field into modes and ordering these structures by frequency. These modes can be identified by their physical characteristics, which include frequency, amplitude and damping and represent a flow system feature. To reduce the computational expense of predicting the flow fields, a subset of the DMD modes is selected to simplify the system, typically by ranking the modes according to their amplitude/energy. This study aims to add insight to the selection process by grouping the modes based on the flow structures using unsupervised classification algorithms. K-means and hierarchical clustering algorithms were compared regarding their capabilities when classifying visualised DMD modes. The Hierarchical clustering algorithm was the better model for this application as it sorted the modes closer to manual classification.

***Keywords*: Wind turbines, Modelling, DMD, Clustering, Classification**

1. Introduction

The introduction and growth of renewable energy has marked a significant transition in global energy systems, driven by concerns over climate change, energy security, and economic viability. Wind power has emerged as a crucial component of the renewable energy landscape. Wind turbines, which convert the kinetic energy of wind into electrical energy, have become increasingly efficient and cost-effective. Modelling the collective effects of wind turbine wakes in a wind farm is useful for designing and predicting the power output of wind farms. In predictive control, accuracy and computational speed are crucial in controlling and optimising the outputs of wind farms, which makes mathematical tools like Reduced Order Models (ROMs) increasingly popular.

1.1. Reduced Order Models

Recently, mathematical techniques like Reduced Order Models (ROM) have been used to identify and trim the number of variables that can represent the characteristics of a system. This reduces the complexity of the scenario while providing reasonably accurate results, these usually require Computational Fluid Dynamics (CFD) models to train the ROMs. Once trained, they can be applied to model similar scenarios outside the initial range, significantly reducing the computational power required to model wind farms on-site. This means results could be provided in real-time and can aid Engineers and Technicians in making decisions and controlling wind farms in reaction to various conditions [1].

Two types of ROMs used in wake modelling are Proper Orthogonal Decomposition (POD) and Dynamic Mode Decomposition (DMD), with each method having multiple variants. POD is used to identify dominant energy modes in the variability in a dataset, this technique decomposes the flow field into temporal and spatial modes arranged by energy [1]. DMD, on the other hand, identifies important coherent structures by decomposing the flow field and ordering these structures by frequency [1].

While POD captures information on dominant structures based on energy levels and spatial correlations, DMD offers additional insights by extracting spatial and temporal orthogonalities, as well as frequency information about the modes. DMD is particularly useful for modelling transient flows and dynamic flow features, making it advantageous for control applications. However, DMD can be more computationally expensive, especially for larger datasets, as it requires more modes to capture the same amount of energy compared to POD and needs more sample time to converge [2][3][4].

Studies on dynamic conditions have been conducted on wind turbines where the pitch or yaw of the upstream turbine blades was changed and variants of DMD were used to predict the effect on downstream turbines [5] [6]. Scenarios like this are useful in control problems as they provide a framework for modelling time-dependent conditions.

While both methods have benefits and shortcomings, DMD is more appropriate for this study, mainly due to its ability to capture dynamic and frequency information that is present in the flow structures found in the wake of a wind turbine.

1.2. Mode selection methods

Selecting the right modes for reconstructing the flow is important for precision and computational efficiency. The modes can be ranked and chosen according to the energy levels. The downside of this is that larger energy/amplitude modes might damp out quicker over time, meaning these modes or modes that have lower energy levels but remain steady over time, could be ranked incorrectly, especially for highly transient or non-linear flows. This approach also fails to consider the nuances in the physical properties of each mode, resulting in the omission of important modes.

In a study by Kou and Zhang [7], selection techniques were analysed and applied to test cases, namely to a subsonic flow around a cylinder in equilibrium and transient states, and an aerofoil in a transonic flow. This approach scaled the amplitude with the eigenvalues and eigenvectors and was shown to work well for both periodic and transient flow systems.

There are a few other selection approaches in literature, including a study by Krake et al, where the traditional DMD modes were redefined before scaling the amplitudes [8]. A recent study by Corrochano et al explored a selection algorithm that clustered variables in combustion by frequencies of the DMD modes and then used an iterative process to reduce the reconstruction of these flow fields [9]. None of these models, however, considers the physics of the flow beyond the frequency information, while the frequencies, amplitudes and damping ratios are important qualities of the modes, the mode structures and flow patterns are also valuable parameters to consider for a physics-based approach to mode selection. This study is the first part of an attempt to develop a physics-informed method to mode selection that groups the modes based on the flow/physical properties before selection. This would allow the user to consider flow feature representation in the selection process, aided by unsupervised grouping of the modes.

1.3. Unsupervised classification algorithms

To streamline grouping modes based on flow features, this would have to be done unsupervised; two algorithms would be considered in this study, K-means and Hierarchical clustering. K-means clustering classifies data points by initially assigning each data point to one group, it then iteratively groups data points within the nearest proximities while updating the centroids of each group (cluster) based on the points [10][11]. Due to its simplicity, K-means is fast, works well with large datasets, and is adaptable to new data points. K-means also works well with spherical clusters of similar sizes [12]. It, however, is sensitive to outliers and initial assignment of the centroids, it also requires that the number of clusters be specified, although there are methods to work around this that will be discussed. This method is also not ideal for non-spherical clusters with varying sizes, and can be susceptible to converging to local optima [13], [14].

Hierarchical clustering classifies datasets by creating a tree-like/dendrogram hierarchy of clusters using either a top-down (divisive) or bottom-up (agglomerative) approach. This classifies the points by quantifying the data points' similarity (or distance) [15]. This method does not require specifying the number of clusters, although it might require the distances to be specified. It is also robust, and the visualisation of the dendrogram can allow the user to trim the results at a reasonable threshold. Hierarchical clustering can be, however, computationally expensive and unsuitable for larger datasets (>100000 points) [16], [17], [18].

Since this study aims for an unsupervised classification algorithm, selecting the number of classes should not be prescribed. Two main methods are used in optimising the number of classes, the elbow and silhouette methods. The elbow method measures the compactness of a cluster using the within-cluster sum of squares (WCSS) and chooses the number of clusters where adding more does not significantly reduce the WCSS which is the "elbow" (Figure 1). The silhouette method measures how well-separated clusters are using a silhouette score. This is based on cohesion, how similar a point is to its cluster, and separation, how it compares to other clusters. This paper compares these two methods to determine which is more suitable for this application.

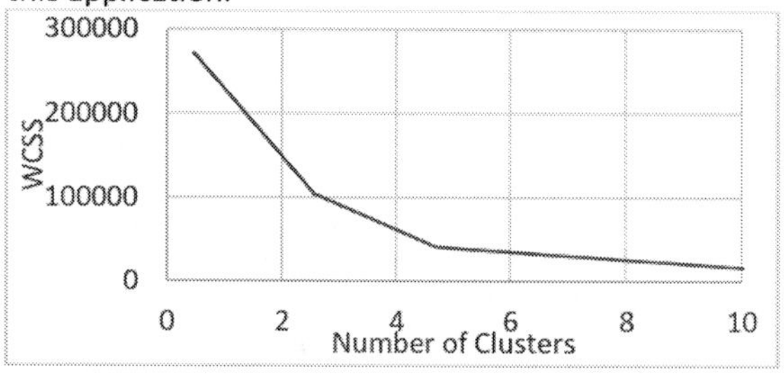

Figure 1: Graph showing a typical elbow plot

2. Methodology

The dataset used is based on a study by Cassamo [5] where LES simulations were run to study the effects of wake steering on downstream wind turbines and the efficacy of DMD in modelling this problem. The following steps were taken to generate results.

1. DMD is performed to obtain the modes' eigenvalues and eigenvectors

2. Normalised velocity fields are reconstructed from DMD mode pairs
3. 2D flow field "slices" were extracted from the 3D flow field at 10 seconds into the simulation
4. The classification algorithm groups the modes based on the visualised slices
5. The number of classes is optimised

Once the results of the classifications are generated, the classes are compared to manual classifications based on the visualisation of the modes. The performance of the modes was judged based on the ratio of misclassified modes to the total.

3. Results and discussion

3.1. Manual classification

The modes were first classified manually based on the top and side views at the rotor centreline. Then, based on the size of the structures visualised, they were categorised into Small, Medium, Large, and Extra-Large. Examples of these are shown in Table 1 (Top view) and Table 2 (Side view). The Extra-Large category was reserved for the mean flow. Most of the modes are easily categorised. However, some borderline cases, such as those in the 3rd column of the medium and large rows (Table 1) that can be subjectively grouped in either the large or medium categories.

3.2. Class number optimisation

The K-means clustering algorithm was chosen as the base to compare the elbow and silhouette optimisations (Table 3). Compared to the manual classification of 4 classes, the elbow method tended to predict more than twice the number of classes. In contrast, the silhouette method predicted the same when using both top and side views. The elbow plot in Figure 2 shows more of a linear relationship than the typical elbow plot, which usually features a distinct bend or 'elbow'. This lack of a clear elbow suggests that the optimal number of clusters is not easily discernible, indicating that this method might not be effective for this application. The silhouette method will be used in further analysis.

3.3. K-means vs Hierarchical clustering

To understand how the algorithms classified the modes, the clustering algorithms were evaluated against manual classifications using a grid (Table 4). Rows show manual classes; columns show algorithm results. The performance of the two classification algorithms was based on the ratio of misclassifications compared to the manual classification, while noting the borderline cases. The hierarchical method misclassified 2 (in orange) out of 7 blocks (ratio 0.29), while K-means misclassified 4 out of 8 (ratio 0.5). The hierarchical method's results align more closely along the table's diagonal, indicating better class separation, whereas K-means deviates more from this pattern. Even though the large group is split into classes 2 and 3 in the K-means method, most of the classifications fall within the diagonal.

Table 1: Samples of modes classified on the top views

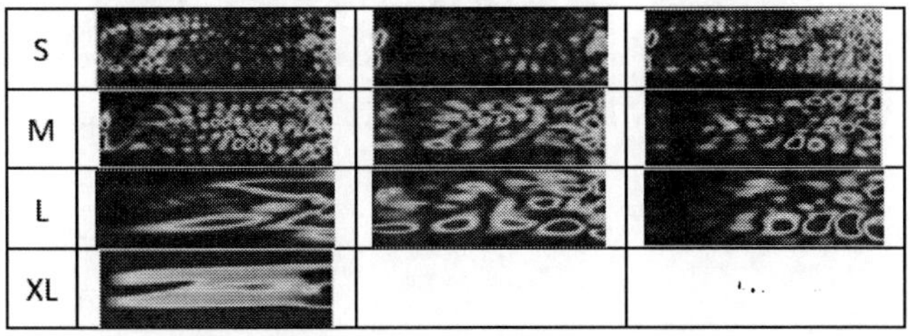

Table 2: Samples of modes classified on the side views

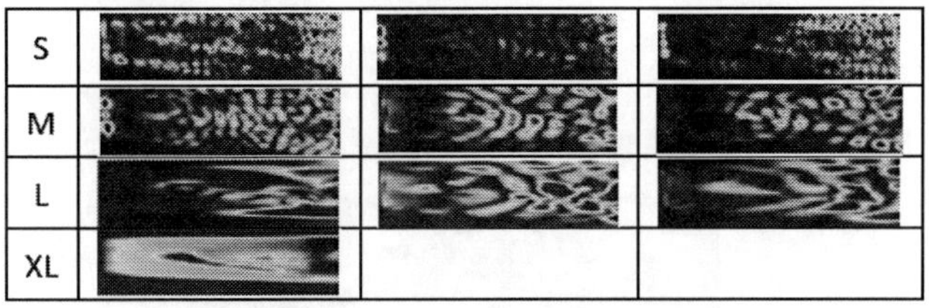

Table 3: Comparing the number of classes predicted by the elbow and silhouette optimisation on the K-means method.

	No of Classes	
View	Elbow	Silhouette
Side	10	4
Top	11	4

Table 4: Comparing hierarchical and K-means methods to the manual classification of top-view mode pairs

	Hierarchical					K-means			
	1	2	3	4	5	1	2	3	4
S				5	30			2	33
M				7	2		1	5	3
L		3	2				1	4	
XL	1					1			

Table 5 lists misclassified modes from the hierarchical clustering algorithm, which are mostly on the borderline between small and medium structures. In the K-means algorithm, mode 3 spans the S, M, and L categories, as shown in Table 6, indicating a wide range of sizes. To clarify this, structure sizes were quantified by normalising the flow field and setting a threshold at 0.7 times the average normalised velocity for each plot to isolate wake structures. The number of white pixels, shown in Figure 3, was counted to determine sizes. However, this method may overestimate average sizes because nearby structures can merge.

Table 5: Sample of misclassified hierarchical modes

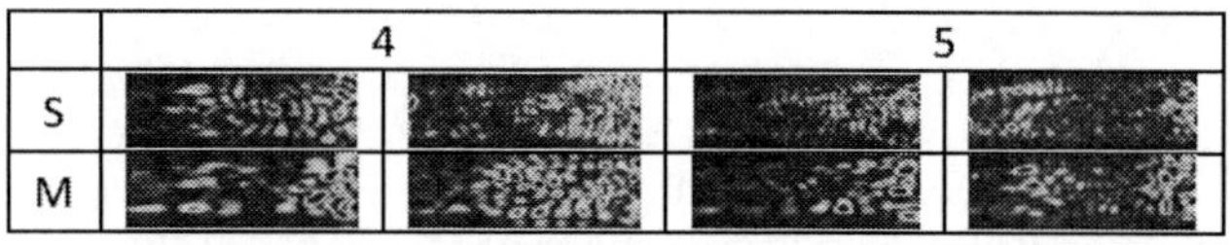

Table 6: Sample of misclassified K-means modes

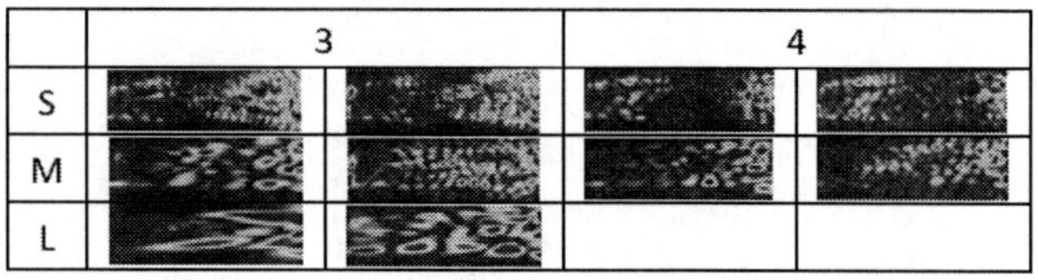

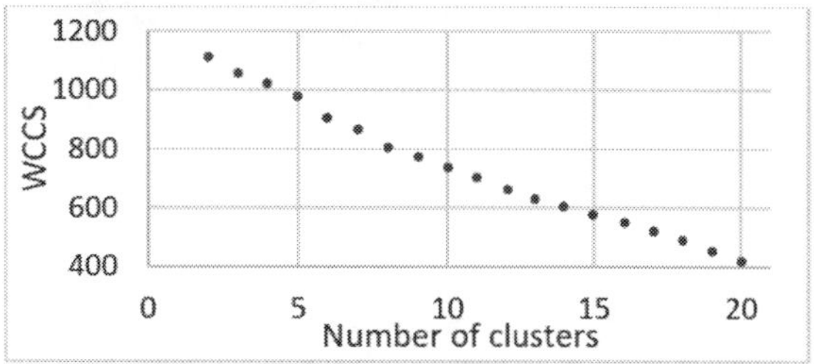

Figure 2: Elbow plot for K-means algorithm using top view

Figure 3: Threshold visualisation for determining the size of the structures

The average structure sizes were plotted against the classes for both the hierarchical and K-means methods in Figure 4. These plots display the variation of sizes across each class and indicate the overlap of average sizes between different classes. The hierarchical method distributes the modes based on size, while the K-means method tends to assign more modes to class 3, which encompasses a wide range of sizes.

The side view classifications in Table 7 show patterns like those seen in the top view. The hierarchical method grouped data along the diagonal, with the main distinction being that the 35 small modes were split into three (with two major) classes. The K-means algorithm showed greater deviation from the diagonal grouping.

For this case, plots of the average sizes show a broad range of sizes in classes 4 and 5 for the hierarchical method, and in classes 2 and 4 for the K-means method, as illustrated in Figure 5. To improve clarity, these graphs exclude the first class. Wide size ranges within certain classes are attributed to the merging of flow structures, which is depicted in Figure 6. Setting the threshold involves balancing the prevention of mergers with excluding some structures that fall below the threshold. This issue can be addressed by taking multiple slices of the views in the domain or at different times. With these considerations, the hierarchical method demonstrates more separation of mode sizes compared to the K-means method for this view.

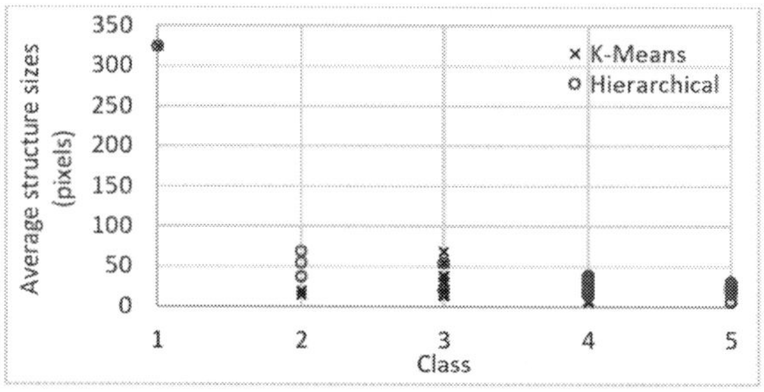

Figure 4: Average flow structure sizes for each mode by class for the Hierarchical and K-Means methods using the top view

Table 7: Comparing hierarchical and K-means methods to the manual classification of side view mode pairs.

	Hierarchical						K-means			
	1	2	3	4	5	6	1	2	3	4
S				1	24	10			4	31
M			1	8	1			9		1
L		2	2					2	2	
XL	1						1			

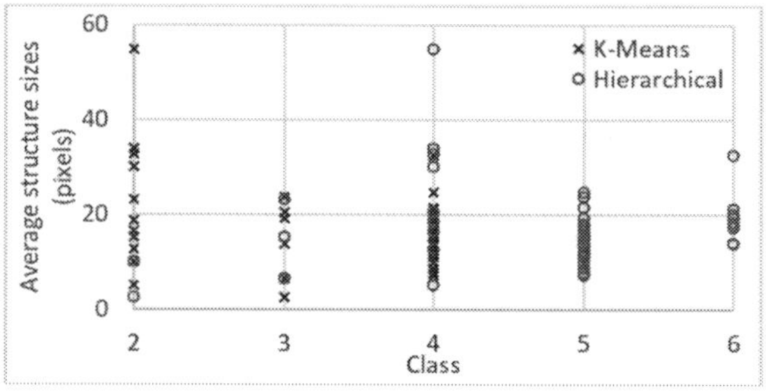

Figure 5: Average flow structure sizes for each mode by class for the Hierarchical and K-Means methods using the side view

According to the misclassification ratios presented in Table 8, the Hierarchical method resulted in fewer misclassifications than manual classification. The cases were also assessed using the ratio of modes misclassified, as shown in the fourth column; this stricter metric revealed similar patterns. Classification using side views also led to a reduction in misclassifications.

Figure 6: Structure merging with image thresholding

Table 8: Misclassification ratio comparisons

Cases	Algorithm	View	Misclass ratio	Misclassed modes ratio
H-Top	Hierarchical	Top	0.29	0.14
K-Top	K-means	Top	0.50	0.20
H-Side	Hierarchical	Side	0.33	0.06
K-Side	K-means	Side	0.43	0.10

4. Conclusions

Both algorithms classified the modes logically, but K-means grouped a broader range of sizes while the hierarchical algorithm created more distinct and numerous classes, closely matching manual results. Despite producing more groups, the hierarchical method still fit the manual sizing guide and outperformed K-means overall. Using multiple 2D spatial slices may further enhance algorithm performance.
The performance of these methods would need to be evaluated within the context of mode selection and the resulting flow-field errors from the resulting DMD reconstructions.

5. References

[1] G. V. Iungo, C. Santoni-Ortiz, M. Abkar, F. Porté-Agel, M. A. Rotea, and S. Leonardi, "Data-driven Reduced Order Model for prediction of wind turbine wakes," in *Journal of Physics: Conference Series*, Institute of Physics Publishing, Jun. 2015. doi: 10.1088/1742-6596/625/1/012009.

[2] Q. Zhang, Y. Liu, and S. Wang, "The identification of coherent structures using proper orthogonal decomposition and dynamic mode decomposition," *J Fluids Struct*, vol. 49, pp. 53–72, 2014, doi: 10.1016/j.jfluidstructs.2014.04.002.

[3] Z. Wu, D. Laurence, S. Utyuzhnikov, and I. Afgan, "Proper orthogonal decomposition and dynamic mode decomposition of jet in channel crossflow," *Nuclear Engineering and Design*, vol. 344, pp. 54–68, Apr. 2019, doi: 10.1016/j.nucengdes.2019.01.015.

[4] T. W. Muld, G. Efraimsson, and D. S. Henningson, "Flow structures around a high-speed train extracted using Proper Orthogonal Decomposition and Dynamic Mode Decomposition," *Comput Fluids*, vol. 57, pp. 87–97, Mar. 2012, doi: 10.1016/j.compfluid.2011.12.012.

[5] N. Cassamo and J. W. Van Wingerden, "On the potential of reduced order models for wind farm control: A koopman dynamic mode decomposition approach," *Energies (Basel)*, vol. 13, no. 24, Dec. 2020, doi: 10.3390/en13246513.

[6] N. Cassamo and J.-W. Van Wingerden, *Model Predictive Control for Wake Redirection in Wind Farms: a Koopman Dynamic Mode Decomposition Approach*. 2021. doi: 10.0/Linux-x86_64.

[7] J. Kou and W. Zhang, "An improved criterion to select dominant modes from dynamic mode decomposition," *European Journal of Mechanics, B/Fluids*, vol. 62, pp. 109–129, Mar. 2017, doi: 10.1016/j.euromechflu.2016.11.015.

[8] T. Krake, S. Reinhardt, M. Hlawatsch, B. Eberhardt, and D. Weiskopf, "Visualization and Selection of Dynamic Mode Decomposition Components for Unsteady Flow," Dec. 2020, [Online]. Available: http://arxiv.org/abs/2012.09633

[9] A. Corrochano, G. D'Alessio, A. Parente, and S. Le Clainche, "Hierarchical higher-order dynamic mode decomposition for clustering and feature selection," *Computers & Mathematics with Applications*, vol. 158, pp. 36–45, 2024.

[10] A. Likas, N. Vlassis, and J. J. Verbeek, "The global k-means clustering algorithm," *Pattern Recognit*, vol. 36, no. 2, pp. 451–461, 2003.

[11] A. Beikmohammadi, "Using MM principles to deal with incomplete data in K-means clustering," Dec. 2022, [Online]. Available: http://arxiv.org/abs/2212.12379

[12] X. Li, L. Yu, L. Hang, and X. Tang, "The parallel implementation and application of an improved K-means algorithm," *J. Univ. Electron. Sci. Technol. China*, vol. 46, no. 1, pp. 61–68, 2017.

[13] T. Kanungo, D. M. Mount, N. S. Netanyahu, C. D. Piatko, R. Silverman, and A. Y. Wu, "An efficient k-means clustering algorithm: Analysis and implementation," *IEEE Trans Pattern Anal Mach Intell*, vol. 24, no. 7, pp. 881–892, 2002.

[14] C. Yuan and H. Yang, "Research on K-value selection method of K-means clustering algorithm," *J (Basel)*, vol. 2, no. 2, pp. 226–235, 2019.

[15] H. Yu and X. Hou, "Hierarchical Clustering in Astronomy," Nov. 2022, doi: 10.1016/j.ascom.2022.100662.

[16] V. Cohen-Addad, V. Kanade, F. Mallmann-Trenn, and C. Mathieu, "Hierarchical clustering: Objective functions and algorithms," *Journal of the ACM (JACM)*, vol.

66, no. 4, pp. 1–42, 2019.

[17] A. K. Jain, M. N. Murty, and P. J. Flynn, "Data clustering: a review," *ACM computing surveys (CSUR)*, vol. 31, no. 3, pp. 264–323, 1999.

[18] G. J. Oyewole and G. A. Thopil, "Data clustering: application and trends," *Artif Intell Rev*, vol. 56, no. 7, pp. 6439–6475, 2023.

CFD Modelling and Experimental Validation of Pre-heating of Manganese Ore with Air to 600 °C in a Packed Bed

Sifiso N. Sambo[a,*], Mopeli Khama[b]

[a,b] *MINTEK, 200 Malibongwe Drive, Randburg, Johannesburg 2194, South Africa*

**sifisos@mintek.co.za*

Abstract

Pyro metallurgical processes often involve thermal treatment of granular mineral ores before smelting processes. The ferroalloys industry has seen increasing interest on pre-heating of ore in packed beds to reduce smelting energy consumption. Packed bed systems take advantage of the high surface area to volume ratio and are convenient for process flow configuration. The use of air as an energy transfer fluid is promising and has attracted interests for potential use on the implementation of renewable energy in minerals processing. Detailed interfacial transport information is crucial in the design and optimization of these systems. The current work-study employs Computational Fluid Dynamics (CFD) and experimental measurements to investigate the interfacial heat transport of a packed bed of lumpy manganese (Mn) ore heated with air to 600 °C. A CFD model was developed and validated for virtual prototyping of the transport process.

Experimentally, it was observed that the Mn ore can be pre-heated to 600 °C with a transfer fluid at 750 °C. Low intra-particle temperature gradients were recorded from experiments and were also predicted by the CFD model. This suggests that the air to Mn ore interfacial heat transfer is efficient. A close agreement between CFD results and experimental measurements was achieved. The CFD model provided detailed information about the transfer process and can be used to simulate interfacial transport processes of randomly packed beds of mineral ores. This will benefit the design, optimization and scale up of air fired pre-heating units.

Keywords: Pre-heating, manganese ore, experimental validation, computational fluid dynamics

1. Introduction

Heat transfer is an essential subject in unit operations across industries, including the chemical and metallurgical industry. Understanding heat transport processes is vital for process control, quality assurance, and safety. Often in the ferroalloys industry, processing of mineral resources through pyro metallurgical techniques involves handling granular materials in packings during pre-treatment stages and smelting. Packed beds provide high surface area to volume ratio [1], [2], and materials pre-treatment units and submerged arc furnace designs take advantage of this phenomenon.

A pre-treatment stage can be introduced to a smelting process to drive off volatiles and reduce feed materials moisture content. This is referred to as pre-heating. Other pre-treatment processes aim to reduce the amount of excess oxygen in the ore feed and is referred to as pre-reduction. Pre-heating and pre-reduction reduces smelting energy requirements. Other pre-treatment processes solve the problem of particle size distribution which may render smelting furnaces inoperable due to safety problems. Agglomeration and or sintering are examples of these pre-treatment processes [3], [4].

Pre-heating of Mn ores before smelting results in reduced smelting energy requirements during the production of Mn ferroalloys. Energy saving of up to 20% can be achieved when Mn ore feed is pre-heated to 600 °C [5]. The feasibility of pre-heating Mn resources with air allows the introduction of concentrated solar power energy (CSP) in minerals processing. Thanks to recent progress reported in studies on the development of air-particle solar receiver technology with working fluid temperatures above 800 °C [6], [7].

CFD models can be used to understand the heat transfer mechanism during pre-heating as they are able to predict the influence of flow field, reactor configuration, bed porosity, bed height and operating protocol on heat transfer. A CFD model that integrates gas flow through porous medium was developed by [8]. They included the convection and radiation models for gas and solid phases and applied the immersed boundary technique for the porous medium inside the computational domain.

Other CFD studies such as the work of [9] couple CFD and discrete element method (DEM) to study the transient heat transfer behaviour in packed beds. Their findings demonstrate that the influence of thermal diffusion within a particle is significant only for particles with a high Biot number (Bi > 0.1).

Given the importance of pre-heating to smelting processes, it is crucial to utilize CFD tools to inform the design and optimization initiatives. This work uses the CFD model that integrates gas flow through a porous medium to analyse heat transfer during pre-heating of Mn ore particles with the view to understand the heat transfer mechanism and ultimately determine the parameters that improve gas-solid heat transfer.

2. Methodology

A bed of Mn ore particles was treated as a homogeneous porous material and bulk properties of the porous material were specified in the model. To allow for an efficient evaluation of the temperature fields, the porous media was included in the main computational domain by employing a special type of Immersed Boundary Method named fictitious domain. In this method, the Mn ore particles are described as a collection of scalar properties such as porosity, density, thermal conductivity and tensor properties such as viscous resistance. A dynamic feed temperature from experiments was used as inlet temperature in the model.

2.1. Governing equations

The Navier Stokes equations describing gas flow through porous Mn ore particles are presented in Equations (1) to (4), which represent momentum, continuity, species conservation and energy balance, respectively. In these equations, C_p is the specific heat, ρ represents density, P is pressure, ϵ is porosity, u is fluid velocity, μ_{eff} is the effective viscosity, D_{eff} is diffusivity, h_{conv} is heat transfer coefficient, S_{Av} is the surface area to volume ratio, and F is the viscous resistance term. The superscript G refers to the gas phase.

$$\frac{\partial}{\partial t}\rho^G u + \nabla.(\rho^G uu) + \varepsilon \nabla P - \nabla.(\mu_{eff}\nabla u) - \rho^G g = -\mu_{eff} D.u - F.u \tag{1}$$

$$\frac{\partial}{\partial t}\varepsilon\rho^G + \nabla.(\rho^G u) = 0 \tag{2}$$

$$\frac{\partial}{\partial t}\varepsilon\rho^G Y_i{}^G + \nabla(\rho^G uY_i{}^G) - \nabla.(\varepsilon\rho^G D_{eff}\nabla Y_i{}^G) = 0 \tag{3}$$

$$\frac{\partial(\varepsilon\rho^G Cp^G T^G)}{\partial t} + \nabla.(\rho^G Cp^G T^G u) - \nabla.\nabla(\varepsilon k_{eff}^G T^G) = -h_{conv}S_{Av}(T^G - T^S) + S^{G,Radiation} \tag{4}$$

The fixed bed of Mn ore particles is described by continuity, species conservation and energy balance equations as shown in Equation (5) to (7). The momentum equation is not included since the solid particles are immobile. The solid and gas phase interact through heat transfer, and the coupling between the phases is achieved through the heat transfer source term.

$$\frac{\partial}{\partial t}(1-\varepsilon)\rho^S = 0 \tag{5}$$

$$\frac{\partial}{\partial t}(1-\varepsilon)Y_k^s\rho^S = 0 \tag{6}$$

$$\frac{\partial}{\partial t}(1-\varepsilon)(\rho^S C_p^s T^S) - \nabla.((1-\varepsilon)\mathbb{K}k_{eff}^s.\nabla T^S) = h_{conv}S_{Av}(T^G - T^S) + S^{S,Radiation} \tag{7}$$

The radiative source term $S^{G,Radiation}$ for the gas phase and the source term $S^{S,Radiation}$ for the solid phase are described by Equation (8) and (9), respectively.

$$S^{G,Radiation} = -4\pi\left(a\frac{\sigma T_g^4}{g}\right) + aG \tag{8}$$

$$S^{S,Radiation} = -4\pi(E_s) + a_s G \tag{9}$$

Where G refers to the incident radiation, σ is Stefan-Boltzmann constant, Es is emission of the solid matrix, a and as are the respective absorption coefficients of the gas mixture and solid matrix. A second order discretization scheme was used and PIMPLE method used for pressure-velocity coupling. A no slip boundary condition was used for both gas and solid at the walls

2.2. Boundary conditions

The convective heat transfer at the porous medium interface is defined by Equation (10). S_{AV} is the surface area to volume ratio, h is the convective heat transfer coefficient. The conductive heat flux at the porous medium interface is zero as observed in Equation (11).

$$q = S_{AV}h(T_g - T_s) \tag{10}$$

$$k\frac{\partial T}{\partial R} = 0 \tag{11}$$

2.3. Geometry and meshing

The geometry and mesh used are presented in Fig.1. The furnace used for pre-heating was 1200 mm long, a radius of 207 mm, and the Mn ore particles were packed to a bed height of 800 mm. The operating conditions are presented in Table 1.

Table 1: Furnace dimensions and operating conditions

Pre-heated gas flow rate (L/min)	1000
Initial temperature of Mn ore pellets (K)	300
Porosity	0.48

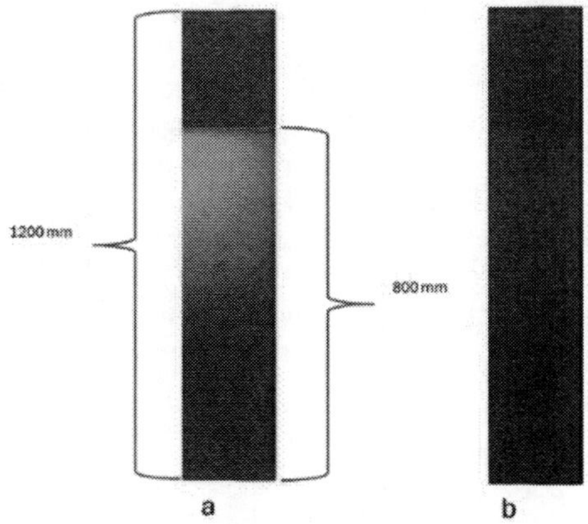

Fig. 1. Geometry and mesh

2.4. Experimental methodology

A shaft column was used to experimentally study the pre-heating process. Compressed air at 2.5 bar was heated to 750 °C in a 30 kW air heater. The air flow rate was varied from 1000-1600 Nl/min at 200 Nl/min intervals for a set of 4 tests. The feed air temperature (T1) was measured at an entry point of the packed bed located at the bottom of the test column. The packed bed temperature was measured at 200 mm lengths at the Centre of the packing of 800 mm height and 207 mm diameter. Core particle temperature was measured on one 20 mm particle located at the 600 mm temperature point. The bed was randomly packed with Mn ore of +6- 20 mm particles with a mean particle diameter of 11 mm.

A 6 mm ceramic fibre paper insulated the inside of a 4.5 mm thick mild steel to make the inner shell of the shaft column. The outside of the shell was insulated with two layers of 50 mm and 1260 °C rated fibre blankets. A 3 mm stainless steel sheet covered the outer surface of the column. The tests commenced with feed air at 25 °C and ramped up to 750 °C. Some hold-up time was afforded until the entire packing was heated to temperatures above 600 °C and the transfer process reached equilibrium. Temperature data was measured with 3 mm K-type thermocouples and transmitted to a DeltaV data acquisition system.

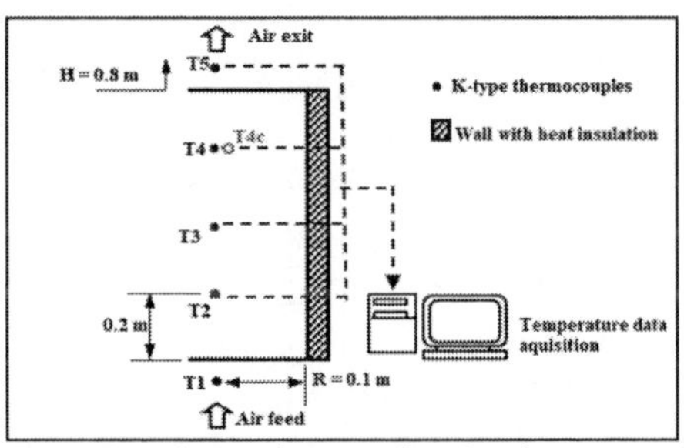

Fig. 2: Experimental set-up showing thermocouple arrangement within the packed bed

3. Results and discussion

3.1. Experimental results

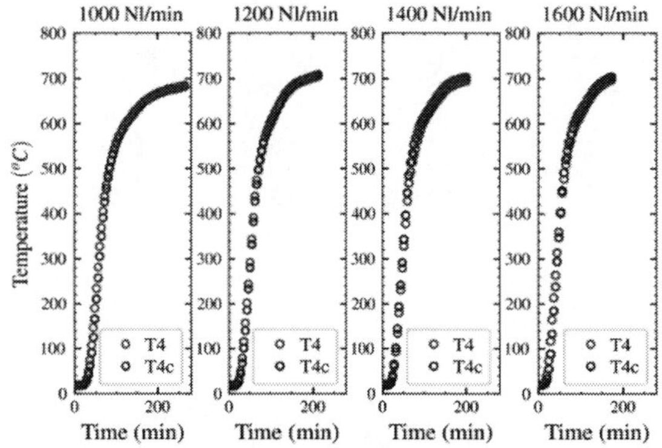

Fig. 3. Fluid temperature (T4) and core particle temperature (T4c) profiles from experiments taken at 600 mm

The feasibility of pre-heating Mn ore to 600 °C was confirmed by the outcome from experiments. The whole bed of Mn ore was pre-heated to +650 °C under a period of 280 minutes for all tests. The only difference between the different tests is the time required to pre-heat the Mn ore when different fluid flow rates are used. A comparison of the solid and the fluid phase temperatures taken at 600 mm is presented in Fig. 3. The results show a small temperature difference between the two phases. This suggests minimal intra-particle temperature gradients within the Mn ore particles.

3.2. Model results

The solid and gas phase temperature fields at 60 and 120 minutes are shown in Fig.4. The findings reveal that even after 120 minutes of pre-heating there exists an axial temperature gradient. However, the pre-heating process reached a targeted temperature within 120 minutes. To understand the heat transfer mechanism, radially averaged temperature distribution at axial temperature points is shown in Fig.5. As observed, during the initial stages of pre-heating, the lower axial positions experienced intensive gas solid heat transfer due to higher temperature gradients between the solid and gas. With time, thermal equilibrium is reached, and the heat transfer intensity shifts to the upper axial positions.

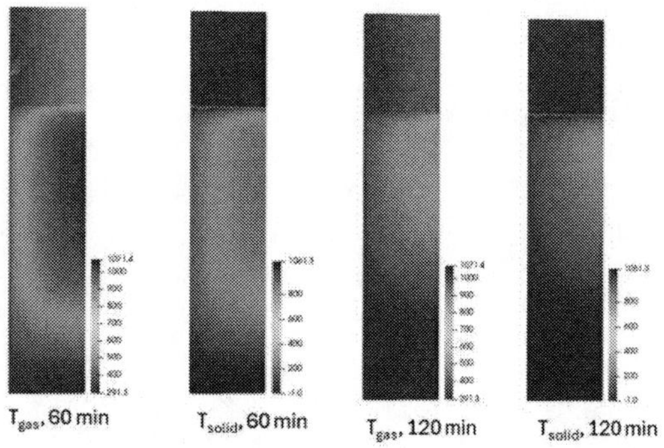

Fig. 4. Gas and solid temperature fields at 3600 and 7200 seconds

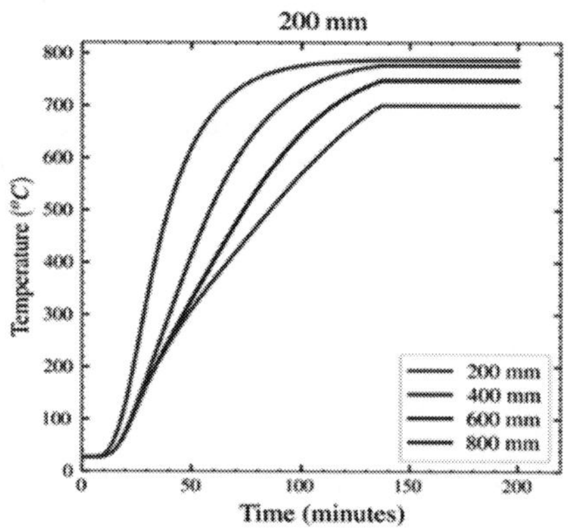

Fig. 5. Solid phase axial temperature profiles from model predictions

The model predictions for solid and gas phase temperatures follow the same trend observed in the experimental measurements as demonstrated in Fig. 6. The radially averaged temperature profiles reveal a small difference between the gas and solid phase temperatures, and this can be attributed to high heat transfer coefficient and efficient heat transfer because of high air flow rates.

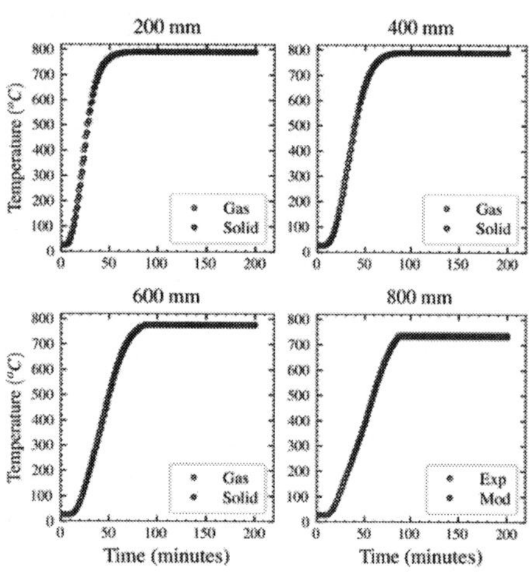

Fig. 6. Fluid and solid temperature profiles for 1000 Nl/min

3.3. Influence of air flow rate on heat transfer

The air flow rate was varied to determine its influence on heat transfer. The findings in Fig.7 indicate that high air flow rates increase the rate of heat transfer, and the targeted temperature is reached faster than when lower flow rates are used. At higher gas temperatures, the thermal boundary layer becomes thinner and heat transfer coefficient increases, resulting in increased convective heat transfer.

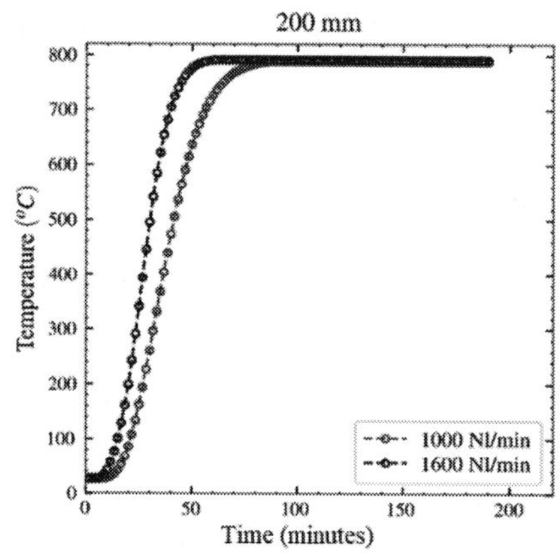

Fig. 7. Temperature profiles at 200 mm with different gas flow rates

3.4. Model validation

The temperature data from the model's predictions was compared to experimentally measured temperature data to assess whether the model predictions were viable. Experimental temperature measurements and model predictions from different axial positions are presented in Fig.8. As observed, the model results and experimental results show a close agreement in the bottom axial positions (200 mm and 400 mm) and the discrepancy manifests towards the upper axial positions.

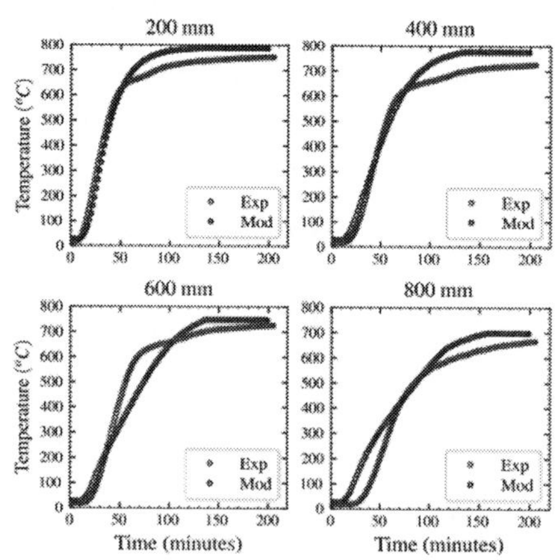

Fig. 8. Model validation results for 1000 Nl/min

4. Conclusions

The CFD model used for studying heat transfer when air is used to pre-heat Mn ore particles was found to be in close agreement with experimental measurements. Both the CFD model and experimental measurements revealed that even after 200 minutes of pre-heating there exists axial temperature gradients. However, the feed fluid temperature of 750 °C was enough to preheat the whole packed bed to temperatures above 600 °C. Low intra-particle temperature gradients were recorded from experiments and were also predicted by the CFD model. This suggests that the air to Mn ore interfacial heat transfer is efficient.

Acknowledgements

This paper is published with the permission of MINTEK. The authors acknowledge the Centre for High Performance Computing (CHPC), South Africa, for providing computational resources to this research project. The authors are grateful for MINTEK for financial support (PDR-24016/MangSolair).

References

[1] S. Bale, S. S. Tiwari, K. Nandakumar, and J. B. Joshi (2019). *Effect of Schmidt number and D/d ratio on mass transfer through gas-solid and liquid-solid packed beds: Direct numerical simulations*. Powder Technol., vol. 354, pp. 529–539. doi: 10.1016/j.powtec.2019.05.067.

[2] J. Yang, J. Wang, S. Bu, M. Zeng, Q. Wang, and A. Nakayama (2012). *Experimental analysis of forced convective heat transfer in novel structured packed beds of particles*. Chem. Eng. Sci., vol. 71, pp. 126–137. doi: 10.1016/j.ces.2011.12.005.

[3] Y. Gordon, J. Nell, and Y. Yaroshenko (2018). *Manganese Ore Thermal Treatment Prior to Smelting*. KnE Eng., vol. 3, no. 5, pp. 71–86. doi: 10.18502/keg.v3i5.2656.

[4] J. D. Steenkamp, D. Chetty, A. Singh, S. A. C. Hockaday, and G. M. Denton (2020), *From Ore Body to High Temperature Processing of Complex Ores: Manganese—A South African Perspective*. JOM, vol. 72, no. 10, pp. 3422–3435. doi: 10.1007/s11837-020-04318-x.

[5] M. Tangstad, K. Ichihara, and E. Ringdalen (2015). *"Pretreatment unit in ferromanganese production*. p. 8.

[6] M. R. I. Sarker, M. Saha, M. S. Rahman, and R. A. Beg (2016). *Recirculating metallic particles for the efficiency enhancement of concentrated solar receivers*. Renew. Energy, vol. 96, pp. 850–862. doi: 10.1016/j.renene.2016.05.047.

[7] A. Calderón et al (2018). *High temperature systems using solid particles as TES and HTF material: A review*. Appl. Energy, vol. 213, pp. 100–111. doi: 10.1016/j.apenergy.2017.12.107.

[8] P. J. Żuk et al. (2022), *OpenFOAM solver for thermal and chemical conversion in porous media*. Comput. Phys. Commun., vol. 278, p. 108407

[9] H. Che, Y. Yue, Z. Jiang, and D. Li (2024). *Numerical investigation of transient gas–solid heat transfer in a packed bed: Impact of intra-particle thermal diffusion*. Particuology, vol. 90, pp. 404–411.

Large Eddy Simulation of Expansion Wave Boundary Layer Interaction

Zahra Shaikh[a,b*], Randall T. Paton[a], Wei Hua Ho[c]

[a]*University of the Witswatersrand, 1 Jan Smuts Ave, Johannesburg 2017, South Africa*

[b]*University of South Africa, Preller St, Pretoria 0002, South Africa*

[c]*University of Cape Town, Rondebosch, Cape Town 7700, South Africa*

**corresponding author email: shaikz@unisa.ac.za*

Abstract

This study uses a Large Eddy Simulation computational model to examine the interaction effect of an externally generated expansion fan on a supersonic boundary layer. An expansion fan generator is placed at a specified distance from the interaction boundary layer. The effects on the boundary layer velocity profiles, turbulence structure and downstream flow field are examined. The boundary layer shows a thinning and subsequent thickening effect in the interaction region. Velocity profiles in the near wall region deviate from the wall log-law profile due to the development of a laminar sub-layer. An increase in upstream Mach number resulted in a steepening of the sub-layer velocity profile. Increasing the expansion wave angle delays the relaminarization effect.

1. Introduction

The interaction of expansion waves with supersonic boundary layers has been seen to considerably influence the boundary layer structure and development. Abrupt acceleration of the boundary layer, due to the favourable pressure gradient of the expansion wave, may induce relaminarization, increasing the chances of separation [1].

This phenomenon comes into play in the design of supersonic aircraft bodies, engine intakes and nozzles. A higher focus being on scramjet intakes (Fig. 1), where an expansion wave is found to occur at the end of the compression ramp interacting with the top wall boundary layer. Limited work has been done to understand the physics of the expansion-boundary layer interaction. An improved understanding of this interaction region could assist in enhancing the design of supersonic intakes and vehicles.

Large Simulation Models have been proposed to further the understanding of such interactions with boundary layers and to enable more accurate prediction of flow field properties [2].

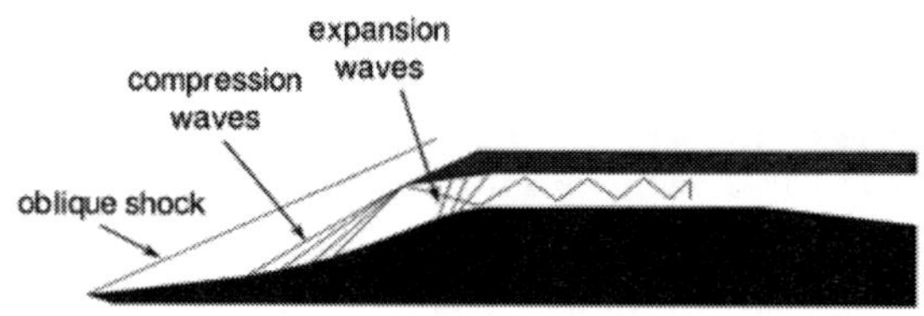

Fig. 1. Sramjet Intake

1.1. Objectives

This study seeks to examine the following:

- Examine the changes in the velocity profiles through the interaction region.
- Examine the effect the expansion wave has on the turbulence characteristics of the boundary layer.
- Study the effect of upstream Mach number (M_1) and expansion generator angle (θ) on the interaction.

1.2. Computational Method

A Large Eddy Simulation (LES) computation fluid mechanics code was used in this study. The commercial Ansys Fluent package was utilized. The Wall Adapting Local Eddy Viscosity SGS model was employed as it shows good prediction in a spatially evolving compressible turbulent boundary with reduced computation cost [3].

Fig. 2. shows the computational domain used in the study. The boundary interaction points with the forward and rearward Mach lines of the expansion wave (l_1 and l_2) and the expected interaction length (l_e) are determined numerically from the Prandl-Meyer expansion wave theory and are a function of the upstream Mach number (M_1) and the expansion generator angle (θ).

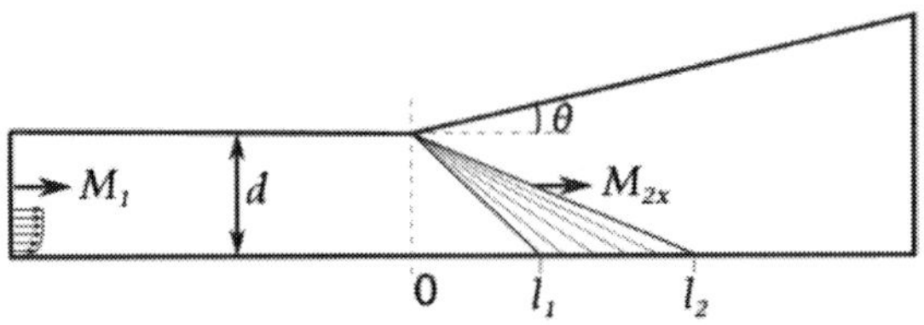

Fig. 2 Computational Domain

The interaction length (l_e), which is the distance between the points l_1 and l_2 is given by the Eqn 1.

$$l_e = d\left(\frac{1}{tan(\mu_2-\theta)} - \frac{1}{tan\mu_1}\right) \quad (1)$$

where μ_1 and μ_2 are the forward and rearward mach angles.

Table 1 shows the grid resolution for the domain for the $\theta = 10^o$ case.

Table 1. Grid Resolution

Area	Resolution $N_x \times N_y \times N_z$	Δy^{+}
Upstream of Corner	500 x 300 x 100	<1
Downstream of Corner	550 x 600 x 100	<1

2. Discussion of Results

2.1. Near Corner Flow Field

Shadowgraph and schlieren photography of expansion wave boundary layer interaction was done in a former study [4]. Fig. 3 shows the interaction region for a 5° expansion generator placed at 7 mm from the wall. The disturbance of the boundary layer is clearly visible in the image showing a thinning effect on the boundary layer followed by a thickening.

Fig. 4 shows a velocity contour of the expansion region. The black lines are the empirically calculated forward and rearward Mach lines. Here the expansion generator is placed at a close distance of 3mm from the interaction wall.

Fig. 3. Schlieren photograph. $\theta = 5^o$, d = 7 mm, Mach 2.5

The tail end of the expansion is disturbed by reflected expansions of the wall. A clearly visible thinning of the boundary layer is seen at the beginning of the expansion followed by a subsequent thickening.

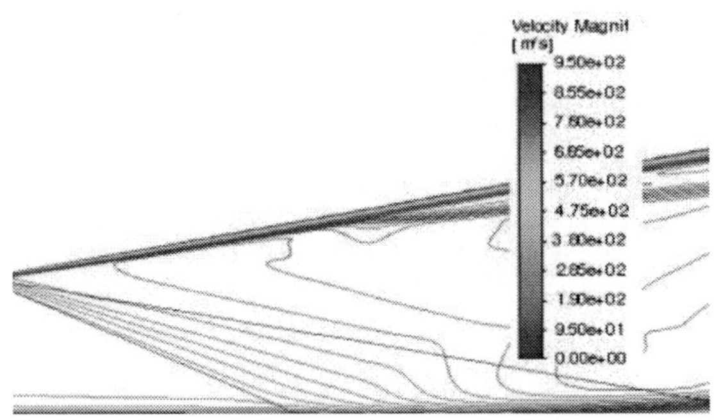

Fig. 4. Expansion region velocity flow field. $\theta = 10^o$, d = 3 mm, Mach 2.5

2.2. Velocity Profiles through the interaction length

The boundary layer velocity profiles through and after the interaction region are plotted in Fig. 5 and can be compared to the upstream profile. The position is given as a ratio of the total interaction length, l_e. The expansion wave is seen to have considerable influence on the near wall boundary layer. Considering that the expansion generator gap was very small, the strength of the expansion could be sufficient to cause relaminarization resulting in a laminar 'sub-layer' in the near wall region.

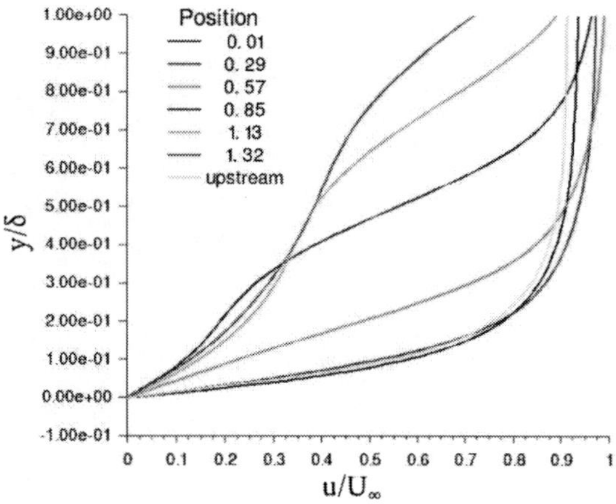

Fig. 5. Boundary layer velocity profiles. Position given as x-l2/le. θ = 10°, d = 3 mm, Mach 2.5

3. Parameter effects on the interaction region

3.1. Effect of Expansion Generator Angle

Fig. 6 shows a plot of the velocity profiles at θ = 10° and θ = 15°. The deviation of the boundary layer is seen to occur earlier on for the 10° case. This is believed to be a result of the difference in the interaction length between the two cases: the empirically calculated interaction lengths being l_e = 10.69 mm for the 10° wedge and l_e = 50.46 mm for the 15° wedge. For lower expansion angles, the effects on the boundary layer are seen to progress faster than at higher angles.

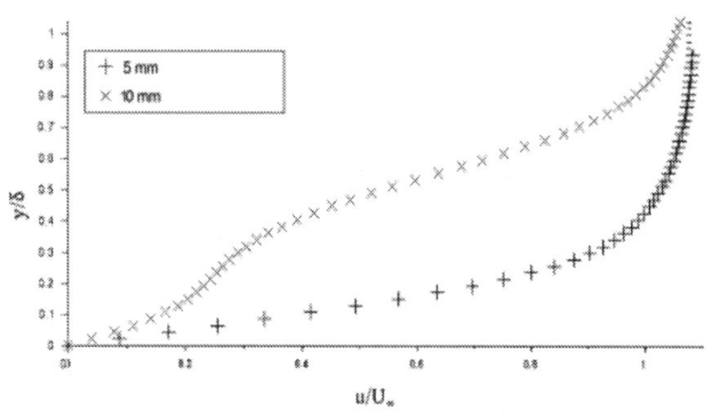

(a)

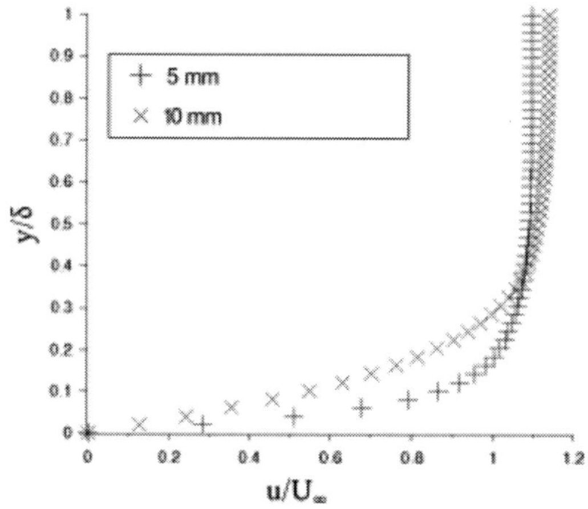

(b)

Fig. 6. Velocity Profiles at 5mm and 10 mm into the interaction region. Mach 2.5, d = 3mm (a) θ = 10°, (b) θ = 15°

3.2. Effect of Upstream Mach Number

The velocity profile plots at M = 3.0 at two locations in the interaction region are shown in Fig. 7. An increase in Mach number is seen to steepen the velocity profile in the laminar sub-boundary layer in the near wall region. A more drastic effect in the boundary layer is seen with an incoming free stream of Mach 3 than at Mach 2.5 in Fig.6b.

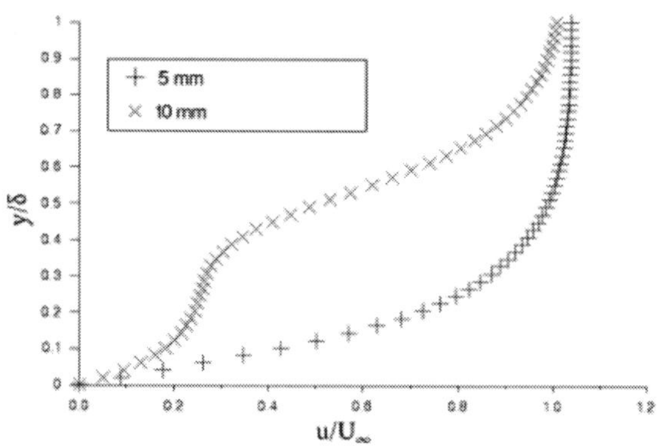

Fig. 7. Velocity Profiles at 5 mm and 10 mm into the interaction region. Mach 3, d = 3mm, θ = 15°

4. Conclusions and Future Work

The study conducted shows the significant effects of relaminarization on a turbulent compressible boundary layer at Mach 2.5 – 3. An impinging expansion wave placed 3mm from the wall shows a significant deviation in the velocity profile from the log-law effect in the near wall region is significant when the expansion wave.

Altering the incoming free stream Mach number as well as the expansion wave angle both influenced the interaction and showed changes in the boundary layer structure and had an effect on the relaminarization of the near wall region.

Future developments to this study would include experimental wind tunnel testing to compliment the numerical work. Additional data will be extracted and examined including velocity fluctuations and skin friction properties. Examining the direct effect of the expansion generator gap from the boundary wall will also be considered.

Acknowledgements

The authors would like to acknowledge the support of the Centre for High Performance Computing (CHPC) for the use of their computing facilites for the purposes of this work and for their ongoing technical support.

References

[1] Goldfeld MA, Nestoulia RV , Starov AV (2000). *The Boundary Layer Interaction with Shock Wave and Expansion Fan.* Journal of Thermal Science, **9**: 9–2.

[2] Dolling DS (2001). *Fifty Years of Shock-Wave/Boundary-Layer Interaction Research: What Next?* AIAA Journal. 39:8

[3] Ben-Nasr O (2017), Assessment of subgrid-scale modeling for large eddy simulation of a spatially-evolving compressible turbulent boundary layer. Computers and Fluids, **151**.

[4] Thomas J, Skews BW, Paton RT. *Expansion Wave/Boundary Layer Interaction.* 30th International Symposium on Shock Waves Proceedings.

Effect of the placing of the coolant risers in a micro nuclear reactor on the thermal-fluid behaviour during a DLOFC

CG du Toit[a,*], W Boyes[b], JF Slabber[b]

[a]*North-West University, Private Bag X6001, Potchefstroom 2520, South Africa*

[b]*University of Pretoria, Private Bag X20, Hatfield 0028, South Africa*

**corresponding author email: jat.dutoit@nwu.ac.za*

Abstract

The thermal-hydraulic behaviour of the Advanced Micro Reactor (AMR) is modelled considering two options using the system code Flownex SE. In option 1 the coolant flows from the inlet to the upper plenum (UP) through risers placed in the outer reflector (OR). In option 2 the coolant flows through the gap between the core barrel (CB) and the reactor pressure vessel (RPV). Under steady-state conditions the maximum fuel and graphite temperatures are 1032.4 °C and 809.0 °C for option 1 respectively, whilst the corresponding temperatures for option 2 are 1012.4 °C and 786.8 °C respectively. The maximum temperatures of the OR, CB and RPV for options 1 and 2 are respectively 397.9 °C, 357.7 °C, 252.4 °C and 593.2 °C, 440.9 °C, 315.6 °C. Under Depressurized Loss of Forced Cooling (DLOFC) conditions that heat rejected by the reactor cavity cooling system (RCCS) in option 1 rises from 86.1 kW to a maximum of 109 kW, whilst in the case of option 2 it rises from 142.6 kW to a maximum of 148.4 kW. The maximum rate at which heat is released by the solids is 47.2 kW in option 1, whilst for option 2 the rate is 78.7 kW.

1. Introduction

Typically, in (V)HTR reactors the coolant flows downwards through the core [1-2]. Entering the reactor vessel at the bottom through the outer annulus of the co-axial duct, the cold coolant can flow upwards either in the gap between the CB and the RPV [1, 3-5], or in riser holes in the OR [2, 6-8]. In none of the reactor analyses a comparison was made of the effect that the placing of the up-flow channel(s) of the coolant has on the thermal-hydraulic behaviour of the reactor. The nearest to that was a study by Kim et al. [9] who considered in addition to the coolant flowing up through riser holes in the OR, a vessel cooling system (VCS) where cold helium flows upwards between the CB and RPV but does not enter the core.

The prismatic block Advanced Micro Reactor (AMR) shown Fig. 1 is a high temperature gas-cooled reactor under design to produce 10MW thermal power [10]. The active core consists of three rings of fuel block assemblies, and nineteen control rod (CR) block assemblies enclosed by a central inner reflector (IR) and an outer reflector (OR). The core contains 1260 fuel assemblies (FAs) each consisting of a silicon carbide tube housing coated particle fuel immersed in a lead-bismuth eutectic alloy (LBE). The fuel and CR assemblies are placed in borings in the graphite (GR) moderator blocks allowing annuli for the helium coolant flow.

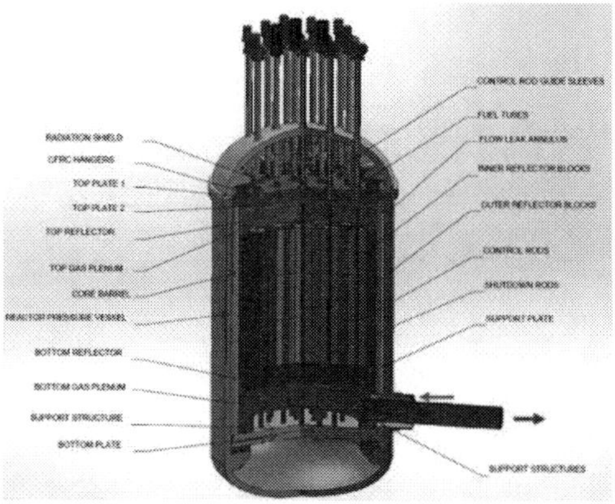

Fig. 1 Cut view through the AMR reactor [10]

The purpose of this paper is to study the effect on the thermal-hydraulic behaviour of the reactor of the coolant flowing through riser holes in the OR, or else flowing in the gap between the CB and RPV from the inlet to the UP.

Under steady-state conditions helium enters the cold inlet, blue arrow in Fig. 1, at a flowrate of 4.44154 kg/s for option 1 and 4.41625 kg/s for option 2 and a temperature of 320 °C. The helium moves up through the riser channels into the UP. The helium flows down through the core where it collects heat, into the lower plenum (LP) and then exits the hot outlet, red arrow, at a temperature of 750 °C and a pressure of 4 MPa.

2. Flownex model of the AMR

A 2D axi-symmetric network model has been set up of the AMR using the system code Flownex SE [11]. The upper and lower domes shown in Fig. 1 are not included in the Flownex model. In the axial direction the reactor has been discretized into 12 increments: top reflector (TR); UP; seven core increments (I – VII), bottom reflector (BR); LP; and bottom support (BS). In the radial direction the reactor has been discretised into 11 rings: Inner reflector (IR + CR); First fuel ring (GR1 + FU1); Second fuel ring (GR2 + FU2 + CR); Third fuel ring (GR3 + FU3 + CR); OR + CR; Gap between OR and CB; CB; Gap between CB and RPV; RPV; Cavity between RPV and RCCS; and RCCS face plate.

The Flownex model consists of solid and fluid nodes (CVs) that account for the conservation of mass, momentum and energy. The nodes are connected by components that account for the transfer of mass, momentum and energy between the nodes. The Flownex model consists of three inter-connected networks simulating the solids, flow and fuel. The details of the geometry of the AMR and the layouts of the networks can be found in Boyes et al. [11].

The temperature at the outer surface of the RCCS face plate was prescribed as 65 °C [12-13]. The power generated in the fuel under steady-state and transient conditions [12] was distributed in the axial direction according to a sine profile. At the beginning of the DLOFC the inlet mass flow rate decreases to 0.0001 kg/s and the outlet pressure decreases to 103 kPa according to a cosine shaped profile over a period of 5 s. The decay power is calculated using the Wigner-Way correlation [14].

Flownex employs a finite volume-based implicit pressure correction method [15] to obtain the steady-state solution of the conservation equations and a non-iterative algorithm to obtain the transient solution.

3. Results

Two sets of results are discussed. Firstly, a comparison is made between the temperatures obtained for options 1 and 2 under steady-state conditions. Secondly, the heat rejected by the RCCS and the heat released by the solids for options 1 and 2 during a DLOFC are discussed.

3.1. Steady-state results

The temperature distributions in the reactor for options 1 and 2 under steady-state conditions are shown in Fig. 2 and Fig. 3 respectively. In the case of the CB and RPV only the temperatures for the inner surfaces are shown. The average temperatures T_ave are the average temperatures of I to BR for the IR, GR1, GR2 and GR3, and the average temperatures of TR to BS for the OR, CB and RPV. The temperature difference (Taj-Tai) is the difference between the average temperature for a radial position and the average temperature of the previous radial position.

It can be observed in Fig. 2 and Fig. 3 that the maximum fuel temperature is 1032.4 °C for option 1 compared to the 1012.4 °C for option 2, and maximum graphite temperature for option 1 is 809.0 °C compared to 786.8 °C for option 2. The maximum temperatures for the OR, CB and RPV for option 2 are respectively 593.2 °C, 440.9 °C and 315.6 °C compared to the corresponding values of 397.9 °C, 357.7 °C and 252.4 °C for option 1.

Fig. 2 and Fig. 3 reveal further that the average IR, GR1 and GR2 temperatures for option 1 are higher than the corresponding temperatures for option 2, whilst the average GR3, OR, CB and RPV temperatures for option 2 are higher than the corresponding temperatures for option 1. Thus, in the case of option 1 the temperatures of the inner part of the reactor are higher than those for option 2, whilst in the case of option 2 the temperatures of the outer part of the reactor are higher than those for option 1. For IR, GR1, GR2 and GR3 from the top of the core (I) to the bottom of the core (VII) the graphite temperature increases on average by 403.6 °C in the case of option 1, whilst for option 2 the corresponding increase is 420.4 °C.

In option 1 the OR transfers 618.5 kW through convection to the coolant flowing up the risers resulting the temperature of the coolant increasing from 320 °C at the inlet to 346.8 °C at the UP. In the case of option 2 the effective amount of convection heat transferred to the coolant is 97.8 kW resulting in the coolant heating up from 320 °C at the inlet to 324.3 °C at the UP.

Because of the large amount of heat that must be transferred to the coolant by the OR, as well as the fact that no heat is generated in the OR, the temperature drop between GR3 and OR in case of option 1 is on average 220.3 °C. This also makes provision for the heat rejected by the RCCS. This leads to the average CB and RPV temperatures of 335.3 °C and 238.6 °C and the amount of 86.1 kW rejected by the RCCS. The coolant flow path therefore results in the lowering of the OR, CB and RPV temperatures.

In the case of option 2 the temperature of the coolant flowing in the gap between the CB and RPV is primarily responsible for the average temperature of 305.2 °C of the RPV. This results in the amount of 142.6 kW of heat being rejected by the RCCS. This amount of heat along with the effective amount of heat transferred to the coolant then leads to the average temperatures of the CB and OR being 391.7 °C and 503.1 °C respectively. The average temperature drop between GR3 and OR in the case of option 2 is therefore 99.4 °C. In this case the coolant flow path results in the higher temperatures of the OR, CB and RPV compared to the corresponding temperatures for option 1.

	IR	GR1	FU1	GR2	FU2	GR3	FU3	OR	CB	RPV
TR	329.9	329.9		329.8		329.7		329.5	300.4	219.2
UP								341.8	311.7	223.2
I	368.3	360.7	406.0	360.8	405.9	358.5	404.9	345.7	313.8	224.1
II	421.4	424.0	605.0	421.9	603.4	415.9	599.3	357.8	325.0	231.7
III	520.1	537.9	836.7	531.6	832.6	517.6	823.4	373.6	338.0	239.8
IV	628.9	665.3	988.8	653.5	981.3	629.7	965.8	388.3	349.8	247.1
V	714.9	764.9	1032.4	747.8	1020.9	714.6	998.4	397.1	356.9	251.6
VI	759.5	809.0	962.5	788.4	947.3	747.0	916.9	397.9	357.7	252.4
VII	774.7	801.6	855.9	774.9	837.2	711.6	797.3	391.4	352.8	249.9
BR	724.6	795.3		764.0		705.4		386.2	347.6	245.1
LP								365.4	334.5	240.2
BS	389.4	388.6		385.3		379.5		369.4	335.1	239.4
T_ave	614.1	644.8		630.3		600.0		370.3	335.3	238.6
Taj-Tai		-30.8		14.5		30.3		220.3	35.1	96.6

Fig. 2. Temperature [°C] distribution in reactor for option 1 under steady-state conditions.

	IR	GR1	FU1	GR2	FU2	GR3	FU3	OR	CB	RPV
TR	346.0	346.1		346.2		346.4		346.7	323.4	262.4
UP								358.2	335.5	293.5
I	346.0	338.3	383.8	338.8	383.8	339.3	383.9	361.6	336.6	306.6
II	399.1	401.2	583.7	400.2	582.9	399.7	582.4	393.8	348.3	314.9
III	497.9	514.7	816.7	510.5	814.0	507.1	812.0	451.9	369.4	315.6
IV	607.1	642.0	969.2	633.6	963.9	626.7	959.8	517.3	406.5	310.6
V	693.5	742.0	1012.4	729.4	1004.2	719.2	997.7	568.0	429.7	311.3
VI	738.4	786.8	941.8	772.0	930.9	758.7	921.5	593.2	440.9	311.3
VII	754.8	781.9	835.4	765.5	823.1	741.9	809.2	592.1	438.7	310.4
BR	702.9	774.6		751.2		727.7		546.9	411.5	311.0
LP								486.1	433.1	308.6
BS	492.5	492.1		489.6		485.1		477.2	426.8	306.4
T_ave	592.5	622.7		612.6		602.5		474.4	391.7	305.2
Taj-Tai		-30.2		10.0		10.1		99.4	82.7	86.5

Fig. 3. Temperature [°C] distribution in reactor for option 2 under steady state conditions.

3.2. DLFOC results

When the initial depressurization is completed after the occurrence of the DLOFC, the heat rejected by the RCCS, $\dot{Q}_{RCCS}$, comprises the decay heat, $\dot{Q}_{DH}$, generated and the accumulated heat, $\dot{Q}_{ST}$, released by the solids. The RCCS heat balances as functions of time for options 1 and 2 are depicted in Fig. 4. The curves for option 1 are shown as solid lines, whereas the curves for option 2 are shown as dashed lines.

In the case of option 1 $\dot{Q}_{RCCS}$ increases to reach 109 kW at 8.680 h, whilst in the case of option 2 $\dot{Q}_{RCCS}$ increases to reach 148.4 kW at 7.842 h, before gradually decreasing.

During the initial depressurization when the value of $\dot{Q}_{DH}$ decreases sharply the solids release heat to the coolant due to the thermal inertia of the solids until the coolant mass flow rate becomes negligible. The solids then accumulate heat again until $\dot{Q}_{RCCS} \geq \dot{Q}_{DH}$, as seen in Fig. 4, after which solids start to release heat as part of the cool down of the reactor.

In the case of option 1 $\dot{Q}_{ST}$ reaches a maximum of 47.2 kW at 12.856 h. It can also be observed that $\dot{Q}_{ST}$ is always less than $\dot{Q}_{DH}$ with its maximum contribution to $\dot{Q}_{RCCS}$ being 45.5%. At 40 h $\dot{Q}_{ST}$ still comprises 41% of $\dot{Q}_{RCCS}$.

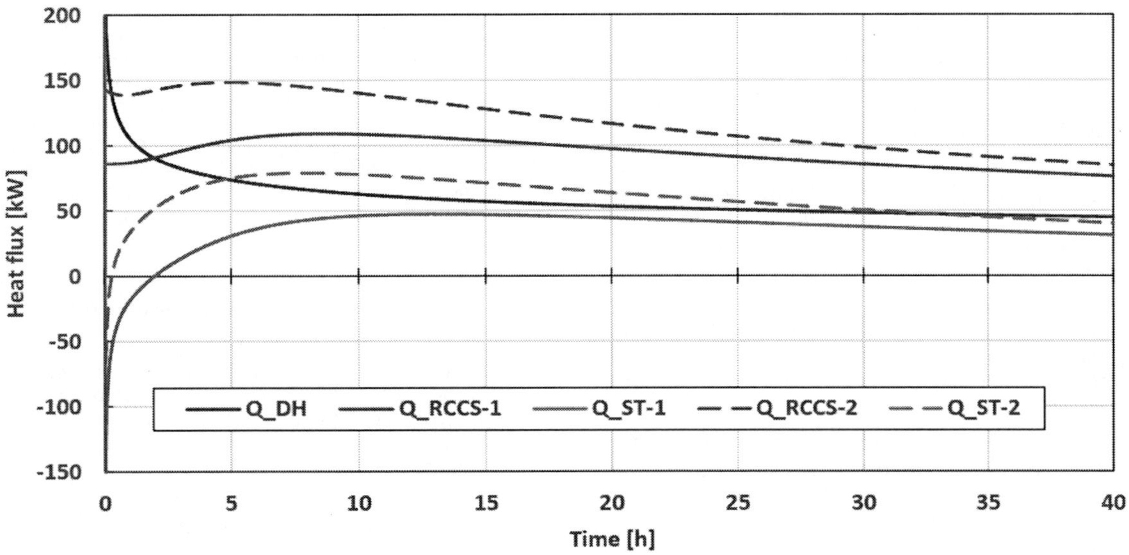

Fig. 4. RCCS heat balances [kW] as functions of time for options 1 and 2.

When considering option 2 it is found that $\dot{Q}_{ST}$ reaches a maximum of 78.7 kW at 7.856 h. Furthermore, $\dot{Q}_{ST}$ exceeds $\dot{Q}_{DH}$ for the period 4.803 h to 32.914 h. This can be attributed to the large amount of heat accumulated in the OR, CB and RPV during steady-state conditions. During this period the maximum contribution of $\dot{Q}_{ST}$ to $\dot{Q}_{RCCS}$ is 55.7%. At 40 h $\dot{Q}_{ST}$ still comprises 47.2% of $\dot{Q}_{RCCS}$. The heat released by the solids, therefore, contributes more to the heat rejected by the RCCS in option 2 compared to option 1.

Table 1. Amounts of heat released by the solids for options 1 and 2 over selected periods.

	Option 1 [kJ]	Option 2 [kJ]
0 < t < 20 h	2426222.6	4897348.4
20 h < t < 40 h	2699346.7	3655259.0

The total amount of heat generated, rejected or released over a given period can be obtained by integrating the heat rate over the relevant period. The heat released by the solids during the periods 0 < t < 20 h and 20 h < t < 40 h for options 1 and 2 are summarized in Table 1. For 0 < t < 20 h the solids in the case of option 2 release 2.02 times more heat than the solids in option 1, whilst for 20 h < t < 40 h the solids in option 2 release 1.35 times more heat than the solids in option 1. Over the same periods the decay heat generated is respectively 4.97 GJ and 3.48 GJ. For 0 < t < 40 h the solids in option 2 contribute 50.3% to the total amount of heat rejected by the RCCS compared to 37.7% in the case of option 1. In both cases the solids thus contribute a significant portion of the heat rejected by the RCCS.

4. Conclusions

In this study the thermal-hydraulic behaviour of the AMR was modelled using Flownex SE. Two options were considered, namely (1) where the coolant flows in risers that are placed in the OR, and (2) where the coolant flows in the gap between the CB and RPV from the inlet to the UP.

Under steady-state conditions the inner part (IR to GR2) is on average hotter for option 1, whilst for option 2 the outer part (GR3 to RPV) is on average hotter.

The DLOFC analysis showed that the solids contribute a significant portion of the heat rejected by the RCCS. In option 2 the amount of heat released by the solids is considerably more compared to the corresponding amount of heat released in option 1.

From the results it can be concluded that the statement by Simoneau et al. [12] that when the coolant flows up to the UP between the CB and the RPV, it limits the heat losses and protects the RPV from excessive temperatures, it not entirely true. In the case of the AMR the configuration where the coolant flows up in the riser holes in the OR is far more optimal.

Acknowledgements

The authors wish to thank the University of Pretoria and the North-West University for the opportunity to perform the study.

References

[1] Kim M-H, Lim H-S, Lee W-J (2008) *A thermal-fluid assessment of a cooled-vessel concept for a VHTR*, Nuclear Engineering & Design, **238**: 3360-3369.

[2] Oh C, Kim E, Schultz R, Patterson M, Petti D (2009) *Thermal—hydraulics of the Very High Temperature Gas Cooled Reactor*, 13th International Topical Meeting on Nuclear Reactor Thermal Hydraulics, Kanazawa, Japan, paper N13P1197.

[3] Tochio D, Nakagawa S (2011) *Numerical simulation of three-diemnsional thermal-hydraulic behavior for HTTR (High Temperature Test Reactor)*, Nuclear Engineering & Design, **241**: 1616-1626.

[4] Gutowska I, Woods B, Cadell SR (2019) *CFD modeling of the OSU High Temperature Test Facility inlet plenum flow distribution during normal operation*, Nuclear Engineering & Design, **353**, 110216.

[5] Wang C, Sun X, Sabharwall P (2020) *CFD Investigation of MHTGR Natural Circulation and Decay Heat Removal in P-LOFC Accident*, Frontiers in Energy Research, **8**, 129.

[6] Zhang Z, Wu Z, Wang D, Xu Y, Sun Y, Li F, Dong Y (2009) *Current status and technical description of Chinese 2x250 MW_{th} HTR-PM demonstration plant*, Nuclear Engineering & Design, **239**: 1212-1219.

[7] Janse van Rensburg JJ, Kleingeld M (2010) *A CFD methid to evaluate the integrated influence of leakage and bypass flows on the PBMR Reactor Unit*, **240**: 3841-3850.

[8] Xhou Y, Zhou K, Ma Y, Sui Z (2013) *Thermal hydraulic simulation of reactor of HTR-PM based on thermal-fluid networdk and SIMPLE algorithm*, Progress in Nuclear Energy, **62**: 83-93.

[9] Kim M-H, Tak N-I, Lim H-S (2010) *Thermal-fluid assessment of the design options for reactor vessel cooling in a prismatic core VHTR*, Annala of Nuclear Engineering, **37**: 1774-1782.

[10] Boyes WA, Slabber J, CU Toit C, Van Heerden F (2023) *Evaluation of the Basic Neutronic and Thermal-Hydraulics for Safety Evaluation of the Advanced Micro Reactor (AMR)*, Nuclear Science, **8**, 8-29.

[11] M-Tech Industrial, Flownex Simulation Environment, http://www.flownex.com, Version 8.16.0.5518, 2024

[12] Simoneau JP, Champigny J, Mays B, Lommers L (2007) *Three-dimensional simulation of the coupled convective, conductive, and radiative heat transfer during decay heat removal in an HTR*, Nuclear Engineering and Design, **237**: 1923-1937.

[13] Lommers LJ, Mays BE, Shahrokhi F (2014) *Passive heat removal impact on AREVA HTR design*, Nuclear Engineering and Design, **271**: 569-577.

[14] Garland WJ, Saunders F (1998) *Decay heat estimates for MNR*, Technical Report 1998-05, McMaster University, Canada.

[15] Greyvenstein GP (2002) *An implicit method for the analysis of transient flows in pipe networks*, International Journal of Numerical Methods in Engineering, **53**: 1127-1143.

A modified effectiveness-NTU method for combined heat and mass transfer in air cooling and dehumidifying coils

Pieter Rousseau[a*], Rashid Haffejee[a], Ryno Laubscher[a]

[a]*Department of Mechanical and Mechatronic Engineering, Stellenbosch University, Stellenbosch 7600, South Africa*

**prousseau@sun.ac.za*

Abstract

Combined heat and mass transfer that takes place in air cooling and dehumidifying coils form an important part of the analysis of heat pumps and heating, ventilating and air conditioning (HVAC) systems. Its design and analysis are typically based on the log mean enthalpy difference (LMHD), as opposed to the log mean temperature difference (LMTD) which is valid in the case of sensible heat transfer only. However, when simulating heat exchangers as part of larger integrated systems, it is more convenient to apply the effectiveness-NTU method since it eliminates the risk of temperature profiles crossing over in between iterations and resulting in unrealistic results. This paper proposes a modified effectiveness-NTU model for combined heat and mass transfer in air cooling and dehumidifying coils. It is based on a modified LMTD approach, rather than LMHD, but while incorporating the effects of mass transfer. The model is first validated and then demonstrated using a case study of a perfectly counterflow heat exchanger configuration.

Keywords: Combined heat and mass transfer, HVAC Heat exchanger modelling, LMTD, Effectiveness-NTU

1. Introduction

Cooling coils are a prominent component in HVAC and heat pump systems. These coils are typically used for air cooling and dehumidification, facilitating heat transfer between air and the fluid in the coils. Operating conditions for cooling coils are dependent on several factors, including air and coolant temperature, air humidity and flow rates. These conditions influence whether the coil operates as a dry, dry and wet, or a purely wet coil. In the wet cases humid air flowing over the coil surface drops below the dew point temperature, resulting in moisture condensation from the humid air in the freestream to the coil surface. Capturing the complex combined heat and mass transfer process is imperative to characterization of the performance of the cooling coil and thus overall system thermal performance.

Detailed design methodologies for cooling coils have been researched extensively and established in sources such as the VDI Heat Atlas [1]. These design methodologies focus on the solution of fundamental governing equations together with correlations for heat and mass transfer. Simplified models have also been proposed, such as that by Elmahdy & Mitalas [2] which is based on the LMHD for modelling the wet sections of the coil including combined heat and mass transfer. For dry sections of the coil, the LMTD is used with only sensible heat transfer taking place. Several modified versions of the LMHD and LMTD methods have also been developed. Liang [3] implemented a modified LMHD accounting for a unit and non-unit Lewis factor. Cui et al. [4] developed a modified LMTD method for indirect evaporative heat exchangers (air to air), based on the wet-bulb temperature and an enthalpy to wet-bulb temperature ratio.

For simulation of heat exchangers in integrated systems, the effectiveness-NTU method may be a more robust method as it eliminates the risk of temperature profiles crossing over (i.e. the temperature of the hot fluid stream dropping below the temperature of the cold fluid stream) in between iterations which can result in unrealistic results. Therefore, using a modified effectiveness-NTU method for combined heat and mass transfer can be advantageous compared to the LMHD. Hasan [5] developed a modified effectiveness-NTU method for indirect evaporative cooling (air to air) applications based on an enthalpy gradient between air in the dry and wet passages. Wang et al. [6] developed a modified effectiveness-NTU method to evaluate the total capacity of chilled-water cooling coils under wet conditions. The formulation was derived based on the enthalpy difference between the inlet moist air and saturated air enthalpy at the water inlet temperature. Braun [7] developed an effectiveness-NTU method for both dry and wet coils also based on an enthalpy difference between the moist air and saturated air at the surface temperature. Gollamudi et al. [8] investigated a coupled effectiveness-NTU method to design and evaluate the performance of liquid-air-membrane energy exchangers with heat and mass transfer. The coupled effectiveness-NTU method solves separate equations for the sensible effectiveness and moisture transfer effectiveness.

This paper proposes a modified effectiveness-NTU method for combined heat and mass transfer in air cooling and dehumidifying coils. The method is based on a modified LMTD approach, rather than the LMHD method. It can incorporate effects of mass transfer for wet sections of the coil and reduces to the conventional LMTD method for dry sections of the coil where mass transfer is not present. The method is derived based on a temperature difference, as opposed to an enthalpy difference typically observed in literature. This allows for application of the method for cooling coils not only using water as coolant, but also other refrigerants. Moreover, the method proposed in this work characterises the combined heat and mass transfer through a single effectiveness value, in contrast to the coupled approach which uses a sensible effectiveness and a moisture transfer effectiveness. The developed methodology is validated using the design methodology applied in the VDI Heat Atlas [1] for a cross-counterflow heat exchanger. Thereafter, the proposed method is applied to a case study of simplified chilled water coils at various inlet water temperatures to analyse dry coil conditions and partially wet-dry coil conditions. The modified effectiveness-NTU method is also compared to the solution that employs the fundamental component characteristic based on the average temperature difference for each increment.

2. Methodology

Figure 1 shows the combined heat and mass flows for a control volume in a perfectly counterflow finned-tube heat exchanger.

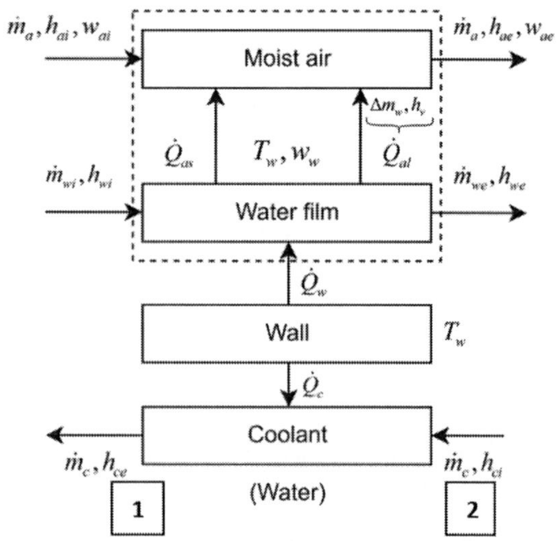

Figure 1: Schematic of combined heat and mass flows for a counterflow heat exchanger

Inlet moist air flows on the finned side counter flow to the coolant flowing inside the tubes, which in this case is chilled water. If the absolute humidity of the incoming air, w_{ai}, is greater than that of the saturated air at the surface temperature, w_{sat,T_w}, condensation occurs and the surface is wet, otherwise it is dry with only sensible heat transfer occurring. For cooling and dehumidification, the actual direction of the heat and mass flows between the moist air and water film shown in Figure 1 are reversed and the values will therefore be negative.

2.1. Governing equations

The steady state dry air, vapour, and water film mass balance equations can be written as:

$$\dot{m}_{ae} = \dot{m}_{ai} = \dot{m}_a \tag{1}$$

$$\dot{m}_a w_{ae} = \dot{m}_a w_{ai} + \Delta\dot{m}_w \tag{2}$$

$$\dot{m}_{we} = \dot{m}_{wi} - \Delta\dot{m}_w \tag{3}$$

The moist air mixture energy balance is given by:

$$\dot{m}_a h_{ae} = \dot{m}_a h_{ai} + \dot{Q}_{as} + \dot{Q}_{al} \tag{4}$$

In Eq. (4), the sum of the sensible and latent heat transfer rates equals the total heat transfer rate, i.e. $\dot{Q}_{as} + \dot{Q}_{al} = \dot{Q}_a$, where:

$$\dot{Q}_{as} = \dot{m}_a c_{pa}(T_{ae} - T_{ai}) \tag{5}$$

$$\dot{Q}_{al} = \Delta\dot{m}_w h_v = \dot{m}_a(w_{ae} - w_{ai})h_v \tag{6}$$

The average enthalpy of the superheated vapour in the humid air mixture as well as the average heat capacity is given as:

$$h_v = ave(h_v |_{T_{ai}}^{w_{ae}}, h_v |_{T_{ai}}^{w_{ai}}) \tag{7}$$

$$c_{pa} = ave(c_{pa} |_{T_{ae}}^{w_{ae}}, c_{pa} |_{T_{ai}}^{w_{ae}}) \tag{8}$$

The energy balance for the water film is:

$$\dot{Q}_w = \dot{Q}_{as} + \Delta\dot{m}_w h_v + \dot{m}_{we} h_{we} - \dot{m}_{wi} h_{wi} \tag{9}$$

For the coolant, the mass and energy balances are given as:

$$\dot{m}_{ce} = \dot{m}_{ci} = \dot{m}_c \tag{10}$$

$$\dot{Q}_c = \dot{m}_c h_{ce} - \dot{m}_c h_{ci} = \dot{m}_c c_{pc}(T_{ce} - T_{ci}) \tag{11}$$

The relationship between $\dot{Q}_w$ and $\dot{Q}_c$ is:

$$\dot{Q}_c = -\dot{Q}_w \tag{12}$$

Component characteristics can be used to determine heat exchanger performance. The sensible heat transfer, $\dot{Q}_{as}$, can be determined using:

$$\dot{Q}_{as} = h_o A_o \Delta T_w^a \tag{13}$$

In Eq. (13), $\Delta T_w^a = T_{w,avg} - T_{a,avg}$, and $h_o A_o$ represents the thermal conductance of the external air flow over the coils. The mass transfer from the saturated humid air mixture next to the water film surface towards the humid air mixture in the free stream is given as:

$$\Delta\dot{m}_w = \rho h_m A_o \Delta w_w^a \tag{14}$$

In Eq. (14), $\Delta w_w^a = w_w - w_a$, with $w_w = w_{sat}|_{T_w}$ By applying the Chilton-Colburn J-factor analogy to obtain a value for the mass transfer coefficient based on the value of the heat transfer coefficient, and by using the definition of the Lewis number, $Le = Sc / Pr$, Eq. (14) can be rewritten as:

$$\Delta \dot{m}_w = h_o A_o \left(c_{pa} Le^{2/3} \right)^{-1} \Delta w_w^a \quad (15)$$

For moist air at typical operating conditions, $Le \approx 1$ [1]. The component characteristic equation for $\dot{Q}_{al}$ can then be written as:

$$\dot{Q}_{al} = h_o A_o (h_v / c_{pa}) \Delta w_w^a \quad (16)$$

Using Eq. (13) and Eq. (16), $\dot{Q}_a$ is then:

$$\dot{Q}_a = h_o A_o (1+X) \Delta T_w^a \quad (17)$$

$$X = (h_v \Delta w_w^a) / (c_{pa} \Delta T_w^a) \quad (18)$$

The component characteristic equation for $\dot{Q}_c$ is:

$$\dot{Q}_c = h_i A_i \Delta T_w^c \quad (19)$$

In Eq. (19), $\Delta T_w^c = T_{w,avg} - T_{c,avg}$ and $h_i A_i$ represents the thermal conductance of the internal coolant flowing within the tubes.

2.2. Modified LMTD method

Considering an incremental control volume of the schematic proposed in Figure 1 with area dA_o, Eqs. (5) and (13) are used to define $d\dot{Q}_{as}$, and similarly Eqs. (6) and (16) are used to define $d\dot{Q}_{al}$, leading to:

$$dw_a / dT_a = \Delta w_w^a / \Delta T_w^a \quad (20)$$

Eq. (20) in combination with the differential forms of Eqs. (5) and (6) leads to $d\dot{Q}_a$ defined as:

$$d\dot{Q}_a = m_a c_{pa} (1+X) dT_a = h_o (1+X) \Delta T_w^a dA_o \quad (21)$$

Defining $A_i = \alpha A_o$, $d\dot{Q}_c$ is defined as:

$$d\dot{Q}_c = \dot{m}_c c_{pc} dT_c = h_i \Delta T_w^c \alpha dA_o \quad (22)$$

Considering the temperature potential between the humid air stream and the coolant stream as $\Delta T_a^c = T_a - T_c$ and using Eqs. (21) and (22) applied to the incremental control volume, $d(\Delta T_a^c)$ can be written as:

$$d(\Delta T_a^c) = \left(\tfrac{h_o}{\dot{m}_a c_{pa}} \Delta T_w^a - \tfrac{\alpha h_i}{\dot{m}_c c_{pc}} \Delta T_w^c \right) dA_o \quad (23)$$

Considering the thermal resistance network, expressions of ΔT_a^w and ΔT_w^c as shown in Appendix A. , and Eq. (23) above:

$$\frac{d(\Delta T_a^c)}{\Delta T_a^c} = \left(\frac{\dfrac{1}{\dot{m}_a c_{pa}(1+X)} + \dfrac{1}{\dot{m}_c c_{pc}}}{\dfrac{1}{h_o(1+X)} + \dfrac{1}{\alpha h_i}} \right) dA_o \quad (24)$$

Integrating the above expression from 1 to 2 as depicted in Figure 1:

$$\ln\left(\frac{\Delta T_a^c|_2}{\Delta T_a^c|_1} \right) = -\left(\frac{1}{h_o A_o (1+X)} + \frac{1}{h_i A_i} \right)^{-1} \left(\frac{1}{\dot{m}_a c_{pa}(1+X)} + \frac{1}{\dot{m}_c c_{pc}} \right) \quad (25)$$

Substituting the integrated form on the left hand side of Eq. (21) and Eq. (22) into the second term on the right hand side of Eq. (25), and applying subscripts 1 and 2 as depicted in Figure 1 for the respective temperatures, $\dot{Q}_a$ can be written as:

$$\dot{Q}_a = -\left(\frac{1}{h_o A_o (1+X)} + \frac{1}{h_i A_i} \right)^{-1} \frac{(T_{a2} - T_{c2}) - (T_{a1} - T_{c1})}{\ln\left(\dfrac{T_{a2} - T_{c2}}{T_{a1} - T_{c1}} \right)} \quad (26)$$

Therefore, the modified UA and modified heat capacity of moist air is given as:

$$UA = \left(\frac{1}{h_o A_o (1+X)} + \frac{1}{h_i A_i} \right)^{-1} \quad (27)$$

$$C_a = \dot{m}_a c_{pa} (1+X) = \dot{m}_a c_{pa} \left(1 + \frac{h_v}{c_{pa}} \frac{\Delta w_w^a}{\Delta T_w^a} \right) \quad (28)$$

The heat capacity for the coolant remains as $C_c = \dot{m}_c c_{pc}$.

The modified UA and C_a derived above can be directly implemented using the effectiveness-NTU method. This paper focuses on the implementation for a counterflow heat exchanger, where:

$$\varepsilon = \frac{1 - \exp(-NTU(1 - C_r))}{1 - C_r \exp(-NTU(1 - C_r))} \quad (29)$$

In Eq. (29), $NTU = UA / C_{min}$, and $C_r = C_{min} / C_{max}$.

2.3. Comparison to VDI Heat Atlas design methodology

The modified $\varepsilon - NTU$ method is first validated using the design methodology proposed in the VDI Heat Atlas [1]. The design methodology is used to size a cross-counterflow heat exchanger, resulting in a six-tube row configuration. The modified $\varepsilon - NTU$ method is used to simulate the sized heat exchanger, and the results are compared to design values from the VDI Heat Atlas, [1] (example 4).

2.4. Case study counterflow heat exchanger

The modified $\varepsilon - NTU$ is applied to chilled water coils for cooling and dehumidification of moist air in a perfect counterflow arrangement using the governing equations described above and the schematic from Figure 1. The heat exchanger is discretized into 100 increments, with governing equations and thermofluid properties solved at each increment.

For verification purposes, the fundamental numerical solution is also found using component characteristic Eqs. (13), (16), and (19) to compute heat transfer rates and compared to the proposed modified $\varepsilon-NTU$ method.

Table 1. Boundary conditions for case study heat exchanger

p_{amb}	101.325 kPa	p_c	300 kPa
$\dot{m}_a$	1.0 kg/s	$\dot{m}_c$	0.60 kg/s
T_{ai}	30 °C	T_{ci}	16°C, 12°C, 8°C
RH	0.4	$\dot{m}_{wi}$	0.0 kg/s
h_oA_o	3.0 kW/K	h_iA_i	40 kW/K

The governing equations, namely mass and energy balances, and component characteristic equations are solved simultaneously using the Engineering Equation Solver (EES). The equations are solved to a maximum variable residual below 1E-3.

Table 1 shows the boundary conditions applied for the case study counterflow heat exchanger. Thermal conductances for both air and water flows are assumed. The inlet water temperature, T_{ci}, is varied for three cases: i) Dry coil operation at 16°C, ii) Partial wet operation at 12°C, and iii) Partial wet operation at 8°C.

3. Results and Discussion

Table 2 shows high-level results comparing the proposed modified $\varepsilon-NTU$ method and the design approach used in the VDI Heat Atlas [1]. Good agreement between results is observed for $\Delta T_a(°C)$, $\Delta T_c(°C)$, and $\dot{Q}_{a,tot}$ (kW). The proposed method under predicts mass transfer by 9% while slightly over predicting sensible heat transfer (3.8%). This may be due to the lack of discretization in the design approach. The number of tube rows determined from the design approach (5.7 tube rows) is also rounded up to the nearest integer, which may contribute a slight error.

Table 2. Validation to VDI Heat Atlas design approach

	$\Delta T_a(°C)$	$\Delta w_a(^{kg}/_{kg})$	$\Delta T_c(°C)$	$\dot{Q}_{a,tot}$ (kW)
VDI Heat Atlas	-15.0	0.0030	7.6	-50.58
Modified $\varepsilon-NTU$	-15.6	0.0027	7.5	-50.59
Rel. difference (%)	3.75	-9.17	0.28	-0.02

Moving on to the case study counterflow heat exchanger, shows a psychrometric chart for all three inlet water temperatures. The inlet air temperature for all three cases is the same.

For the dry coil case at $T_{ci}=16°C$, only sensible heat transfer occurs and thus w_a remains constant. For the wet coil cases at 12°C and 8°C, both sensible and latent heat and mass transfer occur, shown by the decrease in w_a. For the 8°C case with the coldest coil surface temperature, more condensation occurs and also earlier compared to the 12°C case, with a larger decrease in w_a observed over the heat exchanger, and thus a larger Δm_w. The larger temperature differential for the 8°C case between the air and water streams also results in a lower T_{ae}.

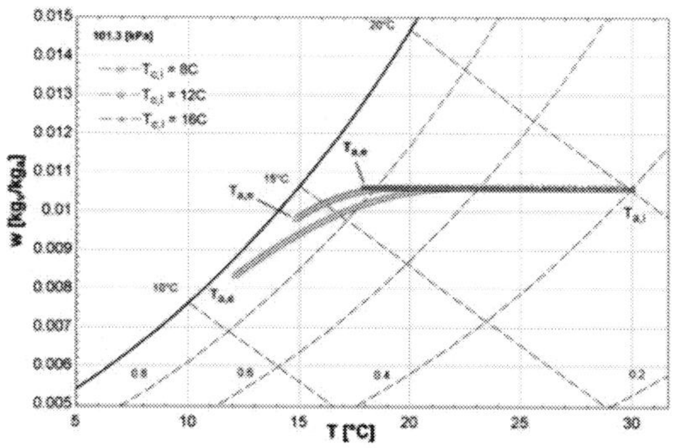

Figure 2: Psychrometric chart for case study heat exchanger

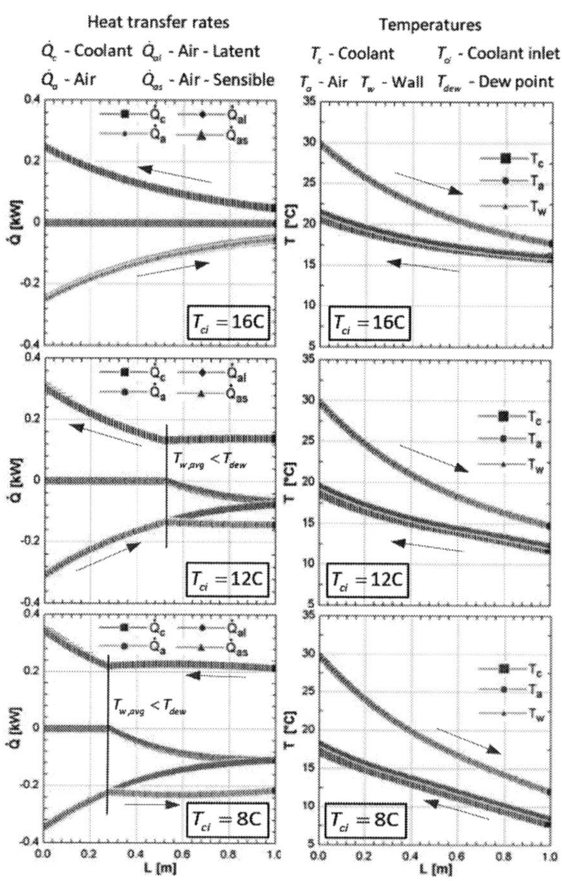

Figure 3: Heat transfer rates and temperatures for each T_{ci}

Figure 3 shows heat transfer rates (left) and temperature profiles (right) for each of the three cases. For dry coil operation ($T_{ci} = 16°C$) only sensible heat transfer occurs with no latent heat or mass transfer. The heat transfer curves therefore represent a standard counterflow heat exchanger with $\dot{Q}_a = \dot{Q}_{as}$.

For the partially wet coil cases, only sensible heat transfer is observed initially. However, once $T_{w.avg} < T_{dew}$, condensation starts to occur with latent heat and mass transfer observed from the humid air to the coil surface. This results in an increase in $\dot{Q}_{al}$ from the point of condensation till the outlet of the heat exchanger. At the outlet, $\dot{Q}_{al}$ makes up 46.8% of $\dot{Q}_a$ at $T_{ci} = 12°C$, and 50% at $T_{ci} = 8°C$. These high percentages indicate the importance of capturing latent heat transfer. Notably, the lower temperature differential between the humid air and coolant towards the tail end of the heat exchanger results in a reduced $\dot{Q}_{as}$, counteracted by the increasing $\dot{Q}_{al}$ which starts to plateau. For $T_{ci} = 8°C$, condensation starts to occur further upstream due to the larger temperature differential between humid air and water in the coils, shown on the right of Figure 3.

The modified $\varepsilon - NTU$ method compares well with the fundamental numerical solution, with the largest observed relative error of 1.3% at the increment where condensation first occurs. This error reduces to within 0.1% for downstream increments. For sensible heat transfer, the results are nearly identical (<0.1% error).

4. Conclusions and Future Work

A modified $\varepsilon - NTU$ method for combined heat and mass transfer is presented and validated using an existing design approach from literature for a cross-counterflow heat exchanger. The presented method is then applied to a case study of chilled water coils in a pure counterflow configuration for both dry and partially wet-dry coils. The method can capture both sensible and latent heat transfer (mass transfer). The results agree well with the fundamental numerical solution. Future work involves detailed validation of the method using experimental data of a chilled water coil for various operating conditions.

5. References

[1] M. Zeller and U. Busweiler, "Humidifying and Drying of Air," in *VDI Heat Atlas*, 2nd Edition., Berlin, Heidelberg: Springer-Verlag, 2010, ch. M8, pp. 1343–1362. doi: 10.1007/978-3-540-77877-6_101.

[2] A. H. Elmahdy and G. P. Mitalas, "A simple model for cooling and dehumidifying coils for use in calculating energy requirements for buildings," *ASHRAE Trans*, vol. 83, no. 2, pp. 103–117, 1977.

[3] X. Liang, "Study on the Heat and Mass Transfer Taking Place in a Direct Expansion (DX) Air Cooling and Dehumidification Coil," The Hong Kong Polytechnic University, 2009. Accessed: Dec. 05, 2024. [Online]. Available: https://theses.lib.polyu.edu.hk/handle/200/5707

[4] X. Cui, K. J. Chua, M. R. Islam, and W. M. Yang, "Fundamental formulation of a modified LMTD method to study indirect evaporative heat exchangers," *Energy Convers Manag*, vol. 88, pp. 372–381, 2014, doi: 10.1016/j.enconman.2014.08.056.

[5] A. Hasan, "Going below the wet-bulb temperature by indirect evaporative cooling: Analysis using a modified ε-NTU method," *Appl Energy*, vol. 89, no. 1, pp. 237–245, 2012, doi: 10.1016/j.apenergy.2011.07.005.

[6] C. Wang, S. Wang, X. Ding, J. An, and X. Fu, "A Simple Explicit Formula for Evaluating the Total Capacity of Chilled-Water Cooling Coils under Wet Conditions," *Buildings*, vol. 14, no. 9, p. 2630, Aug. 2024, doi: 10.3390/buildings14092630.

[7] J. E. Braun, "Methodologies for the Design and Control of Central Cooling Plants," Madison, 1988. Accessed: Jan. 15, 2025. [Online]. Available: https://minds.wisconsin.edu/handle/1793/46694

[8] S. Gollamudi, H. Kamali, M. Fauchoux, E. Krishnan, A. Joseph, and C. Simonson, "Coupled ϵ-NTU Method to Design and Evaluate the Performance of Energy Exchangers With Coupled Heat and Mass Transfer," *ASME Journal of Heat and Mass Transfer*, vol. 147, no. 8, Aug. 2025, doi: 10.1115/1.4068509.

Appendix A. Thermal resistances

Consider the thermal resistance network of the control volume as depicted in Figure 1, with external heat (and mass) transfer from air flow over the coils, and internal heat transfer from water/fluid in the tubes. The thermal resistance network can be used to find the temperature potentials between the air and coil surface (ΔT_a^w), as well as between the coil surface and water/fluid in the tubes (ΔT_w^c). Using the modified *UA* value from Eq. (27):

$$\Delta T_a^w = \frac{(h_o(1+X))^{-1}}{(h_o(1+X))^{-1} + \alpha h_i^{-1}} \Delta T_a^c \tag{30}$$

$$\Delta T_w^c = \frac{(\alpha h_i)^{-1}}{(h_o(1+X))^{-1} + (\alpha h_i)^{-1}} \Delta T_a^c \tag{31}$$

Design and Thermal Analysis of a Retrofitted New Energy Vehicle Battery Pack Module: A Heat Dissipation Augmented Performance Study

Nicholas Xavier[a,*], Antoine F. Mulaba-Bafubiandi[b], Pathmanathan Naidoo[a]

[a]*University of Johannesburg, PO Box 524, Auckland Park, Johannesburg, 2006, South Africa*

[b]*University of Johannesburg, PO Box 2028, Doornfontein, Johannesburg, 2094, South Africa*

**corresponding author email: nicholasixavier317@gmail.com*

Abstract

Global environmental qualms have led to the development and adoption of New Energy Vehicles. Epitomized as the engineering crowns of a sustainable future, homologation requirements have not been neglected. Considering the harsh conditions of charging and discharging of these vehicles, requirements on thermal analysis for safety and optimal operation reasons are mandatory. This work presents a pragmatic design approach for a battery pack of a retrofitted New Energy Vehicle that was sized using vehicle dynamics and further explored for thermal analysis, where a comparative simulation experiment, based on a numerical heat transfer model was adopted to augment its heat dissipation performance. To achieve this, geometric models of three battery modules for the pack are developed using a computer-aided design tool, where heat dissipation fins with embedded serpentine liquid flow minichannels, of different cross-sectional profile shapes, i.e., circular, elliptical, and hexagonal, are arranged parallel to the prismatic batteries. A record of heat dissipation and heat transfer visualization have shown that, the e-fin is best performing followed by the circular, then lastly, the hexagonal. The leading performance of the e-fin shows a heat dissipation performance improvement from the c- and h-fin and will improve the overall battery thermal management system of the designed battery pack. This can be further validated using physical experimentation.

Keywords: NEV, Heat Dissipation, Heat Transfer, Minichannels, Thermal Analysis, Numerical Simulation.

Nomenclature

Symbol

A_{ch}	Cross-sectional area of the minichannel	h	Convective heat transfer coefficient
A_{fin}	Surface area of fin	m	Mass Flow Rate
C_{fluid}	Specific heat of fluid	ΔP_{ch}	Pressure drop in the minichannel
D_h	Hydraulic Diameter	T	Temperature

Subscripts			
ave	Average	C	Circular
in	Inlet	E	Elliptical
out	Outlet	**Greek Symbols**	
max	Maximum	η_{fin}	Fin Efficiency
min	Minimum	ρ_{fin}	Fluid Density

1. Introduction

The adoption and ever-increasing demand for new energy vehicles (NEV) fueled by the global sustainability demands is game changing. To ensure a 'just transition' from conventional vehicles, electrified vehicles must comparatively perform or outperform them to ensure compliance with regulatory and statutory regulations, making them easy to homologate [1]. The performance criteria in Fig. 1 (also in A.1.) summarizes the key factors considered in the overall performance analysis of new energy vehicles.

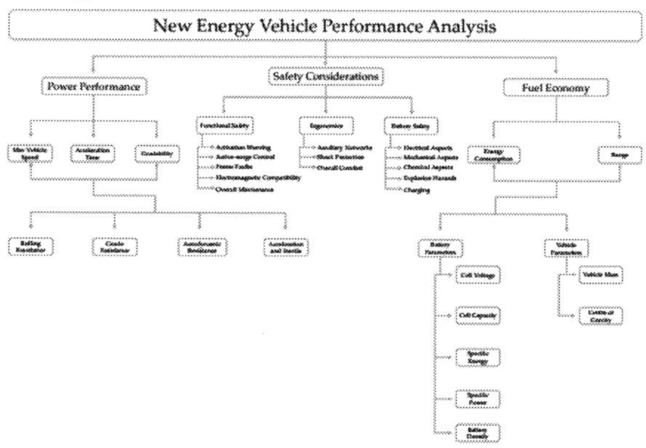

Fig. 1. Performance Analysis Chart for New Energy Vehicles with special attention to Safety

The attention to safety considered in the figure above also includes battery safety, a key homologation factor in vehicular electrification. Engineering design practitioners must consider the following in their designs [2]:

- Electrical Aspects – Where protection against electrocution because of direct and indirect contact with the vehicle user and short-circuits that lead to explosions and fires is considered.
- Mechanical Aspects – Where the spatial design and location of the traction battery in the vehicle is considered to avoid any contact with the user or other power electronic systems in case of an accident. Collision cases are developed, simulated, and practically tested to ensure the overall safety feasibility of the vehicle.

- Chemical Aspects – Where the battery design, electrolyte, and metal elements are considered to avoid overheating of the pack, explosions and fires, fumes to be possibly inhaled, and obviously thermal runaway.
- Battery Charging – Where design considerations using off-board, on-board, and partially on-board battery chargers are used to actively reduce mechanical, electrical, and chemical hazards.

1.1. Background and Motivation

Ambitions to pursue electrification retrofitting in Non-Mediterranean Africa mandate battery thermal analysis requirements for safety and optimal operation. The university's Toyota Avanza vehicle electrification retrofitting use case project (Fig. 2) has reviewed a plethora of retrofitting strategies, where prior works have presented comprehensive powertrain engineering sizing with tallied battery pack sizing with power electronics main and subsystems presented from high-voltage circuits, low-voltage circuits, and communication (CAN) networks systems [3 – 6]. While addressing these critical aspects in vehicular electrification retrofitting powertrain sizing, battery thermal management systems have not been presented. Studies aimed at addressing this gap are important.

Fig. 2. The Toyota Avanza Vehicle Electrification Retrofitting Project ambition undertaken at the University.

1.2. Objective and Scope

This work presents a prismatic lithium-ion battery pack module design and thermal analysis study for a retrofitted new energy vehicle using an active battery thermal management system [7 – 9]. The study uses liquid cooling via a network of thin fins with embedded serpentine liquid flow minichannels [10] of different cross-sectional profile shapes; namely, circular (c-fin), elliptical (e-fin), and hexagonal (h-fin). It seeks to:

- Employ CAD to develop a geometric model design of the prismatic batteries and fins, then the battery pack module,
- Develop a thermal model based on the thermal parameters of the batteries, the fins, and the coolant [10, 11], and
- Employ a heat transfer numerical simulation to study the heat dissipation performance of the different fins studied comparatively for three scenario test cases:
  - ❖ Optimal Battery Performance Temperatures – 20-40 °C [12 – 14]: Chosen Temp = 40 °C
  - ❖ Harsh Lithium-ion Battery Discharge Temperatures – 60-100 °C [13, 15]: Chosen Temp = 100 °C
  - ❖ Harsh Battery DC Fast Charging Temperatures – 150-250 °C [15, 16]: Chosen Temp = 180 °C

This is done to:

- Develop a comparative fin thermal performance analysis for the best performing fin for this battery thermal management system.
- Introducing a comparative heat transfer study where elliptical and circular profile shaped serpentine minichannels were compared to a hexagonal shaped serpentine minichannel for battery thermal management system applications [17, 18]. The comparative approach stems from the possibilities of enhancing heat transfer performance of fins using minichannel geometry optimization techniques summarized in the figure below.

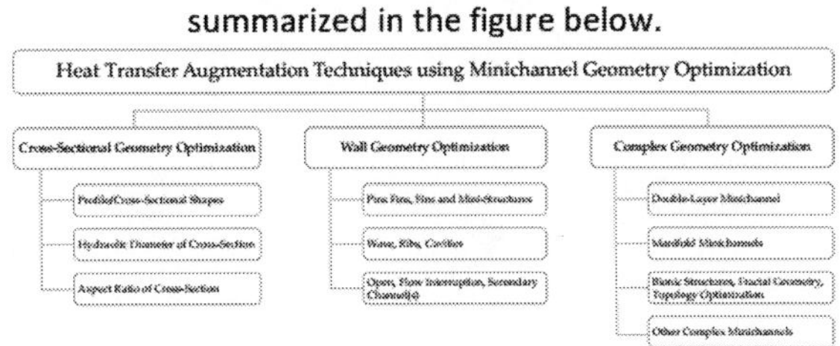

Fig. 3. Geometry Optimization Techniques used to Augment the Heat Transfer in Minichannels

2. Methodology and Computational Approach

Fig. 4 summarizes the methodology employed in this study. The overall approach employs numerical simulation and adopts a comparative approach to interpret and provide conclusions based on the results.

The battery pack sizing of the retrofit vehicle covered in the powertrain sizing study coupled with prior works suggests that integrating battery thermal management systems in electric retrofitting projects may be a challenging endeavor. This follows the homologation mass constraints criteria these vehicles need to comply with prior approval for roadworthiness [19]. In South Africa, the National Regulator for Compulsory Specifications (NRCS) is responsible for this approval [19]. Two major requirements standout for homologation:

- Mass Constraints – The mass of the retrofit vehicle must be lesser or equal to the conventional fuel vehicle's original mass.

- Chassis Modification - Drilling, welding, and cutting techniques are not allowed on the vehicle chassis.

2.1. Geometric Model

The geometric model of the battery thermal management system, combining the batteries and the cooling fins is shown in Fig. 4 of point 2 and the general top and front cross-sectional layout of the fins with their differing profile faces. A.2. shows the geometric parameters of the profile faces of the minichannels. The design of the minichannels follow similar cross-section areas with hydraulic diameters within the required minichannel range.

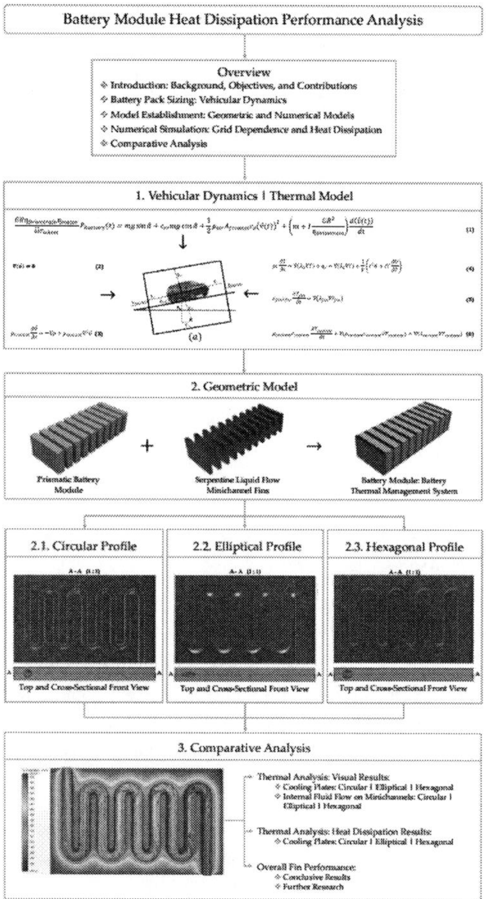

Fig. 4. The methodology summary employed in this study

2.2. Numerical Simulation Approach

The numerical approach simulated the cooling effects of the cooling fins when two batteries (heat sources) are attached to it (shown in B.1.). The technique is set at room temperature for both the fin and the coolant, and the heat transfer effect is set to occur from the contact surfaces of the fin with the batteries to the channels where the heat is transferred to the coolant and then dissipated out of the fin.

The boundary conditions are set at the pre-processing stage of the heat transfer numerical simulation as follows:

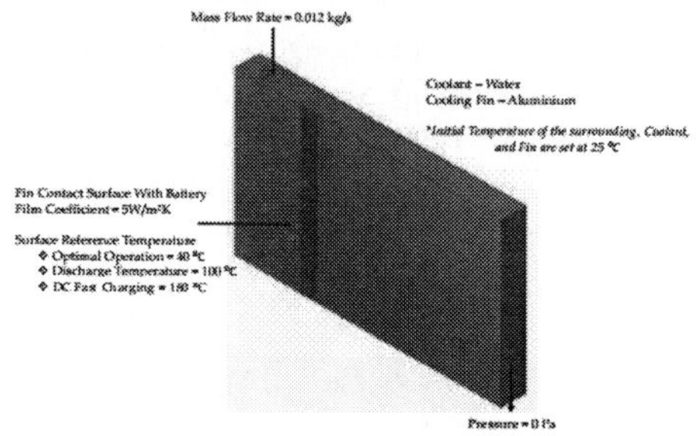

Fig.5. Boundary conditions applied on the cooling fin based on the heat transfer numerical simulation model

The heat transfer numerical simulation processing stage computes the optimal mesh for each of the geometries using automatic adaptive mesh as the simulation iterations progress. Targeted post-processing results are summarized as follows:

- Coolant - Average Pressure, Pressure Drop, Average Temperature, Maximum Temperature
- Cooling Fin - Average Temperature, Average Temperature of Contact Surface

A numerical simulation validation technique is employed against related literature [10, 11] to enhance the validity of the results. Although the heat transfer numerical simulations may be similar in approach, X. Zeng et al. [10] uses CPCM liquid cooling and M. Xiong et al. [11] used pinned fins.

3. Key Results and Discussions

3.1. Heat Transfer Results: Coolant

The following shows the thermal results of the coolant (also in C.1.):

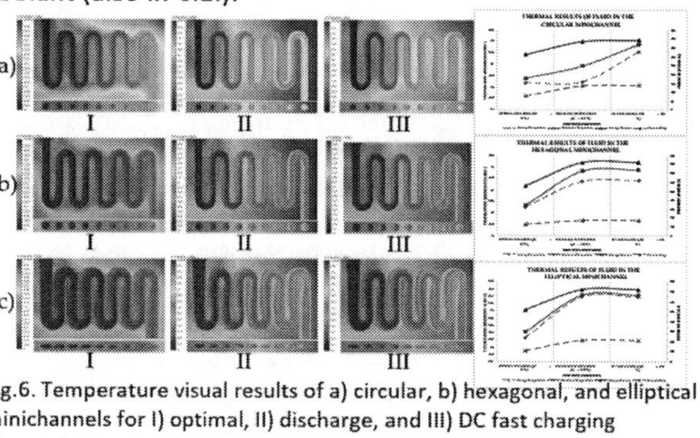

Fig.6. Temperature visual results of a) circular, b) hexagonal, and elliptical minichannels for I) optimal, II) discharge, and III) DC fast charging alongside their graphical thermal results

The graphical results of the coolant as observed in the differing minichannel show comparable pressure drop and temperature results. The temperature ranges about 25 °C, that may suggest the suitability of any of these fins for this project application.

A comparison of the thermal results of the coolant in each of the minichannels is shown below.

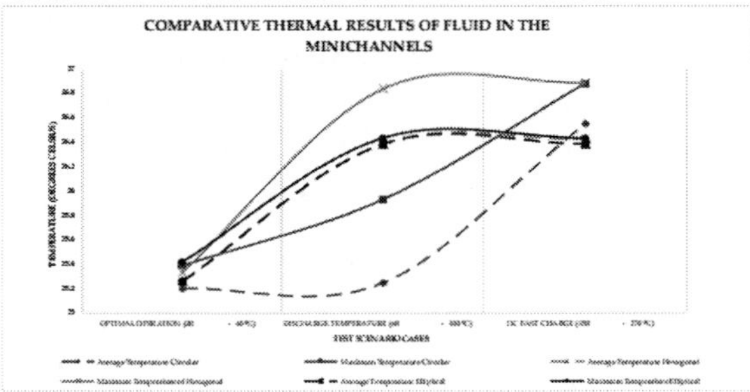

Fig.7. Comparative thermal results of the coolant at differing minichannel profile shapes

Two observations are key to indicate superiority of heat transferred to the coolant: (1) a greater maximum temperature, and (2) closeness in the maximum and average temperatures. The coolant in the elliptical and hexagonal minichannels suggest superior heat transfer based on their maximum temperatures. Although the maximum heat of the coolant in the hexagonal minichannel is greatest compared to the elliptical, the heat distribution (average) results along the coolant in the channel show ineffective heat transfer compared to the elliptical as seen with its maximum results close to its average. The c-fin proves ineffective for harsh discharge conditions showing less heat transferred from the fin to the coolant. Effective heat transfer to the coolant indicates superior heat dissipation as more of the heat will be removed from the fin via the liquid.

3.2. Heat Transfer Results: Cooling Fins

The following shows the thermal results of the cooling fins (also in C.2.):

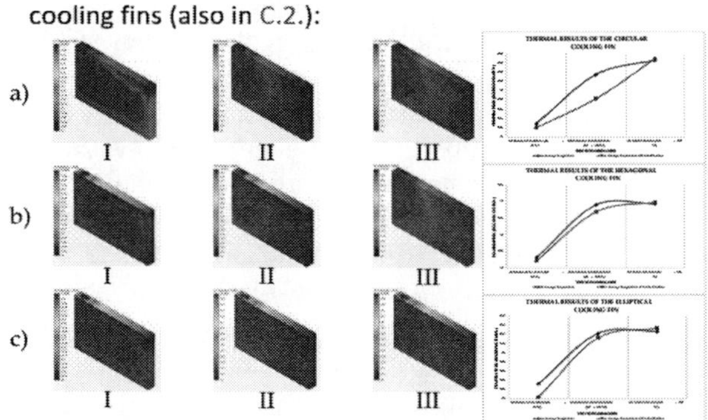

Fig.8. Global temperature visual results a) circular, b) hexagonal, and elliptical minichannels for I) optimal, II) discharge, and III) DC fast charging alongside their graphical thermal results

The graphical results show the average temperatures of the cooling fins and their contact surfaces. Based on the observations, the cooling fins' results show good thermal performance.

A more comparative analysis is adopted in Fig.9 to further explain the results.

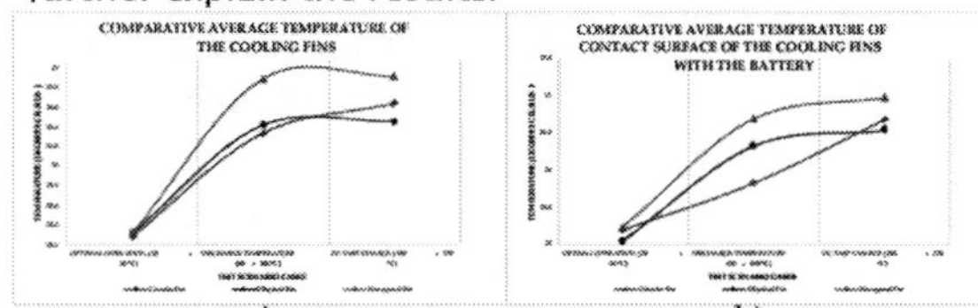

a) b)

Fig.9. The graphs show the (a) average temperature of the fins and the (b) average temperature of the contact surface of the fin

The c- and e-fins appear cooler than the hexagonal fin (Fig.9(a)). The c-fin appears cooler for harsh discharge conditions (Fig.9(a) & (b)); however, the results exhibit a cooler contact surface for optimal and fast charging conditions for the e-fin (Fig.9(b)). This suggests that the e-fin has enhanced heat transfer capabilities than the h- and c- fins, indicating the fin's superior heat dissipation capabilities.

The overall analysis based on the key results and key observations has shown the e-fin as the most performing fin as far as heat dissipation is concerned. The e-fin's minichannel is observed with induced fluid flow mixing after turns in the minichannel (shown in C.3.), where fluid particles in the wall (hot particles) move about the side center to mix with center particles (colder particles) to improve the overall heat transfer from the fin to the liquid coolant, this is observed with the average temperature of the coolant in the e-fin close to its maximum temperature.

Conclusions

A vehicular electrification retrofitting cooling heat transfer numerical simulation was conducted to comparatively analyse the heat dissipation performance of three fins with embedded serpentine minichannels of different profile shapes: namely, Circular, Elliptical, and Hexagonal, based on three scenario cases: i.e., Optimal, Discharge, and DC Fast Charging. The following was observed:

- The coolant results showed superior heat dissipation in the e-fin compared to the circular and the hexagonal.
- The fins' results showed that the c- and e- fins were cooler than the h-fin.
- The overall analysis, including the average temperature of the contact surfaces of the fins showed that the e-fin was cooler, showing enhanced or augmented heat dissipation capabilities, making it the overall ideal choice for this project application.

Further work will explore a 2:1 (battery-to- fin ratio) for the battery pack module, then conduct simulations of the battery module than just the fins. An experimental approach to validate and test these results is practical for further studies.

References

[1] Department of Trade, Industry and Competition, „Electric Vehicles White Paper," DTIC South Africa, Pretoria, November 2023.

[2] P. v. d. Bossche, „Safety Considerations for Electric Vehicles," 27 July 2003.

[3] L. Szabo und V. Lulia, „A Brief History of Electric Vehicles," *Journal of Computer Science and Control Systems,* pp. 19-26, 2022.

[4] S. Z. Rajper und J. Albrecht, „Prospects of Electric Vehicles in the Developing Countries: A Literature Review," *Sustainability,* Bd. 12, 2020.

[5] S. Kaleg, A. Hapid und M. R. Kumia, „Electric Vehicle conversion based on Distance, Speed and Cost Requirements," *Energy Procedia,* Nr. 68, pp. 446-454, 2014.

[6] S. Lacock, A. A. du Plessis und M. J. Booysen, „Using Driving-Cycle Data to Retrofit and Electrify Sub-Saharan Africa's Existing Minibus Taxis for a Circular Economy," *World Electric Vehicle Journal,* Bd. 14, Nr. 296, pp. 1-24, 2023.

[7] C. Zhao, B. Zhang, Y. Zheng, S. Huang, T. Yan und X. Liu, „Hybrid Battery Thermal Management Systemin Electrical Vehicles: A Review," *Energies,* Bd. 13, Nr. 6257, pp. 1-18, 2020.

[8] K. Garud, L. T. Duc, S.-G. Hwang, N. Nghia und M.-Y. Lee, „A Review of Advanced Cooling Strategies for Battery Thermal Management Systems in Electric Vehicles," *Symmetry,* Bd. 7, Nr. 15, 2023.

[9] I. Cetin, E. Sezici, M. Karabulut, E. Avci und F. Polat, „A comprehensive review of battery thermal management systems for electric vehicles," *Proceedings of the Institution of Mechanical Engineers Part E Journal,* Bd. 3, Nr. 237, 2022.

[10] X. Zeng, Z. Men, F. Deng und C. Chen, „Optimization of the Heat Dissipation Performance of a Lithium-Ion Battery Thermal Management System with CPCM/Liquid Cooling," *Materials,* Bd. 15, Nr. 3835, pp. 1-12, 2022.

[11] M. Xiong, N. Wang, W. Li, A. Garg und L. Gao, „Study on the Heat Dissipation Performance of a Liquid Cooling Battery Pack with Different Pin-Fins," *Batteries,* Bd. 9, Nr. 44, pp. 1-20, 2023.

[12] A. Wang, X. Yin, Z. Xin, F. Cao, W. Zan, S. Bengt und D. Xiao, „Performance optimization of electric vehicle battery thermal management based on the transcritical CO2 system," *Energy,* Bd. 266, Nr. 126455, 2023.

[13] D. D. Agwu, F. Opara, N. Chukwuchekwa, D. Dike und L. O. Uzoechi, „Review Of Comparative Battery Energy Storage Systems (Bess) For Energy Storage Applications In Tropical Enviroments," in *2017 IEEE 3rd International Conference on Electro-Technology for National Development (NIGERCON),* Nigeria, 2017.

[14] F. S. Hwang, T. Confery, C. Reidy, D. Picovici, D. Callaghan, D. Culliton und C. Nolan, „Review of battery thermal management systems in electric vehicles," *Renewable and Sustainable Energy Reviews,* Bd. 192, Nr. 114171, pp. 1-22, 2024.

[15] M. Ahnheim, „Thermal management for safe and efficient fast-charging of battery electric vehicles on the road," 2024. [Online]. Available: https://www.cejn.com/articles/thermal-management-for-safe-and-efficient-fast-charging-of-battery-electric-vehicles-on-the-road/.

[16] H. M. Ali, „Thermal management systems for batteries in electric vehicles: A recent review," *Energy Reports,* Bd. 9, pp. 5545-5564, 2023.

[17] E. Sajjad, H. D. B. Mohammad, M. Mohammad, M. Hamid und S. Maisam, „Comparison of Heat Transfer & Fluid Flow in the Micro-channel with Rectangular and Hexagonal Cross Section," *Archives for Technical Sciences,* Bd. 21, Nr. 1, pp. 1-10, 2020.

[18] H. Yu, T. Li, X. Zeng, T. He und N. Mao, „A Critical Review on Geometric Improvements for Heat Transfer Augmentation of Microchannels," *Energies,* Bd. 15, Nr. 9474, pp. 1-45, 2022.

[19] National Regulator for Compulsory Specifications, „AUTOMOTIVE—Homologation of Vehicles," NRCS, Pretoria, South Africa, 2019.

[20] Shenzhen Flyxure Technology Co., Ltd, „CATL NMC 3.7V 72Ah 74Ah 93Ah 95Ah 114Ah 116Ah 117Ah 120Ah 127.5Ah Prismatic Lithium Ion Battery Cells," Shenzhen Flyxure Technology Co., Ltd, 2019. [Online]. Available: https://www.alibaba.com/product-detail/CATL-NMC-3-7V-72Ah-74Ah_1600963094844.html?spm=a2700.details.you_may_like.1.15467755TAhoOP.

[21] J. Kleiner, L. Komsiyska, G. Elger und C. Endisch, „Thermal Modelling of a Prismatic Lithium-Ion Cell in a Battery Electric Vehicle Environment: Influences of the Experimental Validation Setup," *Energies,* Bd. 13, Nr. 62, pp. 1-18, 2019.

[22] J. C. Kurnia, D. C. Lim, L. Chen, L. Jiang und A. P. Sasmito, „Entropy Generation and Heat Transfer Performance in Microchannel Cooling," *Entropy,* Nr. 21, pp. 1-15, 2019.

Appendices

Appendix A. Performance Analysis Chart and Geometric Model(s)

A.1. Performance Analysis Chart of a New Energy Vehicle

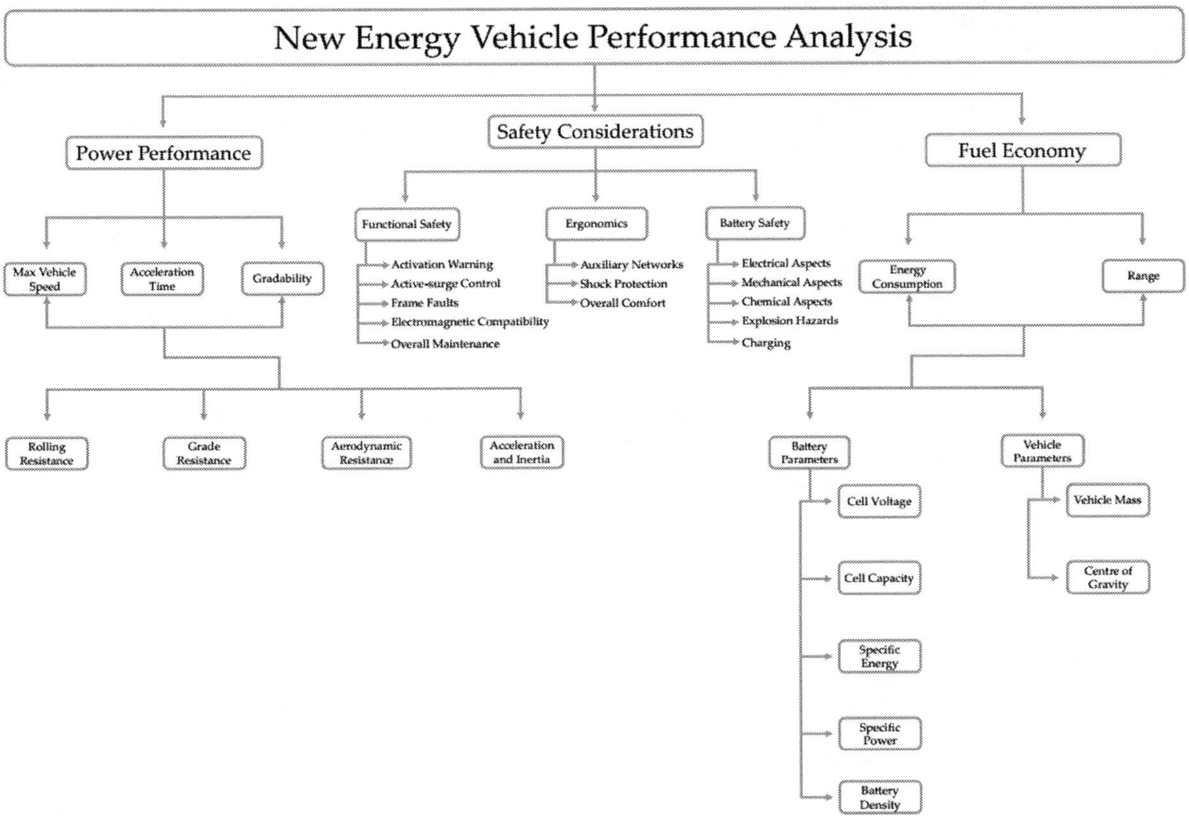

A.2. Geometric Parameters of the Cooling Fin Minichannel Profile

Geometric Name	Geometric Shape	Geometric Parameters					
		Dimensions (mm)	Cross-Section Area Formular	Cross-Section Area (mm²)	Hydraulic Diameter Formular	Hydraulic Diameter (mm)
Circular	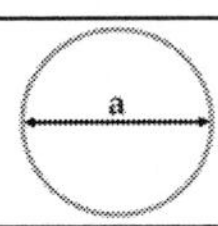	$a = 2$	$A_{ch_C} = \frac{\pi}{4}a^2$	π	$D_{h_C} = a$	$D_{h_C} = 2$
Hexagonal	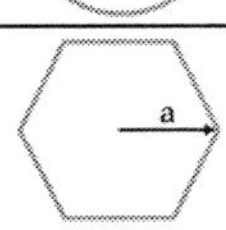	$a = \left(\frac{2\sqrt{3}\pi}{9}\right)^{\frac{1}{2}}$	$A_{ch_H} = \frac{3\sqrt{3}}{2}a^2$	π	$D_{h_H} = 2a$	$D_{h_H} = 2\left(\frac{2\sqrt{3}\pi}{9}\right)^{\frac{1}{2}}$
Elliptical	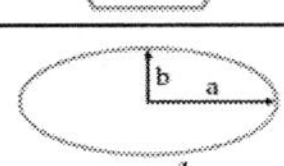 $for\ b = \frac{1}{3}a$	$a = \sqrt{3}$	$A_{ch_E} = \pi ab$	π	$D_{h_E} = \frac{4ab}{a+b} = a$	$D_{h_E} = \sqrt{3}$

A.3. Geometric Parameters of the Battery

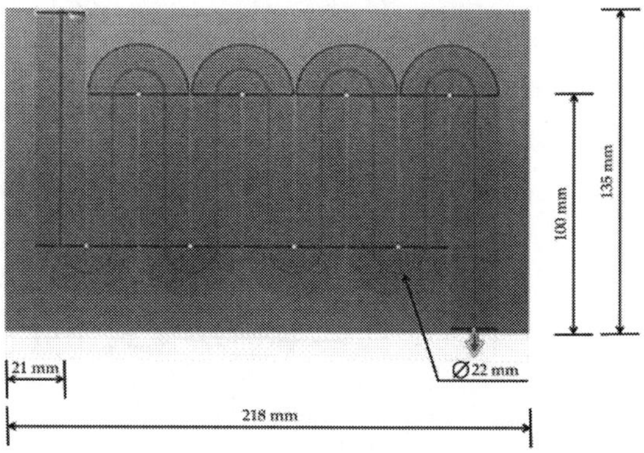

Appendix B. Heat Transfer Numerical Simulation Model

B.1. Numerical Simulation from pre- to post-processing

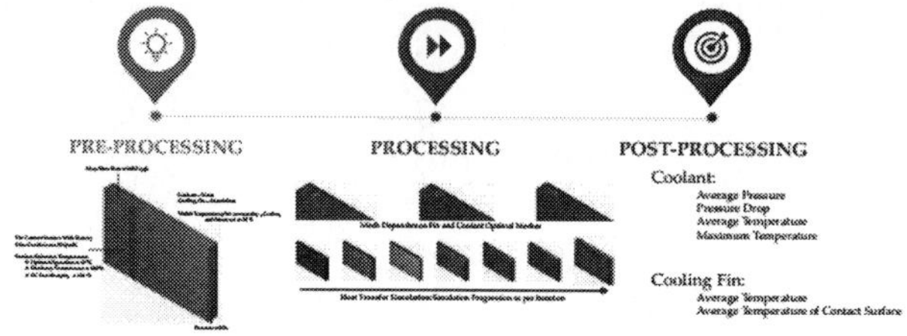

B.2. Thermophysical Properties of the Battery, Cooling Fin, and the Coolant

Component	Material	Thermophysical Properties			
		Specific Heat Capacity (J/kg.K)	Density (kg/m^3)	Thermal Conductivity (W/m.K)	Reference
1. Battery	Combination*	1084.2	1828.7	61.9	Calculated*
1.1. *Electrolyte*	*$LiPF_6$*	134	1290	0.45	[11, 20]
1.2. *Separator*	*Polypropylene (PP)*	1978	492	0.334	[11]
1.3. *Casing*	*Aluminum & Polymer*	900	2700	238	[21]
1.4. *Thermal Pad*	*Connection Cooling*	903	2740	2.22	[21]
1.5. *Cathode*	*NMC*	1270	2200	1.68	[11, 20]
1.6. *Anode*	*Carbon*	1320	1550	129	[11, 20]
2. Cooling Fin	Aluminum	871	2719	202.4	[11]
3. Coolant	Water	4182	998.2	0.6	[11, 17, 22]

**Parameters were averaged*

Appendix C. Key Results - Enlarged

C.1. Thermal results of the liquid flowing inside the minichannels

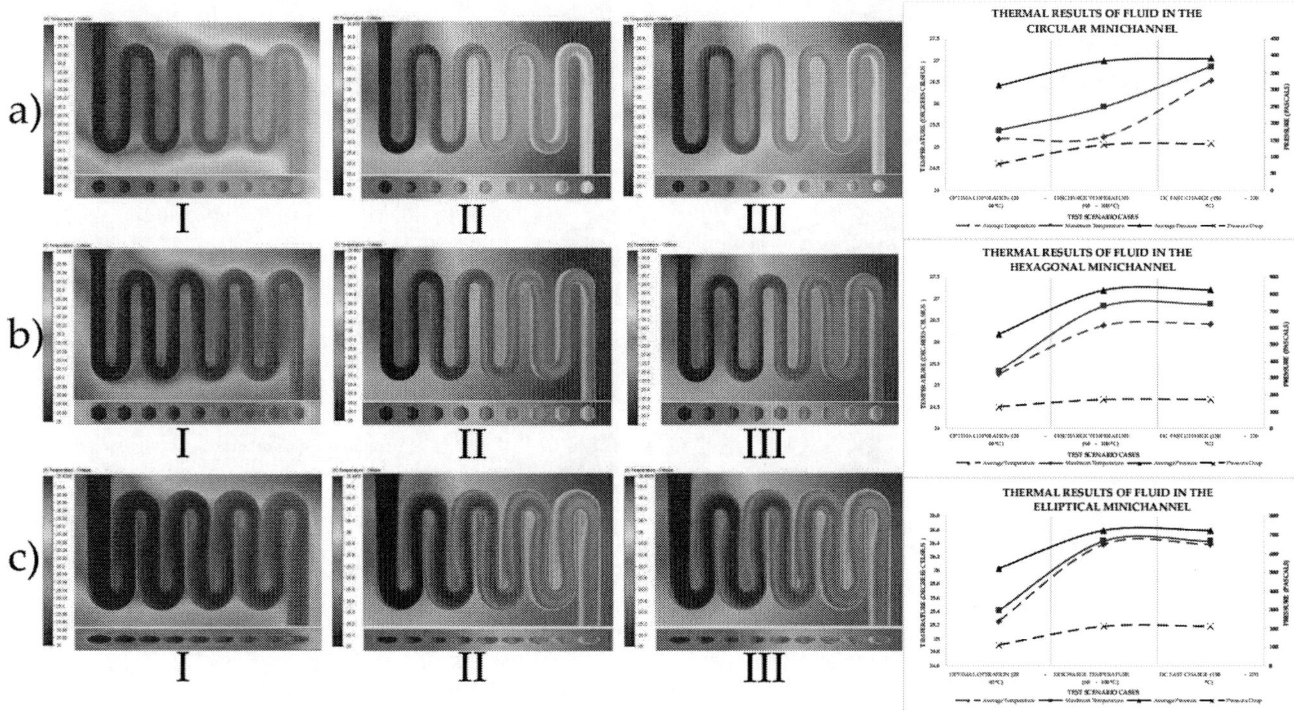

C.2. Thermal results of the fin

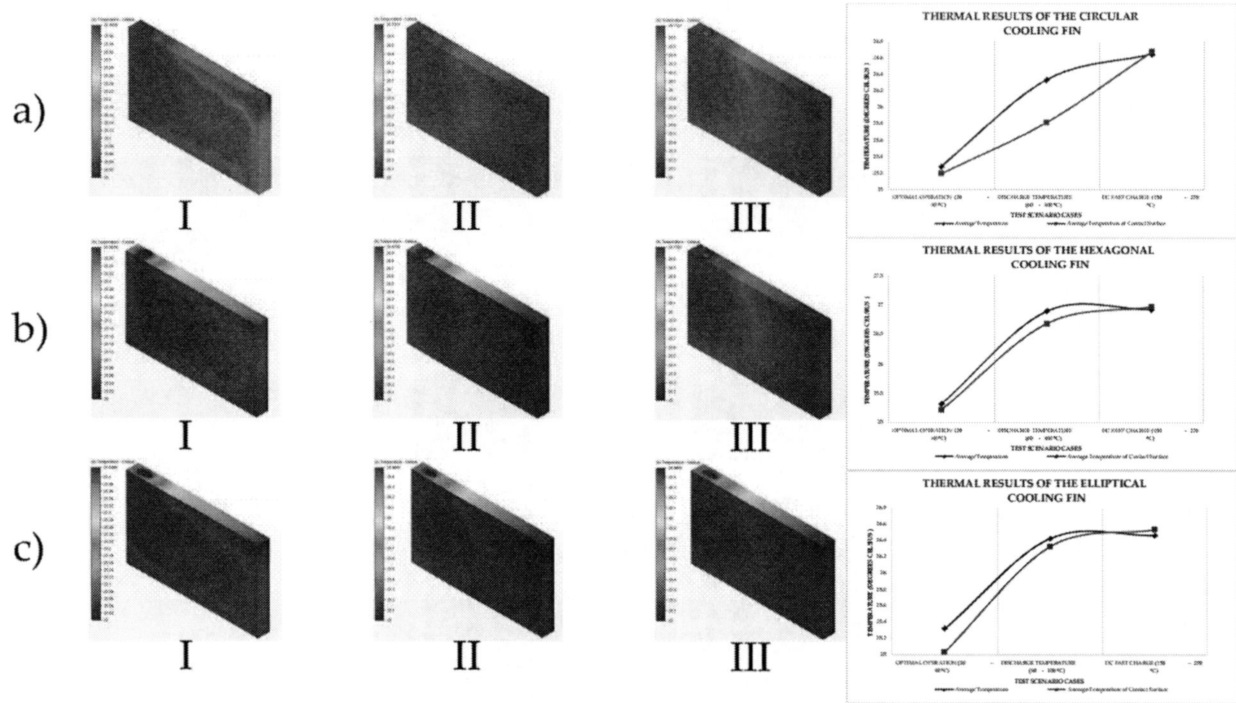

C.3. Thermal results of the fluid flowing inside the elliptical minichannel for particles traced when mixing

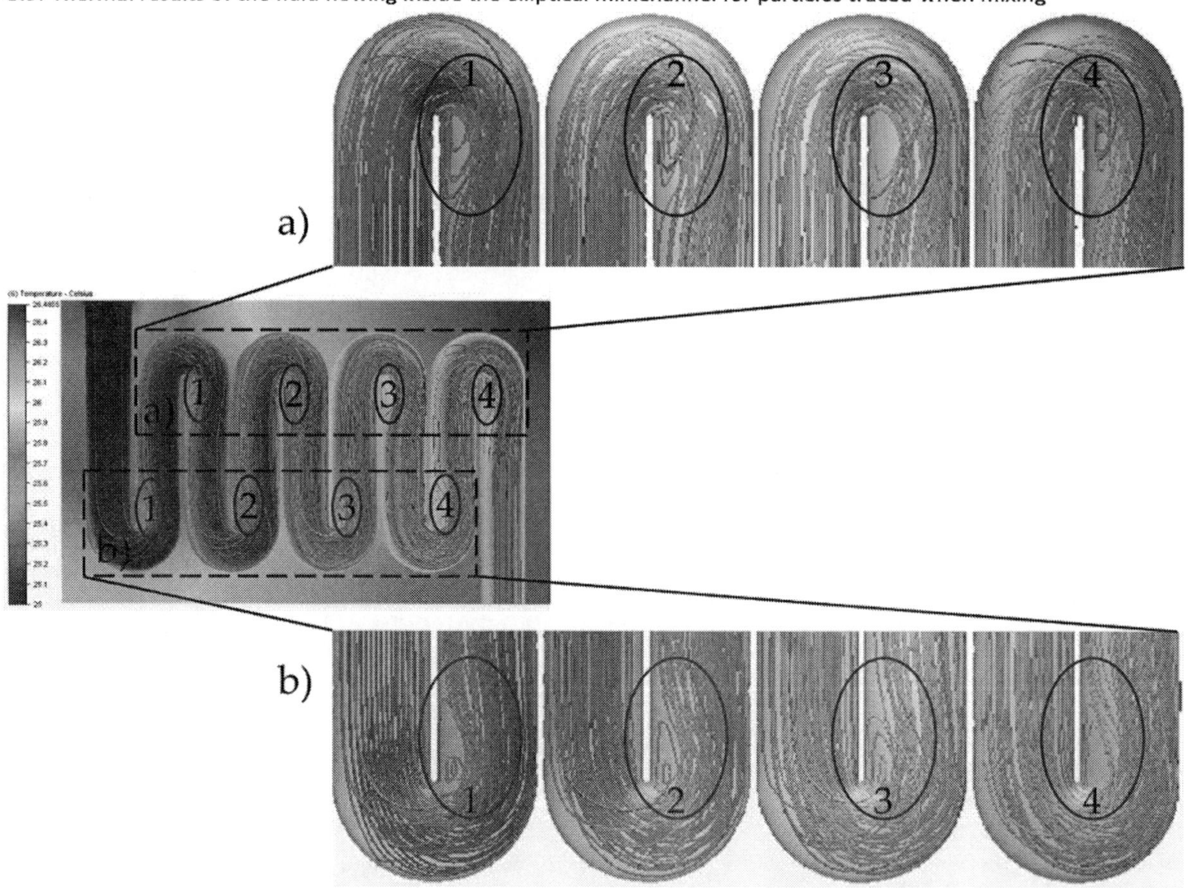

Lyapunov-based stability and control of a receiver aircraft during aerial refuelling

Aarti Panday[1*], Jimoh O. Pedro[1]

[1]University of the Witwatersrand
University of the Witwatersrand, Johannesburg, Private Bag 3, Wits 2050, South Africa
*Aarti.Panday@wits.ac.za

Abstract

The study explores feedback linearization control for aerial refuelling, addressing the nonlinear dynamics between a tanker and receiver aircraft. Stability is ensured through tuning controller gains using a Lyapunov-based cost function. While fuel transfer introduces mechanical energy to the receiver, minimizing the cost function reduces control effort required for the manoeuvre. A proposition is developed to prove that the system energy must be bounded for stability and performance robustness to be maintained. Genetic algorithms are employed to find optimal controller gains that meet performance requirements across all refuelling phases. The controller demonstrates stable, robust performance under disturbances.

Keywords: Nonlinear control, Aerial refuelling, Lyapunov stability, Feedback linearization control.

1. Introduction

Aerial refuelling is a vital part military of operations around the world [1]. The main factors affecting receiver aircraft dynamics in aerial refuelling are the time-varying mass and inertia during refuelling and the tanker aircraft's trailing vortices. These effects can potentially destabilise a receiver aircraft [2]. Due to these complexities, nonlinear control strategies have been proposed for automated aerial refuelling (AAR).

While studies have considered fault tolerant control based on the backstepping [3] and extended state observer-based nonlinear control [4], sliding mode control has been applied in numerous recent research outputs [5], [6], [7]. A major draw- back of sliding mode control is that it is prone to high-frequency oscillations known as chatterring. Manifestation of this phenomenon in control surface deflections for AAR will fall short against the stringent performance and safety requirements for AAR missions. While sliding mode control offers a framework to address external disturbances and parametric uncertainty, the potential of feedback linearization (FBL) has been limited in its exploration. More recently, using additive-state-decomposition-based control, FBL was used to address stabilisation in probe-drogue refuelling during station-keeping [8].

Stability is crucial among control system properties, as it determines convergent behaviour. Lyapunov's direct method gives insight into the stability of a nonlinear system by constructing a scalar "energy-like function" for the system and examining its time variation [9], assuming that dissipated mechanical energy leads the system to an equilibrium. However, in AAR, mechanical energy inherently increases as mass is introduced. This study incorporates the energy increase into the system dynamics rather than treating it as a disturbance, extending Lyapunov's method to such systems. The paper contributes a feedback linearization controller, optimized using a genetic algorithm and a Lyapunov-based cost function, to ensure system stability despite the energy increase.

2. System Modelling

Fig. 1. Tanker and receiver in aerial refuelling

Accounting for time-varying mass in Newton's Second Law, the receiver aircraft nonlinear dynamics during AAR can be described relative to the tanker motion (Fig. 1.) in nonlinear affine-in-control form [10]:

$$\dot{\mathbf{x}}_R = f(\mathbf{x}_R, \mathbf{x}_T, \mathbf{W}, \dot{\mathbf{W}}, \dot{m}) + g(\mathbf{x}_R)u_R \quad (1)$$

$$\mathbf{x}_R = [V_R, \beta_R, \alpha_R, p_{TR}, q_{TR}, r_{TR}, \ldots \\ \phi_{TR}, \theta_{TR}, \psi_{TR}, x_{TR}, y_{TR}, z_{TR}]^T \quad (2)$$

$$\mathbf{y}_R = h(\mathbf{x}_R) = [\phi_{TR}, x_{TR}, y_{TR}, z_{TR}]^T \quad (3)$$

$$\mathbf{u}_R = [\delta_a, \delta_e, \delta_r, \delta_t]^T \quad (4)$$

where $\mathbf{x}_R$ and $\mathbf{x}_T$ refers to the receiver and tanker states respectively in Eq. 1. $\mathbf{W}$ is the wind velocity and u_R is the receiver control vector. The receiver aircraft states in Eq. 2 are speed V_R, sideslip angle β_R, angle of attack α_R, and states relative to the tanker aircraft: roll rate p_{TR}, pitch rate q_{TR}, yaw rate r_{TR}, roll angle ϕ_{TR}, pitch orientation θ_{TR}, yaw angle ψ_{TR}, x- y- and z - separation $[x_{TR}, y_{TR}, z_{TR}]$ respectively. $\mathbf{y}_R$ is the receiver aircraft output states (Eq. 3). The receiver aircraft control surface deflections (Eq. 4) are aileron δ_a, elevator δ_e, rudder δ_r, and throttle δ_t.

The translational dynamics (Eq. 5), rotational dynamics (Eq. 6), translational kinematics (Eq. 7) and rotational kinematics (Eq. 8) are given in compact vector form [10]:

$$\dot{\mathcal{X}}_{\mathcal{R}} = \mathbf{f}_1\dot{\omega}_{B_RB_T} + c_1;\ \mathcal{X}_{\mathcal{R}} = [V_R, \beta_R, \alpha_R]^T \ (5)$$

$$\dot{\omega}_{B_RB_T} = \mathbf{f}_2\dot{\mathcal{X}}_{\mathcal{R}} + c_2\ ;\ \omega_{B_RB_T} = [p_{TR}, q_{TR}, r_{TR}]^T \ (6)$$

$$\dot{\xi} = f_3\left(\boldsymbol{R}^T_{B_RB_T}, V_W\right);\ \xi = \left[x_{TR}, y_{TR}, z_{TR}\right]^T \ (7)$$

$$\dot{\boldsymbol{R}}_{B_RB_T} = \boldsymbol{S}\left(\omega_{B_RB_T}\right)\boldsymbol{R}_{B_RB_T} (8)$$

Functions $\mathbf{f}_1, c_1, \mathbf{f}_2, c_2$ are complex nonlinear relationships linking: fuel flow effects in each tank; changing mass and inertia properties of the receiver aircraft; the tanker motion; the wind field and its spatially-varying action onto the receiver aircraft; and the forces and the moments acting on the receiver aircraft, expressed in the tanker body frame or the receiver wind frame as applicable. The actuator dynamics for each control surface δ_i in Eq. 4 were modelled as a first-order system with time constant τ_{δ_i}.

Furthermore, the system block diagram (Fig. 2) shows the system kinematics and dynamics in two loops. The inner loop (fast dynamics) and outer loop are constituted as follows:

$$\dot{\mathbf{x}}_{\substack{inner\\outer}} = f_{\substack{inner\\outer}}(\mathbf{x}) + g_{\substack{inner\\outer}}(\mathbf{x})\mathbf{u}_{\substack{inner\\outer}} \ (9)$$

$$\mathbf{y}_{\substack{inner\\outer}} = C\mathbf{x}_{\substack{inner\\outer}} \ (10)$$

$$\mathbf{x}_{\text{inner}} = [V_R, p_{TR}, q_{TR}, r_{TR}]^T \ (11)$$

$$\mathbf{u}_{inner} = \mathbf{u}_R = \left[\delta_{a_R}, \delta_{e_R}, \delta_{r_R}, \delta_{T_R}\right]^T (12)$$

$$\mathbf{x}_{\text{outer}} = [x_{TR}, \phi_{TR}, y_{TR}, z_{TR}]^T \ (13)$$

$$\mathbf{u}_{outer} = [V_R, p_{TR}, q_{TR}, r_{TR}]^T (14)$$

3. Controller Implementation

In input-output feedback linearization, a nonlinear transformation is applied to a system, in conjunction with a new input variable v (or pseudo-input variable), such that the nonlinear system behaves linearly from the new input v to the output y, whilst decoupling the system where the ith output depends on the ith external input. The system in Eq. 1 is input-output linearizable if there exists a static state feedback of the form [11]:

$$u_{FBL} = P(x) + Q(x)v(t) \ (15)$$

$P(x) \in \mathbf{R}^m, Q(x) \in \mathbf{R}^{m\times m}$ nonsingular, $v \in \mathbf{R}^m$
Introducing constant-coefficient linear operators of the form:

$$\mathfrak{L}_{\rho_i} = \sum_{k=1}^{\rho_i} \lambda_{ik}\frac{d^k}{dt^k}, i = 1, \ldots, m \ (16)$$

and satisfying the conditions in [11], selection of λ_{ik}, input-output linearisation and decoupling can be achieved [11]:

$$\sum_{i=1}^{m} \sum_{k=0}^{r_i} \lambda_{ik}\frac{d^k y_i}{dt^k} = v \ (17)$$

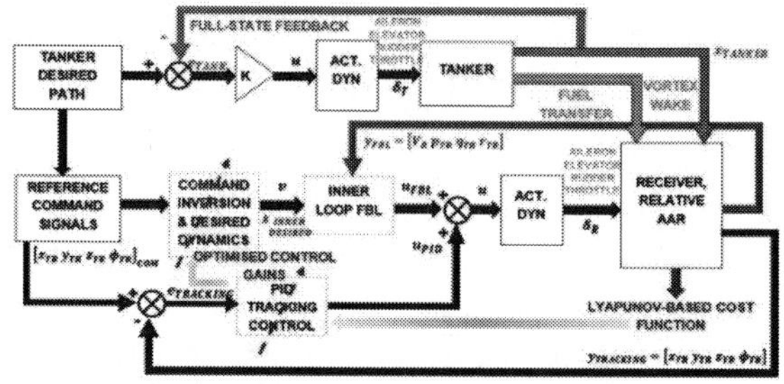

Fig. 2. Closed-loop system block diagram

3.1. Inner-Loop FBL

Applying the Lie derivative, the relative degree is thus: $\{r_1 = 1, r_2 = 1, r_3 = 1, r_4 = 1\}$

Using Eq. 26 with $m = 4$ and $r_i = 1$ according to the relative degree, $v(t)$ is: i goes from 1 to 4

$$v_i = \lambda_{i0}(y_i - y_{id}) + \lambda_{i1}(\dot{y}_i - \dot{y}_{id}) \ (18)$$

3.2. Command Inversion and Outer-Loop Tracking Control

The synthetic input v is shaped for the obtaining the desired tracking performance in the outer loop [12]. The derivative of the original control $\dot{u}$ is taken as the new control inputs - in line with giving the desired dynamics for the inner-loop feedback linearisation. This arrangement also appears to mimic a backstepping controller and so is a "pseudo-backstepping" scheme in that sense. The control law is given in Eq. 19:

$$u = \int_0^t \left[-K_s\left(\frac{\partial v}{\partial u}\right)^T (v - z) - \left(\frac{\partial v}{\partial u}\right)^{-1}\left(\frac{\partial v}{\partial x}\dot{x} - \dot{z}\right)\right] dt \ (19)$$

The variable z is defined in Eq. 20:

$$z = \left[diag[k_1]\right]_{4\times 4}[y_{id} - y_i]_{i=1\ldots4} + [\dot{y}_{id}]_{i=1\ldots4} \ (20)$$

A PID controller is used to close the outer loop and provide feedback control so that the primary controller design requirements are met. The PID controller comprises four control laws for each control surface, given generically in Eq. 21:

$$\delta_{PID} = K_P(y_{com} - y) + K_I\int(y_{com} - y)\,dt + K_D(\dot{y}_{\text{com}} - \dot{y}) \ (21)$$

$$u_\delta = u_{FBL} + \delta_{PID} \ (22)$$

The final control signal in Eq. 22 is of the sum of the feedback-linearised portion and the PID contribution.

3.3. Controller Specifications

Table 1. Refuelling phases

Phase	Description	Time [s]
I	On station, in vortex	0 - 25
II	Fuel transfer: Tank 1 & Tank 2	25 - 85
III	Fuel transfer: Tank 3 & Tank 4	85 - 195
IX	On station, in vortex	195 - 600

The primary requirement of the control design is to maintain a fixed relative separation from the tanker, with zero steady-state error in x_{TR}, y_{TR}, and z_{TR}, while subject to a vortex wind field and varying mass and inertia characteristics in a straight leg refuelling manoeuvre. The transient behaviour expected at the times when fuelling commences and completes (Table 1) should be minimal and not destabilise the aircraft. The receiver aircraft speed, V_R, should be the same as the tanker aircraft throughout the manoeuvre, also with zero steady-state error. The control inputs generated by the controller should not cause significant saturation on the magnitudes and rates of the actuators. Lastly, it was desired that a single set of control gains be used for the entire simulation and that gain scheduling should not be incorporated in the control design. These gains should be applied directly to the nonlinear receiver model and no system linearization should be employed. Robustness to disturbances should be demonstrated: (i) asymmetrical mass flow rate (Phase II at 75% capacity, Phase III, tank 3 at 50%, tank 4 was not filled), and (ii) application of light Dryden turbulence in Phase III.

4. Lyapunov-Based Controller Tuning

In aerial refuelling, the mass cannot increase from its initial value at t_0 to infinity at some time $t \gg t_0$ as there are physical limits imposed on the aircraft. The fuel tank size and location place a physical bound on the maximum energy that the system is able to receive. As mechanical energy is added through fuel transfer, it is important to ascertain stability characteristics throughout the refuelling period, including at the instance when fuel transfer is terminated, as unstable and thus undesirable behaviours would be detrimental to the safety and success of the AAR manoeuvre.

The controller tuning process is initiated with manual tuning of the gains. A cost function J can be formulated such that performance of manually-tuned controller gains can be ascertained and a bounded search space can be found for application of optimisation techniques. In formulating the gain selection as an optimisation problem, an optimal set of control gains can be found within this bounded search space.

A Lyapunov function V for the AAR system can be constructed - whose derivative $\dot{V}$ will be incorporated into the cost function to determine appropriate controller gains to meet stability requirements. In addition to $\dot{V}$, the cost function will include a quadratic element for minimisation of the state deviation from its commanded or nominal values, as well as for minimisation of the control effort:

$$J = \int_0^{t_f} ([(x - x_c)^T \boldsymbol{Q}(x - x_c) + \boldsymbol{u}^T \boldsymbol{R} \boldsymbol{u}] + \dot{V})\, dt \quad (23)$$

Minimising Eq. 23 will lead to the selection of optimal controller gains that ensure system stability, with:

$J \longrightarrow$ constant value K_J as $t \longrightarrow t_f$ and,

$\dot{J}$ driven to $K_{\dot{J}} \leq 0$ as $t \longrightarrow t_f$.

A theorem demonstrating uniform asymptotic stability for the system is provided in the Appendix. A genetic algorithm (GA) was used for the controller gain optimisation using the manually-tuned gains as the initial value within the bounded search space.

5. Results and Discussion

A set of 20 controller gains was tuned manually and the system response and cost function manually checked for performance. When these gains were used in the GA routine, it was found that when applying a high number of generations, the routine provided a divergent solution (Fig. 3). This resulted showed that too much system energy had been shed during optimisation of J (the system energy/cost). A minimum system energy needs to be maintained, as seen in Fig. 4 during optimisation in order for performance and stability characteristics to be maintained. A Proposition was developed in order to determine the minimum energy bound in the system in order for a stable control law to be found using the developed Theorem. The Appendix contains the proposition and proof.

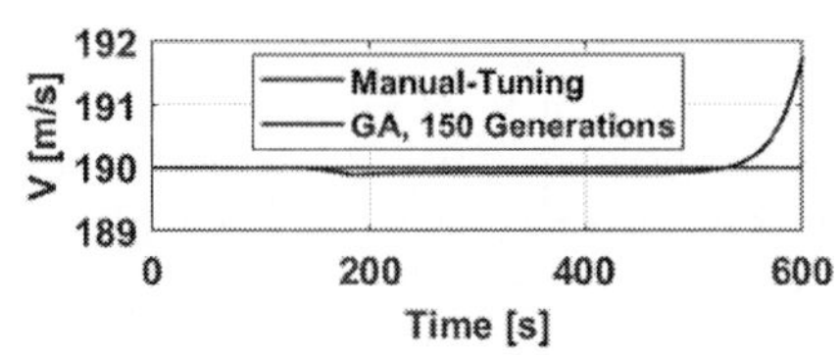

Fig. 3. Divergent results: GA-tuning with 150 generations

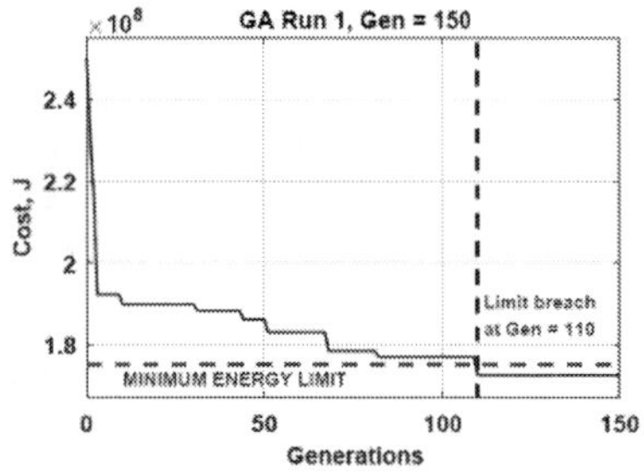

Fig. 4. Minimum energy threshold in J

After applying the bounds determined from the Proposition, the optimisation was able to yield a stable control law that met performance specifications. Table 2 lists the final cost values, showing the GA-optimised controller to be within the determined threshold.

Table 2. Controller cost function

Cost	Manual-Tuned	GA-Optimized
J	2.051e8	1.759e8

In Fig. 5, the underlying principle of the theorem can be seen from the plots of cost function J and $\dot{J}$ in the manually-tuned and optimised gains for the baseline, and two disturbance cases. The cost is driven to a constant value as the refuelling is completed and its derivative is driven to zero at that time, visually showing asymptotic stability.

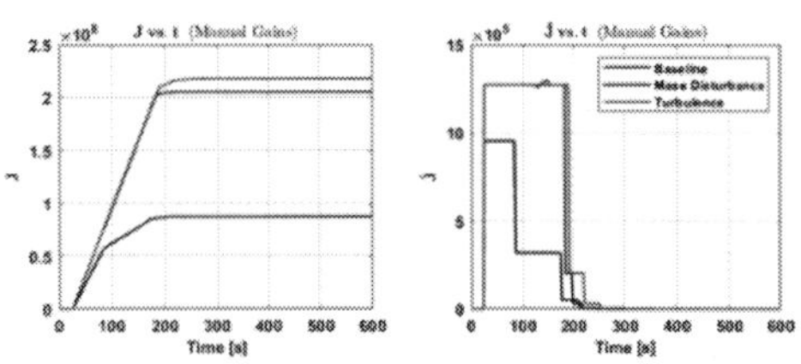

Fig. 5. Minimum energy threshold in J

As expected, since the mass flow rate is reduced in the mass disturbance, cost function J which is also a measure of system energy, is appropriately decreased compared to the baseline. The effect of tanker turbulence on the system is thus to add mechanical energy, since J is larger than that of the baseline case. The difference in J for the baseline and turbulence case becomes more prominent when examining the case where GA-optimised gains have been applied. Since the GA was applied to the baseline non-disturbed case, when the optimised controller gains are then subjected to tanker disturbance, it can be seen that J is distinctly higher than the baseline case. This is attributed to how the cost function J was formulated - as a representation of system energy. Whilst the baseline case was subject to energy being shed through optimisation, the turbulence acted to increase mechanical energy of the system that has not been shed, therefore, the case of the GA-optimised results is indeed larger than the basesline for the turbulence case.

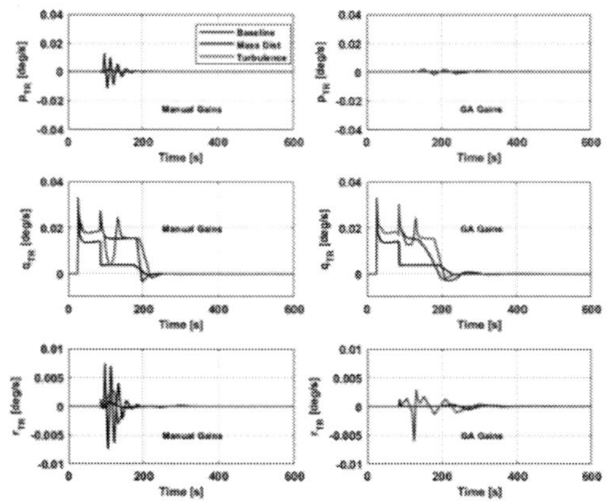

Fig. 6. Time-histories of rotational rates

The inner-loop rotational rates are shown in Figure 6. The lateral-directional roll and yaw rates were sensitive to the addition of fuel to the aft tanks for the manually-tuned system. Upon application of the GA-optimised controller gains, the response was significantly improved with the elimination of this transient behaviour.

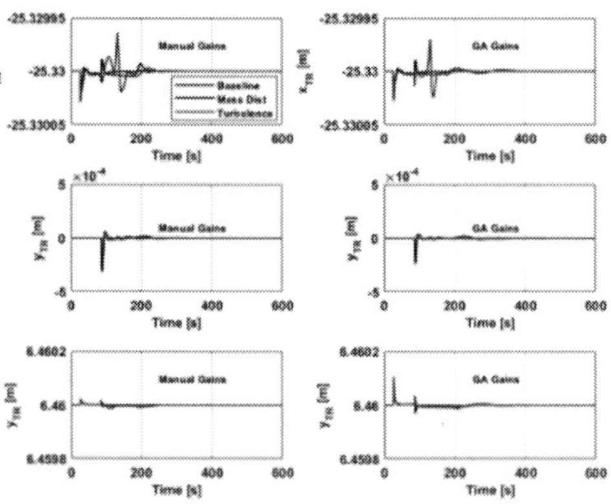

Fig. 7. Time-histories of relative position

The developed control laws were able to maintain the required separation between the two aircraft, even under disturbance conditions (Fig. 7). The baseline performance in x-separation was improved by the GA gain. While turbulence introduced transient behaviour, this was not destabilising and the magnitudes are within the acceptable window of deviation, in the range of 0.0002% from the setpoint.

The control surface deflections required to execute the manoeuvre are shown in Fig. 8. Deflections remained within the aircraft physical limits. The GA-optimised gains were able to reduce the control effort required, showing the impact of the quadratic cost on control within the Lyapunov-based cost function. Turbulence added additional energy, which required higher surface deflections. In the case of the asymmetric mass disturbance, the aileron and rudder deflections were non-zero, and was able to keep the aircraft at its required position.

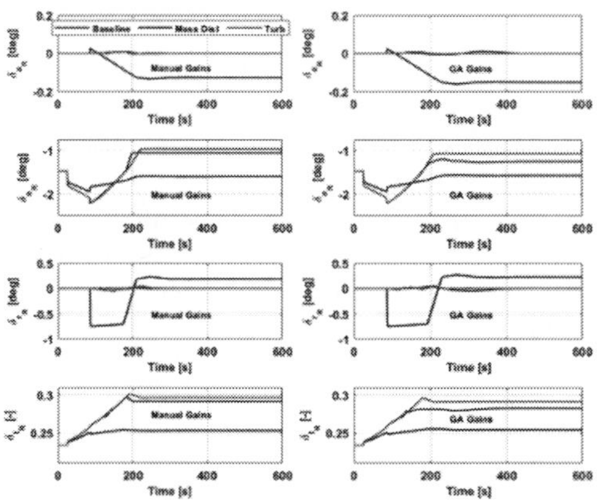

Fig. 8. Control surface deflections

6. Conclusion and Future Work

Lyapunov's Direct Method was successfully extrapolated to address stability of systems with increasing mass. Controller gains were first determined through manual-tuning and then optimised using a genetic algorithm. It was found that for stability to be maintained, a minimum system energy must remain in the system during optimisation. Future work will include modelling of the refuelling boom and comparison of FB sliding mode control. .

References

[1] Walton TA, Clark B (2021). *Resilient Aerial Refueling: Safeguarding the US Military's Global Reach*, Hudson Institute, Washington D.C.

[2] Dogan A, Sato S, Blake WB (2005). *Flight control and simulation for aerial refueling*, in Proceedings of the AIAA Guidance, Navigation, and Control Conference and Exhibit, San Francisco, California, pp1–16,

[3] Yu Z, Zhang Y, and Qu Y (2018). *Fault-tolerant control for autonomous aerial refueling against actuator fault in receiver UAV*, IFAC-PapersOnLine, 51(24):274–279.

[4] Cao H, Xu Y, Wang L (2022). *ESO-based nonlinear flying boom attitude control with the handling qualities requirement,* Aerospace Science and Technology, 120:107235.

[5] Wang L, Cao H, Yue T (2017). *Attitude Control Design for Flying Boom during Air-to-Air Refuelling*, Journal of Systems and Control Engineering, 231(1):812–826.

[6] Yu Z, Qu Y, Zhang Y (2017). *Robust Adaptive Dynamic Surface Control for Receiver UAV During Boom Refueling in the Presence of Vortex,* in Proceedings of 2017 29th Chinese Control and Decision Conference, pp 1798 – 1803

[7] Su Z, Wang H, Li N, Yu Y, Wu J (2018). *Exact docking flight controller for autonomous aerial refueling with back-stepping based high order sliding mode*, Mechanical Systems and Signal Processing, 101:338–360

[8] Ren J, Quan Q, Ma H, Cai K (2021). *Additive-state-decomposition-based station-keeping control for autonomous aerial refueling,* Science China Information Sciences, 64(11):1–3

[9] Slotine JE, Weiping L (1991). *Applied Nonlinear Control.* Prentice Hall, Englewood Cliffs, New Jersey, 1st edition.

[10] Panday A, Pedro JO (2018). *Particle swarm-optimised pid control of a receiver aircraft during aerial refuelling,* in 2018 IEEE Congress on Evolutionary Computation, Rio de Janeiro, Brazil,pp 1–8.

[11] Garces FR, Becerra VM, Kambhampati C, Warwick K (2003). *Strategies for Feedback Linearization: A Dynamic Neural Network Approach*, Oxford Higher Education, London, 1st edition.

[12] Zhang S, Liao F, Cheng Y, He K (2020). *Aircraft trajectory control with feedback linearization for general nonlinear systems,* in 2020 Chinese Control and Decision Conference, Hefei, China, pp 3937–3944.

[13] Marquez HJ (2003). *Nonlinear control systems: analysis and design.* John Wiley & Sons, Hoboken, 1st edition.

Appendix A. Theorem & Proposition

A.1. Theorem

Theorem 1 (Lyapunov stability for a non-autonomous AAR system) *If in a spherical region of the state space initially given by radius R_0, which when subject to addition of mechanical energy continuously increases in radius to R_{final}, there exists a scalar function $V(\boldsymbol{x}, t)$ with continuous first partial derivatives such that:*

- *$V(\boldsymbol{x}, t)$ includes a scalar proxy for system energy*

- *$V(x,t)$ is and remains positive definite as radius R increases from $R_0 \rightarrow R_{final}$*
- *$\dot{V}(x) \leq 0$ is and remains negative semi-definite as radius R increases from $R_0 \rightarrow R_{final}$*

then the equilibrium point $\mathbf{0}$ is stable in the sense of Lyapunov. If furthermore,

- *$V(x,t)$ is decrescent*

then $\mathbf{0}$ is uniformly stable at R_{final}, as well as the instantaneous equilibrium points $\mathbf{0}$ along the trajectory of $R_0 \rightarrow R_{final}$ are also stable. □

Proof of Theorem 1 When an appropriate Lyapunov function has been selected for the AAR system, proof of the proposed theorem corresponds to the proof detailed in [9] and [13]. Applying the analysis at each successive instance as the state-space radius R increases from $R_0 \rightarrow R_{\text{final}}$ as mechanical energy is introduced, uniform asymptotic stability is proven from $R_0 \rightarrow R_{\text{final}}$. □

A.2. Proposition

Proposition 1 When controller gains are optimised using evolutionary algorithms with cost function J defined in Theorem 1, its minimisation will lead to the selection of optimal controller gains that ensure system stability, with $J \longrightarrow$ constant value K_J as $t \longrightarrow t_f$ and $\dot{J}$ being driven to $K_{\dot{J}} \leq 0$ as $t \longrightarrow t_f$.

However, the consequence is that the system energy represented through J, will be reduced compared to the case when gains were determined manually. Under an optimisation scheme, a minimum system energy must be available in order for stability to be maintained, and for a degree of robustness to disturbances to remain, i.e., $J_{\text{manual}} > J > J_{\text{minimum}}$

If $J < J_{\text{minimum}}$, then too much of energy would have been shed during the optimisation process, and there would be insufficient energy to maintain stable migration from system radius R_0 to R_{final}. Thus, there exists a minimum energy threshold, below which, the system becomes unstable. □

Proof of Proposition 1 Proof of this proposition is based on formulating the problem as a nominal motion problem subject to perturbation [9] or as the translation of a non-zero trajectory [13]. Instead of considering stability around an equilibrium point, stability is now examined from the perspective of the motion, i.e., whether a system will remain close to its original motion trajectory if slightly perturbed away from it. In the case of the AAR optimisation, the "perturbation" refers to the motion under optimisation of the control. In Fig. 9, the nominal motion trajectory is labelled $\mathbf{x}(t)$. Thus the "perturbed" motion, i.e., the motion that results from applying the optimised controller gains would correspond to $\mathbf{x}^*(t)$.

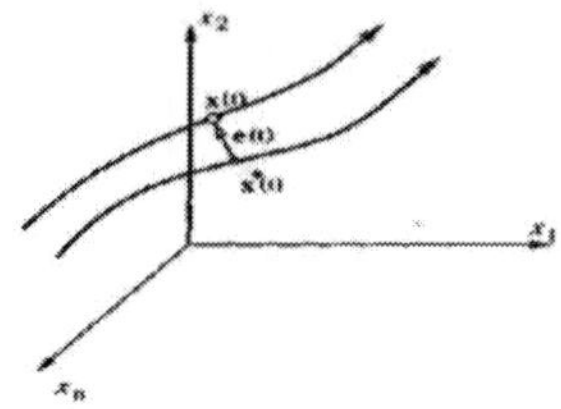

Fig. 9 Nominal motion problem ([9]).

Let $E = \mathbf{x} - \mathbf{x}_{\text{min}}$ and the corresponding Lyapunov function be represented by V_{min} where $V_{\text{min}} = m\mathbf{x}^2_{\text{min}}$. Then the Lyapunov function corresponding to the manually-tuned controller gains, V is larger than the GA-optimised Lyapunov function by a factor α. For ease of representation, $\mathbf{x}$ is written as x in the remainder of this proof.

$$\alpha V_{\text{min}} = V \tag{24}$$

$$\alpha(mx^2_{\text{min}}) = mx^2 \tag{25}$$

$$\alpha(m[x - E]^2) = mx^2 \tag{26}$$

$$E = x\left(1 \pm \frac{\sqrt{2-\alpha}}{2}\right) \tag{27}$$

It can be seen that the parameter α cannot be greater than 2 as the square root would lead to an imaginary solution. Also, the lower limit of α is 1 , meaning that the motion trajectory of the manually-tuned controller gains is the same as the motion trajectory that of the optimised controller gains ($V = V_{\text{min}}$). Now let $\beta = \frac{1}{\alpha}$. Hence:

$$2 > \alpha > 1 \tag{28}$$

$$0.5 < \beta < 1 \tag{29}$$

Therefore in general, J may be optimised upto half of the manually-tuned controller gains system cost, and there exists an optimal value of α and β within their bounds, for the AAR system for which the stability of the motion can be maintained when subject to optimisation. □

While the proposition above shows that the system energy is bound by a factor of two (or half), the optimal factor (α/β) for an AAR system can be found using numerical parameters specific to the system.

A Comparative Study On The Use Of Evolutionary Algorithms For Snake Robot Control

Phumelele S. Magagula[a,*], Aarti Panday[b]

[a]University of the Witwatersrand, 1 Jan Smuts Ave, Johannesburg, 2017, South Africa

Abstract

This paper presents a comparative study on the effectiveness of Evolutionary Algorithms (EAs), including the Genetic Algorithm (GA), Particle Swarm Optimisation (PSO), and Ant Colony Optimisation (ACO), in optimizing the PID gains of a seven Degree-of-Freedom (DOF) wheel-less snake robot modelled in MATLAB/Simulink. This is necessary to improve the control of the multi-link, multi-DOF module, which would make the snake robot useful in hazardous environments where high manoeuvrability is required. The performance of these algorithms was assessed in terms of direction control and control input minimization, with a cost function defined for quantitative analysis. The results show that the GA is most efficient at minimising the cost function, thus has improved overall controller performance compared to the PSO and ACO.

Keywords: *Snake Robot Control, PID Optimisation, MATLAB/Simulink, Evolutionary Algorithms.*

1. Background and Motivation

A snake robot's multi-link, multi-DOF body composed of multiple joints has motion driven by ground friction and changes in internal shape [1]. Its high manoeuvrability capabilities achieve effective navigation [2]. However, that results in more complex control. EAs have been proven to be suitable for solving non-linear constrained problems that have high complexity [3]. Thus, they were proposed to obtain near optimal controller parameters that minimise system costs while achieving the control objective [4].

Many previous studies sought to prove the effectiveness of a control scheme, such as Bhandari et al. [5], who tuned control parameters using the GA for trajectory tracking. Kelasidi et al. [6] paper explored snake robot controller optimisation using the PSO. However, few studies directly compared the performance of different control optimisation schemes. Such as Abut [7], who compares FUZZY-PID and GA-PID to conventional PID. Furthering research on a comparative investigation would be advantageous for choosing the better control scheme.

This study's aim is to compare the effectiveness of three EAs in optimising snake robot control gains obtained using the Ziegler-Nichol (Z-N) classical tuning method, to improve reference trajectory tracking and direction control. To do this, a planar model of a snake robot is completed in MATLAB/Simulink using previously derived expressions, along with an undulating serpentine trajectory profile (sinewave gait). A baseline trajectory tracking controller is developed then optimised using the proposed EAs.

2. Mathematical modelling

The 2D model of the snake robot presented is achieved by approaches similar to the kinematic and dynamic equations introduced in [8], [9], [10], and [11].

2.1. Kinematic equations

The initial positions of the links are as follows:

$$x_i = x_0 + l\left[\cos\theta_i + \sum_{j=1}^{i-1}\cos\theta_j\right] \tag{1}$$

$$y_i = y_0 + l\left[\sin\theta_i + \sum_{j=1}^{i-1}\sin\theta_j\right] \tag{2}$$

Referring to Fig. 1, the position of the centre of mass (CoM) can be written as:

$$\boldsymbol{P} = \begin{bmatrix} P_x \\ P_y \end{bmatrix} = \frac{1}{Nm}\begin{bmatrix} \sum_{i=1}^{N} m x_i \\ \sum_{i=1}^{N} m y_i \end{bmatrix} = \frac{1}{N}\begin{bmatrix} \boldsymbol{e}^T\boldsymbol{X} \\ \boldsymbol{e}^T\boldsymbol{Y} \end{bmatrix} \tag{3}$$

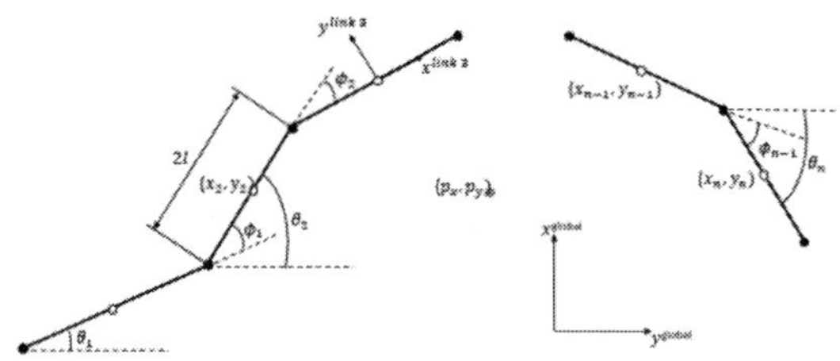

Fig. 1: Snake Robot with N equivalent rigid links

The position of the CoM of each link can be written as:

$$\boldsymbol{X} = -l\boldsymbol{K}^T\cos\boldsymbol{\theta} + \boldsymbol{e}p_x \tag{4}$$

$$\boldsymbol{Y} = -l\boldsymbol{K}^T\sin\boldsymbol{\theta} + \boldsymbol{e}p_y \tag{5}$$

2.2. Anisotropic viscous friction model

The anisotropic viscous friction model is written as:

$$\boldsymbol{f}_R = \begin{bmatrix} f_{R,x} \\ f_{R,y} \end{bmatrix}$$

$$f_R = \begin{bmatrix} c_t(C_\theta)^2 + c_n(S_\theta)^2 & (c_t - c_n)S_\theta C_\theta \\ (c_t - c_n)S_\theta C_\theta & c_n(C_\theta)^2 + c_t(S_\theta)^2 \end{bmatrix} \begin{bmatrix} \dot{X} \\ \dot{Y} \end{bmatrix} \tag{6}$$

2.3. Dynamic equations

The dynamic equations are derived from Fig 2. It is assumed the joints are equipped with motors that apply torque u.

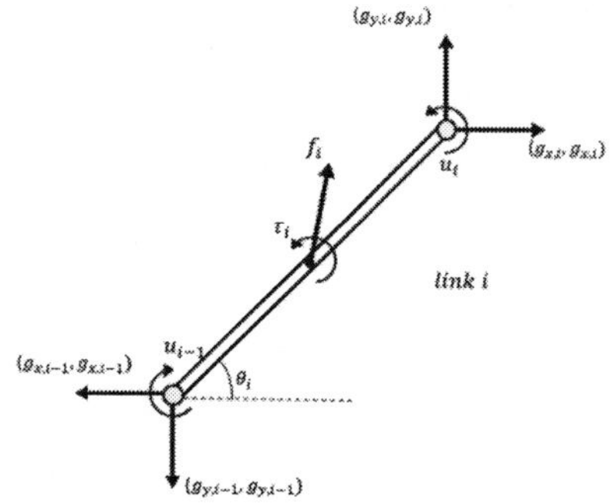

Fig. 2: Free-body diagram of a snake robot link

$$M_\theta \ddot{\theta} = D^T u + \tau - W\dot{\theta}^2 + lS_\theta K f_{R,x} - lC_\theta K f_{R,y} \tag{7}$$

$$\ddot{P}_x = \frac{1}{Nm} e^T f_x \tag{8}$$

$$\ddot{P}_y = \frac{1}{Nm} e^T f_y \tag{9}$$

where:

$$\tau = -c_n J \dot{\theta} \tag{10}$$

$$W = ml^2(S_\theta V S_\theta - C_\theta V C_\theta) \tag{11}$$

$$M_\theta = J I_N + ml^2(S_\theta V S_\theta + C_\theta V C_\theta) \tag{12}$$

2.4. Reference trajectory profile

The parameterised gait reference trajectory can be defined as in equation 13 [12], and the joint angular positions are defined as the difference in the link positions.

$$\theta_{i,ref} = \alpha \sin(\omega t + (i-1)\beta) + \gamma, \qquad i = 1, \dots, N \tag{13}$$

$$\phi_i = \theta_{i+1} - \theta_i \tag{14}$$

2.5. Parameters

Some necessary parameters are listed in the tables below:

Table 1: Snake robot model parameters

Parameter	Value
Number of links	$N = 5$
Link half-length	$l = 0.1\ m$
Mass of each link	$m = 1.5\ kg$
Friction coefficients	$c_t = 0.5; c_n = 3$

Table 2: Reference trajectory parameters

Parameter	Value
Amplitude	$\alpha = 30°$
Angular frequency	$\omega = 3\ rad.s^{-1}$
Phase shift	$\beta = 60°$
Joint offset	$\gamma = 0°$

3. Controller Design

3.1. Controller specifications of snake robot

The design specifications chosen were based on established works that developed and tested snake robot models, including Baysak and Atlas [13], and Ma S, et. al [14]. The snake robot's sinusoidal path will be restricted to an amplitude of $\pm 0.1\ m$, have a minimum joint torque of $1\ Nm$, and a maximum joint angle amplitude of $\emptyset = \pm 30°$. Lastly, the heading angle must degrade to zero over time, to follow a straight-line path along the x-axis, that is $\bar{\theta} = 0°$.

3.2. Joint trajectory law

For optimisation purposes, a standard PID controller to dictate joint control is proposed to be:

$$u_i = k_P e_i + k_I \int e_i dt + k_D \frac{de_i}{dt} \tag{15}$$

where:

$$e_i = \phi_{ref} - \phi \tag{16}$$

$$\phi_{i,ref} = \theta_{i+1,ref} - \theta_{i,ref} \tag{17}$$

3.3. Line-of-sight (LOS) guidance law

To account for path control, the LOS guidance law is used. The snake robot is steered towards a point a distance Δ along the path in front of the robot. The control objective is to make the snake robot converge to a straight-line path. The heading angle is defined as [11]:

$$\bar{\theta}_{ref} = -arctan\left(\frac{P_y}{\Delta}\right) \tag{18}$$

where P_y is the cross-tracking error. The look ahead distance is twice the length of the robot, such that:

$$\Delta = 2Nl \tag{19}$$

To achieve path following control, the LOS controller is applied as an offset such that:

$$\boldsymbol{\gamma = k_\theta(\bar{\theta} - \bar{\theta}_{ref})} \tag{20}$$

3.4. Complete snake robot model with Z-N tuning

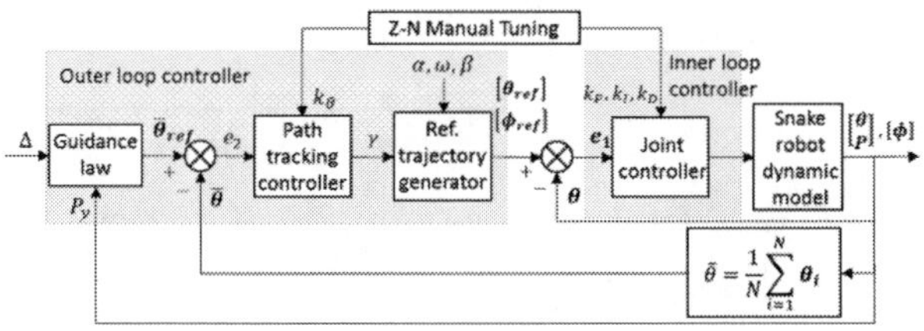

Fig. 3: Complete snake robot model including manual tuning of the PID gains

The control block diagram for applying it is given as above. To obtain the necessary PID gains the Ziegler-Nichols (Z-N) manual tuning method was first applied only as a starting point for gain optimisation.

3.5. Control Optimisation

A quadratic cost function in the form of the integral time square error (ITSE) indicates how well the system achieves its control objectives. It may be expressed as:

$$J = ITSE = \frac{1}{T}\int_0^T \left[\frac{(P_y - P_{y,ref})}{P_{max}^2} + (\boldsymbol{u^T Q u})^T + \frac{([\phi-\phi_{ref}]^T Q[\phi-\phi_{ref}]^T)}{\phi_{max}^2} + (\bar{\boldsymbol{\theta}} - \bar{\boldsymbol{\theta}}_{ref})\right] \tag{21}$$

The aim of the controller design is to minimise the error in the position, angular position of the joints and control effort (actuation). With optimisation applied the system block diagram is represented as:

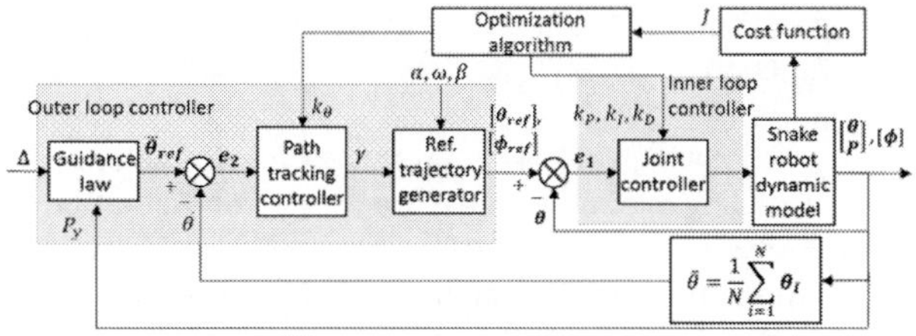

Fig. 4: Complete snake robot model including the PID gain optimisation with EAs

4. Simulation Results and Discussion

4.1. Gains computed through the suggested algorithm

Fig. 5 shows the superimposed convergence history of the EAs, and the resulting gains are displayed in Table 3. The GA produced a superior cost function value, which can be attributed to its efficient balance between exploration and exploitation capabilities when searching the search space. The algorithm retains the fittest individuals to be passed on to the next iteration, which ensures they are always available for reproduction. The ACO performed the worst, which can be explained by its low exploration capabilities as the pheromone trail increases the probability of choosing a local minima that may not be the global minima. Resetting the pheromone trail improved the algorithm but that resulted in the slow convergence observed.

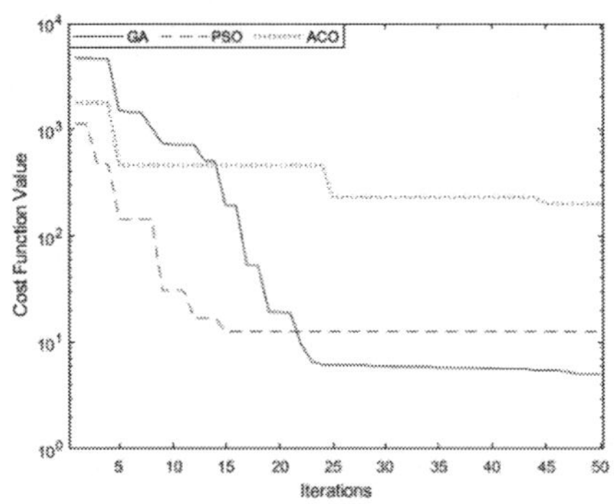

Fig. 5: Convergence history of the various proposed search algorithms

4.2. Analysing performance of optimized gains

The ACO's gains resulted in the snake robot's CoM slipping beyond the acceptable bounds of $\pm\ 0.1\ m$, as observed in Fig. 6. The manually tuned gains result in straight-line motion within the acceptable bounds but utilised the highest control effort.

Table 3 shows the GA has the lowest cost function value, indicating improved overall system performance. It and the PSO both prioritised minimising control effort over path following with the gains achieved over the iterations to minimise the cost function.

Observing Table 4 shows the PSO performed best regarding minimising cross-tracking error. Actuation torque contributes significantly to the cost function value, thus gains that improve on control effort produce a low cost function value. The performance of the GA and PSO reflect this well. The GA uses the lowest control effort, as shown by its RMS and peak value.

Table 3: Gains computed using EAs

Control loop	Outer loop	Inner loop			J
Gain	K_θ	K_P	K_I	K_D	
Manual	0.60110	18.0000	16.3636	4.95000	7004.60
GA	0.60489	12.47039	13.01570	0.25089	5.01
PSO	0.72368	24.51434	5.69261	0.47144	12.62
ACO	0.60156	10.15602	30.46805	2.03120	200.94

Table 4: Performance of the plant with optimised control parameters

Performance criteria	RMS	Peak	RMS	Peak
Cases	Cross-tracking error (m)		Heading error (deg)	
GA	0.0491	0.0917	33.56	-168.10
PSO	0.0487	0.0867	23.31	-179.12
ACO	0.0582	0.1254	30.38	-170.94
Cased	Actuator torque (Nm)		Joint-tracking error (deg)	
GA	4.36	17.03	10.52	27.91
PSO	7.66	32.49	8.45	27.82
ACO	15.86	96.71	6.78	27.55
Cases	Transient time (s)		Settling time	
GA	4.71		24.96	
PSO	4.08		24.98	
ACO	3.31		24.97	

By observing Fig. 7, an initial transient response with a high start-up torque is noted. This is when the plant is transitioning from its initial position (i.e. $\phi_i = 0$) into a serpentine gait before steady state. The settling times achieved by all optimisation methods are effectively the same, as shown in Fig. 8. An optimal control system minimises the transient time. Among the EAs the PSO performed the best in this aspect.

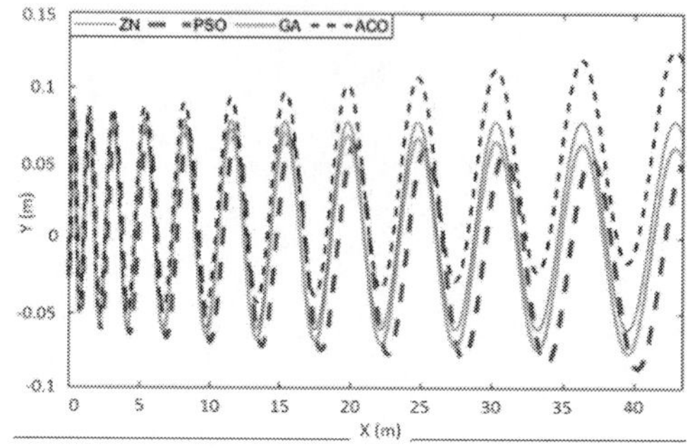

Fig. 6: CoM position of a snake robot with optimised control parameters

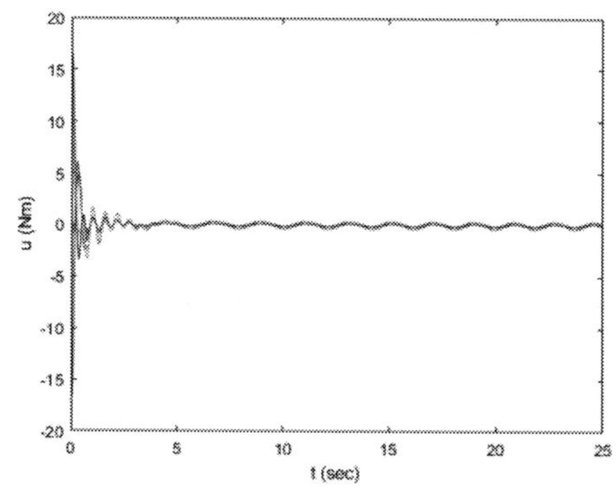

Fig. 7: GA based plant actuator torque

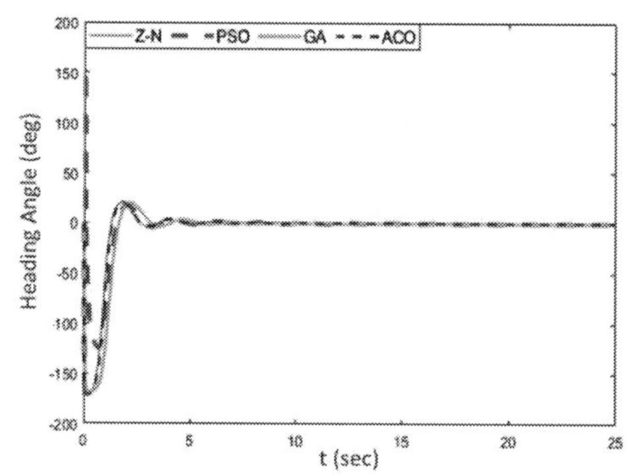

Fig. 8: Heading error of plant with optimised control parameters

5. Conclusion

In this paper a comparison was carried out between different proposed EAs in tuning the snake robot PID controller gains, namely the GA, PSO, and ACO. The GA resulted in the best overall performance with the lowest cost function. Regarding direction control, among the EAs the GA performed the best and ACO the worst. The GA performed best in terms of lowering the control effort. For future work, it is recommended to assign a higher weight to the cross-tracking error to prioritise direction control over other control parameters.

References

[1] T. Dear, B. Buchanan, R. Abrajan-Guerrero, S. D. Kelly, M. Travers, and H. Choset, "Locomotion of a multi-link non-holonomic snake robot with passive joints," *Int J Rob Res*, vol. 39, no. 5, pp. 598–616, Apr. 2020, doi: 10.1177/0278364919898503.

[2] A. Sibilska-Mroziewicz, A. Hameed, J. Możaryn, A. Ordys, and K. Sibilski, "Analysis of the Snake Robot Kinematics with Virtual Reality Visualisation," *Sensors 2023, Vol. 23, Page 3262*, vol. 23, no. 6, p. 3262, Mar. 2023, doi: 10.3390/S23063262.

[3] S. Ventura, M. Luna, and J. M. Moyano, *Genetic Algorithms*. IntechOpen, 2022. doi: 10.5772/intechopen.94664.

[4] T. Hägglund, "The one-third rule for PI controller tuning," *Comput Chem Eng*, vol. 127, pp. 25–30, Aug. 2019, doi: 10.1016/J.COMPCHEMENG.2019.03.027.

[5] G. Bhandari, R. Raj, P. M. Pathak, and J. M. Yang, "Robust control of a planar snake robot based on interval type-2 Takagi–Sugeno fuzzy control using genetic algorithm," *Eng Appl Artif Intell*, vol. 116, p. 105437, Nov. 2022, doi: 10.1016/J.ENGAPPAI.2022.105437.

[6] E. Kelasidi, M. Jesmani, K. Y. Pettersen, and J. T. Gravdahl, "Locomotion Efficiency Optimization of Biologically Inspired Snake Robots," *Applied Sciences 2018, Vol. 8, Page 80*, vol. 8, no. 1, p. 80, Jan. 2018, doi: 10.3390/APP8010080.

[7] T. Abut, "Dynamic Model And Optimal Control Of A Snake Robot: TAROBOT-1 Tayfun Abut," *International Journal of Scientific & Technology Research*, vol. 4, no. 11, 2015, [Online]. Available: www.ijstr.org

[8] P. Liljeback, K. Y. Pettersen, O. Stavdahl, and J. T. Gravdahl, "A simplified model of planar snake robot locomotion," in *2010 IEEE/RSJ International Conference on Intelligent Robots and Systems*, IEEE, Oct. 2010, pp. 2868–2875. doi: 10.1109/IROS.2010.5649110.

[9] A. J. Gonzalez and D. D. Dankel, "Biologically Inspired Robots: Snake-Like Locomotors and Manipulators by Shigeo Hirose Oxford University Press, Oxford, 1993, 220pages, incl. index (£40)," *Robotica*, vol. 12, no. 3, pp. 282–282, May 1994, doi: 10.1017/S0263574700017264.

[10] M. Saito, M. Fukaya, and T. Iwasaki, "Serpentine Locomotion with Robotic Snakes," *IEEE Control Syst*, vol. 22, no. 1, pp. 64–81, 2002, doi: 10.1109/37.980248.

[11] K. Y. Pettersen, "Snake robots," *Annu Rev Control*, vol. 44, pp. 19–44, 2017, doi: 10.1016/J.ARCONTROL.2017.09.006.

[12] D. Rollinson and H. Choset, "Pipe Network Locomotion with a Snake Robot," *J Field Robot*, vol. 33, no. 3, pp. 322–336, May 2016, doi: 10.1002/ROB.21549.

[13] Y. A. Baysal and I. H. Altas, "Modelling and Simulation of a Wheel-Less Snake Robot," *2020 7th International Conference on Electrical and Electronics Engineering, ICEEE 2020*, pp. 285–289, Apr. 2020, doi: 10.1109/ICEEE49618.2020.9102599.

[14] S. Ma *et al.*, "Self-adaptive rolling motion for snake robots in unstructured environments based on torque control," *Biomimetic Intelligence and Robotics*, vol. 3, no. 3, p. 100117, Sep. 2023, doi: 10.1016/j.birob.2023.100117.

Investigating the effects of reconfiguration on a 5R planar parallel manipulator

Junior Ranthethi[a], Wesley Dharmalingum[a,*]

[a]The University of the Witwatersrand, 1 Jan Smuts Avenue, Johannesburg, 2017, South Africa

*corresponding author email: wesley.dharmalingum@wits.ac.za

Abstract

This study investigated the reconfiguration of the link lengths of a 5R Planar Parallel Manipular (PPM) and observed the effects on workspace, Maximal Inscribed Circle (MIC) area and velocity. The "R" refers to revolute joint. The proximal, distal and base links were scaled individually and then altogether. Exponential relationships were discovered when the entire PPM was scaled concerning its relationship to the workspace and the MIC area. The velocity of the PPM was dependent on the initial posture and the actuation rates of the proximal links. This study demonstrated that altering the architectural parameters such as the links lengths to obtain the required performance from a 5R PPM is not a straightforward task.

Keywords: Reconfigurability, parallel kinematic manipulator, robotics, workspace, singularities.

1. Background and Motivation

Parallel Kinematic Manipulators (PKMs) have received substantial research interest over the past few decades due to their attractive mechanical properties such as high stiffness and payload to weight ratio. Due to their excellent stiffness properties, the end effector of a PKM can perform movements with high accuracy and repeatability. Certain PKM designs minimise moving masses by situating the motors on the base link hence reducing the inertia of the links. PKMs are characterised by multiple closed-loop kinematic chains. Each of which connects the base to the end effector. Despite their attractive properties, PKMs suffer drawbacks such as relatively small workspaces and complicated forward kinematics and calibration procedures [1-3]. A Planar Parallel Manipular (PPM) is a type of PKM in which the motion of the end effector is constrained to move only along a plane.

Reconfigurable robots can perform a wider variety of tasks in comparison to traditional robots therefore increasing the versatility and usefulness of the robot. Reconfigurability is usually enabled by designing the robot as a collection of modules that can be added or removed. Reconfigurable robots can change their shape and form to adapt to flexible industrial requirements [4, 5]. These robots have been adopted in machine tools, gripper design, additive manufacturing amongst others. Despite their advantages, reconfigurable robots generally require manual intervention and could potentially lead to longer cycle times. Reconfigurable robots can also be expensive to maintain [6, 7].

Clark and Rojas [4] developed a serial 2-DOF malleable reconfigurable robot with continuum links. The robot could achieve numerous topological configurations. Kereluk and Emami [5] investigated the task-based optimisation of reconfigurable manipulators. Through the optimisation process, 6 manipulator configurations were achieved and successfully implemented. Reconfiguration was imparted to a 5R PPM in a study conducted by Venkataramanujam and Larochelle [6]. The 2-DOF PPM was converted into four different 1-DOF 4-bar mechanisms by locking the revolute joints individually in separate instances. This produced different motion paths. Huang et al [7] designed a novel reconfigurable overconstrained spatial PKM. The PKM could alter its performance indices by changing its structural parameters. A fuzzy-PID controller was successfully developed to control the PKM.

Marchi et al [8] produced the kinematic equations and determined the singularity loci of a class of PPMs. The reconfigurable end effector enabled the robot to grasp objects and place them in a predetermined location. A reconfigurable PKM for injured knee rehabilitation tasks was designed and developed by Valero et al [9]. Reconfiguration of the mounting points of the limbs on the base and end effector were implemented to avoid singularities during the movement of the end effector. Actuator force minimisation was also investigated. A novel 6-DOF spatial PKM was developed by You et al [10] that had three different topological configurations and lockable prismatic joints. The stiffness index showed that the PKM in its locked states performed better than when in the unlocked states.

The investigation of robot reconfigurability naturally leads to the exploration of multi-objective optimisation. Various optimal configurations are found for a given set of objectives, constraints and variable limits [11]. The final selection of the most suitable PKM structure is then determined by the user. A structural optimisation of a 5R PPM was conducted by Kavala Sen et al [12]. The multi-objective optimisation considered the shaking forces and moments. Le et al [13] optimised a 5R PPM's link lengths to obtain the maximum usable workspace inside of which there were no singularities. The configuration for the maximum condition number was determined. Optimisation to determine singularity-free workspaces was also investigated by Erwin-Christian et al [14] on a 5R PPM for predefined shapes. Manipulability, the condition number and stiffness were some of the indices that were addressed. Kelaiaia et al [15] produced the Pareto Front to optimise the kinematic performance and workspace of a 5R PPM. The 5R PPM is also commonly referred to as a 5-bar PPM.

From literature, the link lengths are either the only variables considered or are part of a set of variables when investigating reconfiguration and optimisation of the 5R PKM. Reconfiguration of the link lengths is therefore a significant research area. This paper presents the effects of reconfigurability on the workspace, singularities and end effector velocity by scaling the link lengths altogether and individually by various scaling factors.

The following chapters present the objectives and scope of the study in Section 2. The preliminary theory describing the PPM's geometry and singularities are discussed in Section 3. Section 4 presents the methodology. Section 5 documents the key results and analysis. Section 6 concludes the paper and provides future research directions.

2. Objectives and Scope

The objective of this research was to investigate the PKM's performance characteristics concerning workspace, singularities and end effector velocity when reconfiguration is imparted by scaling the link lengths. The scope of the investigation was limited to a symmetrical 5R planar PKM only and focussed on investigating the workspace, MIC area and velocity. Other performance indices such as stiffness, acceleration, manipulability and dexterity, amongst others, were not investigated.

3. Preliminaries

3.1. Description of architecture

The 5R PPM is comprised of two limbs. Each limb is comprised of two links and three revolute joints as shown in Fig 1. The terminal revolute joint in each limb is shared at G which is the end effector. Therefore, there are five revolute joints which are placed at E_1, E_2, F_1, F_2 and G. In this symmetrical 5R PPM, the link length ℓ_1 describes the length from E_1 to F_1 and similarly from E_2 to F_2. These links are referred to as the proximal links. The link length ℓ_2 describes the length from F_1 to G and similarly from F_2 to G. These links are known as the distal links. The length ℓ_3 describes the distance from the origin, point O, to E_1 and similarly to E_2. The joints E_1 and E_2 are actuated through angles θ_1 and θ_2 respectively. The actuated joints are identified in grey. E_1, E_2 and the origin are colinear. The end effector, joint G, is capable of 2 DOFs namely: translation in the x and y axes.

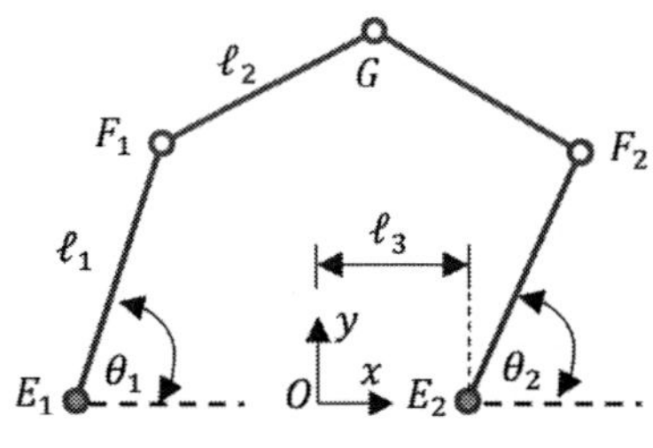

Fig. 1. The 5R PPM

3.2. Inverse kinematic analysis

The Inverse Kinematic (IK) analysis determines the actuation angles for a given end effector position and orientation. Vector loops are usually used to solve the IK problem. The vector loop for limb 1 is shown below.

$$\overrightarrow{OF_1} = \overrightarrow{OE_1} + \overrightarrow{E_1F_1} \tag{1}$$

$$\overrightarrow{OG} = \overrightarrow{OF_1} + \overrightarrow{F_1G} \tag{2}$$

$$\therefore \begin{bmatrix} G_x \\ G_y \end{bmatrix} = \overrightarrow{OE_1} + \overrightarrow{E_1F_1} + \overrightarrow{F_1G} \tag{3}$$

By also formulating the vector loop for limb 2 and expanding and simplifying the equations, the following equations are developed:

$$\ell_2{}^2 = (G_x - \ell_1 \cos\theta_1 + \ell_3)^2 + (G_y - \ell_1 \sin\theta_1)^2 \tag{4}$$

$$\ell_2{}^2 = (G_x - \ell_1 \cos\theta_2 - \ell_3)^2 + (G_y - \ell_1 \sin\theta_2)^2 \tag{5}$$

Equation 4 and 5 are solved using the quadratic equation and half angle trigonometric identities. Both Equation 4 and 5 result in two solutions each therefore there is a total of four solutions to the IK problem. This means that for a given end effector position, the limbs can be arranged in one of four possible combinations that are referred to as Working Modes (WMs). Each WM is identified with a signs namely: "+ +", "+ −", "− +" and "− −". The "+ −" WM is shown in Fig 1 whilst the other postures for the same end effector position is presented in Appendix A.

3.3. Forward kinematic analysis

The Forward Kinematic (FK) analysis determines the location of the end effector for a given combination of actuation angles. Equation 4 and 5 are solved simultaneously and eventually using the quadratic formula. This produces two possible end effector locations. An example of such is shown in Fig 1 (up configuration) and the alternative end effector position (down configuration) is shown in Appendix A. These are referred to as different Assembly Modes (AMs).

3.4. Workspace analysis

The workspace of the PPM is all possible points that can be reached by the end effector. In this case, this refers to the maximal workspace since there is no rotation of the end effector [16]. This workspace can also be referred to as the maximal theoretical workspace since the analyses presented do not consider link interference with each other. When θ_1 and θ_2 are specified in Equation 4 and 5, the workspace can be determined. The workspace is an intersection of the circular annuluses that are created by Equation 4 and 5 for rotation angles that range from 0 to 2π. The circle centers and radii are dependent on the link lengths. The discretization method was used to generate the workspace in MATLAB.

3.5. Singularity conditions

Equation 4 and 5 are differentiated to obtain the velocity equations. For PKMs, it is convenient to split the Jacobian matrix, J, into two separate matrices concerned with the input and output joint rates. Equation 6 presents the relationship between the input actuation rates and output end effector velocities.

$$J_{out}\dot{\boldsymbol{d}} = J_{in}\dot{\boldsymbol{e}} \qquad (6)$$

Where J_{out} and J_{in} refer to the Jacobian matrices of the forward kinematics and inverse kinematics respectively. The column vectors and $\dot{\mathbf{e}}$ and $\dot{\mathbf{d}}$ refer to the input actuation rates and output end effector velocities respectively. Singularities occur when the determinant of J_{out} and/or J_{in} is zero. The postures that describe these instances are shown in Appendix B. Only IK and FK singularities were applicable to this study and not the combined singularity when both J_{out} and J_{in} are simultaneously zero.

4. Methodology and Computational Approach

The methodology and computational approach are shown in Fig 2. The analysis was first conducted on the baseline model then the reconfiguration was implemented.

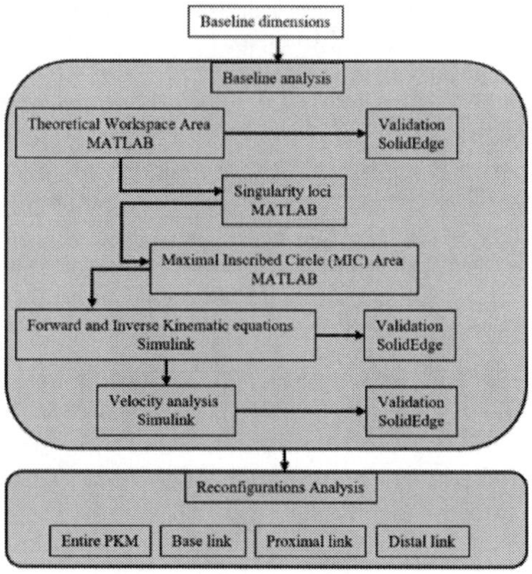

Fig. 2. Methodology and computational approach

5. Key Results and Discussion

The baseline link lengths were set to 230 mm for ℓ_1 and ℓ_2 whilst ℓ_3 was set to 137.5 mm as broadly guided by [17].

5.1. Workspace area analysis

The effects of link length scaling on workspace area are shown in Fig 3. When only scaling the proximal link length, there is a directly proportional linear relationship to the workspace area. The same trend is observed when only scaling the distal link. However, when the base link (ℓ_3) is only scaled, the workspace decreases. When all links are scaled together, an exponential relationship is observed as show in in blue. Selected workspace areas and a table of numerical values are shown in Appendix C.

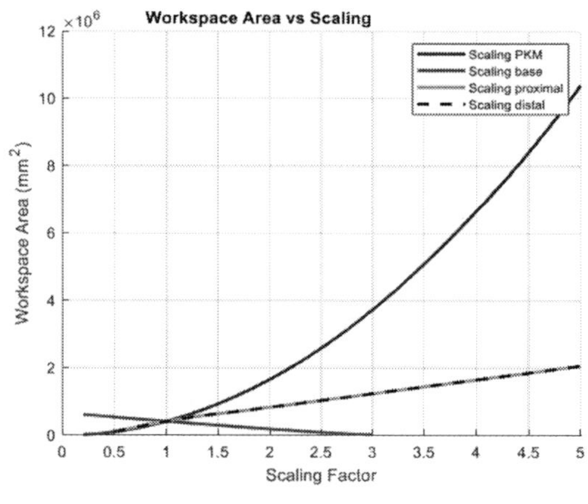

Fig. 3. The effects of link length scaling on workspace area

5.2. Maximal inscribed circle area analysis

Singularities are locations and postures of instability and unpredictability of a PKM. Such locations should be avoided. The Maximal Inscribed Circle (MIC) area refers to the largest circular area in which no singularities exist. Fig 4 presents the effects of link length scaling on MIC area. Scaling up the entire mechanism exponentially increases the MIC area. Selected MIC areas and a table of numerical values are shown in Appendix C. The scaling of the base has a negative effect on the MIC area as it reduces the overlapping annuluses that create the workspace. Scaling the proximal and distal links individually peaks and then stabilizes since the distribution of the singularities also change as the links are scaled and limit the MIC area.

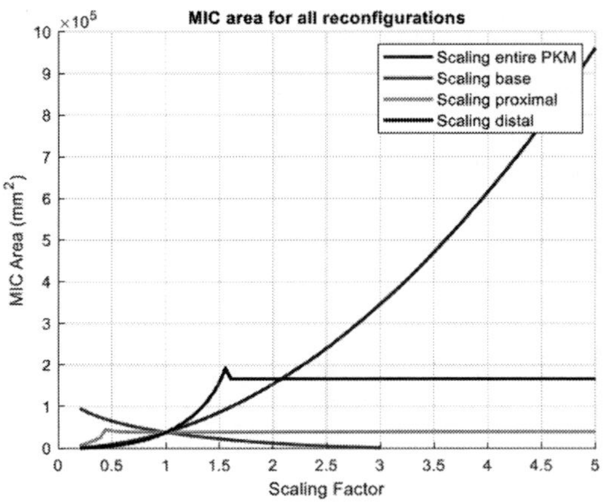

Fig. 4. The effects of link length scaling on MIC area

5.3. Velocity analysis

The velocity analysis was conducted where both the proximal links were vertically upright for its initial posture and both proximal links were actuated arbitrarily at a rate of 3 rpm over 20 seconds. This allowed the end effector to trace a perfect circle as confirmed by the sinusoidal response shown in Fig 5 and 6. When the proximal link lengths were increased, this resulted in an increase in the end effector's velocities. When the base and distal link lengths were scaled individually by different values, this had no effect on the velocity of the end effector. In this initial posture and due to both proximal links being actuated at identical rates, scaling the distal links only changes the location of the traced circle but does not alter the radius. Therefore, the distance travelled by the end effector over the 20 seconds does not change. The same argument follows for the scaling of the base. This indicates that the initial posture and joint rates affect the trajectory and velocity of the end effector. The effect of scaling the proximal link lengths on the maximum velocities is shown in Appendix C.

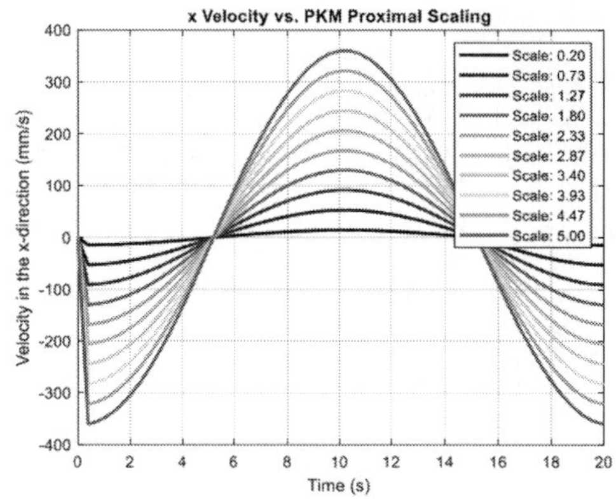

Fig. 5. The effects of link length scaling on x velocity

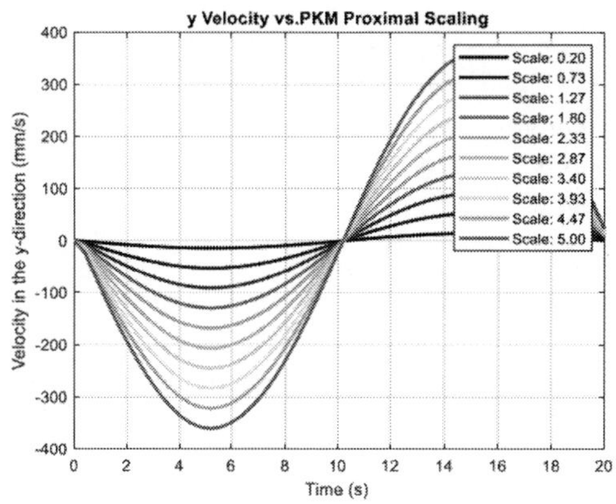

Fig. 6. The effects of link length scaling on y velocity

6. Conclusion and Future Work

This research successfully investigated reconfiguring the link lengths and observing the effects on workspace, MIC area and velocity. Simply scaling the architectural parameters to obtain the different performances is not a straightforward task. The most effective method to increase the workspace of the 5R PPM is to scale the entire PPM. This should be cautiously implemented since changing link lengths also create voids and increases the complexity of the trajectory planning. Similarly, the MIC area is increased most effectively by scaling the entire PPM. Although not investigated, the results suggest that the MIC area and workspace can be increased by only scaling up the proximal and distal links. The velocity analysis is more sensitive to initial posture and actuation rates and careful consideration is required to alter the end effector's velocity. Future research directions include investigating different initial postures and actuation rates, isolating and investigating the different WMs, exploring other performance indices, considering the effects of link interference and optimization of antagonistic performance indices.

References

[1] Rosyid A, El-Khasawneh B, Alazzam A (2020). *Review article: Performance measures of parallel kinematics manipulators*. Mechanical Sciences, **11**: 49-73.

[2] Lorenz M, Brinker J, Prause I, Corves B (2016). *Power manipulability analysis of redundantly actuated parallel kinematic manipulators with different types of actuators*, 2016 IEEE International Conference on Robotics and Automation (ICRA). Stockholm, Sweden, pp 2129-2136.

[3] Pandilov Z, Dukovski V (2014). *COMPARISON OF THE CHARACTERISTICS BETWEEN SERIAL AND PARALLEL ROBOTS*. Acta Technica Corviniensis-Bulletin of Engineering, **7 (1)**: 143–160.

[4] Clark AB, Rojas N (2022). *Malleable Robots: Reconfigurable Robotic Arms With Continuum Links of Variable Stiffness*. IEEE Transactions on Robotics, **38 (6)**: 3832-3849.

[5] Kereluk JA, Emami MR (2017). *Task-based optimization of reconfigurable robot manipulators*. Advanced Robotics, **31 (16)**: 836-850.

[6] Venkataramanujam V, Larochelle PM (2016). *Design and development of planar reconfigurable motion generators*. Mechanics Based Design of Structures and Machines, **44 (4)**: 426-439.

[7] Huang G, Zhang D, Tang H, Kong L, Song S (2020). *Analysis and control for a new reconfigurable parallel mechanism*. International Journal of Advanced Robotic Systems, **17 (5)**: 1-20.

[8] Marchi T, Mottola G, Porta JM, Thomas F, Carricato M (2021). *Position and Singularity Analysis of a Class of Planar Parallel Manipulators with a Reconfigurable End-Effector*. Machines, **9 (1)**: 7.

[9] Valero F, Díaz-Rodríguez M, Vallés M, Besa A, Bernabéu E, Valera A (2020). *Reconfiguration of a parallel kinematic manipulator with 2T2R motions for avoiding singularities through minimizing actuator forces*. Mechatronics, **69**: 102382.

[10] You J, Xi F, Shen H, Wang J, Yang X (2021). *A novel Stewart-type parallel mechanism with topological reconfiguration: Design, kinematics and stiffness evaluation*. Mechanism and Machine Theory, **162**: 104329.

[11] Yang C, Ye W, Li Q (2022). *Review of the performance optimization of parallel manipulators*. Mechanism and Machine Theory, **170**: 104725.

[12] Kavala Sen D, Yildiz A, Kopmaz O (2022). *Optimal Design of a Five-Bar Planar Manipulator and Its Controller by Using Different Algorithms for Minimum Shaking Forces and Moments for the Largest Trajectory in a Usable Workspace*. Machines, **10**: 971.

[13] Le TD, Kang H-J, Doan QV (2013). *A method for optimal kinematic design of five-bar planar parallel manipulators*, 2013 International Conference on Control, Automation and Information Sciences (ICCAIS). Nha Trang, Vietnam, pp 7-11.

[14] Erwin-Christian L, Valentin C, Tivadar D, Alexandru O, Elida-Gabriela T, Melania-Olivia S (2024). *Optimal Synthesis of Five-Bar Linkage Based on Singularity-Free Workspaces with Predefined Shapes*. Robotics, **13**: 173.

[15] Kelaiaia R, Chemori A, Brahmia A, Kerboua A, Zaatri A, Company O (2023). *Optimal dimensional design of parallel manipulators with an illustrative case study: A review*. Mechanism and Machine Theory, **188**: 105390.

[16] Merlet J-P (2006). *Parallel robots*. 2nd ed. Springer Science & Business Media.

[17] Figielski A, Bonev IA, Bigras P (2007). *Towards development of a 2-DOF Planar Parallel Robot with optimal workspace use*, 2007 IEEE International Conference on Systems, Man and Cybernetics. Montreal, QC, Canada, pp. 1562-1566.

Appendix A. Working modes and assembly modes

A.1. Working modes

Fig 7 shows three WMs of the 5R PPM in addition to the WM shown in Fig 1.

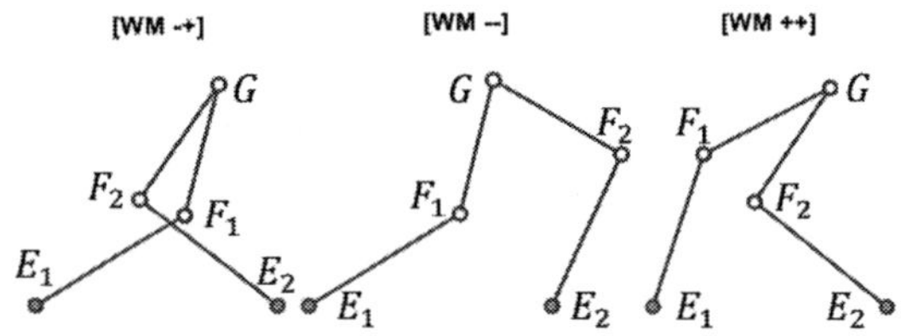

Fig. 7. Different WMs of the 5R PPM

A.2. Assembly modes

Fig 8 shows the down configuration AM whilst Fig 1 shows the up configuration.

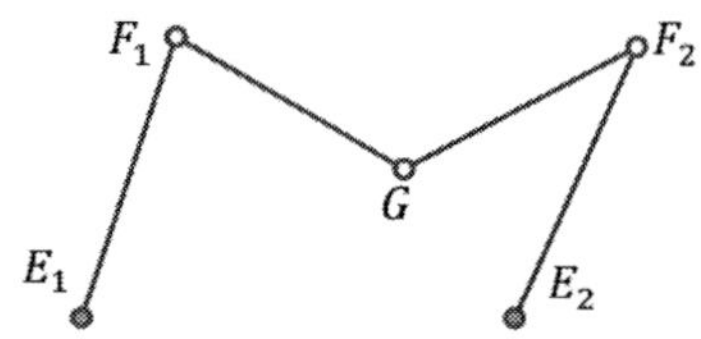

Fig. 8. Down configuration AM

Appendix B. Singularity postures

Fig 9 depicts two IK singularity postures. In one instance, one of the limbs is fully stretched out whilst in the other posture, one of the limbs is folded on itself. Fig 10 presents an FK singularity in which F_1, G and F_2 are colinear. Fig 11 shows two additional FK singularity postures in which F_1 and F_2 are coincident. In the other case, E_1 and E_2 are coincident and F_1 and F_2 are coincident.

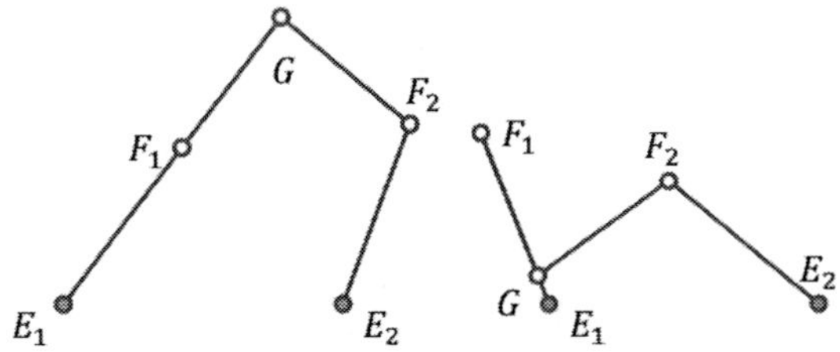

Fig. 9. IK singularity postures

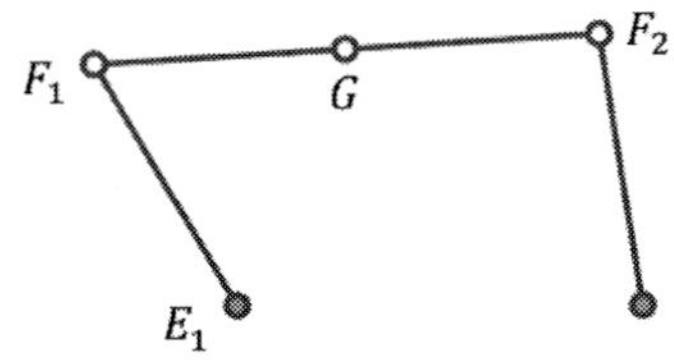

Fig. 10. An example of an FK singularity posture

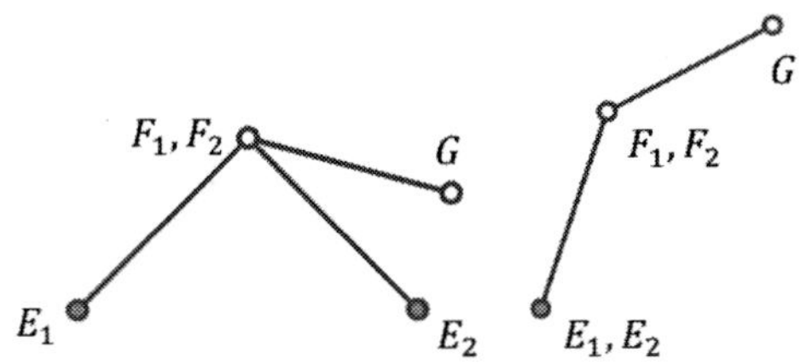

Fig. 11. Other FK singularity postures

Appendix C. Simulation results

C.1. Workspace

Fig 12-14 depicts the workspace area for different scaling factors of the proximal link. Table 1 presents selected numerical values concerning the relationship between workspace area and link scaling.

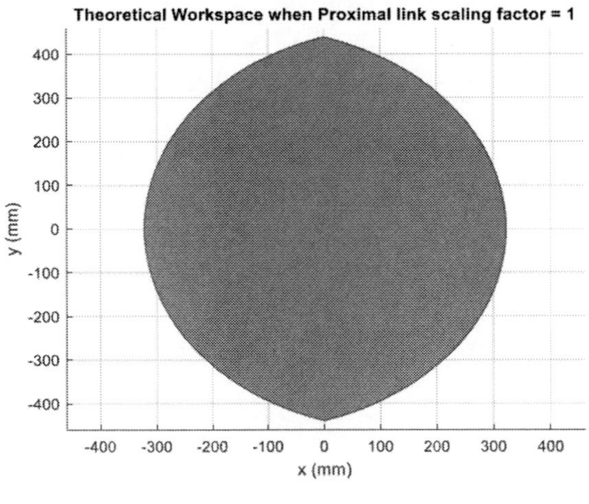

Fig. 12. Workspace with proximal link scaling of 1 (baseline model)

Table 1. Relationship between workspace area and link scaling

Links scaled	Scaling factor	Workspace Area ($\times 10^6\ mm^2$)
All links	0.5	0.10
	1	0.42
	1.5	0.94
	2.5	2.60
Proximal only	0.5	0.11
	1	0.42
	1.5	0.64
	2.5	1.04
	5	2.06
Distal only	0.5	0.11
	1	0.42
	1.5	0.64
	2.5	1.04
	5	2.06
Base only	0.5	0.54
	1	0.42
	1.5	0.30
	2.5	0.10

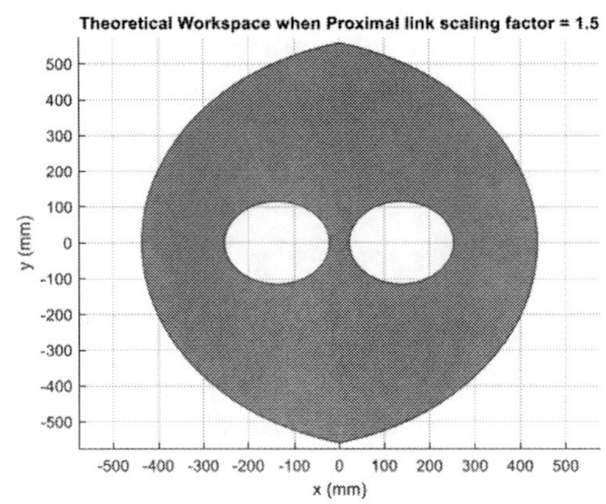

Fig. 13. Workspace with proximal link scaling of 1.5

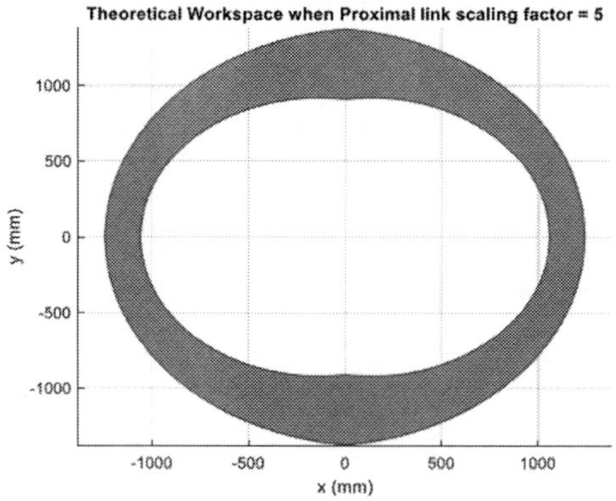

Fig. 14. Workspace with proximal link scaling of 5

C.2. MIC Area

Fig 15-17 depicts the MIC area for different scaling factors of the proximal link. Table 2 presents selected numerical values concerning the relationship between MIC area and link scaling.

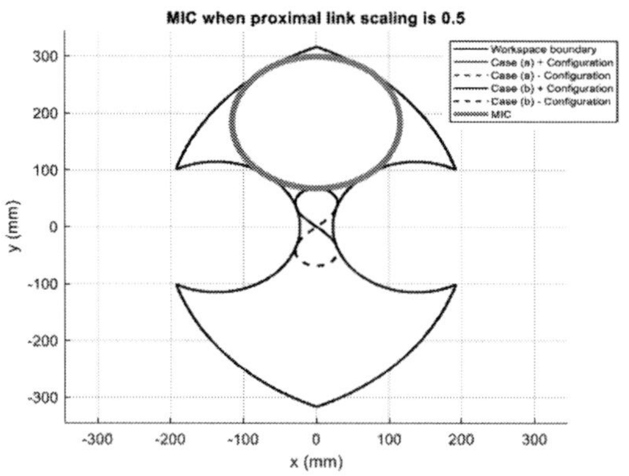

Fig. 15. MIC area with proximal link scaling of 0.5

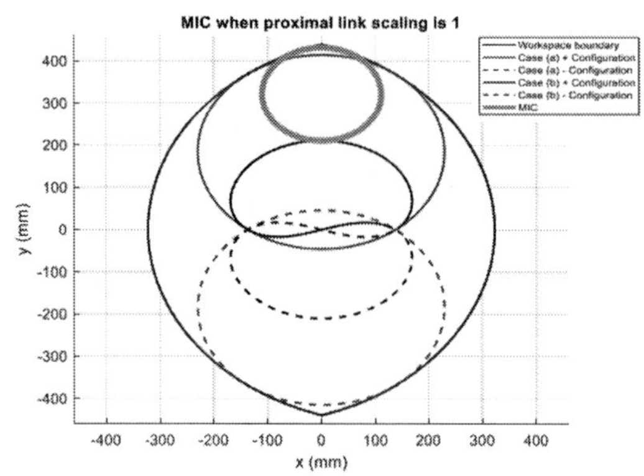

Fig. 16. MIC area with proximal link scaling of 1 (baseline model).

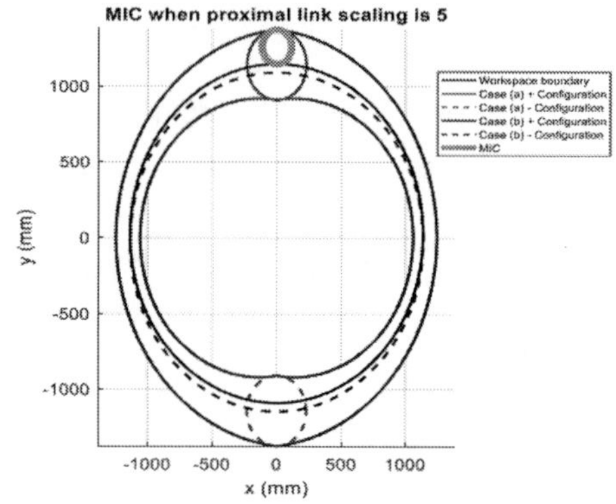

Fig. 17. MIC area with proximal link scaling of 5

Table 2. Relationship between MIC area and scaling factors

Links scaled	Scaling factor	Workspace Area ($\times 10^5\ mm^2$)
All links	0.5	0.10
	1	0.38
	1.5	0.86
	2.5	2.40
Proximal only	0.5	0.42
	1	0.38
	1.5	0.39
	2.5	0.39
	5	0.40
Distal only	0.5	0.05
	1	0.38
	1.5	1.57
	2.5	1.66
	5	1.66
Base only	0.5	0.66
	1	0.38
	1.5	0.23
	2.5	0.05

C.3. Velocity

Fig 18 shows the effects of scaling the proximal links on the maximum velocity in the x and y directions.

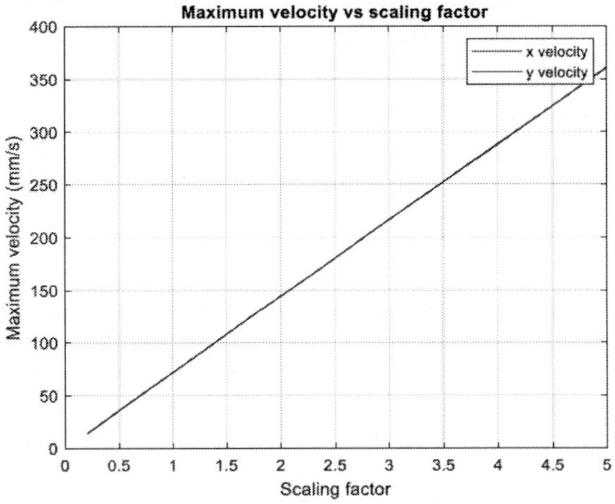

Fig. 18. The effects of scaling the proximal links on the maximum velocity

Rapid and Robust Electric Vehicle Velocity Profile Generation: Balancing Energy Efficiency, Logistical Constraints and Computational Time

Riaan Meeser[a*], Daniel N. Wilke[a]

[a]*Department of Mechanical and Aeronautical Engineering, University of Pretoria, Pretoria, 0086, South Africa*

**corresponding author email: riaan.meeser@up.ac.za*

Abstract

Reducing energy wastage is vital for sustainable and efficient transportation. This study presents a rapid method to determine near-optimal velocity profiles for vehicles on known routes, minimising energy consumption while meeting logistical constraints. By optimising speed based on route characteristics and regenerative capabilities, the approach reduces operational costs and environmental impact, addressing challenges in modern transportation systems.

We propose a fast, efficient method leveraging route topography to obtain an estimate for an energy-efficient velocity profile to overcome the computational demands of a full optimisation. Our approach delivers usable velocity profiles within 20 seconds, achieving energy consumption within 10% of the optimal solution, which typically requires 3–4 hours of computation. Simulations show optimal energy savings of up to 33% compared to uninformed driving. This enables real-time implementation of energy-efficient velocity profiles, balancing efficiency and real-time applicability, without compromising logistical constraints.

Keywords: Velocity Profile, Energy, Efficiency, Optimisation, Electric Vehicle

1. Introduction and Background

Improving machine efficiency can increase profitability, reduce resource wastage, and lower carbon footprints, which is critical for sustainable development [1], [2]. This study proposes a method to rapidly approximate an optimal velocity profile for vehicles travelling known routes, minimising tractive energy consumption while adhering to logistical and regulatory constraints. The approach enables real-time application by addressing the computational challenges of detailed optimisation. The current state of the art is to make use of masses of data to train stochastic or neural network methods to obtain models capable of performing real-time optimisation of velocity profiles. A drawback of these methods is that they require vast amounts of data to train them, use simplifications to the vehicle's energy estimations or rely on off-site computation in a cloud-based server which can perform massive computations in a shorter timeframe [3], [4]. Adaptability of these models to variations in input parameters make their use restricted, if a significant amount of training data is not available, as, for example, a change in mass or unexpected traffic situations may fall outside the trained data, rendering the solution unusable [5], [6]. First-principles based models exist that aim to accurately model all of these variables in a useful manner, making use of gradient-based solvers to find the minima of the cost function. These work very well in offline applications, but require significant computational time to converge to a solution, if any, also being very sensitive to the choice of initial guess used in the optimisation [7]. With little data available, neural network and similar large-data models are infeasible, but real-time application of first principles models also make their use restricted, which led to the work presented in this paper.

A first-principles model is developed to accurately simulate vehicle dynamics and energy requirements on a known route, unlike assumption-heavy methods [2], [8], this model incorporates key system variables in its calculations.. To facilitate comparison of optimised energy consumptions for simulated results with real-world non-optimal data, a sensor-equipped test vehicle was used to record velocity and route topography data. The output from this study will then be an advisory system that can inform vehicle controllers, be it human or computer based, on energy-efficient velocity profiles.

The novelty of the proposed strategy lies in a fast, low-complexity method to approximate optimal velocity profiles. Using a simplified optimisation process, the approach presented reduces simulation time from 3 hours to 20 seconds. The solution achieves energy consumption within 14% of high-fidelity optimisation and offers 22% improvement over an uninformed real person driving. This study demonstrates substantial energy savings with minimal computational overhead, contributing to practical and sustainable vehicle energy management.

2. Vehicle Energy Model

Understanding and characterising the factors influencing energy consumption is essential for optimising vehicular energy use. Key contributors include the vehicle type, total mass, route characteristics, and velocity profiles, all affecting the energy required to complete a route under specified logistical constraints [9]. Figure 1 illustrates these relationships.

Framing the problem in terms of energy simplifies modelling by representing propulsion power, vehicle acceleration, and route topography through basic potential and kinetic energy forms, eliminating the need to model each component explicitly [2]. This approach estimates the energy required to traverse the route (wheel energy) rather than the energy drawn from the source, which would necessitate detailed modelling of internal factors such as motor and transmission efficiencies.

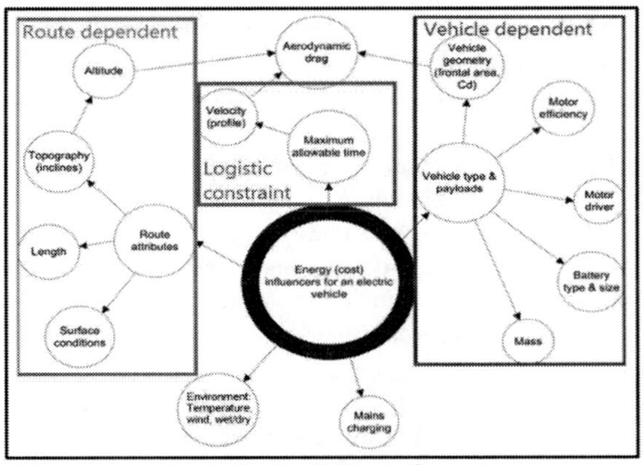

Figure 1: Key factors influencing energy consumption in vehicles are categorised by route-dependent (green), logistical (purple), and vehicle-dependent (blue) constraints.

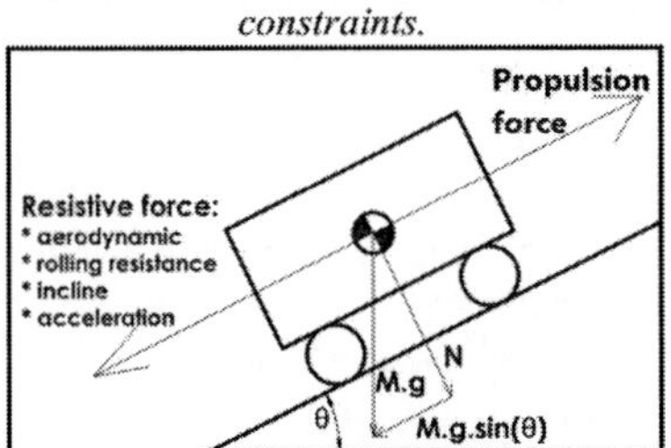

Figure 2: Free-body diagram of a vehicle on an incline, illustrating forces influencing motion and energy consumption.

Figure 2 presents a free-body diagram of a vehicle travelling on an incline, serving as the foundation for the relevant equations. Certain parameters, such as mass, affect multiple factors, including the incline and rolling resistance terms. This interdependence introduces complexity, making it challenging to decouple and solve the model during optimisation.

2.1 Parameters used in energy estimation

A Matlab/Simulink [10] first principles model was constructed based on an energy model, with the advantage that many of the applicable equations may be simplified and accounted for indirectly and summed easily to yield a single value for the energy consumed to complete the route. The product of force and velocity yields power, with the integral of power being energy. A schematic of the Simulink model is shown in Figure 3. The parasitic power losses due to aerodynamic and rolling resistance forces are determined using equation 1 and 3 respectively. The velocity profile for a known route can thus be input, with the resultant route completion energy being the model's output. The regeneration model allows potential energy (*E=MgH*) and kinetic (*E= ½MV²*) to be exchanged both ways without penalty, only applying an efficiency/penalty when converting any of those to another stored form of energy, as with regeneration (labelled as "Regen model" in the figure).

$$P_{Aero} = F_{Aero}V = {}^{1}\!/_{2}\,\rho C_d A_F V^3 \qquad [6](1)$$

$$F_{roll-basic} = C_{rr}N \qquad (2)$$

$$P_{roll} = \left(\left[Constant1\left(1+\frac{Velocity}{Constant2}\right)\right]N\right)V \qquad [6](3)$$

$$** \; (C_{rr} = 0.012 - 0.015 \; on \; tar \; surfaces)$$

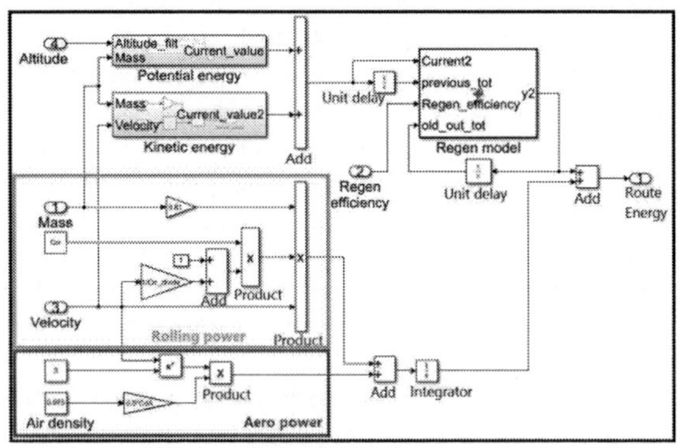

Figure 3: Simulink energy consumption model: incorporating potential and kinetic energy and integrating parasitic power losses from rolling and aerodynamic forces to estimate the total energy consumed.

2.2 Optimisation

It is essential to define the optimisation problem before approximations to the optimal velocity profile for energy-efficient vehicle operation can be generated. This problem is defined as a constrained nonlinear, non-separable, single-objective cost function that minimises overall energy consumption, subject to constraints like speed limits and maximum completion time for a known route. The objective function for the optimisation thus is the route's overall energy consumption, determined using the Matlab/Simulink based first-principles model shown in Figure 3. The fmincon local minima solver function was used, with the following settings: MaxFunctionEvaluations set to 3×10^4, MaxIterations set to 1×10^4 and the "active-set" algorithm used, based on the recommendation in the fmincon documentation on the large steps that this algorithm can take in problems with nonsmooth constraints, which was the case for this study. The output-variable of interest to this optimisation is the velocity as a function of distance. A higher resolution to this velocity profile will facilitate smoother tracing of the optimal solution, resulting in a better solution. This resolution relates to the model fidelity; more velocity increments to solve however increases the solution complexity drastically. When utilising the fmincon function for optimisation, the choice of initial guess significantly impacts the likelihood of finding the global minimum. This study introduces a rapid method to generate a high-quality initial guess, also serving as an effective standalone approximation of the velocity profile.

Before developing this approximation, it's essential to analyse the cost function to assess the influence of various initial guesses, their ranges, sensitivity to variations, simulation times, and typical outputs from the local minimum solver. Achieving predictable and repeatable solutions is vital, as some problems may lack feasible solutions. A systematic approach begins with constant initial guess values for a low-fidelity optimisation problem on basic test routes with constant inclines and declines. Increasing the fidelity may yield better solutions; however, exhaustively exploring every element with sufficient resolution becomes impractical due to computational demands.

Developing a method to provide a good initial guess makes the optimisation process more efficient and effective, enhancing the practical application of energy-efficient velocity profiles in real-world scenarios. The test routes included flat terrain, 2% and 5% inclines and declines, 10% inclines and declines, and combined 5% incline-decline and decline-incline sections. Each spanned 10 km and took 933 seconds to complete, matching the distance and completion time for real-world Route 1, for verification purposes.

The top of Figure 4 displays the optimal velocity profiles derived from a constant initial guess, resulting in nearly uniform velocity vectors across all routes. Notably, the 5% incline-decline and decline-incline routes produced similar constant velocities, contrary to intuitive expectations of inverse profiles due to their opposing topographies. Additionally, constant velocities on incline routes indicate a missed opportunity to leverage topography for energy efficiency. These outcomes suggest that a constant-velocity initial guess leads the optimiser to local minima near the initial values, hindering optimal solutions.

In contrast, the bottom of Figure 4 illustrates optimal velocity profiles obtained using random initial guesses, yielding improved energy outcomes in some cases but worse in others, with a fidelity of 10 applied for practicality. Despite achieving better final energies for certain routes, the resulting velocity profiles were erratic, potentially causing uncomfortable driving and indicating the possibility of more optimal solutions. These findings highlight the importance of selecting suitable initial guesses to achieve realistic and efficient velocity profiles in optimisation processes. The zero final energy values in Figure 4 result from excluding regenerative braking, allowing steep descents to convert potential energy into kinetic energy, enabling route traversal without additional energy consumption. A 2% decline approaches the threshold where a route transitions from requiring to not requiring energy from the vehicle's powertrain.

As previously applied, optimal velocity profiles for the real-world test Route 1 were determined using the best of 10 random initial guesses. Optimisations were conducted with fidelities ranging from 1 to 50 blocks. These are shown in Figure 5, with the right-hand axis and black line depicting the route's topographic profile.

Analysis of the optimal velocity profiles for real-world test routes reveals an inverse relationship with topography: velocity decreases on uphill segments and increases on downhill segments. This pattern aligns with findings by Johannesson et al. [11] and suggests that energy-efficient driving mimics a rollercoaster, efficiently converting kinetic and potential energy. Leveraging this inverse relationship can improve initial guesses in optimisation processes.

Increasing fidelity enhances the solver's ability to trace topographic changes, yielding lower minimum route energy. However, higher fidelity also increases simulation time and may introduce noise into the solution. As an example, a fidelity of 10 takes around 5 minutes to find a solution, whereas a fidelity of 50 takes 3-5 hours, quickly rendering higher fidelities impractical.

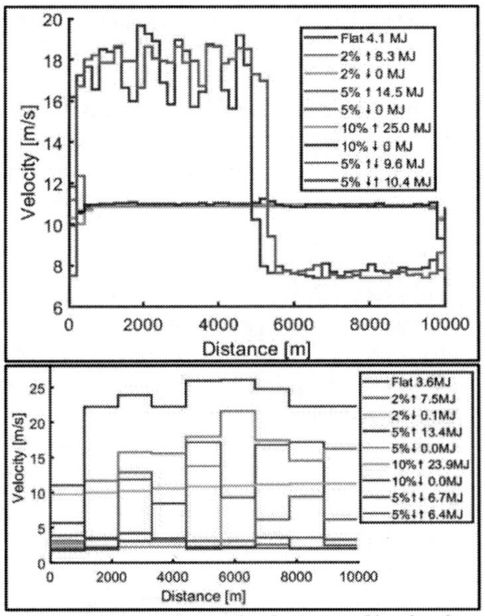

Figure 4: Optimal velocity profiles for the test routes using a constant initial guess, fidelity of 50 (Top) and random initial guess for a fidelity of 10 (Bottom).

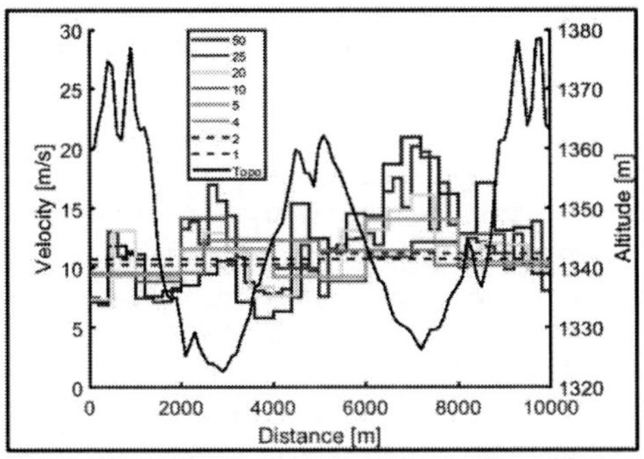

Figure 5: Optimal velocity vectors for real-world Route 1 using the best of 10 random initial guesses.

The choice of initial guess significantly influences convergence and solution quality. A constant initial guess offers simplicity but may lead the solver to local minima, potentially missing the global optimum. Conversely, random initial guesses introduce variability, enhancing the likelihood of escaping local minima and finding the global solution. However, random guesses can result in erratic convergence and increased computational effort. Therefore, selecting an appropriate initial guess balances solution quality and computational efficiency.

3. Initial guess: Inverted topographic

Assuming an inverse relationship between route topography and the velocity profile enables constructing an effective initial guess for optimisation, leading to more robust results and reduced computational requirements. By inverting, scaling, and shifting the known topographic profile, a tailored initial guess can be generated using fewer design variables. This method delivers a high-quality initial guess in approximately 20 seconds without losing topographic resolution. The equation for converting from the topographic profile to the initial guess for the velocity profile is performed using a simple linear equation of the for $y = -m(x)+c$, with y being the velocity vector, x being the topographic profile, m being the scaling factor and c being an offset. This means that a simple optimisation with a fidelity of 2 can be performed to find the optimal values of m and c, to appropriately scale and shift the topographic profile yielding the velocity profile. The minus before the scale factor is used to invert the profile, so that the vehicle gains velocity when going downhill and reduces when going uphill.

The contribution of real-world testing in this work was to obtain realistic energy consumption values to prove model validity facilitate comparisons between the energy consumed by an uninformed driver vs. that of an optimal solution, though these could not be used directly in verifying the proposed optimal solutions, as the travel of a real-world route is too unpredictable, due to traffic etc., making travel at the optimal pace close to impossible. The whole purpose of developing a real-time capable solution was for this reason; so that it can propose an optimal solution in the midst of constantly changing driving parameters. Real-world testing on Route 1 showed that an uninformed driver consumed 8.5 MJ of energy. Applying the optimised inverted topographic velocity profile alone reduced energy consumption to 6.67 MJ—a 22% improvement. Further optimisation with a fidelity of 10, using this improved initial guess, achieved 6.40 MJ. At a fidelity of 50, the energy consumption decreased to 5.73 MJ, representing a 33% total improvement but requiring 2 to 4 hours of computation.

This approach is insensitive to route complexity and length, maintaining effectiveness without needing convenient route segmentations. The top of Figure 6 compares the velocity profiles of the optimised inverted topographic method and the complete optimisation solution, while the bottom of Figure 6 illustrates their respective energy consumptions throughout the route.

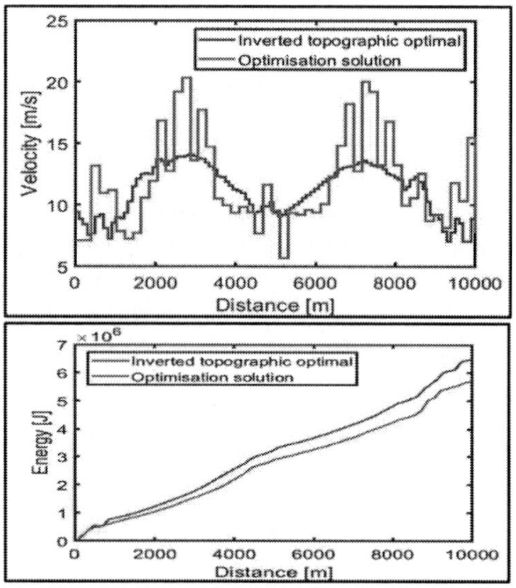

Figure 6: Inverted topographic approximation vs. the complete optimal solution. Velocity profile (Top), Energy consumed (Bottom).

4. Conclusions

This study introduces a method to rapidly generate an initial guess for vehicle velocity optimisation by inverting, scaling, and shifting the route's topographic profile. This approach not only accelerates the optimisation process but also serves as a close approximation to the optimal solution. A first-principles model estimates energy consumption based on route topography, length, vehicle type, load, and velocity profile, revealing a correlation between topography and optimal velocity. Optimised in under a minute, the low-fidelity cost function provides an initial guess that drastically reduces simulation time, enhancing real-time applicability. This method achieves energy consumption within 14% of the optimal solution and offers a 22% improvement over an uninformed driver. However, it may not elegantly handle local speed limits, potentially limiting its applicability in certain scenarios.

5. References

[1] A.D. Bosman, 'Investigating Load Shift and Energy Efficiency of New Technology Loco Battery Chargers', North-West University, South Africa, 2006. Accessed: Nov. 29, 2023. [Online]. Available: https://repository.nwu.ac.za/handle/10394/1057

[2] E. Hellström, J. Åslund, and L. Nielsen, 'Design of an efficient algorithm for fuel-optimal look-ahead control', *Control Engineering Practice*, vol. 18, no. 11, pp. 1318–1327, Nov. 2010, doi: 10.1016/j.conengprac.2009.12.008.

[3] W. Shabbir and S. A. Evangelou, 'Real-time control strategy to maximize hybrid electric vehicle powertrain efficiency', *Applied Energy*, vol. 135, pp. 512–522, Dec. 2014, doi: 10.1016/j.apenergy.2014.08.083.

[4] P. Shen, Z. Zhao, X. Zhan, J. Li, and Q. Guo, 'Optimal energy management strategy for a plug-in hybrid electric commercial vehicle based on velocity prediction', *Energy*, vol. 155, pp. 838–852, Jul. 2018, doi: 10.1016/j.energy.2018.05.064.

[5] S. F. Tie and C. W. Tan, 'A review of energy sources and energy management system in electric vehicles', *Renewable and Sustainable Energy Reviews*, vol. 20, pp. 82–102, Apr. 2013, doi: 10.1016/j.rser.2012.11.077.

[6] M. Back, M. Simons, F. Kirschaum, and V. Krebs, 'PREDICTIVE CONTROL OF DRIVETRAINS', *IFAC Proceedings Volumes*, vol. 35, no. 1, pp. 241–246, 2002, doi: 10.3182/20020721-6-ES-1901.01508.

[7] Snyman, JA, 'Practical mathematical optimization: an introduction to basic optimization theory and classical and new gradient-based algorithms', *Choice Reviews Online*, vol. 44, no. 05, pp. 44-2754-44–2754, Jan. 2007, doi: 10.5860/CHOICE.44-2754.

[8] K. N. Genikomsakis and G. Mitrentsis, 'A computationally efficient simulation model for estimating energy consumption of electric vehicles in the context of route planning applications', *Transportation Research Part D: Transport and Environment*, vol. 50, pp. 98–118, Jan. 2017, doi: 10.1016/j.trd.2016.10.014.

[9] R. Basso, B. Kulcsár, B. Egardt, P. Lindroth, and I. Sanchez-Diaz, 'Energy consumption estimation integrated into the Electric Vehicle Routing Problem', *Transportation Research Part D: Transport and Environment*, vol. 69, pp. 141–167, Apr. 2019, doi: 10.1016/j.trd.2019.01.006.

[10] The MathWorks Inc., *Matlab*. (2021). The MathWorks Inc. [Online]. Available: https://www.mathworks.com

[11] L. Johannesson, N. Murgovski, E. Jonasson, J. Hellgren, and B. Egardt, 'Predictive energy management of hybrid long-haul trucks', *Control Engineering Practice*, vol. 41, pp. 83–97, Aug. 2015, doi: 10.1016/j.conengprac.2015.04.014.

Neural Network-Based PID Tuning for a Ball and Plate System

Aarti Panday[a,*], Xolani Radebe[a]

[a] *University of the Witwatersrand*

University of the Witwatersrand, Johannesburg, Private Bag 3, Wits 2050, South Africa.

**Aarti.Panday@wits.ac.za*

Abstract

The ball and plate balancing system is a multi-input, multi-output (MIMO) system where two variables—the ball's position along the X and Y axes—are controlled through two inputs: the plate's rotations around these axes. Despite its simple conceptual design, the system exhibits nonlinear dynamics, inherent instability, and cross-coupled behaviour between the axes. This study investigates the performance of a Neural Network-based PID controller compared to the MATLAB Autotuned PID tuner in controlling the system. Results demonstrate that the Neural Network-based controller outperforms the MATLAB Autotuned PID, particularly in handling nonlinear models.

Keywords: Ball Balancing Plate, Neural Network, PID gains.

1. Background and Motivation

Traditional controllers struggle to adapt to nonlinearities and dynamic changes in system parameters. When complex mechanical systems are linearized, modelled numerically, and simulated, their responses often differ significantly from experimental results. [1] [2]

There are alternatives that have been developed to deal with the problem identified, some of the alternatives are: Switch control approach [3], Model Predictive control, [4] Fuzzy Logic Control [5], Adaptive Control etc. [6]

These models work as current systems use them, and they are functional. However, it is not always the case that the system parameters changing are going to be projected, predicted or expected. Therefore, this research explores the use of AI in tuning PID gains for a system, and the AI model to be used is a Neural Network (NN).

The system to be analyzed is a ball-balancing plate. This model is complex because the ball is not constrained to the plate, requiring the inclusion of various system variables in the analysis, such as friction, drag, Coriolis force, and others. [7]

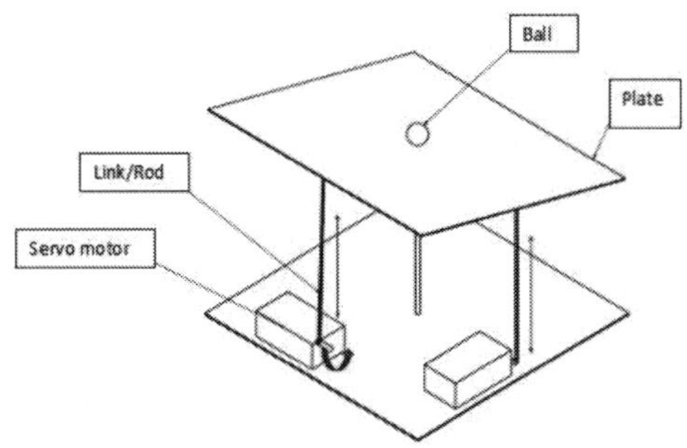

Fig. 1. 3D sketch of the ball balancing plate model.

2. Objectives and Scope

The primary motivation for the research is to prove that systems stabilized using PID tuning alone do not work well when the system is not linear and to investigate if Neural Networks can be used as a PID tuner.

The main objectives of this research are:

- To numerically model the system (ball balancing plate).
- Linearize the modeled system.
- Use MATLAB Autotune to find the PID gains for the model with the shortest response and settling time.
- Develop a Neural Network PID tuner and investigate how the results compare to that of the autotuned model.
- Change the system parameter (ball mass) and investigate how the model system responds.

3. System Modelling

3.1. System Modeling

3.1.1. Newton-Euler mathematical model derivation

The mathematical model is designed to capture the interplay of forces. This serves as a foundation for further analysis and numerical simulations of the system. The model of 1 axis is the same as the model of the other axis, therefore the mathematical derivation is done for 1 axis and for the other axis it is derived from the first axis. Figure 2 below shows the relationship between the motor linkage and the plate angle of the x axis. The model is derived from [7].

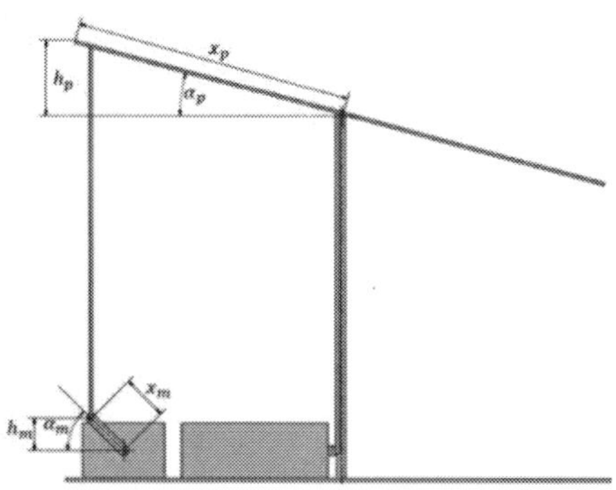

Fig. 2. Relationship between the motor angle and the plate angle on the x axis.

Final mathematical model equations combining ball on plate, plate model (Eq 1-4) and, voltage driving servo motor to actuators (Eq 5 – 6) [7]:

$$\left(1+\frac{J_b}{m_b r_b^2}\right)\frac{d^2x}{dt^2}+\frac{dx}{dt}\left(\frac{k_b}{m_b r_b}+\frac{k_{cx}}{m_b}\left|\frac{dx}{dt}\right|\right)+x\left(\frac{d\beta}{dt}\right)^2=-g\sin(\beta) \qquad (1)$$

$$\left(1+\frac{J_b}{m_b r_b^2}\right)\frac{d^2y}{dt^2}+\frac{dy}{dt}\left(\frac{k_b}{m_b r_b}+\frac{k_{cy}}{m_b}\left|\frac{dy}{dt}\right|\right)+x\left(\frac{d\alpha}{dt}\right)^2=-g\sin(\alpha) \qquad (2)$$

$$\left(\frac{J_p+J_{Gx}+J_mG^2}{m_b}+x^2\right)\frac{d^2\beta}{dt^2}+\left(2x\frac{dx}{dt}+\frac{k_{py}}{m_b}+\frac{k_0G^2}{m_b}\right)\frac{d\beta}{dt} = \frac{G.k_m\phi}{m_b}i_y-x.g.\cos(\beta) \qquad (3)$$

$$\left(\frac{J_p+J_{Gx}+J_mG^2}{m_b}+y^2\right)\frac{d^2\alpha}{dt^2}+\left(2y\frac{dy}{dt}+\frac{k_{px}}{m_b}+\frac{k_0G^2}{m_b}\right)\frac{d\alpha}{dt} = \frac{G.k_m\phi}{m_b}i_x-y.g.\cos(\alpha) \qquad (4)$$

$$L\frac{di_y}{dt}+k_u\phi G\frac{d\beta}{dt}+(R_z+R)i_y+R_zi_x=u_xU_0 \qquad (5)$$

$$L\frac{di_x}{dt}+k_u\phi G\frac{d\alpha}{dt}+(R_z+R)i_x+R_zi_y=u_xU_0 \qquad (6)$$

From the Eqs. 1 - 6 above, m_b is the ball's mass, r_b its radius, and J_b its moment of inertia. k_b and k_c are the coefficients of friction and environmental resistance. J_p, J_{Gx}, and J_m are the moments of inertia of the plate, drive, and rotor. k_{py}, k_{px} represent system losses, and G is the gear ratio. L, R, R_z, k_m, k_u, and k_0 are the motor's inductance, winding resistance, source resistance, torque constant, rate constant, and loss approximation. ϕ is the magnetic flux, i_x, i_y are current components, α, β are angular positions, u_x is the control voltage, and g is gravitational acceleration.

3.1.2. De-coupled Linearized model

Using small angle approximations and flat table as initial condition, the decoupled linearized model for α and x is given in Eqs. 7 – 9. Decoupling was possible due to examining the order of magnitude in the cross-coupled motor parameters.

$$\left(1+\frac{J_b}{m_b r_b^2}\right)\frac{d^2y}{dt^2}+\frac{dy}{dt}\left(\frac{k_b}{m_b r_b}+\frac{k_{cy}}{m_b}\left|\frac{dy}{dt}\right|\right)+x\left(\frac{d\alpha}{dt}\right)^2=0 \qquad (7)$$

$$\left(2y\frac{dy}{dt}+\frac{k_{px}}{m_b}+\frac{k_0G^2}{m_b}\right)\frac{d\alpha}{dt}=\frac{G.k_m\phi}{m_b}i_x \qquad (8)$$

$$L\frac{di_x}{dt}+k_u\phi G\frac{d\alpha}{dt}=u_xU_0 \qquad (9)$$

4. Control Law Design

The PID control law equation is given in Eq. 10:

$$u(t)=K_pe(t)+K_i\int e(t)\,dt+K_d\frac{d}{dt}e(t) \qquad (10)$$

The error is the different distance between the ball position and the origin. The primary plant control law is implemented as per Figure 3, with the Autotuned PID Controller. The control law that features the use of the Ziegler Nicholes method of tuning is shown in Figure 4 with ZN as the tuning method of the PID.

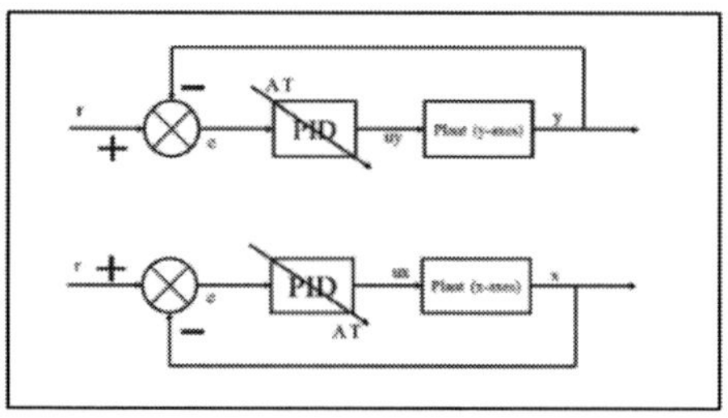

Fig. 3. Autotuned Control law for the plant.

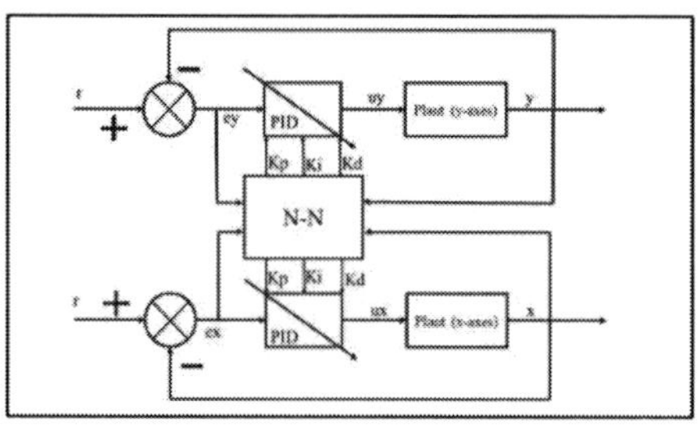

Fig. 4. The control law with the Neural Network as a tuner.

5. Neural Network

The Neural Network (NN) used in this research is the NARX NN. It is important to note that the NARX Neural Network approach was used purely on the simulation and was not implemented experimentally. [9] The development and training of the model was as follows:

5.1.1. Data generation

The model was trained using data generated from the plant model, where various PID gain values served as inputs and system responses over 20-second intervals were recorded as outputs. Each response, tied to specific PID gains, was stored in arrays and used to train the NARX Network.

5.1.2. Model training

The model, summarized in Table 1 and detailed in Appendix B, was trained with 1001 data points. It featured a layer size of 10 and a time delay of 2 seconds.

Table 1. Summary of training model details.

Training Algorithm			
Data Division	Random		
Training Algorithm	Levenberg-Marquardt		
Performance	Mean Square Error		
	Observation	MSE	R
Training	699	0.0948	0.9898
Validation	150	0.1166	0.9874
Test	150	0.0981	0.9894

For a well-trained NARX Neural Network, the mean square error (MSE) should approach zero, and the regression coefficient (R) should be as close as possible to 1 [10]. During the model's training, the MSE and R convergence value was 0.099411 and 0.989, indicating effective network training.

6. Experimental Setup

The experimental setup uses a smartphone camera and Droidcam software to monitor the ball's position in real time, transmitting video to Python via Jupyter Notebook. The ball's position is identified using OpenCV, and its coordinates are sent to an Arduino, which calculates error distances and controls servos using PID gains. Distance calibration converts video coordinates into actual distances through predefined measurements. A second Arduino reads IMU sensor data to compare plate angular displacement with numerical predictions.

The test rig design and schematics are included in Appendix A.

7. Results and Discussion

The Simulation results are as shown in Table 2.

Table 2. Summary of simulated experiment results.

	Autotune	NN Settled
Kp	-1.8803	-1.85
Ki	-0.0868	-0.25
Kd	-7.2164	-4.75
% overshoot	36.2	20
Settling position (x & y)	0.1m	0.1m
Settling time	16 s	35s

From the results tabulated above the PID gains do not differ too much. The settled PID gains refers to the PID gains that resulted at once the system had reached the error. This is shown in the Figure 5 below.

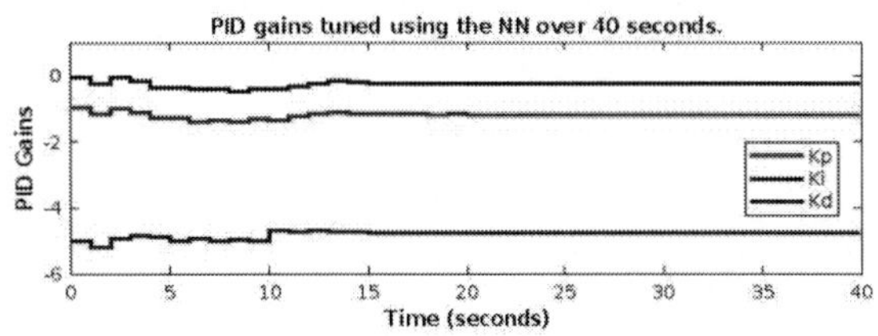

Fig. 5 PID gains tuned using the NN over 40 seconds.

The system response for the linear and nonlinear model, where the PID gains used on the nonlinear model were the PID gains tuned for the linear model is shown in Figure 6. Also seen on the table above, there is a trade-off between settling time and overshoot. This means that this would have to be considered when designing and training a NN. The higher the settling time the lower the overshoot.

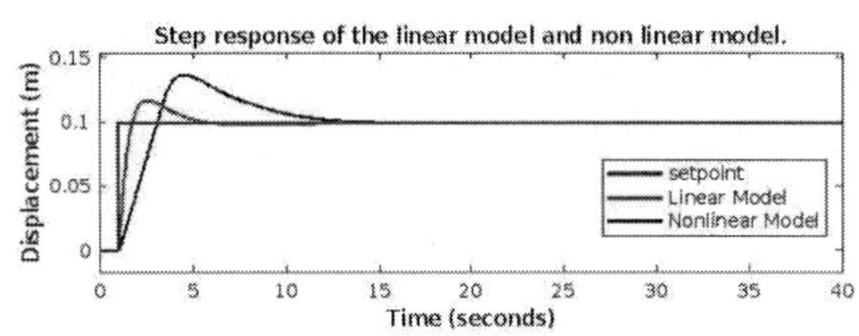

Fig. 6. Step response for the linearized model and un-linearized model tuned using MATLAB Autotune.

This variation in the results show that the changes in system properties affect the system response, potentially causing instability. The Neural Network system response is shown in the Figure 7 below.

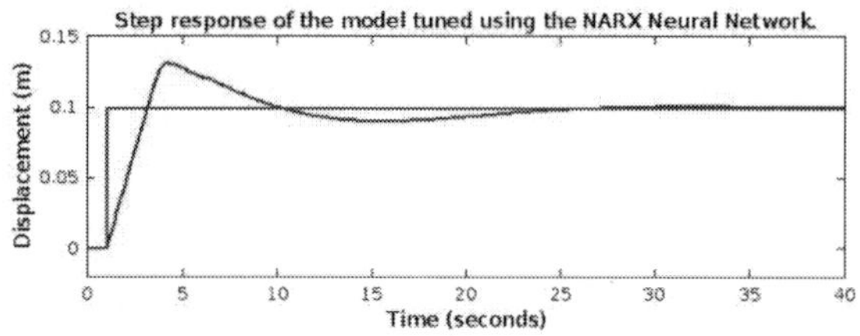

Fig. 7. Step response of the model tuned using the NARX Neural Network.

For the experiment, the PID gains determined using the MATLAB PID tuner were used and the system became unstable. Through trial and error, the PID gains were tuned to; Kp = 0.039, Ki = 0.088, and Kd = 0.7 and the system responded as shown in Figure 8.

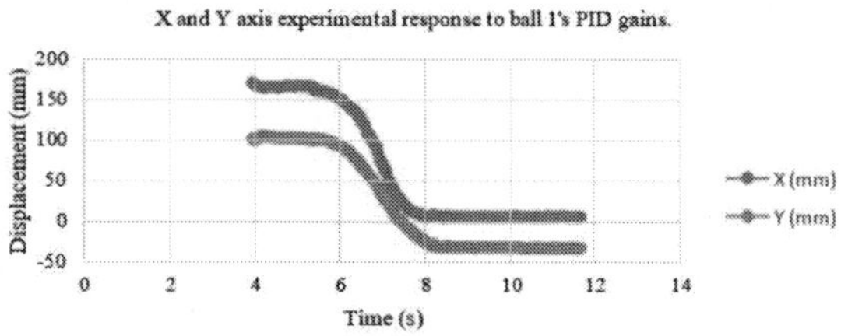

Fig. 8. Experimental results from using PID gains determined through trial and error.

Simulated results confirm that Neural Networks are effective in tuning PID gains.

The Neural Network's ability in handling non-linearized models and parameter variations was shown through simulation. A trained model for one mass, 18.2g, was tested by altering the mass to 14g at 40s and initializing a new set point.

For the constant gains obtained using the autotuner, the nonlinear system becomes unstable when the mass is changed. This demonstrates the limitations of linearization because the PID gains were determined based on this model. The corresponding result graphs Figure 9, 10 and 11 are shown below.

In contrast, when using the NARX Neural Network as the PID tuner, the model eventually settles after some time. The graph below, shown on figure 10, displays the output of the model.

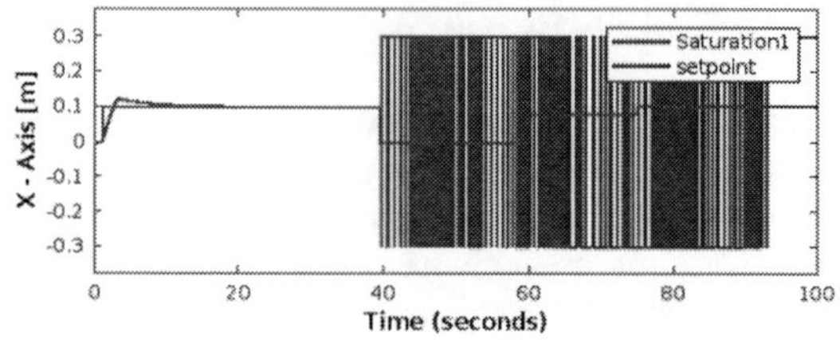

Fig. 9. Model response using static PID gains tuned using MATLAB Autotune.

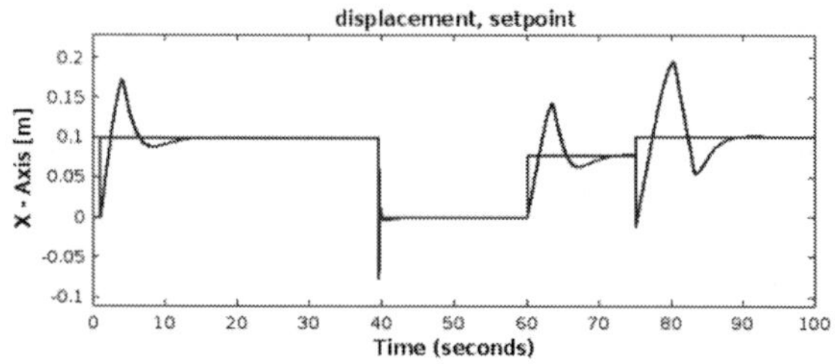

Fig. 10. Displacement results when Neural Network over 100 seconds when the mass changes along with the setpoint after 40 seconds

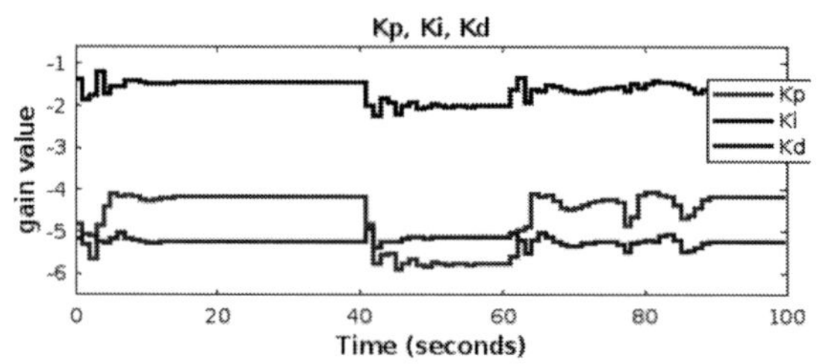

Fig. 11. PID gains (Kp, Ki, Kd) over 100 seconds for the Neural Network changing with the mass change and setpoint change.

The PID gains are recalculated based on the new set point, as the model aims to optimize the PID for minimal overshoot and the quickest settling time. The figure 11 above illustrates the PID outputs for the NARX model.

8. Conclusion and Future Works

The PID gains obtained from tuning the linearized model fails to stabilize the nonlinear system with changing parameters. However, Neural Networks effectively address this challenge and can be used as PID tuners, adapting to changes in the system model and parameters. The difference in PID gains used in simulation and experimentation can be attributed to factors such as the delay between the camera reading the ball's position and the actuators moving the plate, which was not accounted for in the mathematical model. Future work should include live experimental testing.

Acknowledgements

The authors would like to acknowledge Dr. M. Atkins, chair of the School's Lab Committee for funding the research equipment. ChatGPT was used to refine the grammar of the first paragraph of this paper.

References

[1] Nguyen, D.H., Lowenberg, M.H. & Neild, S.A. Identifying limits of linear control design validity in nonlinear systems: a continuation-based approach. *Nonlinear Dyn* **104**, 901–921 (2021). https://doi.org/10.1007/s11071-021-06341-2

[2] H. Nijmeijer, "Control of nonlinear mechanical systems," *Final Program and Abstracts on Information, Decision and Control*, Adelaide, SA, Australia, 2002, pp. 29-29, doi: 10.1109/IDC.2002.995381.

[3] Massachusetts Institute of Technology, 2007. *Thesis (Ph.D.)—Massachusetts Institute of Technology, Dept. of Mechanical Engineering, 2007.* [online] Available at: http://hdl.handle.net/1721.1/39897 [Accessed 15 Jan. 2025].

[4] H. Nijmeijer, "Control of nonlinear mechanical systems," *Final Program and Abstracts on Information, Decision and Control*, Adelaide, SA, Australia, 2002, pp. 29-29, doi: 10.1109/IDC.2002.995381.

[5] Tang, H. H., & Ahmad, N. S. (2024). Fuzzy logic approach for controlling uncertain and nonlinear systems: a comprehensive review of applications and advances. *Systems Science & Control Engineering*, *12*(1). https://doi.org/10.1080/21642583.2024.2394429

[6] Jastrzębski, M., Kabziński, J. Adaptive control of magnetic levitation system based on fuzzy inversion. *Sci Rep* **14**, 24815 (2024). https://doi.org/10.1038/s41598-024-76067-9

[7] Dušek, F., Honc, D., & Sharma, K. R. (2017). Modelling of ball and plate system based on first principle model and optimal control. 2017 21st International Conference on Process Control (PC), Strbske Pleso, Slovakia, 216-221.

[8] Arali, R., Halawal, V. and Naik, A., 2021. Colour Detection from Image. International Journal of Advances in Engineering and Management (IJAEM), 3(2164), pp.5252-030721642171.

[9] MathWorks, n.d. NARXNET. [online] Available at: https://www.mathworks.com/help/deeplearning/ref/narxnet.html [Accessed 15 Jan. 2025].

[10] Alallaf Almustawi, Omaima & Abdulqader, Shahla. (2011). Nonlinear Autoregressive neural network for estimation soil temperature: a comparison of different optimization neural network algorithms. Ubiquitous Computing and Communication Journal (UBICC), Journal issue for ICIT2011. 6. 43-51.

Appendix A. Assembled Experimental Rig

Fig. 12. Experimental Test Rig setup.

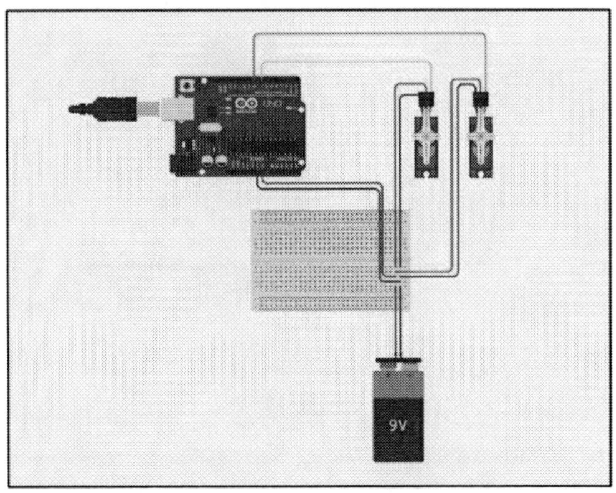

Fig. 13. Electronics schematics for the experimental setup.

Appendix B. Training results for the NARX network 1.

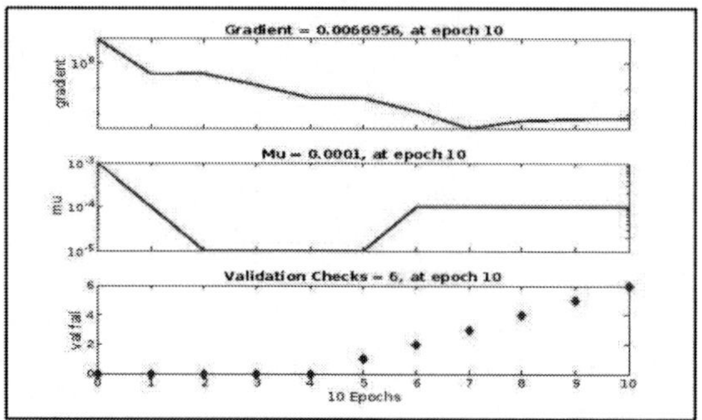

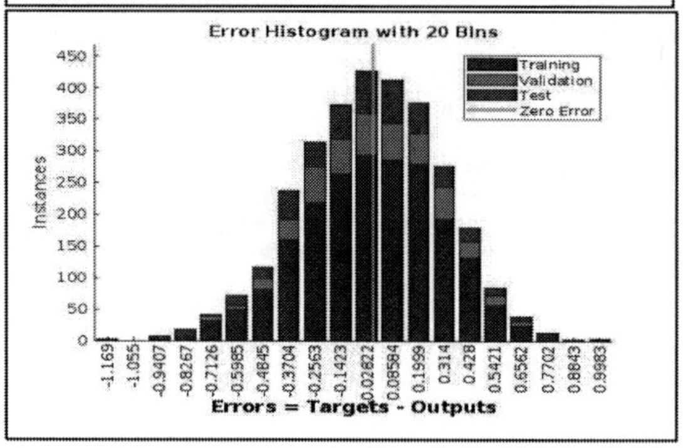

Fig. 14. Error Histogram training graph results for the NARX.

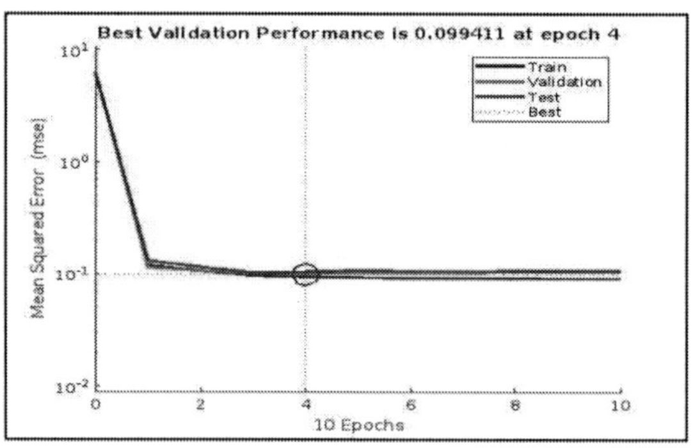

Fig. 15. Mean Square Error Training Results for the NARX network.

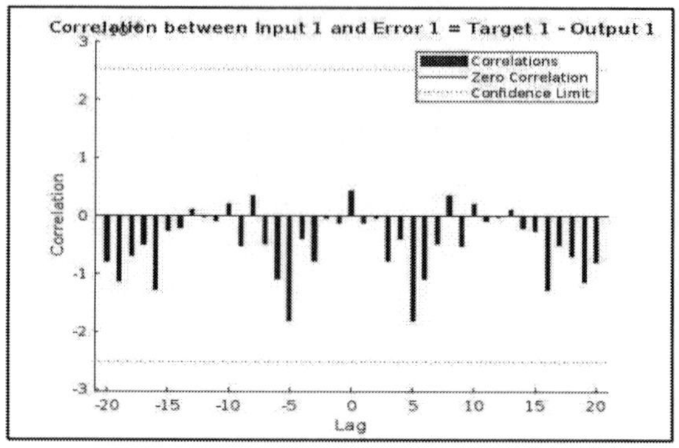

Fig. 16. Correlation Results for the NARX network.

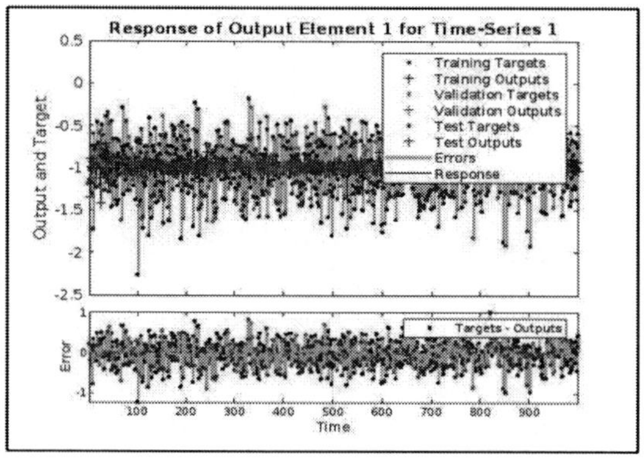

Fig. 17 Output and Target training Results for the NARX network.

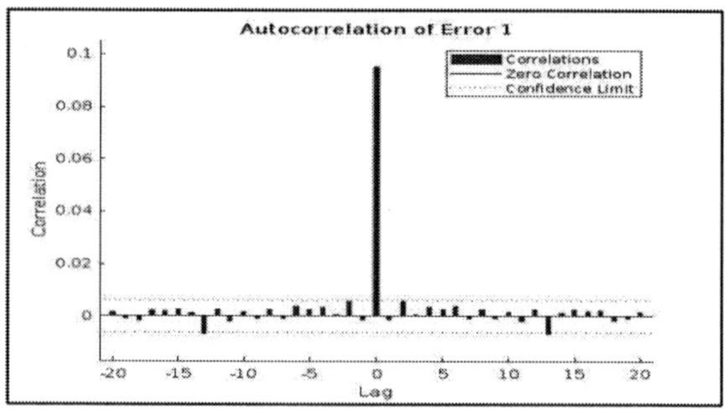

Fig. 18 Autocorrelation of Error 1 results for the NARX network.

14th South African Conference on Computational and Applied Mechanics (SACAM 2025)
21-23 January 2025, Johannesburg, South Africa

A Weibull Distribution-Based Approach for Rapid Ranking of Mechatronic Systems

Tinashe Erwin Vhurumuku[a*], Dr. Aarti Panday[a], Dr. Chioniso Kuchwa-Dube[a]

[a]*University of the Witwatersrand, Johannesburg*

1 Jan Smuts Avenue, Braamfontein, 2000, Johannesburg, South Africa

**tinashe.vhurumuku@wits.ac.za*

Abstract

This paper proposes a novel technique for fast evaluation and scoring of mechatronic systems by combining the Weibull distribution with feature selection based on machine learning algorithms. The purpose of this tool is to provide a rapid comparison of mechatronic systems that belong in the same domain. It is important to evaluate their individual subsystems and their total system performance as well as integration of mechanical, electrical and control components, which are called mechatronics. At the present time these methods can be slow, often unstandardized and do not deliver complete and comparable representations of reliability and efficiency aspects of a given system. Due to its flexibility and versatility when it comes to modeling failure behaviors the Weibull distribution can be effectively employed in ranking systems [1]

In this study, historical performance data is used to model a ranking system through the Weibull distribution [2], allowing for the evaluation of relative failures, long-term wear-out patterns, and general performance. Machine learning techniques are employed to identify the most critical features contributing to system performance, ensuring that the model prioritizes key metrics across the entire system.

The outcome of this work is a rapid ranking tool that will evaluate mechatronic systems and rank them on a rating scale from 1 to 10, based on their reliability, efficiency, and performance relative to their counterparts. This tool will provide decision-makers with a quick, data-driven method for comparing and optimizing mechatronic systems.

1. Background and Motivation

Evaluating mechatronic systems—integrating mechanical, electrical, and control components—requires standardized frameworks for benchmarking reliability, performance, and efficiency. Traditional methods focus on meeting design specifications but lack comparative perspectives against historical data or industry standards, limiting informed decision-making. Benchmarking tools like OEE in manufacturing or performance metrics in sports analytics provide useful rankings but remain domain-specific and unstandardized for mechatronic systems. Existing methods often neglect the variability and wear-out behaviours crucial to engineering contexts.

Weibull distributions are well-suited for modelling reliability and performance variability in engineering due to their flexibility in representing asymmetric failure and wear-out patterns, unlike Gaussian distributions. They have been successfully applied to life data and reliability analysis, providing a robust statistical foundation for benchmarking [3] This study leverages Weibull modelling combined with feature weighting and machine learning to standardize rankings across diverse domains, including robotics, manufacturing, and aerospace.

The proposed methodology incorporates historical data, evaluates Key Performance Indicators (KPIs), and aligns composite scores to a Weibull distribution for intuitive and realistic performance benchmarking. By employing percentile-based ranking and statistical rigor, this approach addresses gaps in existing methods, offering a rapid and interpretable system for ranking mechatronic systems across domains.

2. Objectives and Scope

2.1. Objectives

The primary objectives of this study are to:

1. Develop a robust ranking methodology using the Weibull distribution and weighted percentile scoring to evaluate mechatronic systems.
2. Provide standardized benchmarking tools to compare mechatronic systems against historical data
3. Simplify performance evaluation through an intuitive scoring scale from 0 to 10
4. Validate the methodology through illustrative examples and case studies

2.2. Scope

1. **Ranking Framework:** Application of Weibull distribution for realistic modelling of performance variability.
2. **System Versatility:** Adaptation of the methodology for mechatronic systems in robotics, automation, and aerospace engineering
3. **Decision Support:** Providing engineers with data-driven insights for benchmarking and optimizing system designs

3. Methodology and Computational Approach

The proposed methodology ranks mechatronic systems by leveraging percentiles of Key Performance Indicators (KPIs), weighted according to their importance, to compute a total score. The resulting scores are aligned with a Weibull distribution for realistic performance modelling and benchmarking. It is important to clarify that the use of the Weibull distribution is a normative modelling choice, not an assumption that the scores will naturally follow this distribution. While the Central Limit Theorem (CLT) suggests that the sum of many random variables will tend toward a normal distribution, the KPIs in a complex mechatronic system are neither independent (e.g., engine performance and fuel efficiency are correlated) nor identically distributed. Since the conditions for the CLT are not met, there is no theoretical basis to expect a normal distribution. Instead, we engineer the final score distribution to align with a target Weibull distribution. This choice is justified by the Weibull distribution's flexibility and its widespread use in reliability engineering to model performance data, which is often skewed and not well-represented by a symmetric normal distribution.

3.1. Step-by-Step Overview of Methodology

1. **Collect Historical Data**: Gather historical data containing n KPIs for systems within the same domain. Each row in the dataset represents a unique system, and each column corresponds to a KPI such as rise time, settling time, or reliability metrics.
2. **Data Preprocessing**: Clean and preprocess the data to address missing or inconsistent values. Ensure KPIs are expressed on a consistent scale (e.g., seconds for timing metrics, percentages for reliability metrics
3. **Select KPIs**: Key Performance Indicators (KPIs) and their weights are identified using statistical tools and machine learning algorithms [4], prioritizing those with the highest impact based on frequency, classification performance, correlation with target metrics, and domain knowledge.

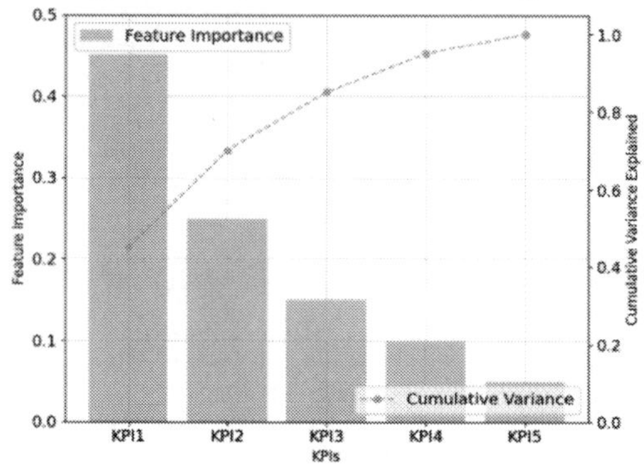

Figure 1 KPI importance using machine learning: Illustrative example from generated data

1. Figure 1 shows an illustration of KPI importance using machine learning.
4. **Assign KPI Weights:** Assign weights C_i to each KPI based on its relative importance. The sum of all weights must equal 100 to ensure the total score is capped at 100:

$$\sum_{i=1}^{n} C_i = 100 \qquad (1)$$

5. **Calculate Percentile Scores:** For each KPI Ki, compute the percentile rank $P_{i,j}$ of each system j relative to the dataset. Higher percentiles correspond to better performance:
6.

$$P_{i,j} = \frac{\text{Number of systems with } K_i \le K_{i,j}}{\text{Total number of systems}} \times 10 \qquad (2)$$

7. **Compute Composite Scores:** Calculate the total score Sj for each system j as the weighted sum of its percentile scores:

$$S_j = \sum_{i=1}^{n} C_i \cdot \frac{P_{i,j}}{100} \qquad (3)$$

Here, $(\frac{P_{i,j}}{100})$ converts the percentile to a decimal value (e.g., 95th percentile becomes 0.95).

8. **Weibull Distribution Alignment**: Adjust the weights Ci iteratively to ensure the scores S_j align with the desired Weibull distribution. This optimisation task is non-linear, so instead of gradient-based methods that risk converging to local minima, we use Particle Swarm Optimization (PSO), a metaheuristic algorithm. PSO effectively explores the complex search space to find weights that minimize the divergence between the score

distribution and the target Weibull model, proving more robust than methods like Maximum Likelihood Estimation (MLE) for this application. The Weibull cumulative distribution function (CDF) is given by:

$$F(x;\lambda,k) = 1 - e^{-\left(\frac{x}{\lambda}\right)^{k}} \quad (4)$$

Where:

- (λ) (scale parameter) determines the spread of the distribution,
- (k) (shape parameter) governs the failure behaviour (e.g., early-life, random, or wear-out failures).

Weights are adjusted iteratively using a cost function to minimise the divergence between the observed distribution of scores and the Weibull target distribution:

$$\text{Cost Function} = \sum_{j=1}^{m} \left(F(S_j;\lambda,k) - F_{\text{target}}(S_j)\right)^2 \quad (5)$$

F_{target} represents the target cumulative distribution function (CDF) value from the predefined Weibull distribution for a given score S_j. The choice of Weibull parameters and determines the scenario and desired characteristics of mechatronic system ratings. For this work we set $(k = 6)$ (shape parameter) and $(\lambda = 3)$ (scale parameter) to define our desired distribution.

These values result in a dataset where the mean rating lies between 5 and 6, fitting a psychologically and statistically intuitive framework. In a rating system ranging from 1 to 10, scores around 5 are naturally interpreted as average, aligning with human expectations of a midpoint in a scoring scale.
The mean of approximately 5 also corresponds to a statistically average value, making the distribution relatable and easy to interpret. This aligns with the convention that 50% represents average in many evaluation frameworks. The resultant distribution is in

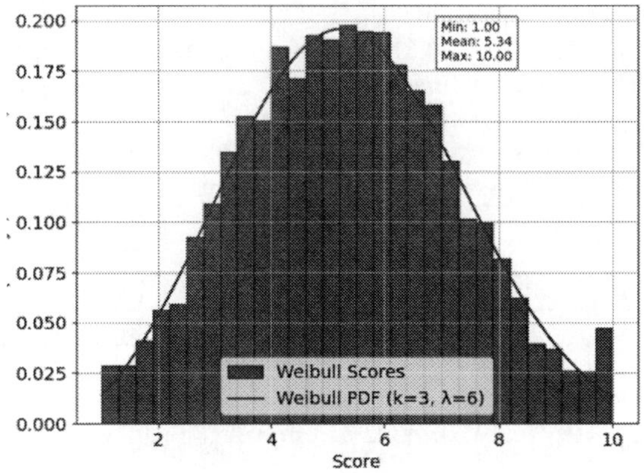

Figure 2 Desired Distribution

9. **Convert Scores to a 0–10 Scale:** After the scores are aligned with the Weibull distribution, convert them to a 0–10 scale by dividing the total score by 10:

$$S_j^{\text{Final}} = \frac{S_j}{10} \quad (6)$$

10. Model Finalisation: Save the final weights (C_i) and Weibull parameters (λ, k) as the evaluation model. These parameters enable rapid evaluation of new systems.
11. New System Evaluation: For a new system, compute its KPIs, determine the weighted percentile scores, and calculate the final score using the saved model. The score benchmarks the new system against historical data.

3.2. Comprehensive Worked Example

To illustrate the full 10-step methodology, consider a small, hypothetical dataset of four mechatronic systems (A, B,C,D), evaluate on three KPIs: Efficiency (%), Response Time (s) and Reliability (MTBF hrs).

Step 1: Collect Historical Data

System	Efficiency (%)	Response Time (s)	Reliability (MTBF hrs.)
A	85	0.5	1200
B	92	0.3	1500
C	78	0.8	950
D	88	0.4	1350

Step 2: Data Preprocessing

Data is clean. Note: lower values are better for Response Time. This will be handled by ranking KPIs in ascending order for beneficial metrics and descending order for detrimental ones.
Step 3 & 4:
Select KPIs & Assign Initial Weights Assume PCA

yields initial weights:
C_1(Efficiency)= 40, C_2(Time)= 30, C_3(Reliability)= 30 with weights summing to 100.

Step 5: Calculate Percentile Scores

Percentile scores for System B are $P_{1,B} = 100$, $P_{2,B} = 100$, are $P_{3,B} = 100$. Percentile scores for all systems:

System	P1 (Eff.)	P2 (Time)	P3 (Rel.)
A	50	50	50
B	100	100	100
C	25	25	25
D	75	75	75

Step 6: Compute Composite Scores (Initial)

$$S_B = 40\left(\frac{100}{100}\right) + 30\left(\frac{100}{100}\right) + 30\left(\frac{100}{100}\right) = 100$$

The full set of initial scores is $S_A = 50$, SB = 100, SC = 25, SD = 75.

Step 7: Weibull Distribution Alignment

The initial scores {25, 50, 75, 100} are uniformly distributed. The target distribution is a Weibull with shape k = 6 and scale λ = 3. An optimization algorithm (e.g., Particle Swarm Optimization) yields new weights:

$C_1 = 50$, $C_2 = 20$, $C_3 = 30$

Step 8: Recompute Composite Scores (Optimized)

$S_B = 50\cdot1+20\cdot1+30\cdot1 = 100$

The updated scores are {28.75, 57.5, 86.25, 100}, which are better aligned with the target Weibull shape.

Step 9: Convert Scores to 0–10 Scale

$$S_B = \frac{10}{100} S_B = 10.0$$

Step 10: Model Finalization

The final model includes the optimized weights and Weibull parameters:

(C1, C2, C3) =(50,20,30), Weibull parameters: k = 6, λ = 3

4. Key Results and Discussion

The method was applied to a car performance data set with several KPIs [5]. The feature importance analysis was performed using Principal Component Analysis (PCA) to identify the most significant features across the dataset. Initial weights are further refined using a cost function to align the system scores with the desired Weibull distribution. The initial weights were optimised (Table 1) to meet the desired distribution in **Error! Reference source not found.**.

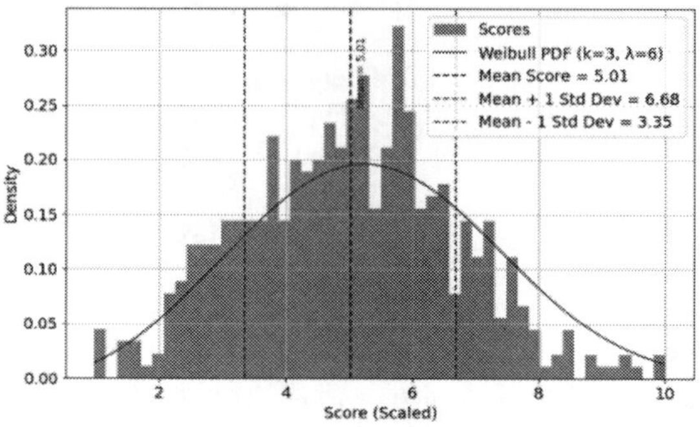

Figure 3 Alignment of Final Scores with Weibull Distribution

Table 1 Comparison of Initial and Optimised Weights

Feature	Initial Weight (%)	Optimized Weight (%)
Fuel Efficiency	11.54	12.78
Safety Rating	9.68	11.99
Engine Performance	10.45	10.68
Braking Performance	11.91	7.73
Comfort Rating	10.31	10.31
Emissions	11.55	11.55
Acceleration	10.43	10.43
Suspension Performance	11.17	13.95
Scenario Adaptability	10.66	10.54

The optimized scoring process ranked the cars based on their relative performance. Table 2 Top 10 Ranked Cars Based on Final Scores shows the top 10 ranked cars, providing insights into their performance.

Table 2 Top 10 Ranked Cars Based on Final Scores

Car Model	Final Score (Scaled)	Rank	Performance Tier
Corolla, Toyota	10.00	1	Exceptional
E-Class, Mercedes	9.50	2	Exceptional
Model 3, Tesla	9.30	3	Exceptional
Sonata, Hyundai	9.28	4	Exceptional
GLE,	9.04	5	Exceptional

Mercedes			
Civic, Honda	8.90	6	High
Camry, Toyota	8.75	7	High
Altima, Nissan	8.66	8	High
Model S, Tesla	8.49	9	High
3 Series, BMW	8.46	10	High
Ford Escape	6.85	25	Average
Jeep Cherokee	6.10	38	Average
Chevrolet Trax	5.32	51	Below Average
Mitsubishi Mirage	4.77	62	Below Average
Dodge Journey	3.91	75	Poor

The proposed Weibull distribution-based ranking methodology effectively evaluates mechatronic systems by integrating weighted percentile scoring, feature selection, and Weibull alignment, as shown in the illustrative vehicle performance data. The methodology highlights the importance of historical performance data, though availability may be limited in domains like robotics and manufacturing. Synthetic data can help in data-scarce environments but risks introducing biases. Domain-specific variability necessitates tailoring KPIs, such as path-following accuracy for robotics or throughput for manufacturing, with standardized definitions to ensure meaningful comparisons.

Ranking interpretability is influenced by Weibull parameters, which must align with the system context, such as prioritizing early-life failures for consumer robotics. Additionally, KPI selection and weighting are critical, as irrelevant or inconsistent metrics can skew results. Despite these challenges, the methodology demonstrates significant potential for robust and reliable rankings, provided issues like data availability, KPI customization, and parameter sensitivity are addressed.

5. Conclusion and Future Work

This paper introduces a Weibull distribution-based methodology for evaluating and ranking mechatronic systems using historical data, weighted percentile scoring, and feature selection to create interpretable rankings aligned with realistic statistical distributions. While effective in domains like robotics, manufacturing, and avionics, the methodology faces challenges such as reliance on historical data, domain-specific KPI selection, and sensitivity to Weibull parameters. Recommendations for improvement include developing domain-specific datasets, standardizing KPIs, refining weighting processes with expert input, validating with real-world data, and tailoring the framework to specific applications. Future integration with design tools, maintenance workflows, and machine learning could enhance adaptability and utility, making it a robust foundation for evaluating diverse mechatronic systems.

6. References

[1] A. Ramesh Kumar and V. Krishnan., "A study on system reliability in Weibull distribution". International Journal of Innovative Research in Electrical, Electronics, Instrumentation and Control Engineering, 5(3):38, 2017.

[2] D. Chin-Diew Lai, "Weibull distributions and their applications," in *Springer Handbook of Engineering Statistics*, 2006, pp. Chapter 3:63-78.

[3] J. Frost, "Weibull distribution: Uses, parameters and examples," 2023Paper.

[4] I. Guyon. and A. Elisseeff, "An introduction to variable and feature selection.," *Journal of Machine Learning and Research,* vol. 3, p. 1157–1182, 2003.

[5] Ziya, "Car Performance Dataset," 2024. [Online]. Available: https://www.kaggle.com/datasets/ziya07/car-performance-dataset. [Accessed 30 November 2024].

Mechanical analysis of the Hoyt Xceed archery riser

C. Lopes[1]*and F. Pietra[1]

[1]School of Mechanical, Industrial and Aeronautical Engineering, University of the Witwatersrand
1 Jan Smuts Ave, Braamfontein, Johannesburg, 2017

*carinadalopes@gmail.com
Francesco.Pietra@wits.ac.za

ABSTRACT

The riser is the central rigid part of the recurve bow connected to the limbs, and its mechanical performance is crucial to the archer's success. The riser should be as stiff as possible with minimal bending to support the limbs and ensure an accurate shot.The Hoyt Xceed riser has gained popularity amongst archery experts and all male medallists in the 2024 Paris Olympic games [1]. This is an interesting statistic as it is outdated and commercialised as inferior to the 2020 Formula Xi.

The aim of this investigation was to conduct Finite Element Analysis (FEA) and compare the mechanical performance of the two risers to determine if there are any geometrical features and mechanical properties that indicate why the Xceed is the preferred choice. Furthermore, the design of the Xceed was evaluated to determine if it was based on engineering rationale by comparing it against a generic riser that was topologically optimised to have the same mass as the former. An initial 3D scanning of the Xceed riser was obtained and a Computer Aided Design (CAD) version of the riser was created. The CAD was analysed using ANSYS to evaluate the life, damage, stress and strain state of the respective risers under the shooting loading conditions.

The investigations demonstrated that the Formula Xi outperformed the Xceed in every aspect of the comparative analysis aside from twist. The FEA validated the statements of Hoyt, suggesting that favouring the Hoyt Xceed over the Formula Xi riser was based on personal preference rather than mechanical performance. Despite this, the FEA demonstrated that the Hoyt Xceed excelled for a riser of its mass as it performed significantly better than the generic riser. Furthermore, the comparative analysis emphasised that cutouts are an effective and practical way to optimise risers.

Keywords: Recurve Bow, FEA Analysis, Reverse Engineering, 3D Scanning

1. Introduction

The archery riser is the central, rigid part of the recurve bow. It plays a significant role as it is the foundation to which all the parts are connected to. Aside from affecting yielding and fatigue, the riser's mechanical performance is crucial to the archer's success within this sport of accuracy. The riser should be as stiff as possible to minimise twist and bending. Significant riser twist will cause the arrow to be propelled in both the axial and transverse direction [2]. This is not ideal as the arrow will skid to the side rather than being shot accurately towards the target. The limbs are the flexible parts that store potential energy which is converted to kinetic energy to propel the arrow in flight. The limbs are connected to and apply a bending moment on the riser. The riser should have insignificant bending to structurally support the limbs and ensure they will adequately bend to store sufficient potential energy to shoot the arrow the required distance to the target.

Furthermore, the riser's mass plays a crucial role in the archer's success. Archers can be expected to shoot up to 200 times per day to prepare and train for competitions [2]. This can be demanding as they are expected to support the bow outstretched in front of them while pulling a significant draw weight (22 kg – men's division or 16 kg – women's division) with each shot [3], [4]. Thus, the riser's mass should be minimised to ensure the conventional position can be maintained by the archer without experiencing fatigue.

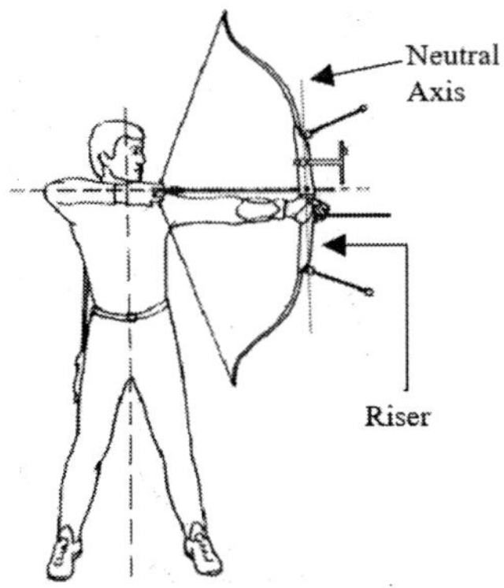

Figure 1: Image demonstrating the conventional archer position [5].

In addition, the conventional technique (Figure 1), practiced by archer's to hold the bow, results in the recurve bow rotating once the arrow is shot [5]. Riser's are typically designed to have a centre of gravity as low and as forward, to the neutral axis (Figure 1), as possible such that it has the tendency to rotate forward and away from the archer's face while still mainitaining stable condition. This ensures the riser returns to the upright position after rotation.

There are two types of limb connection systems available in the market, namely: the Interntaional Limb Fitting (ILF) system and the Formula system. The most significant difference between the two systems is that the distance between the u-shaped cutout and dovetail is much greater for the Formula in comparison to ILF (Figure 2) [6], [7]. Furthermore, the bushing on Formula limbs allows a damper to be connected directly onto the limbs. This is not the case for ILF limbs.

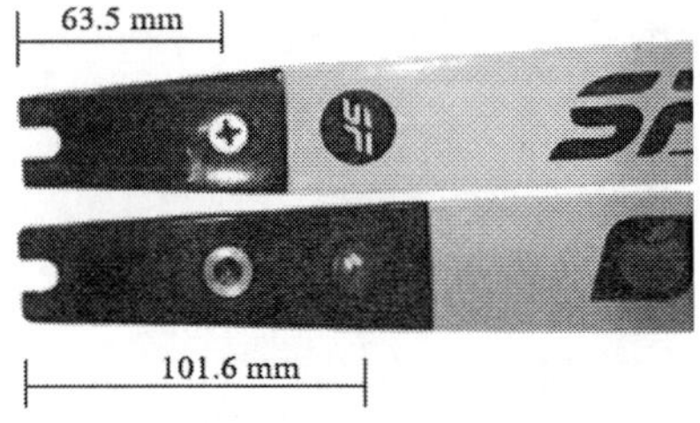

Figure 2: Image of the ILF limbs (top) and Formula limbs (bottom) [6], [7].

One particular riser that has performed significantly well is the Hoyt Xceed 25-inch riser, which has the ILF connection system (Figure 3). It was the preferred choice for all the men's podium finishers at the 2024 Olympic games and has been described as a breakthrough model, providing a testament of the riser's popularity amongst experts in archery [1].

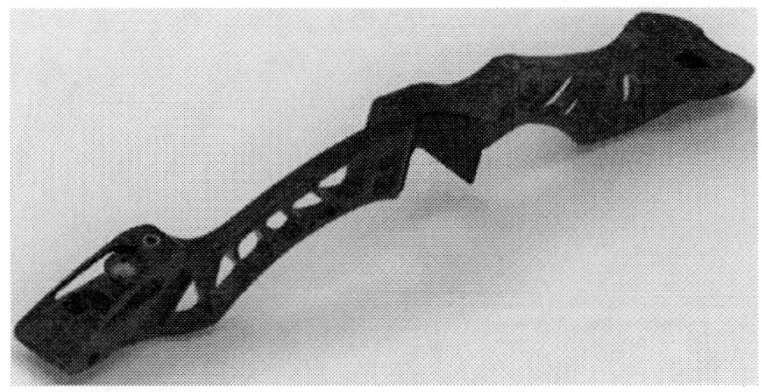

Figure 3: Image of the Hoyt Xceed riser.

This is a very interesting statistic for multiple reasons: the Xceed is a riser designed in 2020 and it doesn't have the latest technological advances of the next generation of risers, such as integrated dampening or revised geometries to optimise the stiffness-to-weight ratio [8]. Furthermore, Hoyt commercialises and claims that the Xceed design and ILF connection system are inferior to the Formula design and connection system in terms of stability and dampening [9], [10]. This paper aimed to investigate why the Xceed is the most favoured riser by top archers despite being labelled as outdated and inferior to the Formula risers. This was done by comparing the mechanical performance, and geometrical characteristics of the Xceed against its counterpart, the Formula Xi which was also released in 2020 and has the Formula limb connection system (Figure 4).

Figure 4: Image of the Formula Xi riser.

In addition, cutouts are either introduced onto structures based on engineering rationale, with intent to optimise the mass of geometries, or for aesthetic purposes. This paper aimed to investigate the rationale behind the inclusion of the cutouts on the Xceed and to evaluate what geometrical features of the Xceed contribute to it excelling within the industry. This was done by comparing its design and mechanical performance against a generic riser which was optimised, to have the same mass, with the loading conditions and engineering analysis as the primary focus (Figure 5).

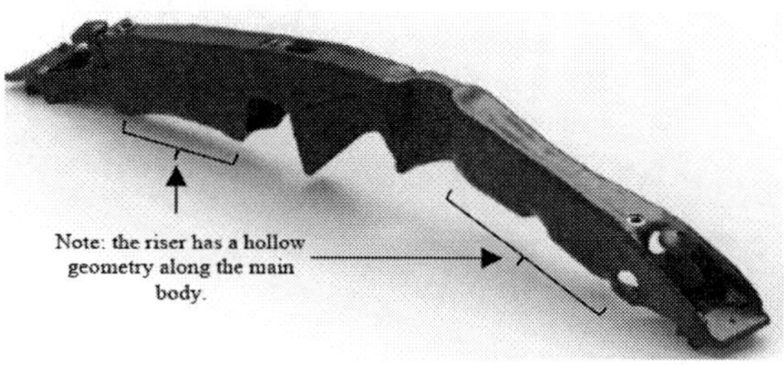

Figure 5: Image of the generic riser.

2. Methodology

FEA was completed in ANSYS 2024 R2 to obtain and compare the mechanical performances of the respective risers, including deformation, degree of twist, equivalent von Mises stress and shots to failure. Risers are typically complex in geometry to ensure they are sufficiently stiff to withstand the applied loads while maintaintaining a relatively low mass. Thus, it was chosen to implement a reverse engineering approach to obtain the CAD models for the FEA. Both the Hoyt Xceed and Formula Xi risers were scanned using an Artec Eva 3D scanner. The generic riser was created using Artec Studio 17, GeoMagic Design X 2022, SolidWorks 2019 and SolidEdge 2021. The softwares were used to add material onto the previously scanned xceed riser to remove its pre-existing shape and create a box-like generic design as it is shown in Figure 6. After which, the topological optimisation tool in ANSYS 2024 R2 was used to refine the shape by removing material and obtaining geometry, according to the FEA results, that had the same mass as the Xceed riser (1.27 kg). The stresses experienced by a structure influence how it behaves—how it deforms, rotates, and ultimately where and how it may fail [11]. Thus, the objective was to reduce the mass while maintaining the stress distribution, and the locations in which the stress is concentrated, with the intention that the mechanical performance would be maintained too.

Figure 6: Image of the generic riser before topological optimisation.

The Xceed and generic risers were defined to be made of aluminium 6081 while the Formula Xi was defined to be made of aluminium 2024 [12]. Their properties are outlined in Table 1 [13], [14], [15], [16]. The handles were defined to be rigid grade PVC, and the additional components, such as the bolts and screws, were defined to be steel.

Table 1: Table of material properties [13]-[16].

Property	***Magnitude***	
Type of aluminium alloy	6081	2024
Young's Modulus (E) (GPa)	69	72.4
Density (ρ) (kg/m³)	2700	2780
Poisson Ratio (ν)	0.33	0.33
Yield Strength (σ_y) (MPa)	270	345
Ultimate Tensile Strength (σ_{uts}) (MPa)	310	427

Two separate analyses, static and fatigue, were conducted for each riser to evaluate the relative mechanical performance. The static analysis evaluated the deformation (mm) and equivalent von Mises stress (MPa) of each riser. The fatigue analysis evaluated the life (number of shots) and damage of each riser. The damage is a dimensionless value representing the ratio of the intended design life to the available life. For this project, the intended design life was estimated to be 624 000 shots, assuming a professional archer training program was followed for a period of 10 years [2]. Each approach will be discussed subsequently in more detail.

2.1 Static Analysis

The loads used for all the static analyses are provided in Figure 7 and Table 2. They were obtained from a literature paper regarding FEA of a recurve bow [17]. An extra force, F_{By}, was added to the simulation and estimated to be 10% of the x – component of force at the dovetail in the global coordinate system. It was assumed that the bow string would not be drawn perfectly back due to imperfect technique, thus, there would be a small component of force perpendicular to the side of the riser. The loads F_{Hy}, F_{Hs} and F_{Hp} represent the forces of the archer's hand to support and prevent the bow from moving. The archer's hand was simulated as displacement contraints in the simulations, as it is shown in Figure 8.

Table 2: Table providing the loads applied on a riser [17] .

Description	*Load*	*Magnitude (N)*
Limb force on main bolt.	F_{As}	1440.70
Limb force on dovetail.	F_{Bs}	1144.89
	F_{Bp}	161.74
	F_{By}	35.00

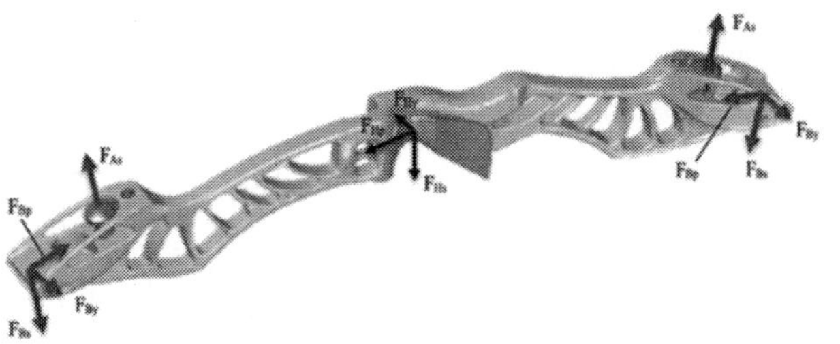

Figure 7: Diagram demonstrating the loads applied on a riser [17].

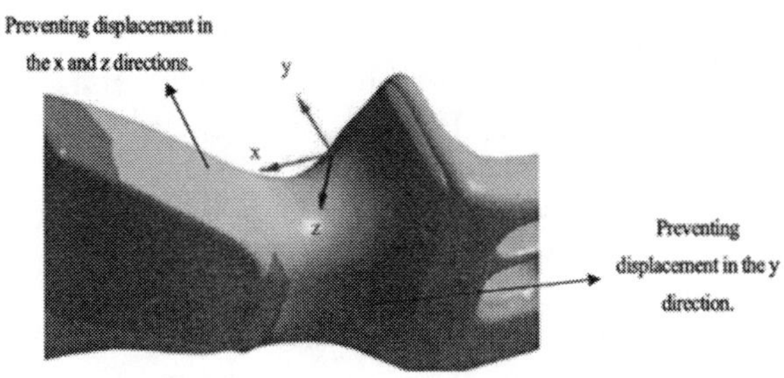

Figure 8: Diagram demonstrating the displacement constraints on the handle of the risers.

2.2 *Fatigue Analysis*

The dynamic loading was obtained using a reference shot. It was acquired by conducting physical testing in the Archery Laboratory of the University of the Witwatersrand (Figure 9). Five tests were conducted on the Hoyt Xceed 25-inch riser, in its naked configuration, and the test that best described the average shot was used as the reference shot (Figure 10). The naked configuration refers to the riser in its most basic form, that is, with no stabilisers, additional weights, or damping components attatched.

Figure 9: Image of the riser during the test set up.

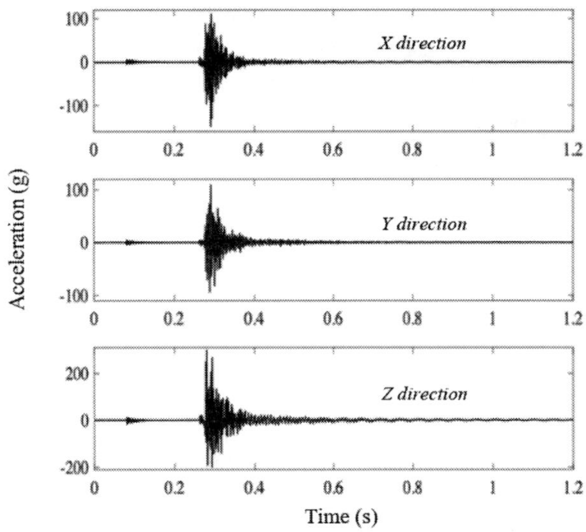

Figure 10: Dynamic loading for the fatigue analyses in x (top), y (middle) and z (bottom) directions.

The loading conditions for the fatigue analyses differed for each riser to accommodate the dynamic loading. The maximum forces in each direction were obtained by calculating forces equivalent to 1 g of acceleration according to the mass of the respective risers. The remaining forces in each direction were then calculated using ratios provided in Figure 7. This approach ensured the data was converted to m/s^2, while still considering the dynamic aspect of the loading and the exact locations where the accelerations were applied onto the risers. Once more, the force from the archer's hand was represented as displacement constraints on the handles of the risers. The static loads for each riser are provided in Figures 11-13.

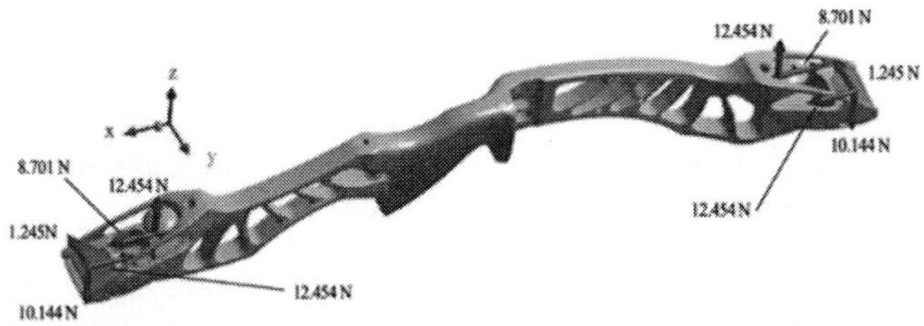

Figure 11: Diagram demonstrating the static loading conditions of the Xceed riser.

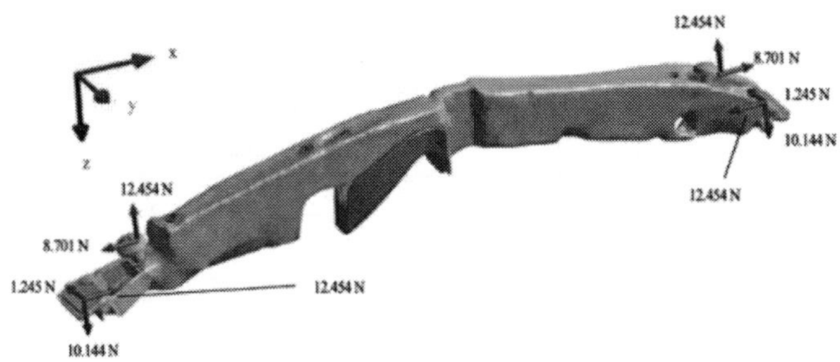

Figure 12: Diagram demonstrating the static loading conditions of the generic riser.

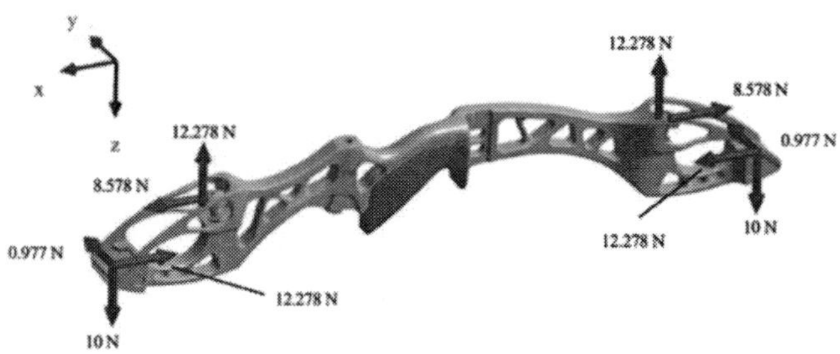

Figure 13: Diagram demonstrating the static loading conditions of the Formula Xi riser.

Table 3 provides the settings used throughout the fatigue analyses to set up the dynamic loading [18], [19].

Table 3: Table providing the settings used in the fatigue simulations [14], [15].

	Outcome	***Reasoning***
Stress Theory	*Solderberg*	Most conservative theory for aluminium parts.
Analysis Type	*Stress Life*	Risers are assumed to be at the low cycle of their lives (after manufacturing).
Scale Factor	*1.143*	Scaling the data to have a draw weight of 18.144 kg to coincide with literature.
Fatigue Strength Factor (k_f)	*0.896 (Xceed & Generic)* *1.12 (Formula Xi)*	$k_f = k_s k_d k_t k_e$ • k_s = 0.986 (machined surface) • k_d = 2.271 (equivalent diameter = 7.227) • $k_t = 1$ (primarily bending) • k_e = 0.4 → Xceed and generic (σ_{uts} < 328 MPa) • k_e = 0.5 → Formula Xi (σ_{uts} > 328 MPa)

3. Results

Before the results were analysed, a mesh independency test was used to validate the results of the simulations. The built-in convergency test of ANSYS 2024 R2 could not be used to validate the results as there were stress concentrations in the geometry, mainly due to the post-processing of reverse engineering, which caused the tests to fail. Thus, the mesh element size of each riser was changed to determine if the equivalent von Mises stress was independent of the mesh size to determine the reliability of the results. The general rule of thumb is as follows: the stresses would be deemed mesh independent if they did not change by more than 5%.

Table 4: Table providing the maximum equivalent von Mises stress (MPa) for different element sizes.

	Maximum Equivalent Von Mises Stress (MPa)		
Element Size (mm)	*Hoyt Xceed riser*	*Formula Xi riser*	*Generic riser*
5	66.27	75.53	125.21
2.5	67.28	78.51	133.02
1.25	67.24	77.21	141.17

The change in stresses did not exceed the 5% threshold. This was with exception to the generic riser, which had changes of 5.87% and 6.13%. However, these values were considerably close to the threshold, thus the results were deemed mesh independent.

3.1 Static Analyses

3.1.1 Total deformation

The total deformation, in mm, has been provided in Figures 14-16. There were a few insights in the results. The largest deformations occurred at the limb pockets and the least amount of deformation was experienced at the handle. This concurred with what was expected as the riser was supported by the archer at the handle while accelerations were induced, by the limbs, at the limb pockets. Furthermore, the upper limb pocket consistently deformed more for all the risers.

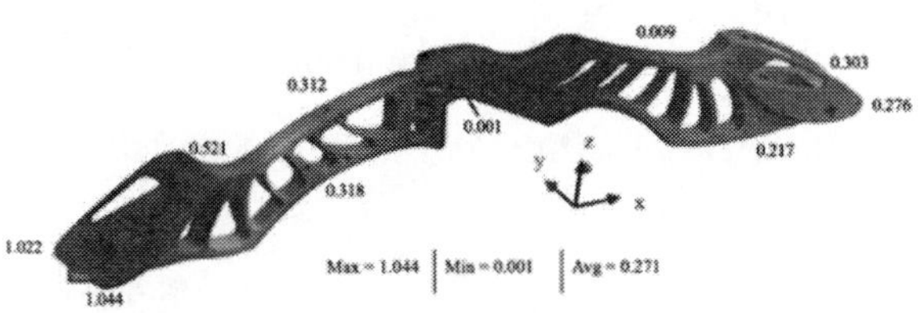

Figure 14: Diagram demonstrating the deformation (mm) of the Hoyt Xceed riser.

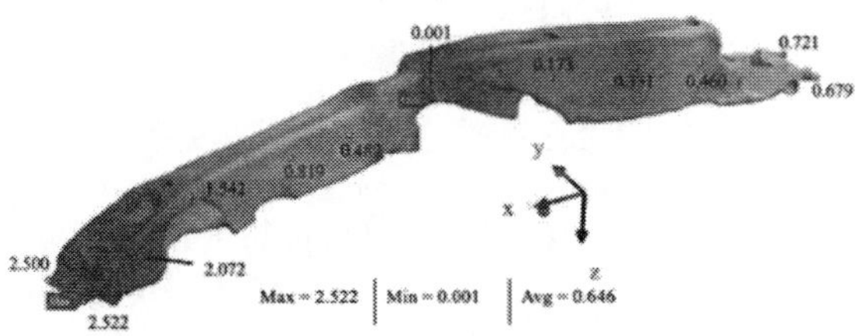

Figure 15: Diagram demonstrating the deformation (mm) of the generic riser.

Despite being optimised with the fatigue analysis as the primary focus, the generic riser experienced the worst total deformation. The Formula Xi had the least deformation.

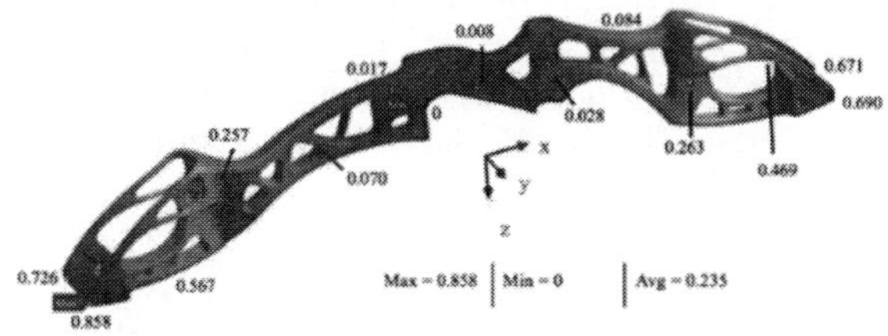

Figure 16: Diagram demonstrating the deformation (mm) of the Formula Xi riser.

3.1.2 Degree of Twist

In addition to the deformation, a dimensionless value was calculated to determine the angular rotation of the limb pockets or degree of twist (T_w). It was calculated using equation 1, where d_1 and d_2 are the total deformations, in mm, of probes at the corner of each limb pocket and l is the distance between the probes in mm. The results are provided in Table 5.

$$T_w = \frac{d_1 - d_2}{l} * 1000 \tag{1}$$

Table 5: Table providing the degrees of twist for the respective risers.

	Degree of Twist	
	Upper Limb Pocket	*Lower Limb Pocket*
Hoyt Xceed riser	0.71	0.57
Formula Xi riser	3.46	0.42
Generic riser	1.10	0.58

Despite having the least amount of deformation, the Formula Xi had the greatest degree of rotation at the limbs. The Hoyt Xceed experienced the smallest degree of twist overall.

3.1.3 Challenges of reverse engineering

Due to the emergent and novel nature of the reverse engineering process, specifically for post-processing of topologically optimised or 3D scanned componets, the 3D CAD models obtained had some defects. This included uneven surfacing, non-manifiold geometry or unintended sharp edges. These defects made it a bit more complex to evaluate the results for the equivalent von Mises stress as they acted as stress concentrators but were not a true reflection of the risers' mechanical performances. Most of the geomteries were cleaned up and defects removed prior to the analysis. However, due to the limitation of resources and time, the models could not be processed more than their current state. Despite this, the results were deemed accurate to present as it was possible to distinguish between what was a true reflection of the performance compared to what was simply caused by a bad node / geometry. In many instances, the defects caused the maximum and average stress to be higher than its true value. Thus, in each instance a magnified view has been provided to show the defect and how the stress has been concentrated due to the defect. Through critical analysis, the accurate maximum stress has been provided in the tables within the figures below the risers. Due to the defects acting as stress concentrators, they also influenced the damage and life results. The same approach has been followed for these sections, in which the defects are highlighted and the true value is provided below.

3.1.4 Equivalent von Mises stress

The distribution of stresses for each riser are provided in Figures 17-19. None of the risers exceeded their yield strength, therefore will not fail due to yielding.

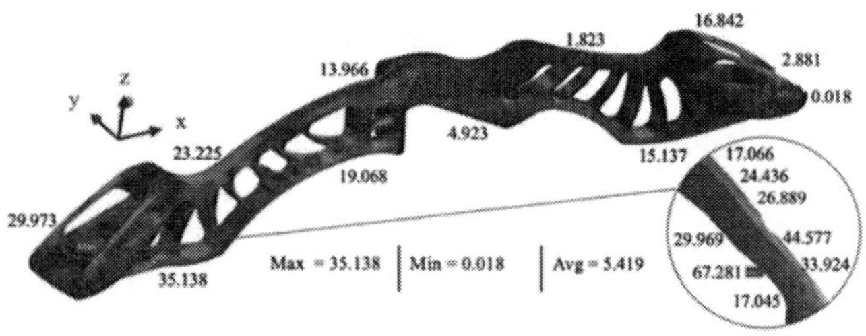

Figure 17: Diagram demonstrating the equivalent von Mises stress (MPa) of the Hoyt Xceed riser.

From Figure 17, the maximum stress is stated to be 35.1 MPa. Aside from the defect, the ribs at the limb pockets are typically regions of high stress. This is expected due to the small cross sectional areas and close proximity to the applied forces.

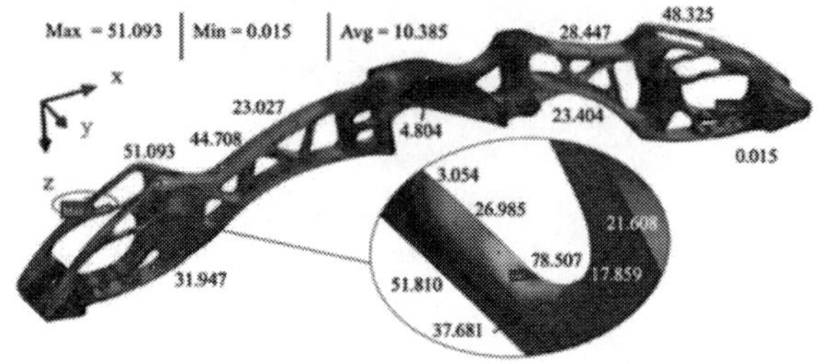

Figure 18: Diagram demonstrating the equivalent von Mises stress (MPa) of the Formula Xi riser.

Evident in Figure 18, the ribs at the limbs pockets once more experienced high stresses. However, the stresses were more evenly distributed along the main body of the Formula Xi in comparison to the Xceed. The maximum stress was evaluated to be 51.1 MPa.

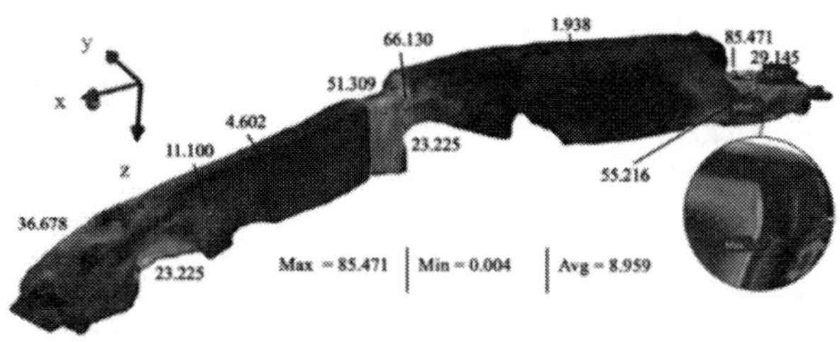

Figure 19: Diagram demonstrating the equivalent von Mises stress (MPa) of the generic riser..

The maximum stress was evaluated to be 85.5 MPa and was elevated in comparison to the other models. Despite this, the equivalent von Mises stress was well below the yield strength of aluminium 6081 (270 MPa) and would not fail due to yielding [14]. In contrast to the Xceed and Formula Xi, it is important to note that the stress was more concentrated at the clicker region for the generic riser. When topologically optimising the riser, the objective was to maintain the stress distribution of the generic riser. The two risers have similar stress distributions (Figure 19 and 20), in which the stress is concentrated at the clicker region (shown in Figure 20), due to the smaller cross-sectional area. The generic riser, before topological optimisation, experienced minimal stress, deformation and twist due to its bulky nature and larger mass moments of inertia. Thus, the intention behind maintaining the stress distribution was to maintain the mechanical properties of a bulky generic riser while reducing the mass. The stress distribution heavily influenced the box-like and hollow shape of the generic riser. This also provided the opportunity to critically evaluate the cutouts of Xceed, and most commercially available risers, to determine if the cutouts are the most effective way to minimse mass.

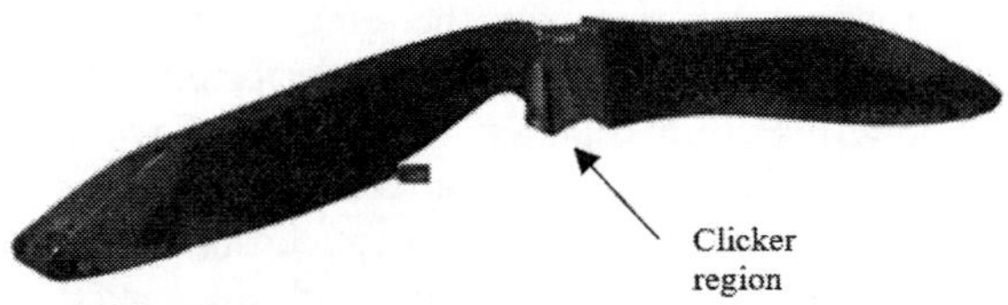

Figure 20: Diagram demonstrating the distribution of equivalent von Mises stress of the generic riser.

3.2 Fatigue Analyses

3.2.1 Life and Damage

The distribution of combined damage has been provided from Figures 21-23. The minimum life in each direction and combined minimum life are provided in Table 6. The combined results represent the life and damage of the respective risers due to the combination of loadings in each direction. One important insight was that none of the risers would exceed their intended design life of 624 000 shots.

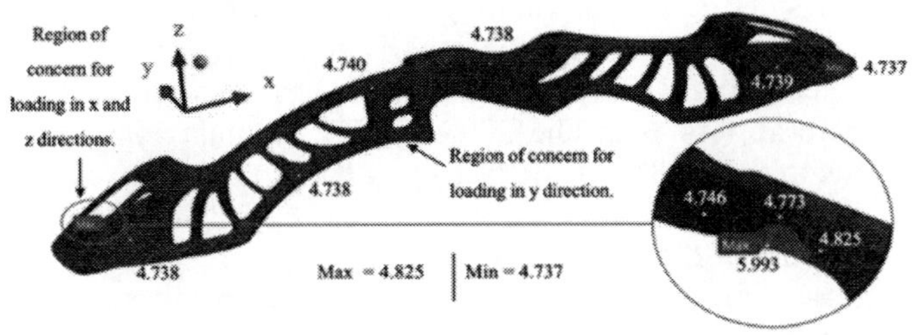

Figure 21: Diagram demonstrating the damage of the Hoyt Xceed riser.

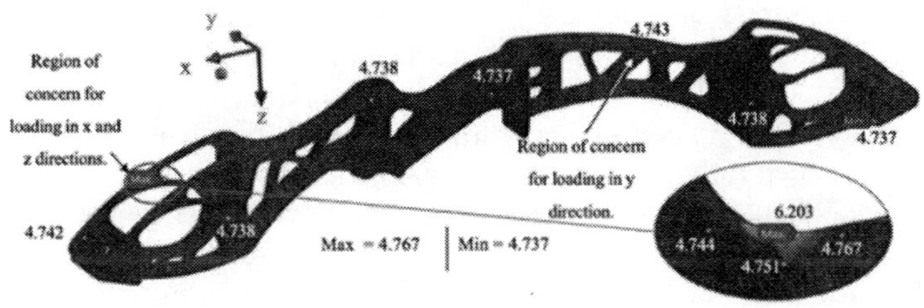

Figure 22: Diagram demonstrating the damage of the Formula Xi riser.

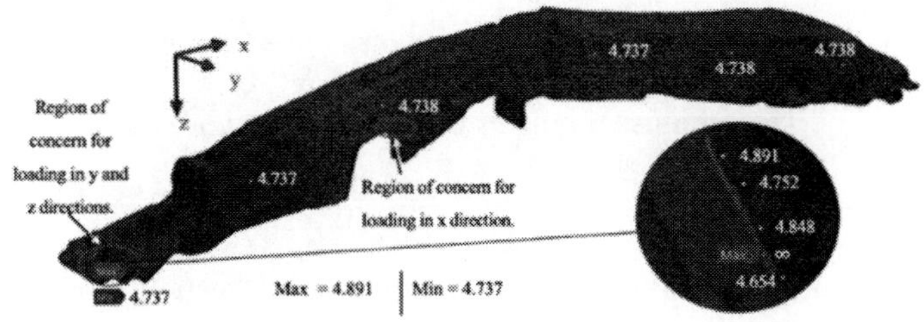

Figure 23: Diagram demonstrating the damage of the generic riser.

Table 6: Table providing the minimum life for the respective risers in each direction.

	Minimum Life (no. of shots)			
	X	*Y*	*Z*	*Combined*
Hoyt Xceed riser	3.15e5	3.72e5	5.56e5	1.29e5
Formula Xi riser	3.16e5	3.75e5	5.55e5	1.31e5
Generic riser	3.17e5	3.75e5	5.56e5	1.28e5

Fatigue is inevitable for aluminium as it does not have an endurance limit [18]. It is ideal to have a high minimum life and low damage as it indicates the riser has a greater ability to postpone fatigue. The Formula Xi performed the best with respect to life and damage. However, it was not by a significant amount. In addition, all the risers performed the worst in the x-direction.

3.3 Geometrical Properties

3.3.1 Mass Moment of Inertia

The higher the mass moment of inertia, the greater the riser's ability to resist bending about a given axis. Thus, risers should be designed to have large mass moments of inertia to minimise bending and twist, and ensure an accurate shot. The mass moments of inertia for the respective risers are provided in Table 7.

Table 7: Table providing the mass moments of inertia for the respective risers about the given axes.

	Mass moments of inertia (kg.m^2)		
	I_{xx}	I_{yy}	I_{zz}
Hoyt Xceed riser	0.001	0.038	0.037
Formula Xi riser	0.001	0.033	0.032
Generic riser	0.001	0.033	0.033

All the risers had the lowest mass moments of inertia about the x – axis. The Hoyt Xceed had the largest mass moments of inertia about both the y – and z – axes.

3.3.2 Centre of Gravity

As mentioned in the introductory section, the position of the riser's centre of gravity (CG) plays a significant role in the functionality of the bow. As a general rule of thumb, the CG should be as low and as forward, compared to the neutral axis (Figure 1) as possible, to ensure the bow rotates away from the acher's face and independently reverts to the upright position. The position of the CG for the respective risers have been compared in Figure 24.

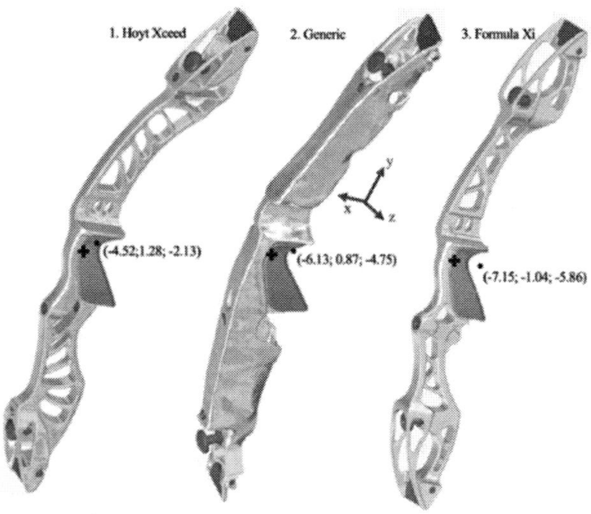

Figure 24: Diagram demonstrating the CG for the respective risers relative to the origin (+).

The placement for the CG of the Formula Xi was the most favourable out of the three risers due to its tendency to rotate away from the archer's face. A high CG, as seen with the generic riser, would be beneficial if the CG was positioned significantly forward (beyond the neutral axis). However, this was not the case. In its current design, the generic riser will have the tendency to rotate into the archer's face and is thus, the least favourable riser in terms of CG positioning.

4. Discussion

Prior to discussing the individual risers, there were some important trends that need to be discussed that were relevant to all the risers. The foremost was that none of the risers would exceed their intended design life of 624 000 shots. They would all have to be checked after approximately 2 years if a professional archer training program was followed. This is particularly concerning, especially for athletes training for the Olympics that occurs every 4 years, as they may not be able to use the same riser for the entirety of their training. Furthermore, the risers performed the best, in terms of life, in the z – direction and worst in the x – direction. This agrees with what is expected as risers are primarily designed to withstand bending. This was supported by the mass moment of inertia about the x – axis being consistenly lower in comparison to the other axes for all the respective risers.

The mechanical performance of the Hoyt Xceed was compared against its 2020 counterpart, the Formula Xi, to determine if it had any merits that would suggest or indicate why the former is the preferred choice by top archers. While it was not significant, in many cases, the Formula Xi narrowly edged out and performed better than the Xceed, as Hoyt commericalised. This was specifically true for its position of centre of gravity, minimum life and total deformation.

Considering that the Formula Xi had lower mass moments of inertia in comparison to the Hoyt Xceed, it was not expected that the former would deform less than the Xceed. The Formula Xi deformed 17.8% (0.186 mm) less than the Xceed, at the upper limb pocket which experienced maximum deformation for both models. The first reason for the lower maximum deformation was that the deformation was more evenly distributed between the upper and lower limbs pockets in the Formula Xi. Although it deformed less at the upper limb pocket in comparison to the Xceed, it deformed 56.1% (0.387 mm) more at the lower limb pocket. Thus, demonstrating that the design of the Formula Xi distributed the deformation more evenly between the two limb pockets rather than concentrating the deformation to the upper limb pocket, as it is with the Hoyt Xceed. The deformation of the riser affects its ability to support the limbs and the deformation of the limbs. Thus, it would be more favourable and stable, when shooting the arrow, to have more evenly distributed deformation between the upper and lower limb pockets of the riser to ensure the limbs deform by the same proportion and store approximately the same amount of potential energy. Disparities in energy between the limbs may result in asynchronous limb motion when the arrow is released, causing the arrow to deflect vertically. This is not ideal as the shot accuracy and stability will be compromised. In addition, the Formula Xi experienced a greater amount of stress along its main body. This difference in distribution and magnitude of stress would also cause the deformation decrease despite the lower mass moments of inertia. The lower maximum deformation and more evenly distributed deformation at the limb pockets are beneficial qualities of the Formula Xi design compared to the Xceed.

On the other hand, the Formula Xi had significantly greater degrees of rotation at the limbs than the the Xceed. The Xceed has a greater distribution of material at the limb pockets which would prevent the limb pockets from rotating.

The results demonstrated that each model had their own strengths and downfalls. However, the Formula Xi typically outperformed the Xceed, supporting the claims of Hoyt. Aside from its ability to prevent rotation at the limb pockets, there is no apparent evidence, from its mechanical properties and performance, to suggest why the Xceed is the preferred choice by top archers and experts in comparison to the Formula Xi.

The generic riser was far superior to the Hoyt Xceed, in terms of mechanical performance, before it was optimised [20]. The aim was to investigate if it was possible to maintain the riser's excellent mechanical performance, by maintaining the stress distribution, while optimising its mass to be identical to that of the Xceed. From the results, it was apparent that the mechanical performance could not be maintained as the Hoyt Xceed performed significantly better to the generic riser. The Hoyt Xceed experienced 58.6% (1.48 mm) and 35.5% (0.387) less deformation and twist, respectively, in comparison to the generic riser. This was not expected to be the case since the modified design had been developed from and had a similar stress distribution to the excelling generic riser before it was modified.

The poor performance indicated that the box-like geometry only yielded good mechanical properties when it had a significant amount of material. The results for the generic riser, before it was optimised, were completed and presented in prior research [20]. The mass moments of inertia for the generic riser were lower than that of the Xceed indicating why it deformed and twisted more. Thus, more material would have to be added to ensure the riser was comparable in performance to the Xceed, and sufficiently resisted bending and twisting. As it had been established, this is not desirable as the mass of the risers should be minimised to prevent archer fatigue. The Hoyt Xceed is far superior to the generic riser, in terms of mechanical performance, thus suggesting that the commercially available riser was designed with engineering rationale and mechanical performance as a primary focus. Furthermore, the excelling results demonstrated that the cutouts were an effective way to remove material and optimise the mass of the riser without hindering mechanical performance.

Although, mechanical performance significantly impacts the accuracy of the archer's shot and the riser's ability to resist failure or fatigue, it is not the sole determinant for the success of the riser. The success of a design entails a lot more including functionality, practicality and aesthetics. In most cases, risers are manufactured using high precision CNC machining. The generic riser would be more challenging, and thus more expensive, to manufacture in comparison to the commercially available design. Furthermore, the Hoyt Xceed is more aesthetic than the generic riser, which would make it more appealing to archers and potential customers. In addition, the Hoyt Xceed has a greater tendency to rotate away from the archer's face as a result of the lower centre of gravity. The commercially available riser, the Hoyt Xceed, has been designed for performance, manufacture, functionality and aesthetics. Thus, its holistic design and cutout approach leads it to be a far superior design in comparison to the box-like hollow design of the generic riser.

5. Conclusion

The Hoyt Xceed has gained significant popularity and praise by top archers and experts in the field of archery despite being commercialised by Hoyt as inferior to the Formula Xi. The aim of this research was to investigate if there were any geometrical features or mechanical properties that would indicate or suggest why the Xceed is the preferred choice. The investigations demonstrated that the Formula Xi narrowly edged out the Xceed in all areas of mechanical performance studied in this research aside from degree of twist (limb pocket rotation). There is no apparent evidence to suggest why the Xceed is favoured over its 2020 counterpart, the Formula Xi. Rather, the research validated the claims of Hoyt and suggested favouring the Xceed, over the Formula Xi, was based on a property that is yet to be established or on personal preference rather than engineering rationale and mechanical performance.

Although the Hoyt Xceed riser did not achieve mechanical performance credible enough to deem the model superior to the Formula Xi, it did achieve significantly better results in comparion to the generic riser of the same mass. The analysis indicated that the riser achieved noteworthy results in comparison to the latter. Besides confirming that the Xceed design had been designed based on engineering rationale and mechanical performance, the comparison process demonstrated that the riser was far superior in many factors of the design process, including manufacturing, functionality and aesthetics. It emphasised the effectiveness of the cutout approach to minimise mass and optimise riser geometry. Overall, the research demonstrated that the Hoyt Xceed riser has a well-refined, holistic design, rendering it as a favourable option for a riser of its mass.

REFERENCES

[1] "Hoyt Exceed Riser at Paris 2024 all three mens medals." Accessed: Sep. 20, 2024. [Online]. Available: https://www.archerytalk.com/threads/hoyt-exceed-riser-at-paris-2024-all-three-mens-medals.6267787/

[2] A. Jordan, S. Vajna, S. Clement, M. Heller, and J. Edelmann-Nusser, "Lightweight Design Optimization of a Bow Riser in Olympic Archery Applying Evolutionary Computing," *Eur J Sport Sci*, vol. 4, no. 3, 2005.

[3] NBC Olympics, "Archery 101: Olympic equipment." Accessed: Sep. 09, 2024. [Online]. Available: https://www.nbcolympics.com/news/archery-101-olympic-equipment#:~:text=Olympic%20recurve%20bows%20are%20comprised,the%20handle%20to%20the%20tip.

[4] World Archery, "Archery Sport ." Accessed: May 16, 2024. [Online]. Available: https://www.worldarchery.sport/sport

[5] I. Zanevskyy, "Modeling and Computer Simulation of Bow Stabilization in the Vertical Plane," Jan. 2008.

[6] "Hoyt Grand Prix, ILF, Hoyt Formula – What Is the Difference?" Accessed: Dec. 01, 2024. [Online]. Available: https://archeryparent.wordpress.com/2017/09/24/hoyt-grand-prix-ilf-hoyt-formula-what-is-the-difference/

[7] F. Pietra, P. W. Loveday, and R. Reid, "Investigation into the correlations between engineering measurements and archers' perceptions of different limbs in a recurve bow," p. 4, Jul. 2024.

[8] Archery Attack, "Hoyt GMX 3 Grand Prix 27 2024." Accessed: Oct. 09, 2024. [Online]. Available: Hoyt GMX 3 Grand Prix 27 2024

[9] T. van Rooijen, "Which is better: Formula or ILF?" Accessed: Jan. 10, 2024. [Online]. Available: https://improveyourarchery.com/which-is-better-formula-or-ilf/

[10] D. Denton, "Hoyt ILF vs. Formula." Accessed: Dec. 09, 2024. [Online]. Available: https://www.archerytalk.com/threads/hoyt-ilf-vs-formula.6097320

[11] R. C. Hibbeler, *Mechanics of Materials in SI Units*, 10th ed. Harlow, UK: Pearson , 2018.

[12] DiSport, "Hoyt Riser Xceed GP ILF 2023." Accessed: Sep. 20, 2024. [Online]. Available: https://www.disport.it/en/product/118982-1000/hoyt-riser-xceed-gp-ilf-2023

[13] GRANTA, "Material Database."

[14] "6081 (6081-T6, AlSi0.9MgMn) Aluminum." Accessed: Sep. 22, 2024. [Online]. Available: https://www.makeitfrom.com/material-properties/6081-6081-T6-AlSi0.9MgMn-Aluminum

[15] Federal Aviation Administration, *Metallic Materials Properties Development and Standardization (MMPDS)*. 2019.

[16] "Aluminium 2024 - T6." Accessed: Nov. 10, 2024. [Online]. Available: https://asm.matweb.com/search/SpecificMaterial.asp?bassnum=ma2024t6

[17] Z. Ahmad, B. Ponnusamy, and Z. Ghazalli, "Finite element analysis approach on 3D modeling of archery bow design," Mar. 2016.

[18] J. K. Nisbett and R. G. Budynas, *Shigley's Mechanical Engineering Design* , 10th ed. New-York: McGraw-Hill, 2015.

[19] N. E. Dowling, *Mechanical Behaviour of Materials: Engineering Methods for Deformation, Fracture, and Fatigue*, 4th ed. Pearson, 2013.

[20] C. Lopes, "Mass reduction of a commercial archery riser using the topological optimisation approach," University of the Witwatersrand , Johannesburg , 2024.

Evaluating digital image correlation speckle pattern quality using finite element-based displacement fields

Noko Phukubye[a*], Gerhard Venter[a], Melody Neaves[a]

[a]Department of Mechanical and Mechatronic Engineering
Stellenbosch University, Stellenbosch, South Africa
*Author email: nokophukubye24@gmail.com

Abstract

This study contributes to the field of digital image correlation (DIC) speckle pattern quality evaluation by employing finite element methods to generate numerically deformed images, which serve as ground truth models for assessing DIC accuracy. The results are then compared against established speckle pattern quality metrics, and sub-pixel translation tests are conducted to further supplement the analysis. By exploring the relationship between speckle pattern quality and measurement accuracy, this research seeks to support the optimisation of speckle patterns using existing metrics, while also investigating the potential development of new ones. The goal is to improve the accuracy and robustness of DIC in engineering applications. The findings suggest that some existing global metrics may not consistently predict speckle pattern quality, highlighting the potential value of further research into more localised assessment criteria.

Keywords – *Digital image correlation, Finite element methods, Numerical image deformation, Speckle pattern*

1. DIC background

Measuring surface deformation in materials and structures under different loading conditions is an important aspect of experimental mechanics [1]. While the traditional strain gauge method provides point-wise measurements, advancements have led to the development of various full-field, non-contact optical techniques for this purpose [2]. Among these, digital image correlation (DIC) has emerged as a popular method.

DIC is an image-based technique for measuring deformations, relying on both image processing and numerical computation. The core principle involves tracking or matching identical points (or pixels) between a reference image and its deformed counterpart [2]. Among the various DIC approaches, the subset-based technique is the most commonly used. In this method, a square subset is selected from the reference image, and its corresponding location in the deformed image is identified using a correlation algorithm. This process produces a mapping vector that consists of displacements which are used to generate a displacement field [2].

2. Speckle patterns

To enable the tracking of image subsets, a speckle pattern is applied to the specimen surface to generate random grey level variations [3]. Numerous studies [1, 4, 5, 6] have demonstrated that the quality of the speckle pattern significantly influences the measurement accuracy of DIC techniques. Consequently, substantial research has been devoted to understanding this relationship, leading to the development of various pattern assessment criteria and a range of pattern quality measurement parameters being proposed in the literature. However, [3] highlighted that some global metrics may not adequately assess pattern quality.

This study assesses the reliability of speckle pattern quality metrics as predictors of DIC measurement accuracy. To achieve this, six speckle patterns were considered: five were artificially generated in Python and the sixth was selected from the Digital Image Correlation Society's (iDICs) DIC challenge 2.0 image set. The patterns were analysed using the chosen metrics, and sub-pixel translation tests, similar to those conducted in [5], are then performed. Finally, the finite element (FE) method was used to generate ground truth models using the same patterns for comparison with DIC results.

3. Methodology

3.1. Digital image correlation software

For this study, the open-source DIC software, SUN-DIC, developed at Stellenbosch University [7], was used for all tests.

3.2. FE-based image deformation

The use of FE-methods to simulate specimen deformations for assessing DIC accuracy had been explored in previous studies [4, 8]. For this study, the

2023 version of Hexagon's MSC Apex software was used as the FE solver [9]. Image deformation based on an FE simulation was performed using Python libraries, including PyNastran (for importing FE data) and SciPy's ndimage (for remapping images). The deformation process began with the creation of a 2D model of the reference image, which was simulated in MSC Apex, generating two separate files: an input text file and a corresponding binary results file. These files were then imported into a Python environment, where the deformed mesh was constructed using the data. The mesh was subsequently interpolated using thin plate spline radial basis functions. The resulting deformation field was applied to the reference image, and a new image was created by remapping the image coordinates according to the displacements from the deformed FE mesh. The imposed FE mesh for this study was a 2000 by 500 mm 2D mesh of a rectangular plate, consisting of 30 mm quadrilateral elements. The rectangular specimen was simply supported on one end and displaced axially on the other by 1 mm. The selected mesh density ensured minimal DIC measurement errors associated with the interpolation of the FE displacement field were present. The image deformation process is illustrated in Figure 1.

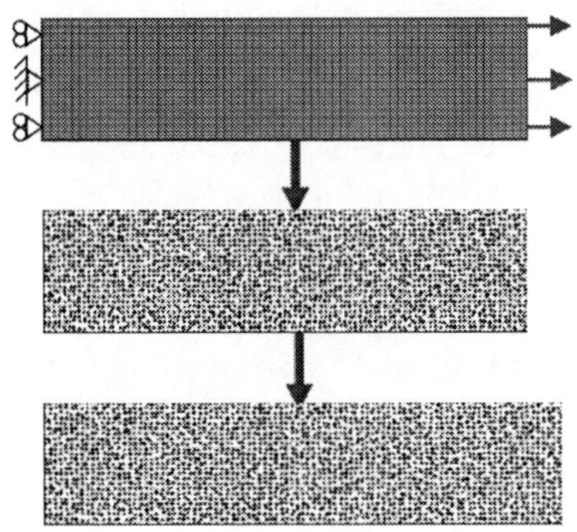

Fig. 1. Image deformation process flow. A deformed mesh is imposed on a reference image which results in a deformed image.

3.3. Sub-pixel translation tests

Sub-pixel translation tests were conducted to complement the FE-based technique. In this process, a series of translated images was generated by applying a translation motion to speckle pattern images. A set of six speckle patterns was considered, and for each pattern, the x-coordinates of the pixels were displaced by multiples of 0.05 pixels. This resulted in a sequence of images translated from 0 to 1 pixel. The purpose of the translation tests is to investigate the bias that results from the DIC process. It has been shown that the resulting bias depends on both the pixel interpolation method and the speckle pattern gradient [10]. The resulting errors were compared with the speckle pattern quality metrics.

4. Speckle pattern generation and evaluation

4.1. Pattern generation

A set of six noise-free speckle patterns was prepared for the study. The first five patterns were generated using a speckle pattern generator [11], while the sixth was a noiseless speckle pattern published by the iDICs as part of the DIC Challenge 2.0 image set [12]. This last pattern is included for performance comparison. All the patterns have image dimensions of 2000 by 500 pixels. The images are shown in Appendix A.1.

4.2. Pattern quality evaluation

To assess the quality of the speckle patterns used in this study, several well-known global assessment metrics from the literature were employed. These included the Mean Intensity Gradient (MIG), the Mean Subset Fluctuation (MSF), Shannon Entropy and the so-called E_f parameter. The following section provides a brief overview of each of these metrics.

The Mean Intensity Gradient (MIG), denoted by δf, was introduced by [2] as a global parameter for evaluating the overall quality of a speckle pattern. Both the mean bias error and the standard deviation of the measured displacements are influenced by the MIG of the speckle pattern. A speckle pattern with a high MIG typically results in smaller displacement measurement errors [2]. The MIG is defined as follows:

$$\delta_f = \sum_{i=1}^{W}\sum_{j=1}^{H}\frac{|\nabla f(x_i, y_i)|}{W \times H} \tag{1}$$

where W and H are the image width and height, $|\nabla f(x_i, y_i)|$ represents the magnitude of the local intensity gradient.

The mean subset fluctuation (MSF) parameter, introduced by [1], is another global pattern quality assessment metric. Shown in equation 2, the metric considers a pixel of interest, with the intensity $f(x_2, y_2)$, and its immediate neighbours as a subset, as shown in Figure 2.

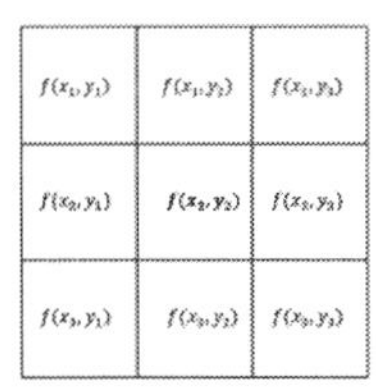

Fig. 2. Pixel of interest, bold, and surrounding pixels.

The subset fluctuation, S_p, is calculated as the sum of the differences between the intensity of each pixel $f(x_i, y_i)$, and the mean intensity of the subset, $\bar{f}$, as expressed in the following formulation:

$$S_p = \sum_{i=1}^{3}\sum_{j=1}^{3}|f(x_i, y_i) - \bar{f}| \quad (2)$$

The fluctuation values are aggregated across the image and normalised by image dimensions, W and H, to arrive at the global MSF value

$$S_f = \frac{\sum_{p\in S} S_p}{W \times H} \quad (3)$$

where S is the class of points in the image. Patterns with a higher MSF value, according to [1], tend to result in lower bias errors and thus more accurate DIC measurements.

Shannon entropy was first introduced as a pattern assessment metric in [13]. It quantifies the information content of an image, with each pixel representing a unit of information. For an 8-bit image, entropy is calculated as:

$$H(Y) = -\sum_{j=0}^{2^{\beta}-1} p(a_j)\log\left(p(a_j)\right) \quad (4)$$

In this equation, $p(a_j)$ represents the normalised probability of each grey level, a_j, which can be computed using the image's histogram. β is the pixel depth of the image. The original authors demonstrated that patterns with higher entropy values tended to exhibit smaller measurement errors.

The final metric was introduced by [14] as a global quality metric. The so-called E_f parameter considers both the first order and second order gradients of the speckle pattern's intensity values as shown by the following formulation:

$$k_x = \frac{\sum_{i=1}^{W}\sum_{j=1}^{H}|\nabla^2 f_x(x_i, y_j)\nabla f_x(x_i, y_j)|}{\sum_{i=1}^{W}\sum_{j=1}^{H}|\nabla f_x(x_i, y_j)|^2} \quad (5)$$

$$k_y = \frac{\sum_{i=1}^{W}\sum_{j=1}^{H}|\nabla^2 f_y(x_i, y_j)\nabla f_y(x_i, y_j)|}{\sum_{i=1}^{W}\sum_{j=1}^{H}|\nabla f_y(x_i, y_j)|^2} \quad (6)$$

In these equations, W and H are the image width and height, respectively, and $|\nabla^2 f_x(x_i, y_j)\nabla f_x(x_i, y_j)|$ is the magnitude of the product of the first order and second order intensity gradients of the speckle pattern. The components are combined to calculate the final global parameter

$$E_f = \sqrt{k_x^2 + k_y^2} \quad (7)$$

Each pattern was analysed using the metrics described above, and the respective values are presented in Table 1.

Table 1. Pattern quality metric values

Pattern	MIG	MSF	EF	Shannon
1	26.3	180	0.570	4.06
2	19.4	123	0.386	6.71
3	47.3*	330*	0.635*	5.17
4	33.1	292	0.563	5.59
5	34.9	229	0.554	6.81
6	21.5	139	0.632	7.13*

*Symbol indicates the metric value that implies the best pattern quality

5. Results and discussion

5.1. Sub-pixel translation test results

The results from the sub-pixel test are shown in Figure 3. This test applies rigid translations to each image and isolates bias error, representing the simplified conditions under which many speckle pattern quality metrics were originally developed [1, 13, 14]. For this reason, it is presented first to evaluate whether the metrics align with measurement accuracy under ideal conditions. Interestingly, the relationship between measurement accuracy and pattern quality appears inconsistent in this case. Patterns 2, 4, and 5, which exhibited the lowest bias errors, did not correspond with the expected best metric values. Likewise, patterns 1 and 3, which had significantly different metric values, performed similarly with large bias errors.

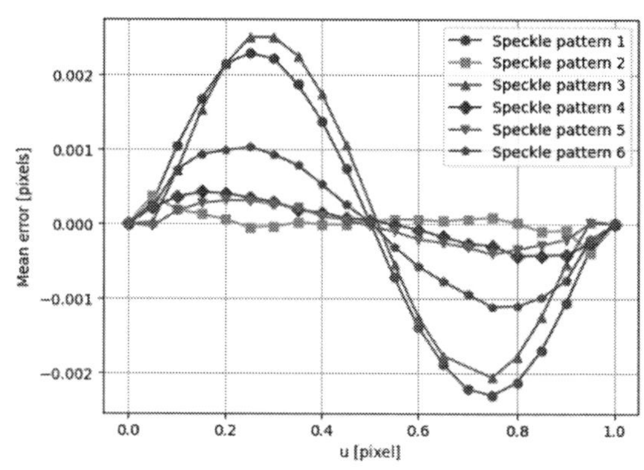

Fig. 4. Sub-pixel translation results showing bias error

5.2. Interpolation effects on sub-pixel test results

It is important to note that the SUN DIC interpolation method differs from typical spline-based approaches, using local Taylor expansions with finite difference derivatives instead. A cubic spline interpolation behaviour was assumed. This assumption was validated using the sinusoidal approximation of bias error, presented in [15], with a cubic spline interpolation function. The theoretical bias error results are shown in Figure 4. Although the SUN-DIC result had a smaller amplitude, the overall trend was consistent (except for pattern 4) which supports the assumption.

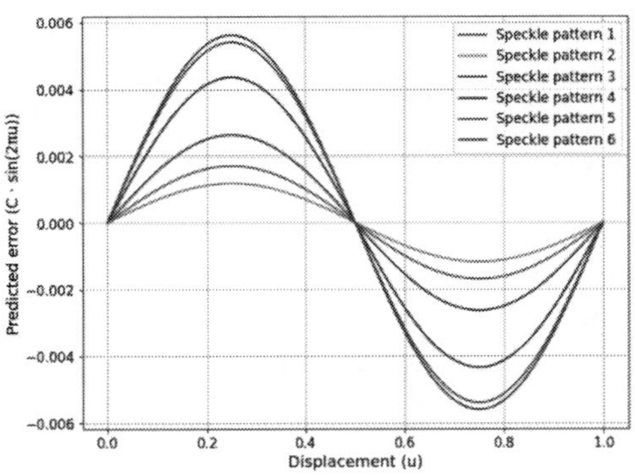

Fig. 4. Theoretical error predictions based on sinusoidal approximation presented in [15]

5.3. FE-based deformation results

Following the analysis of the speckle patterns and sub-pixel translation tests, the results from the DIC analysis of the patterns deformed using an FE mesh are shown by the histograms in Figure 5. The FE displacement fields were verified to be free of artifacts that could influence results. Thus, they are unlikely to impact the relative performance trends observed between speckle patterns. The histograms represent the distribution of total displacement errors, calculated by interpolating the FE-imposed displacements onto the DIC grid and subtracting them from the DIC-measured values. A narrower error distribution with higher peaks indicates better alignment between the measured results and the ground truth, reflecting higher accuracy. Unlike Figure 3, which isolates bias under rigid translation, Figure 5 captures both bias and random error under more realistic, non-rigid deformation. The patterns that had lower bias errors in the sub-pixel test also exhibited better performance in the FE-based deformation analysis. Notably, patterns 2, 4, and 5 performed particularly well, despite having significant variations in their assessment values across the different metrics, while patterns 1 and 3 showed poor performance, consistent with the results from the sub-pixel test.

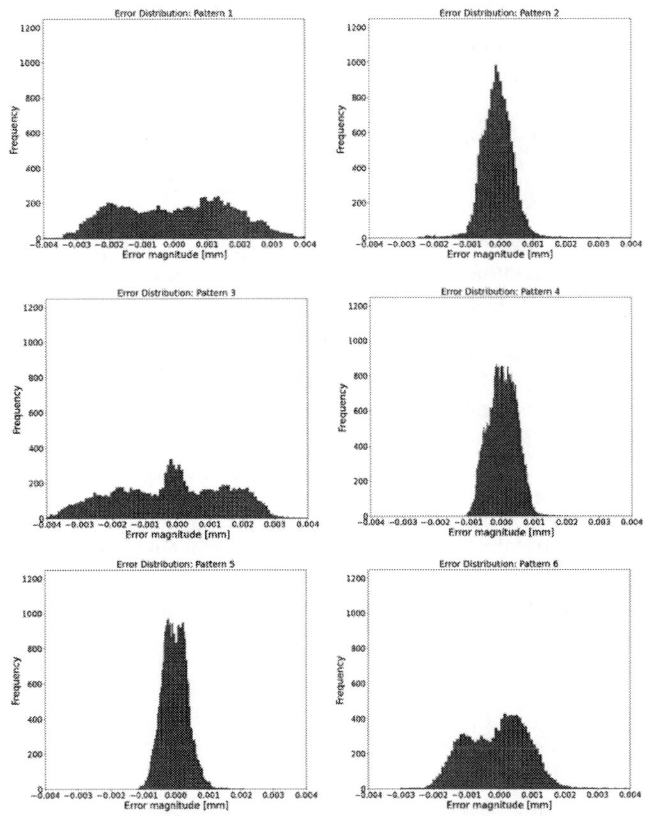

Fig. 5. Error histograms showing total error after FE-based deformation

6. Conclusion and future work

This study assesses the reliability of speckle pattern quality metrics as predictors of DIC measurement accuracy. The results suggest that some global assessment parameters may not reliably reflect actual DIC errors, as indicated by the lack of consistent trends across both rigid translation and FE-based deformation tests. Global quality metrics may not fully capture all aspects of a pattern's quality. As noted by [3], many of these metrics do not account for within-pattern variability and may overlook morphometric properties (such as speckle size, shape, and distribution).

Further research and new metrics may be needed to better understand the relationship between speckle patterns and error. While global metrics sometimes correlate with bias error, optimising patterns solely on this basis may be problematic due to the issues discussed earlier. Future work could focus on developing localised pattern analysis metrics to enable more targeted optimisation and reduce disparities.

References

[1] Hua T, Xie H, Wang S, Hu Z, Chen P, Zhang Q. Evaluation of the quality of a speckle pattern in the digital image correlation method by mean subset fluctuation. Optics & Laser Technology. 2011 Feb 1;43(1):9-13.

[2] Pan B, Qian K, Xie H, Asundi A. Two-dimensional digital image correlation for in-plane displacement and strain measurement: a review. Measurement science and technology. 2009 Apr 27;20(6):062001.

[3] Crammond G, Boyd SW, Dulieu-Barton JM. Speckle pattern quality assessment for digital image correlation. Optics and Lasers in Engineering. 2013 Dec 1;51(12):1368-78.

[4] Lecompte D, Smits AS, Bossuyt S, Sol H, Vantomme J, Van Hemelrijck D, Habraken AM. Quality assessment of speckle patterns for digital image correlation. Optics and lasers in Engineering. 2006 Nov 1;44(11):1132-45.

[5] Pan B, Lu Z, Xie H. Mean intensity gradient: an effective global parameter for quality assessment of the speckle patterns used in digital image correlation. Optics and Lasers in Engineering. 2010 Apr 1;48(4):469-77.

[6] Liu XY, Li RL, Zhao HW, Cheng TH, Cui GJ, Tan QC, Meng GW. Quality assessment of speckle patterns for digital image correlation by Shannon entropy. Optik. 2015 Dec 1;126(23):4206-11.

[7] Venter G. SUN-DIC [GitHub repository]. GitHub; 2024. Available from: https://github.com/gventer/SUN-DIC

[8] Lava P, Cooreman S, Coppieters S, De Strycker M, Debruyne D. Assessment of measuring errors in DIC using deformation fields generated by plastic FEA. Optics and Lasers in Engineering. 2009 Jul 1;47(7-8):747-53.

[9] Hexagon. *MSC Apex, Version 2023* [software]. Hexagon; 2024. Available from: https://hexagon.com/products/productgroups/computer-aided-engineering-software/msc-apex

[10] Wang YQ, Sutton MA, Bruck HA, Schreier HW. Quantitative error assessment in pattern matching: effects of intensity pattern noise, interpolation, strain and image contrast on motion measurements. Strain. 2009 Apr;45(2):160-78.

[11] Gorjup D. speckle_pattern [GitHub repository]. GitHub; 2024. Available from: https://github.com/ladisk/speckle_pattern

[12] Reu PL, Blaysat B, Andò E, Bhattacharya K, Couture C, Couty V, Deb D, Fayad SS, Iadicola MA, Jaminion S, Klein M. DIC challenge 2.0: developing images and guidelines for evaluating accuracy and resolution of 2D analyses: focus on the metrological efficiency indicator. Experimental Mechanics. 2022 Apr;62(4):639-54.

[13] Liu, X.Y., Li, R.L., Zhao, H.W., Cheng, T.H., Cui, G.J., Tan, Q.C. and Meng, G.W., 2015. *Quality assessment of speckle patterns for digital image correlation by Shannon entropy*. *Optik*, *126*(23), pp.4206-4211.

[14] Hu X, Xie Z, Liu F. Assessment of speckle pattern quality in digital image correlation from the perspective of mean bias error. Measurement. 2021 Mar 1;173:108618.

[15] Su Y, Zhang Q, Gao Z, Xu X, Wu X. Fourier-based interpolation bias prediction in digital image correlation. Optics express. 2015 Jul 16;23(15):19242-60.

Appendix A.

A.1. Speckle patterns

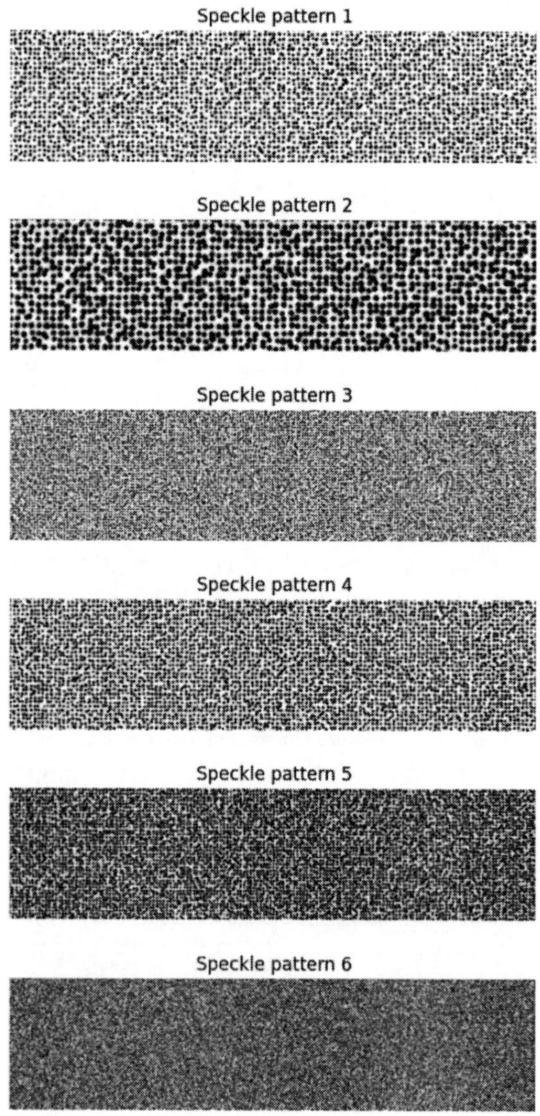

Fig. 5. Speckle pattern images used in study

Mechanical properties of epoxy resins under various loading conditions – Comparison between bio- and petroleum- based resins commercially available in South Africa

Ntokozo T. Sindane[a,*], Sherlyn Gabriel[a,b], James A. Dicks[a]

[a]*Centre of Materials (CME), University of Cape Town 217 Library Rd, Rondebosch, Cape Town, 7701*

[b]*Blast Impact Survivability Research Centre (BISRU) University of Cape Town, -0A Madiba Cir, Rondebosch, Cape Town, 7700*

**corresponding author email: SNDNTO007@myuct.ac.za*

Abstract

Fibre reinforced polymer matrix composites using an epoxy matrix are commonly used in a variety of industries, including within automotive and aerospace fields. Considering the rising environmental concerns and climate change, biobased epoxy resins that are partially synthesised from renewable resources are proposed to be a better alternative to petroleum-derived epoxy resins [1-3]. However, before designing composites using these biobased epoxy resins, the mechanical properties and failure mechanisms associated with their constituents must be experimentally investigated and understood to ensure sustainability without compromising structural integrity. In this study, the quasistatic mechanical properties of commercially available biobased epoxy resins, Sicomin InfuGreen 810 (38% biobased carbon content [4]) and Gurit BioPrime 37 (38% biobased carbon content [5]), were combined with appropriate hardeners and compared to the petroleum-derived epoxy resins, Gurit Prime 27 and Gurit Prime 37. Uniaxial tensile, uniaxial compression, 3-point bending flexural, and hardness tests were performed following ASTM standards. The mechanical properties suggested that BioPrime 37 could adequately replace its petroleum-derived counterparts, presenting an encouraging future for implementation within sustainable composite materials.

Keywords: biobased epoxy; fibre reinforced polymer matrix composites; mechanical properties

1. Introduction

Epoxy resins are widely used to develop high performance fibre reinforced polymer matrix (FRPM) composites due to their advantageous mechanical and chemical properties. Traditionally, epoxy resin monomers are petroleum-derived, however, as the demand for sustainable materials grow, more ecofriendly alternatives have become commercially available [1-3]. Most petroleum-derived epoxy resins are based on bisphenol derivatives (e.g., diglycidyl ether of bisphenol-A, DGEBA), which are functionalised with oxirane ring moieties [6]. Epoxies used in composite applications are typically copolymerised using multifunctional amine monomers (e.g., isophorone diamine, IPDA) to form thermosetting polymers via step-growth polymerisation [7]. Here, both the epoxy monomer and amine comonomer present opportunity for biobased constituents to be incorporated into their structure. Firstly, this has been achieved through the replacement of petroleum derived epichlorohydrin with biobased Epicerol® through the chlorination of glycerol [8, 9]. Secondly, biobased aliphatic diols (e.g., butanediol/hexanediol diglycidyl ether) and cycloaliphatic terpenes (e.g., α-pinene oxide) have also been incorporated for reactive diluent components [10]. Thirdly, the use of biobased components have also been incorporated either partially or fully into amine monomers [11]. Considering that fully biobased (i.e., completely derived from renewable resources) are not currently commercially achievable, a biobased carbon content >25% is generally accepted to classify a resin system as "biobased" [12]. For example, both Sicomin InfuGreen 810 and Gurit BioPrime 37 epoxy resins contain the same major constituents (DGEBA, butanediol diglycidyl ether, and the reaction product of 2-(chloromethyl)oxirane, formaldehyde, and phenol), whilst BioPrime 37 additionally contains a minor fraction (<10%) of C12–C14 alcohol glycidyl ethers [4, 5]. With regards to the curing agents, Sicomin SD8824 contains methylpentane diamine, *m*-xylenediamine, and amine-terminated polypropylene glycol trimethylolpropane ether [13]. On the other hand, Gurit Ampreg 3X STD contains IPDA, propylene glycol diamine, and a mixture of aliphatic amines derived from poly- and triethylene [14]. Whilst it is evident that these resin systems have incorporated biobased components into their formulations, one of the challenges faced by the development of these biobased epoxy resins is the ability to effectively substitute components of the monomers with renewable feedstocks without compromising mechanical performance [15,16]. This study therefore investigates the quasistatic mechanical properties of four epoxy resins, namely two bio-based epoxy resins (Sicomin InfuGreen 810, Gurit BioPrime

37) and two petroleum derived epoxy resins (Prime 37 and Prime 27), to ascertain whether the biobased epoxy resins display comparable mechanical properties for direct replacement of their petroleum derived counterparts for used within high performance composite applications. These resins are widely used in South African manufacturing industries, notably in vacuum infusion processes owing to their low viscosity, making them the most pragmatically viable options for replacement for key engineering industries within South Africa.

2. Methods

2.1. *Specimen Preparation*

For the preparation of all polymers, the manufacturing method outlined by Jost and Pardoen [17] and the manufacturers' guidelines were followed. Sicomin InfuGreen 810 was copolymerised with Sicomin SD8824, whilst all Gurit epoxy resins were copolymerised with Gurit Ampreg 3X STD using curing cycles based on the technical datasheets provided for each resin [4, 5, 18, 19].

Table 1. Copolymerisation of epoxy polymers.

Epoxy	Hardener	Ratio	Cure cycle
Prime 27	Ampreg 3X STD	100:29	RT 24 h, 50°C 5h
Prime 37	Ampreg 3X STD	100:29	RT 24 h, 50°C 5h
BioPrime 37	Ampreg 3X STD	100:29	RT 24 h, 50°C 5h
InfuGreen 810	Sicomin SD8824	100:22	RT 24 h, 50°C 5 h

The epoxy resins were mixed with their respective hardeners and mechanically stirred for 5 minutes to ensure a homogeneous blend. The resin was then placed in a vacuum chamber for 20 minutes to degas the mixture. Following degassing, the mixture was poured into silicon molds with dimensions specified by ASTM mechanical testing standards. The resins were allowed to cure for 24 hours at room temperature (RT) followed by an elevated temperature post-curing cycle specified by the manufacturer's technical datasheets (Table 1).

2.2. *Mechanical Testing*

Uniaxial tensile testing was performed in accordance with ASTM D638-22 using Type V specimen geometries on a universal testing machine (Zwick 1484, Zwick-Roell, Germany) equipped with a 10 kN loadcell. Testing was performed using a crosshead speed of 5 mm/min and displacement was measured using an integrated video extensometer (LIMESS Messtechnik und Software GmbH, Germany). At least 5 specimens were tested and reported as the arithmetic mean with standard error.

Uniaxial compression testing was performed in accordance with ASTM D695-23 on a universal testing machine (Instron 3365, Norwood, MA, USA) equipped with a 5 kN loadcell. Testing was performed using a crosshead speed of 1.3 mm/min. At least 5 specimens were tested and reported as the arithmetic mean with standard error.

3-point flexural testing was performed in accordance with ASTM D790-17 on a universal testing machine (Instron 3365, Norwood, MA, USA) using a 5 kN loadcell. Testing was performed using a crosshead speed of 1.3 mm/min. At least 5 specimens were tested and reported as the arithmetic mean with standard error.

Shore D hardness test type D was performed in accordance with ASTM D2240-15 using a digital durometer (Durotech M202, Durotech, Hampden Test Equipment Limited, Kettering, UK). At least 12 specimens were tested and reported as the arithmetic mean with standard error.

Statistical analysis was performed using IBM SPSS Statistics v.30. Results were checked for normality using Shapiro-Wilk test, followed by one-way analysis of variance and pot-hoc Tukey honestly significant difference, with statistical significance considered for $\alpha < 0.05$ and graphically annotated by an asterisk.

3. Results and discussion

3.1. *Uniaxial tensile properties*

All four epoxy polymers displayed uniaxial tensile behaviour typical of rigid thermosetting materials, as seen in the representative stress-strain curves in Figure 1, with the tensile properties summarised in Table 2.

Table 2. Tensile properties of epoxy polymers.

Resin	E_t (GPa)	σ_t (MPa)	ε_t (%)
Prime 27	2.79±0.12	83.7±2.8	5.4±0.4
Prime 37	3.01±0.08	85.7±1.5	4.9±0.3
BioPrime 37	2.88±0.15	81.6±4.0	4.8±0.3
InfuGreen 810	2.74±0.09	74.7±0.9	5.8±0.4

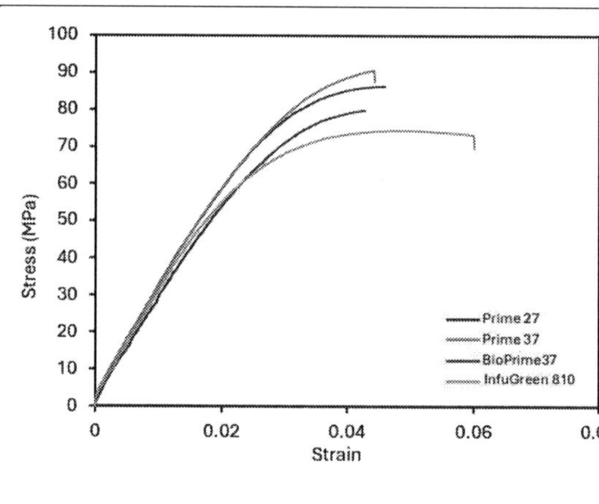

Figure 1: Representative tensile stress-strain curves of epoxy polymers.

The Young's modulus (E_t), ultimate tensile strength (σ_t), and elongation at break (ε_t) results for each polymer were compared for statistical differences. Firstly, no significant difference in E_t was apparent between any of the polymer comparisons. Secondly, only InfuGreen 810 displayed significantly lower σ_t compared to Prime 37 ($p = 0.001$), Prime 27 ($p = 0.014$), and BioPrime 37 ($p < 0.001$), whilst all other polymer comparisons of σ_t were not significantly different to one another (Figure 2). Thirdly, there was no significant difference in ε_t between any of the polymers.

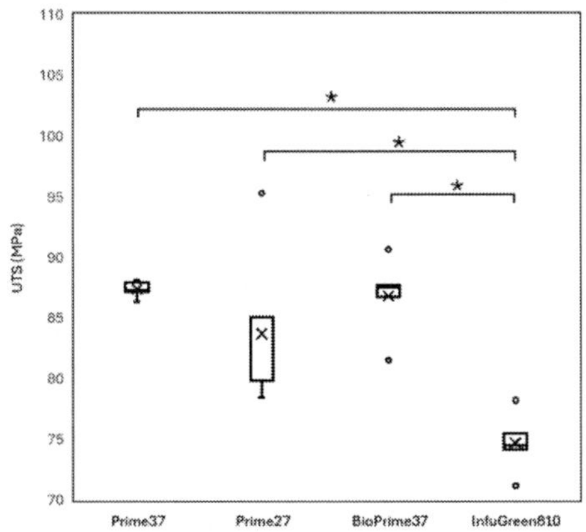

Figure 2: Ultimate tensile strength of epoxy polymers.

3.2. *Uniaxial compression properties*

The uniaxial compressive modulus (E_c) of the polymers was 1.65±0.015 GPa for Prime 37, 1.68±0.014 GPa for Prime 27, 1.68±0.022 GPa for BioPrime 37, and 1.48±0.025 GPa for InfuGreen 810. Of the polymers, only InfuGreen 810 displayed a significantly lower E_c compared to all other polymers ($p < 0.001$ in all cases, Figure 3), whilst all other comparisons between the polymers were not significantly different.

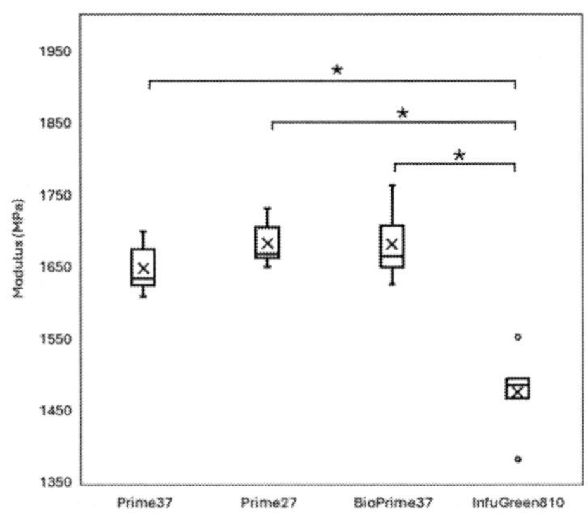

Figure 3: Compressive modulus of epoxy polymers.

3.3. *Three point bending properties*

The three-point flexural behaviour of the polymers generally supported those of the tensile properties, with representative stress-strain curves presented in Figure 4, and flexural properties summarised in Table 3.

Table 3. Flexural properties of epoxy polymers.

Resin	E_f (GPa)	σ_f (MPa)	ε_f (%)
Prime 27	2.55±0.13	94.5±0.13	8.5±0.9
Prime 37	2.58±0.08	97.0±0.11	8.3±0.04
BioPrime 37	2.90±0.55	100.6±0.1	5.4±0.2
InfuGreen 810	2.06±0.08	82.9±0.05	10.9±0.2

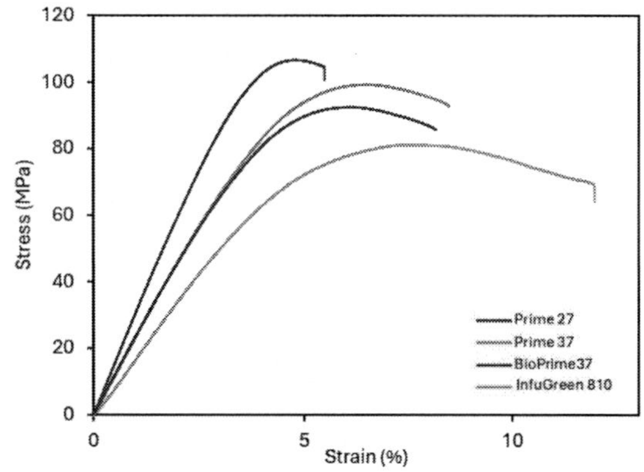

Figure 4: Representative flexural stress-strain curves of epoxy polymers.

The flexural modulus (E_f) did not display significant difference between the petroleum derived and biobased polymers, and there was only a significant difference between the BioPrime 37 and InfuGreen 810 ($p = 0.01$). Similarly, the only difference in flexural strength (σ_f) was between BioPrime 37 and InfuGreen 810 ($p = 0.014$), with no significant differences between the petroleum derived and biobased polymers. Lastly, the flexural elongation at break (ε_f) was significantly lower between BioPrime 37 and Prime 27 ($p = 0.042$), as well as InfuGreen 810 ($p < 0.001$), yet was not significantly different compared to Prime 37.

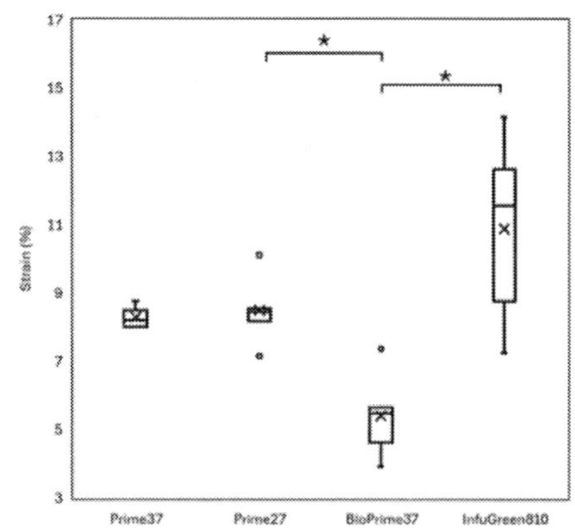
Figure 5: Flexural elongation at break of epoxy polymers.

3.4. *Shore D hardness*

The Shore D hardness of the polymers was 83±0.4 for Prime 37, 82±0.4 for Prime 27, 82±0.3 for BioPrime 37, and 81±0.3 for InfuGreen 810. Of the polymers, only InfuGreen 810 displayed a significantly lower Shore D hardness compared to Prime 37 ($p = 0.001$), whilst all other comparisons between the polymers were not significantly different.

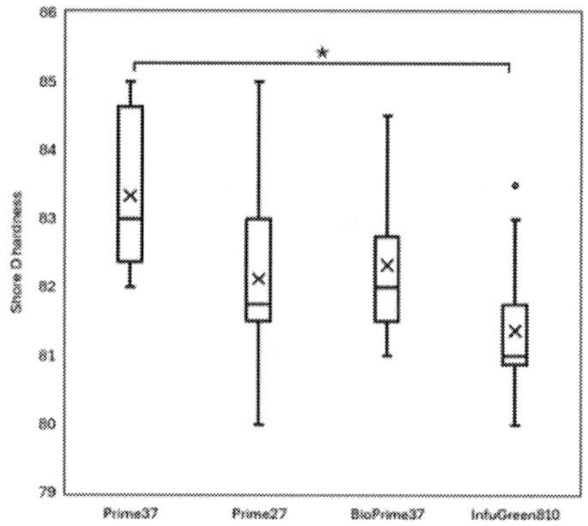

Figure 6: Shore D hardness of epoxy polymers.

4. Conclusions

As industries move towards sustainable materials, understanding the mechanical properties of bio-based alternatives assists in providing confidence to transfer over to these systems for drop-in replacement of petroleum derived systems. For this reason, we compared the quasistatic mechanical properties of two commonly used petroleum derived epoxy systems (Prime 37 and Prime 27) and two biobased epoxy system alternatives (BioPrime 37 and InfuGreen 810). From these results, there was no significant difference in the tensile, compressive, flexural, and hardness properties between either Prime 37 or Prime 27 with BioPrime 37, setting an initial precedent of feasibility to directly replace Prime 37 or Prime 27 with the biobased counterpart. On the other hand, significant differences were displayed between the petroleum derived polymers and InfuGreen 810, establishing that it would not be suitable for direct replacement. Given the promising results of BioPrime 37, it would now be of value to ascertain whether significant differences are observed within FRPM composites with regards to interfacial effects and efficacy in load transfer.

Acknowledgements

The authors wish to graciously acknowledge the Ada & Bertie Levenstein Bursary, Algaeholix THRIP Bursary, and the UCT Emerging Researchers Programme for their financial support.

References

[1] Kumar, S., Krishnan, S., Mohanty, S., Nayak, S. K. (2018). Synthesis and characterization of petroleum and biobased epoxy resins: a review. Polymer International, 67 (7), 815-839.

[2] Zhang, Y., Liu, X., Wan, M., Zhu, Y., Zhang, K. (2024), From Renewable biomass to bio-based epoxy monomers and bio-based epoxy curing agents: synthesis and performance. Polymer Degradation and Stability, 110988.

[3] Gonçalves, F. A., Santos, M., Cernadas, T., Ferreira, P., Alves, P. (2022). Advances in the development of biobased epoxy resins: insight into more sustainable materials and future applications. International Materials Reviews, 67 (2), 119-149.

[4] Sicomin Epoxy Systems, SR InfuGreen 810/SD 882X Green Epoxy Systems for Injection and Infusion, https://sicomin.com/datasheets/srInfuGreen810_sd882x.pdf (Accessed July 2023).

[5] Gurit, Prime 37 Bio Epoxy Infusion System, www.gurit.com/wp-content/uploads/bsk-pdfmanager/2024/06/PDS-PRIME-37-4-0124.pdf (Accessed 29 August 2024).

[6] May, C. (Ed.). (2018), Epoxy Resins: Chemistry and Technology. CRC Press.

[7] Jin, F.L., Li, X., Park, S.J. (2015). Synthesis and application of epoxy resins: A review. Journal of Industrial and Engineering Chemistry, 29, 1-11.

[8] Ondrey, G. (2006). A more-direct route to epichlorohydrin makes its commercial debut. Chem. Eng. 2006, 113 (3), 14-15.

[9] Ricciardi, M., Cespi, D., Celentano, M., Genga, A., Malitesta, C., Proto, A., Capacchione, C., Cucciniello, R. (2017). Biopropylene glycol as value-added product from Epicerol® process. Sustainable Chemistry and Pharmacy, 6, 10-13.

[10] Terry, J. S., Taylor, A. C. (2021). The properties and suitability of commercial bio-based epoxies for use in fiber-reinforced composites. Journal of Applied Polymer Science, 138 (20), 50417.

[11] Froidevaux, V., Negrell, C., Caillol, S., Pascault, J.-P., Boutevin, B. (2016). Biobased Amines: From Synthesis to Polymers, Present and Future. Chemical Reviews, 116 (22), 14181-14224.

[12] Daystar, J., Handfield, R., Golden, J. S., McConnell, E., Pascual-Gonzalez, J. (2021). An economic impact analysis of the US biobased products industry. Industrial Biotechnology, 17 (5), 259-270.

[13] Ampreg™ 3X low toxicity hardeners, https://gurit.com/wpcontent/uploads/2022/12/Ampreg-3X-Systems.pdf (Accessed 15 January 2025).

[14] SD 882X Infusion System, https://sicomin.com/datasheets/sr8100_sd882x.pdf (Accessed 15 January 2025)

[15] Gabriel, S., Langdon, G.S., von Klemperer, C.J., Kim Yuen, S.C. (2022). Blast behaviour of fibre reinforced polymers containing sustainable constituents. Journal of Reinforced Plastics and Composites, 41(19-20), 771-90.

[16] von Klemperer, CJ, Langdon, GS, Gabriel, S, Kim Yuen, SC., (2019), Blast response of panels containing sustainable materials, ICCM22 2019. Melbourne, VIC: Engineers Australia.
4723-33.

[17] Jost, N. and Pardoen, T. (2020). Mechanical properties of a biosourced epoxy resin for composite applications. Masters thesis, Université Catholique de Louvain.

[18] Gurit Prime 27 Epoxy Infusion System, https://www.gurit.com/wp-content/uploads/2022/12/prime-27.pdf (Accessed 15 January 2025).

[19] Prime 37 Low toxicity Epoxy Infusion System, https://www.gurit.com/wp-content/uploads/bsk-pdfhttps://www.gurit.com/wp-content/uploads/bsk-pdf-manager/2022/08/PRIME-37-1.pdfmanager/2022/08/PRIME-37-1.pdf (Accessed 15 January 2025)

Evaluation of the relationships between measured archery bow performance and archer sensations

Francesco Pietra[a,*]

[a]*School of Mechanical, Industrial and Aeronautical Engineering, University of the Witwatersrand*
1 Jan Smuts Ave, 2001 Braamfontein, Johannesburg
**corresponding author email: Francesco.Pietra@wits.ac.za*

Abstract

Archery is a precision sport in which the athlete is required to hit the centre of the target with accuracy and consistency. In archery, the interaction between the archer and the bow is of fundamental importance to guarantee performance. Unfortunately, there is no literature to support archers in selecting their bows. It is not always clear if a new designs improve performance, due to the complex interaction between the archer and the bow, and it is not simple to measure if the archer-bow combination is better. In this paper, a mixed research methodology is presented to evaluate the relationships between the performance, the design and the athletes' sensations in using archery equipment. The proposed methodology is divided into four steps: selection of the design details to analyse; definition of a set of measurable quantities to evaluate change in performance due to change in design (objective metrics); definition of a set of metrics related to the archer experience with the various designs (subjective metrics); and analysis of the correlations between the defined metrics. The methodology was applied in three investigations, which were published as journal articles. The first paper explored the best distribution of stabilising weights for the barebow style, the second paper studied the effect of the introduction of damping on the shooting experience of the archers, and the third paper analysed the problem of the performance of different sets of archery limbs from objective and subjective points of view. In each paper, the methodology demonstrated its capability of analysing complex problems that involve design, performance and athletes' sensations in using sports equipment. The study of the correlations between objective and subjective metrics led to a deep understanding of the analysed problem, suggesting less-than-obvious relationships between metrics. The methodology utilised in this paper may be applied to investigating other topics in archery and can be adapted to other sports where there is a necessity to understand how a change in the design of sports equipment can impact athletes' performance and athletes' perception of the sports equipment.

Keywords

Mixed research methodology, sportive equipment, performance, sensations.

1. Background and Motivation

The problem of evaluating performance in sports and sports equipment is of great interest: athletes are continuously trying to evaluate and improve their performance, while sports brands are promising an increase in performance with the adoption of the latest technology. It is often not clear if the new technologies are capable of improving performance, or if the promise of better performance is simply a marketing strategy to sell new sports equipment. In addition to the problem of evaluating performance, there is the problem of how the change in design is perceived by the athletes. The relationship between perception/emotions and actual performance is complex. In some sports an increase in the performance of sports equipment corresponds directly with an increase in athletes' performance (for example cycling) while, in other sports, this is not the case: the athletes might not be capable of using the new design to its best, or the feeling of the new design might not be right, reducing athletes' performance.

In this paper, the study of the relationship between sports equipment performance and the athlete's perception of this performance will be presented. The problem will be analysed using a mixed research methodology [1, 2], evaluating the effect of a new design on the performance of sports equipment (quantitative study), and evaluating if these changes in performance are perceived by the athletes and how the athletes rate this new design (qualitative study). From the study of the relationships between sports equipment performance and athletes' perceptions, a better understanding of the impact of changes in sports equipment design is expected.

The sport selected for this investigation is archery. Archery is a precision sport in which the athlete is required to hit the centre of the target with accuracy and consistency, and where the interaction between the archer and the bow is of fundamental importance to guarantee performance. Moreover, the modern bow is highly customisable, allowing for adjustments tailored to the specific requirements of the archer. The tuning phase could be based on performance: the standard approach suggested in the archery manuals is to try something different on your bow (a new design feature), score thirty arrows with the new configuration and check if the score and the grouping are better than with the previous configuration [3]. This quantitative approach has been contested by Park as the score of thirty arrows is not statistically significant enough to decide upon the new configuration [4, 5]. The other common approach is the qualitative approach, a typical 'club' approach. The archer is requested to try the new configuration and if it 'feels good', then adopt it as it is better than the previous one (coach Dean McHendry, Archers of the Zoo Lake, Johannesburg). Since it cannot be quickly established that a change in equipment improves precision, the archer's perceptions become the most important criterion in the selection of equipment. Understanding the relationships between archery equipment design and archers' perceptions is therefore fundamental to the development of new designs focused on increasing precision and athletes' compatibility with the archery equipment.

1.1. Modern archery

Modern Western archery can be dated back to the 1780s when the English longbow enjoyed a renaissance among the English aristocracy. Historical archery became a fashionable recreation for both men and women and a cultural way to connect with English traditions. Later, during 1819–1845, longbow archery developed into a mass movement, as it spread initially into the English middle class [6, 7], and then into the United States of America (USA). The first scientific study about archery is dated 1937 by C. N. Hickman [8] and, successively, by P. E. Klopsteg in 1943 [9] who demonstrated that the longbow D-shaped limb cross-section is not the best from the stress distribution point of view, starting the process that created the modern bow. Klopsteg was also the first to solve the Archer's Paradox using speed-flash photography to describe the oscillatory behaviour of the arrow upon shooting. In 1969, the first scientific description of the working recurve bow by Schuster [10] was published. The name *working recurve bow* refers to the classification of the modern Olympic recurve bow: the limbs are initially curved toward the archer and then curved again away from the archer (re-curve, Figure 1), and the tip of the limbs is flexible (working) in contrast with the rigid tips of the composite bows.

The modern Olympic recurve bow design is a hybrid between the English longbow and the traditional composite bow (Turkish bow): while the tips of the limbs resemble the traditional composite bow being initially bent in the opposite direction of the draw, the recurve bow is longer than the traditional composite bow and similar in size to the English longbow [11].

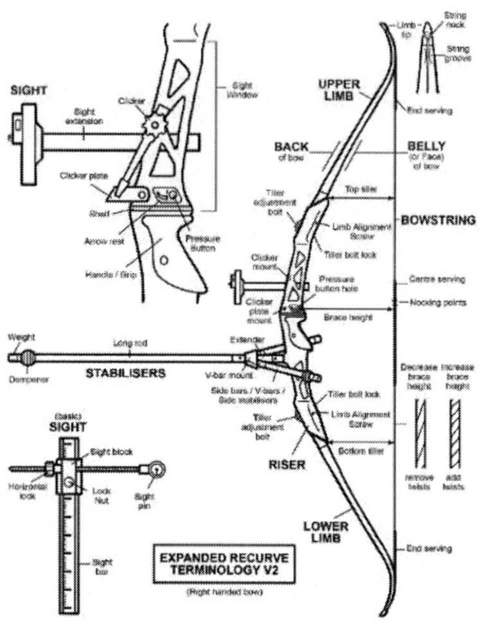

Fig. 1. Bow structure in the recurve style [12].

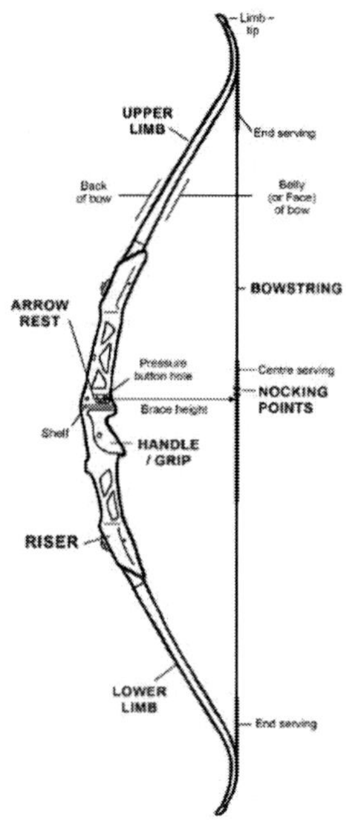

Fig. 2. Bow structure in the barebow style [12].

The World Archery Federation (WA, also formerly known as FITA from the French Fédération Internationale de Tir à l'Arc) is the governing body of the sport of archery. WA is responsible for the Olympic and Paralympic sport of archery, and for organising international events, including a world championship in archery [13]. WA recognises three main categories of bow: the recurve bow, which is the style of bow used at the Olympic Games; the compound bow; and the barebow [14]. The barebow (Figure 2) and the recurve bow (Figure 1) use the same type of bow, divided into a rigid centre called the *riser*, two flexible parts called *limbs* and a *string* to connect the ends of the limbs. The riser is made of aluminium alloy or composite materials; the limbs are made of a combination of wood, glass and carbon fibres. In the recurve style, the archer is allowed to use a stabilisation system, a sight and a clicker. In the barebow style, sight and clicker are not allowed, and the stabilisation system is limited to weights attached to the riser [14].

2. Objective and Scope

As previously discussed, both the qualitative and the quantitative approaches to the study of performance in archery exhibit certain limitations. The historical approach to the problem was a score-based method to assess improvements resulting from modifications to the bow configuration. This quantitative approach is outlined in various manuals [3, 15, 16, 17] and is normally accepted within the archery community as a standard practice. The approach suggests shooting thirty arrows, using the score as an indicator of performance differences between the tested configurations. The problem with this approach is the number of arrows necessary to achieve a score with sufficient statistical significance to differentiate between bow configurations. As demonstrated by Park in [4, 5], even for an Olympic-level archer, evaluating performance changes due to modifications in the bow requires approximately 100 ends (600 arrows) to achieve sufficient statistical significance. Nevertheless, the quantitative methodology can serve as a viable approach to studying archery performance, provided that an appropriate and carefully selected set of variables is used.

The qualitative approach is commonly used by clubs to evaluate different bow configurations. In this method, the archer shoots a limited number of arrows and provides subjective feedback on whether the new configuration "feels better" than the previous one. If the archer perceives an improvement, they are encouraged to adopt the new setup, based on the assumption that a more positive sensation with the bow will translate into better performance (coach Dean McHendry, Archers of the Zoo Lake, Johannesburg). This approach has been extensively tested in the field by the author (coach Francesco Pietra, Archers of the Zoo Lake, Johannesburg) and has proven quite effective, particularly with novice and beginner archers. However, its primary limitation is that archers unconsciously integrate various sensory inputs to form an overall impression of the shot, making it difficult to identify specific feelings or sensations. Before these studies, it was unclear what specific feelings should be targeted—what exactly an archer perceives during arrow release—and whether there is a measurable correlation between these sensations and actual performance, or if archers can even perceive differences attributable to the design of the bow.

Considering the strengths and limitations of both qualitative and quantitative research methodologies, it was decided to study the problem with a mixed research methodology, aiming to harness the advantages of both and achieve a more comprehensive understanding of the problem. This paper will present the results of the use of a mixed research methodology to evaluate the relationship between performance (engineering measurements, quantitative studies), equipment design and athlete sensation in the sport of archery (qualitative studies).

3. Methodology

Mixed methods research involves collecting quantitative and qualitative data and combining or integrating the data to yield insights or inferences from the combined data. Qualitative data typically involves open-ended responses that allow participants to share their perspectives, providing detailed insights into the problem. In contrast, quantitative data usually comprises data from tests or closed-ended responses, such as those found on questionnaires with predefined options, enabling numerical analysis and statistical testing [1, 2].

For the quantitative methods, a particular design feature of the bow is considered for testing. Subsequently, an engineering approach is developed which considers the measurable effects of the different configurations to test the effect of that design feature on performance. Objective metrics (measurable quantities) are defined, and tests are performed to quantify the change in performance relative to the various configurations. For the qualitative methods, an archer-centred approach is developed to evaluate the impact of the tested configuration on the archers' shooting experience. The subjective metrics (archers' experience-related quantities) are defined to quantify the sensations of the archer using the tested configurations. The subjective metrics are collected by interviewing the archers and asking closed-ended questions. The answers are collected with the use of a questionnaire tailored to the specific study. The subjective metrics are then coded [18, 19] and compared to the objective metrics (the integration phase of the methodology [1, 2]), evaluating the Pearson correlation coefficient r and the coefficient of determination $V = r^2$ between the defined metrics (Figure 3). The study of the correlation between objective and subjective metrics (metainferences [1, 2]) gives insight into the problem by seeking relationships between what is measured (objective metrics) and what is perceived (subjective metrics) by the archers.

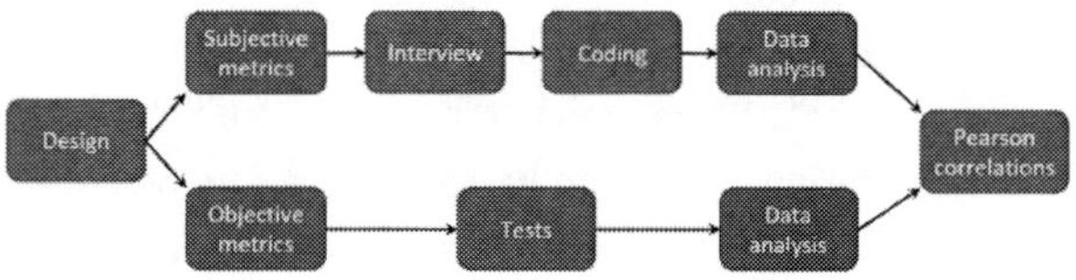

Fig. 3. Graphical representation of the mixed research methodology followed in this study.

Various research designs were used to evaluate the best approach to the problem. In the first paper [20], an exploratory sequential mixed methods design was used, starting the research with a qualitative research phase exploring the views of participants. The data were then analysed, and the information was used to build a second, quantitative phase. In the second paper [21], an explanatory sequential mixed methods design was attempted. With this approach, the researcher first conducts quantitative research, analyses the results, and then builds on the results to explain them in more detail with qualitative research. It is considered explanatory because the initial quantitative data results are explained further with the qualitative data. In the third paper [22], a convergent mixed methods design was used. With this approach, the researcher converges or merges quantitative and qualitative data to provide a comprehensive analysis of the research problem. In this design, the researcher typically collects both forms of data at roughly the same time and then integrates the information in the interpretation of the overall results [1, 2].

A longitudinal approach [1, 2] was also considered, but eventually discarded as not pertinent to this work. Allowing the archer to get used to the new configuration of the bow is necessary for improving the archer's score [23, 24]. However, this acclimatisation process is not necessary when assessing sensations related to a new design. Based on the author's experience as both an archer and coach, just a few shots are sufficient to evaluate the characteristics and feel of a new design, aligning with the club's approach to assessing performance through subjective perception.

4. Key Results and Discussion

The first case study [20] analysed the problem of the best weight distribution for barebow archery style. The problem was complex due to the vague definition of 'best' (best as considering archery accuracy? best for the archer?) and the absence of any reference literature on the subject. The exploratory mixed methods design employed in the first study proved to be highly effective in reducing the amount of qualitative data considered in the study. It allowed for the immediate identification of the most preferred weight configurations based on the archers' feedback, while the quantitative data was primarily used to investigate the underlying reasons behind these choices.

The second case study [21] analysed the effect of the introduction of damping in the barebow archery style. While the measurements of the effects of damping were complex from the testing point of view but straightforward in defining the objective metrics, the problem was to evaluate if the archer could feel any difference for the various configurations and to what extent the damping changed their experience of shooting. The second study employed the explanatory mixed methods design. This approach is particularly effective for guiding the selection of qualitative data, as collecting quantitative data first naturally informs and shapes the subsequent qualitative data collection. A potential risk of this approach is that the quantitative data are typically limited to the capabilities of the available testing equipment, which may lead to the omission of some essential qualitative metrics. However, this was not an issue in the second study, as it built upon the first. The subjective metrics in the second study were informed and refined based on the findings from the initial research, ensuring a comprehensive assessment.

The third case study [22] analysed the effect on the archer of different manufacturing technologies for the production of archery limbs. The problem was complex due to the variability in the limbs (poundage, manufacturing technology and connection system) and the absence of any reference literature on the subject. The third study effectively employed the convergent mixed methods design, which allowed for the simultaneous analysis of multiple variables and facilitated the identification of meaningful correlations between objective and subjective metrics. In the absence of a predefined sequence in selecting these metrics (whether quantitative to qualitative or vice versa), it is the researcher's responsibility to ensure that a comprehensive set of metrics is included to thoroughly investigate the problem. Without a deep understanding of the problem, there is a risk of overlooking fundamental metrics that are essential for the validity and completeness of the research.

The three published papers document the evolution of the application of the mixed research methodology to the sport of archery. The applied methodology has proven to be a powerful research tool for understanding complex problems involving design, equipment performance and athletes' perception in the sport of archery. Although this methodology allows significant freedom in the definition of the previous quantities, the following procedure is suggested:

a) Definition of the problem
In this phase, the researcher decides the design feature of interest and determines the various configurations that will be considered in the study. A basic understanding of the problem is required to define meaningful research questions, but it has been demonstrated that different types of problems can be analysed, from broadly defined questions ('What is the best weight distribution?'), to specific questions ('What is the effect of damping?'), to the evaluation of the impact of manufacturing technologies (limb analysis).

b) Definition of the research design approach
In this phase, the researcher decides which design approach to use in the research: an exploratory, an explanatory, or a convergent mixed methods design. All three methods were used with success to analyse the proposed problem. The researcher also needs to decide if a longitudinal method is necessary to better follow the possible temporal evolution of the problem.

c) Definition of objective metrics
In this phase, the researcher decides what can be measured and how it will be measured. This phase is normally constrained by the instruments available to the researcher, but the broader the spectrum of data available, the better the understanding of the problem.

d) Definition of subjective metrics
In this phase, the researcher decides what questions will be asked of the athletes. The questionnaire is prepared. This is, in the experience of the author, the most critical phase of the entire procedure. The questions have to be designed to help the athlete in the process of quantifying a sensation and distinguishing different aspects of a sensation. Humans are quite good at averaging different inputs, so creating a questionnaire that asks the right questions can be challenging. It is suggested that a good understanding of the analysed sport is necessary. It is also suggested to carefully plan the data acquisition to minimise the effect of external and confounding variables (for example, weather and environmental conditions).

e) Data analysis and study of the correlations
In this phase, the researcher analyses the data relative to the objective metrics and codes the subjective metrics to allow the study of the correlations between metrics. Also in this phase, a good understanding of the analysed sport is necessary to understand the physical meaning of the correlations between subjective and objective metrics and to gain a deeper understanding of the problem.

Expanding on the fourth phase of the procedure, the definition of the subjective metrics is based on the knowledge of the researcher about the analysed sport: what can be sensed with a sufficient resolution is an experience that comes from practising the sport. If the researcher is not experienced in the sport, then it will be necessary to collaborate with an expert athlete to define a meaningful set of subjective metrics. The same problem is encountered in the creation of the questionnaire. The researcher needs to know the sport-specific terms to formulate questions that have the same meaning for all athletes. Questions formulated using a language non-specific to the sport will normally cause misunderstanding and, possibly, random answers from the athletes.

The applied methodology has some limitations: it is not possible to study the correlations between effects that cannot be measured or sensed. It is also not possible to study the correlations between effects that cannot be sensed with sufficient resolution by the athletes, or that cannot be separated from other sensations. All three studies have a very small statistical population, but this is typical of research based on athletes. Moreover, it is very complicated to guarantee a homogeneous population: athletes do not have the same level of performance but, more importantly, they do not have the same level of sensitivity. If it is relatively simple to guarantee a common level of performance (for example defining a common level of expertise), it is more complicated to guarantee a common level of sensitivity. As reported in the three studies, the variability of the replies among the archers is sometimes significant, meaning that the archers had different levels of sensitivity and different ways of interpreting the same sensation. Both problems, small statistical population and variability in the answers are unavoidable and accepted by the research community as normal in this type of research.

6. Conclusion and Future Work

This study was successful in using a mixed research methodology to evaluate the relationships between performance, design details and user sensations in archery equipment. The three case studies successfully analyse three different aspects of the sport of archery and report the evolution of the application of the methodology. In each of the presented case studies, the applied methodology was capable of providing a deep understanding of the problem, revealing correlations between metrics that were not immediately apparent.

The three case studies are just the first steps in studying the sport of archery. The applied methodology is perfectly suited to studying the riser (the central, rigid part of the bow) and how a change in the design can affect the archer. This problem is particularly complex due to the complexity of the riser and the multitude of design details that could make a difference. Arrows are another interesting field of research due to the complexity of the arrow movements in flight. New technologies and materials are continuously developed and offered to the archery community without any technical explanation. The applied methodology is again perfectly suited to evaluate the impact of these new technologies on the archers and their search for performance.

As a possible development of these studies, more sports could be considered to evaluate if the mixed research methodology could be used as the reference methodological framework to analyse the effects of new designs on athletes' performance. Also, the three archery studies raised some very interesting questions about what athletes can sense in practising their sport and, more generally, what a person can sense and how they sense it. Research focused on these questions would complement the presented studies and offer new points of view for future developments.

References

[1] P. D. Leedy and J. E. Ormrod, Practical Research, Planning and Design, New York: Pearson Education, Inc. ISBN 13: 978-0-13-477565-4, 2019.

[2] J. C. Creswell and J. D. Creswell, Research Design - Qualitative, Quantitative and Mixed Methods Approaches, Thousand Oaks, California: SAGE Publications, Inc., 2023.

[3] Easton, "ARROW TUNING AND MAINTENANCE GUIDE," 04 1999. [Online]. Available: http://eastonarchery.com/wp-content/uploads/2019/08/TuningGuideEaston.pdf. [Accessed 15 07 2025].

[4] J. L. Park, "Modelling archers' scores at different distances to quantify score loss due to equipment selection and technique errors," Proc IMechE Part P: J Sports Engineering and Technology, vol. 228(4), no. DOI: 10.1177/1754337114539308., p. 250–258, 2014.

[5] J. L. Park, "Using group sizes to optimise archery equipment," Proc IMechE Part P: J Sports Engineering and Technology, vol. 1–8, no. DOI: 10.1177/17543371211053716., 2021.

[6] H. Arnstad, "The Amazon Archers of England. Longbows, Gender and English Nationalism 1780-1845," 2019. [Online]. Available: https://www.academia.edu/39147585/The_Amazon_Archers_of_England_Longbows_gender_and_English_nationalism_1780_1845. [Accessed 15 07 2025].

[7] M. Johnes, Archery, Romance and Elite Culture in England and Wales, c. 1780-1840, Oxford OX4 2DQ, UK: Blackwell Publishing Ltd., 2004.

[8] C. Hickman, "The dynamics of a bow and arrow," Journal of Applied Physics, vol. 8, no. DOI:10.1063/1.1710314, p. 404, June 1937.

[9] P. Klopsteg, "Physics of Bows and Arrows," American Journal of Physics, vol. 11, no. DOI:10.1119/1.1990474, p. 175, August 1943.

[10] B. Schuster, "Ballistic of the Modern-Working Recurve Bow and Arrow," American Journal of Physics, vol. 37(4), no. DOI:10.1119/1.1975578, pp. 364-373, 1969.

[11] H. O. Meyer, "Applications of Physics to Archery," 6 11 2015. [Online]. Available: DOI:10.48550/arXiv.1511.02250. [Accessed 15 07 2025].

[12] J. Emmett, "Archery Beginners Recurve Diagrams," 2000-2023 Jessica Emmett, 28 08 2024. [Online]. Available: https://www.jessica-emmett.com/blog/archery-beginners-recurve-diagrams/. [Accessed 15 07 2025].

[13] World Archery Federation, "About us," World Archery Federation, 2025. [Online]. Available: https://www.worldarchery.sport/about-us. [Accessed 15 07 2025].

[14] World Archery Federation, "CHAPTER 11 ATHLETES EQUIPMENT," 2025. [Online]. Available: https://www.worldarchery.sport/rulebook/article/793. [Accessed 15 07 2025].

[15] R. McKinney, The simple art of winning, Tokyo: Leo Planning Inc., 1996.

[16] H. T. Kim, Kim, Hyung Tak Archery, 175 Seo bu li Goe san ub Goe san Gun Chung Buk: Crapas, 2015.

[17] J. Kaminski, Tuning for performance: a comprehensive Olympic recurve archery tuning guide, USA: Independently Published, 2020.

[18] F. Pietra, P. Loveday and R. Reid, "Analysis of preferred weight configuration for barebow shooting style," Proceedings of the Institution of Mechanical Engineers, Part P: Journal of Sports Engineering and Technology, vol. 0, no. DOI:10.1177/17543371231155510, p. 0, 2023.

[19] S. J. Tracy, Qualitative Research Methods - Collecting evidence, crafting analysis, communicating impact, Hoboken, USA: John Wiley and Sons, Inc., 2020.

[20] N. K. Denzin, Y. S. Lincoln, M. D. Giardina and G. S. Cannella, The SAGE Handbook of Qualitative Research, USA: SAGE Publications, Inc., 2024.

[21] F. Pietra, P. Loveday and R. Reid, "Objective and subjective measurements of the influence of dampers on the shooting experience of barebow archers," Proceedings of the Institution of Mechanical Engineers, Part P: Journal of Sports Engineering and Technology, vol. 0, no. DOI:10.1177/17543371231213420, p. 0, 2024.

[22] F. Pietra, P. Loveday and R. Reid, "Investigation into the correlations between engineering measurements and archers' perceptions of different limbs in a recurve bow," Proceedings of the Institution of Mechanical Engineers, Part P: Journal of Sports Engineering and Technology, vol. 0, no. DOI:10.1177/17543371241258542, p. 0, 2024.

[23] K. M. Haywood and C. F. Lewis, Archery - Steps to Success, Champaign, IL 61825-5076: Human Kinetics, 2014.

[24] K. Lee and T. Benner, Total Archery - Inside the Archer, Chula Vista, CA 91915: Astra LLC, 2009.

Analysing the effects of cyclic loading on corrugated paperboard boxes

Tiaan Cronjé [a,*], Corné J. Coetzee [a]

[a] *Department of Mechanical and Mechatronic Engineering, Stellenbosch University, Joubert Street, Stellenbosch 7600, Republic of South Africa*

* *Correspondence: 22625224@sun.ac.za*

Abstract

Corrugated paperboard packaging, which is commonly used to store and transport horticultural products, is susceptible to failure due to cyclic loading experienced during transportation [1] and relatively humid conditions commonly found in cold storage [2]. In this study, field tests were conducted, with a load measuring device, on boxes used to export Cape flora to determine the loads and accelerations a palletised stack of boxes would typically experience during transportation on a road vehicle. This data was used to set up repeatable and reliable acceleration tests on a PLAT (pallet lateral acceleration tester); a machine that simulates the lateral motion of an accelerating road vehicle on a palletised stack of boxes [3]. Using motion tracking and BCT's (box compression tests), boxes from these tests were analysed to determine how pallet orientation, humidity, shock loading, and cyclic loading may affect the stability of a stack and the strength of a box. Additionally, BCT's were conducted on undamaged standard and conditioned boxes to determine the effects transportation and/or cold storage may have on their structural integrity. High humidity and prolonged cyclic loading independently reduced the strength of a box, but a combination of these conditions improved it. Cyclic-BCT's, a modified version of BCT's where the compressive load cycles between minimum and maximum load values for a set number of cycles, were conducted on the boxes using a combination of BCT-standards and load data extracted from the acceleration tests. These tests gave more detail on what happens to the strength of boxes during repetitive acceleration, especially at lower cycles. At lower cycles a box quickly loses strength and regains it after some time, making boxes more susceptible to failure at lower cycles due to shock loads.

Keywords: corrugated paperboard boxes, cyclic loading

1. Introduction

An important part of the farm-to-consumer supply chain is the packaging of fresh produce, due to the protection it proves against external physical damage and spoilage. Different types of materials are used for this purpose, of which corrugated paperboard is the most common. This is because of its good strength-to-weight ratio, abundance, and high recyclability. However, this material is susceptible to factors such as static and dynamic loads introduced during the transport of the product, as well as relatively humid environments found in cold storage rooms. A reduction in the material's structural integrity leads to the failure of the packaging, damaging the product in the process. [1, 2, 4]

Cape Flora SA makes use of palletised S22 boxes (Figure 1a) to export Cape flora products from South Africa. The flowers need to be kept in cold storage between 2 °C and 4 °C whilst being transported by road vehicle and sea or air freight [6]. The combination of dynamic loads and humid environments causes about 5 % to 13 % of the boxes to fail during transport (Figure 1b), rendering the product unsuitable for the market [5].

(a) (b)

Figure 1: (a) Palletised stack of S22 boxes and (b) failure of the bottom layer boxes [5]

This study aims at determining the effects of transport and storage conditions on the strength of boxes by: determining the loads and accelerations experienced by a palletised stack of boxes, transported by a road vehicle, using a load measuring device; replicating the motion of the vehicle in a controlled environment using a PLAT (pallet lateral acceleration tester) and determining the effects of cyclic loading on the box's structural integrity using BCT's (box compression test); using C-BCT's (cyclic box compression test) to simulate the dynamic loads applied to the boxes during transportation at low cycle counts.

2. Materials and methodology

2.1. Box samples

The S22 corrugated paperboard box (Figure 2) was the focus of this study. It featured a telescopic design with a tray (200/160C/200 g/m^2) and lid (175/155B/175 g/m^2), and was of external dimensions 1002×390×207 mm. The internal structure of the tray consisted of double panels on the longer sides and tri-corners for additional vertical strength. It should be noted that the lid was not the same height as the tray, causing the bottom 67 mm of the tray to be unsupported. [7]

For the purposes of this study, the boxes were produced in a factory and folded by the same person to minimise variance. All the samples were conditioned in accordance with the BS EN ISO 187:2022 standard [8] for 48 h at either near-standard ($T_{db} = 22.5$ °C, $\phi = 61$ %) or cold room ($T_{db} = 3.0$ °C, $\phi = 85$ %) conditions, respectively. Two different batches of production boxes were obtained due to a lack in supply.

Figure 2: S22 box (left) and its components (right)

2.2. Field tests

To measure the loads and accelerations experienced by a palletised box, a load measuring device (Figure 3) was designed and manufactured. This device featured four compressive load cells in the corners, each with a capacity of 2 kN, as well as a three-axis accelerometer. The device had a similar weight (19.3 kg) and external dimensions to that of a filled S22 box being investigated. The load measuring device was calibrated to have a constant error of less than 1 % for loads above 180 N.

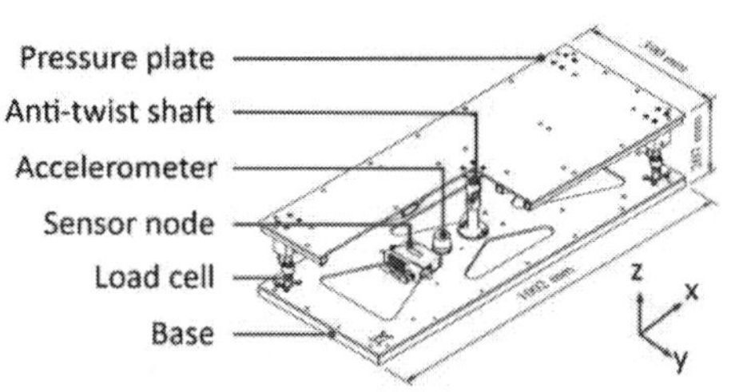

Figure 3: Diagram of load measuring device

This device was placed in the corner of the stack's bottom layer (Figure 4) with the x- and y-axes facing away from the centre of the stack. The stack consisted of 33 S14 boxes (similar footprint to that of S22, with a height of 140 mm) and five S22 boxes with an average weight of 10.8 kg ±2.35 kg and 19.3 kg ±0.07 kg each, respectively. The stack was loaded onto a truck with the x-axis facing the rear. The test was conducted in Matroosfontein, Cape Town, with a speed limit of 60 km/h and a duration of around 23 min.

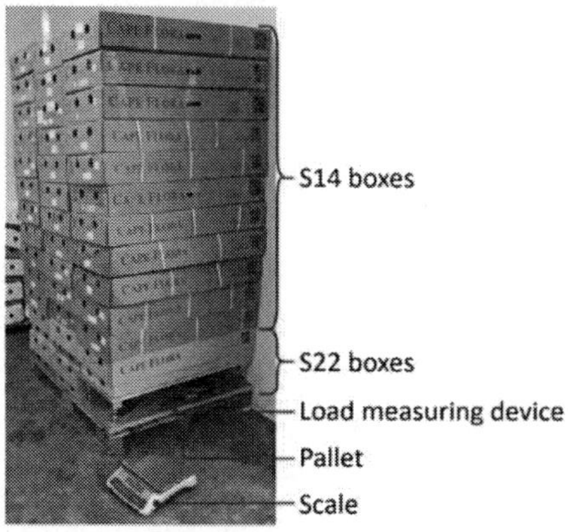

Figure 4: Palletised stack of boxes used in field tests

2.3. Box compression tests

BCT's (box compression test) were conducted on a 5982 Instron Universal Tester with platens, having a capacity of 10 kN (Figure 5), to estimate the strength of new, conditioned, and damaged boxes. These were done at a rate of 10 mm/min and a pre-load of 250 N, in accordance with the BS EN ISO 12048:2001 standard [9].

Figure 5: BCT setup

2.4. Cyclic box compression tests

C-BCT's (cyclic box compression test) were set up similar to that of normal BCT's in terms of travel rate and pre-loading. The key difference is that the compressive load was cycled between minimum and maximum load values (for 1, 5, 10, 30, 70, or 100 cycles) determined through field and experimental cyclic loading. Before and after cycling, a load equal to the static load of the pallet on the lower layer of boxes was applied to the box for 10 s. These tests were conducted to determine the deflection of the boxes during cyclic loading, as well as the effects of cyclic loading on the strength of the boxes (through subsequent BCT's). Three specimens were tested per sample.

2.5. Pallet lateral acceleration tests

It can be difficult and expensive to conduct repeated field experiments to determine the effects of cyclic loading on the strength of palletised boxes. For this reason, cyclic loading is applied to the boxes using a PLAT (Figure 6); a machine that can simulate the lateral acceleration and deceleration of a road vehicle on a palletised stack of boxes [3]. The machine allowed for the tracking of the lateral displacement of each layer, in accordance with the EUMOS 40509:2020 standard [10], and, with the help of the load measuring device, measured the loads on a palletised box for different acceleration profiles. Additionally, the PLAT was used to apply cyclic loading on conditioned boxes (bottom layer) which were subsequently subjected to BCT's. The machine was calibrated to operate at 0.1 G, 0.2 G, and 0.3 G acceleration for stack orientations x- and y-directions.

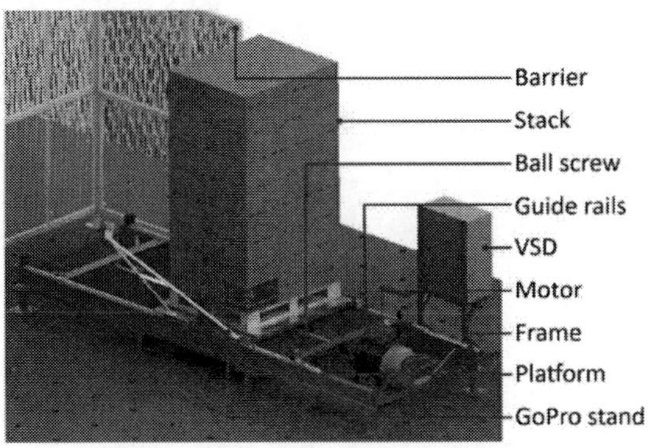

Figure 6: Pallet lateral acceleration tester [3]

3. Results and discussion

3.1. Field tests

During field tests the load measuring device measured loads between 1.7 kN and 4.0 kN (Figure 7). These loads correlated very well with the magnitude of the vehicle's acceleration (Figure 8), showing that there is a relationship between the loads experienced by a palletised box and the acceleration of the stack. From the acceleration data, it was evident that the maximum mean acceleration, of the top 5 % acceleration data, occurred at 3.13 m/s^2 to the right and the majority of the top 5 % acceleration data occurred at 1.5 m/s^2.

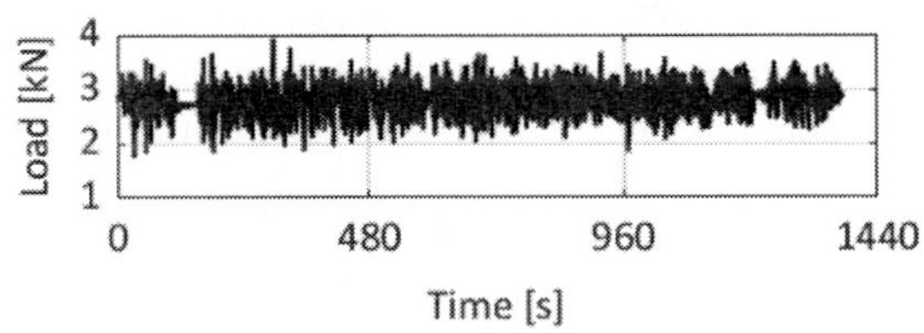

Figure 7: Field test results - Total load

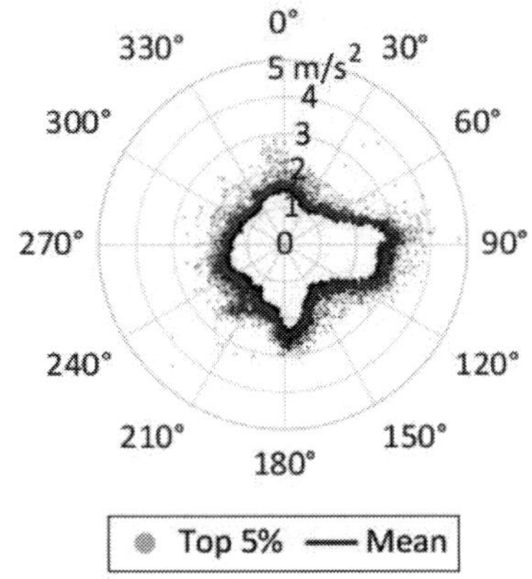

Figure 8: Field test results - Acceleration

3.2. Cyclic performance of boxes

PLAT tests in the x-direction resulted in multiple displacement cycles during a single acceleration-deceleration test run. Tests in the y-direction at 0.3 G revealed that the stack is unstable (stack displacement exceeds 220 mm). Additionally, when conducting cyclic loading on cold room conditioned boxes at 0.2 G in the y-direction, the bottom layer failed in shear after seven cycles, with the entire stack collapsing after eight cycles (Figure 9). Cold room conditioned boxes survived at least 100 cycles in tests conducted at 0.1 G in the y-direction, which forms the new basis for cyclic testing.

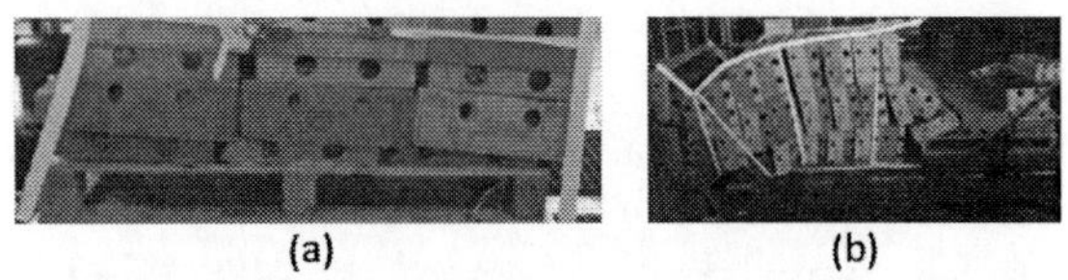

Figure 9: (a) Critical and (b) catastrophic failure of boxes (Batch 1) during cyclic-PLAT tests (0.2 G, 3 °C, 85 % RH)

BCT's were conducted on boxes subjected to various conditions and/or cyclic-PLAT tests, of which the results are shown and summarised in Figure 10 and Table 1, respectively. From these results, it is evident that relatively humid environments and prolongued cyclic loading independently reduced the box's strength but a combination of both improved it.

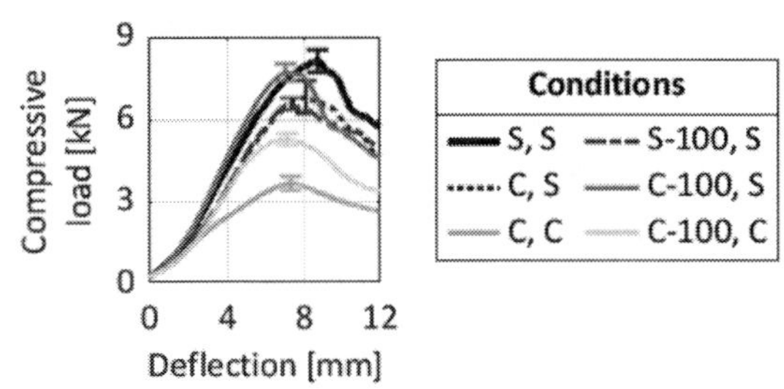

Figure 10: BCT of boxes (Batch 1) subjected to various conditions (i.e., *S-100, S* indicates standard pre-conditioning, followed by 100 PLAT cycles, and a BCT at standard conditions)

Table 1: Maximum BCT values of boxes (Batch 1) subjected to various conditions

Pre-conditioning	Test case PLAT cycles	Post-conditioning	Maximum BCT load [kN]
Standard	0 (New)	Standard	8.12
Standard	100	Standard	6.56
Cold	0 (New)	Cold	3.65
Cold	0 (New)	Standard	6.80
Cold	100	Cold	5.24
Cold	100	Standard	7.71

To further investigate the effects of cyclic loading on cold room conditioned boxes, cyclic-PLAT tests were conducted for 10, 30, and 70 cycles. The BCT results of these boxes, in addition to the new and 100 cycle boxes, are shown in Figure 11 (top). It can be observed that the cyclic strength of the cold room conditioned box increased asymptotically to a maximum of 5.29 kN.

However, this trend did not intercept with the data for new (undamaged) boxes, causing some uncertainty as to what happens at lower cycle counts. Conducting cyclic tests on the PLAT for smaller cycle counts could result in erroneous data due to on- and offloading of the stack. For this reason, the BCT setup was modified to conduct C-BCT's between minimum (1.31 kN) and maximum (2.56 kN) load values, with a static load (1.83 kN) applied before and after cycling. These loads were measured on the plat using the load measuring device for an acceleration of 0.1 G in the y-direction. This was done for 1, 5, 10, 30, 70, and 100 cycles, with cyclic-PLAT tests done for 10, 30, and 100 cycles to validate the results (Figure 11, bottom). Similar to the cyclic-PLAT tests, the C-BCT's revealed that cyclic loading asymptotically increased the strength of the cold room conditioned boxes, after 10 cycles, to a maximum of 5.87 kN. However, below 10 cycles the strength decreased to a minimum of 5.32 kN.

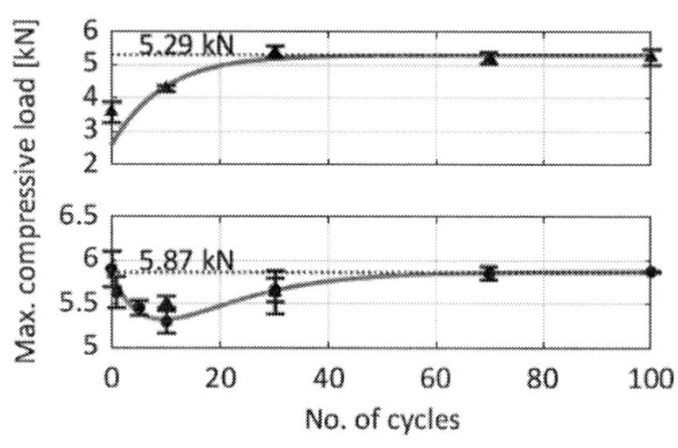

Figure 11: Cyclic strength of boxes (Batch 1 in top, Batch 2 in bottom) subjected to C-BCT's (3 °C, 85 % RH)

The C-BCT deflection data at the minimum and maximum load values for all the tests was extracted and transformed into the deflection rate (Figure 12); the difference in deflection values between two subsequent cycle counts. It shows an asymptotic trend, similar to the cyclic strength of the boxes. If this trend was summated in per-cycle increments, it would give a possible prediction for the deflection of the box during cyclic loading (3.74 mm after 100 cycles, Figure 13), which is not significantly different from the mean deflection data ($R^2 = 0.92$). Future studies with boxes of other geometrical properties are recommended for further proof of these observations.

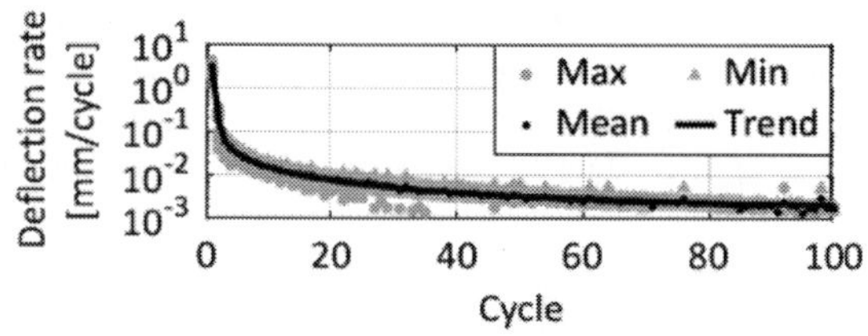

Figure 12: Deflection rate of boxes (Batch 2) subjected to C-BCT's (3 °C, 85 % RH)

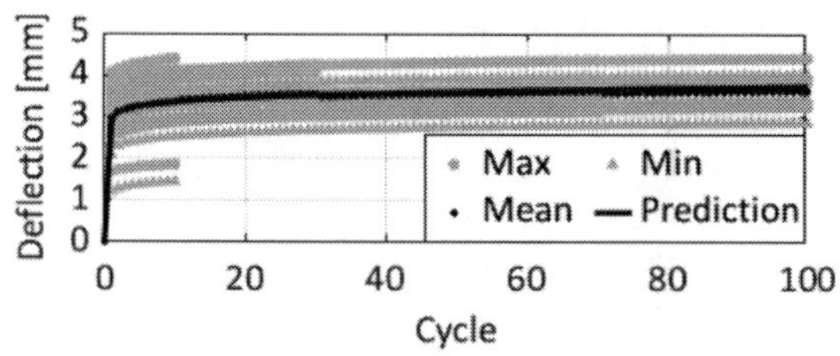

Figure 13: Deflection of boxes (Batch 2) subjected to C-BCT's (3 °C, 85 % RH)

4. Conclusion

Cape Flora SA uses corrugated paperboard boxes to export their products but found that some of the boxes fail during transport. Field tests were conducted on a palletised stack of S22 boxes to determine the loads typically experienced by a palletised box. PLAT tests were conducted on a stack to examine the stability of the individual layers and the loads required for C-BCT's. BCT's conducted on boxes for various conditions revealed that relatively humid environments reduced the strength by up to 55 %, whereas a combination of these conditions and cyclic loading only reduced it by 35 %. This means that cyclic loading increased the strength of cold conditioned boxes by 44 %. Conducting C-BCT's on the boxes for conditions similar to 0.1 G acceleration and cold room storage revealed that the strength reduced by 9 % after 10 cycles before retaining it again after 100 cycles, which is acceptable as the mass of a palletised stack's column is around 174 kg. It also showed that the box can deflect by 3.74 mm after 100 cycles.

Acknowledgements

This work is based on the research supported in part by the National Research Foundation of South Africa (Grant Numbers: 137952). This study was possible due to the partnership between the Department of Mechanical and Mechatronic Engineering, Stellenbosch University, and Cape Flora South Africa (CFSA) with the Post-Harvest Innovation (PHI) Programme and all relative partners. It was also done under supervision and guidance of Prof. Corné Coetzee as part of a thesis presented in partial fulfilment of the requirements for the degree of Master of Engineering (Mechanical).

References

[1] Fadiji, T., Berry, T.M., Coetzee, C.J. and Opara, U.L. (2018). Mechanical design and performance testing of corrugated paperboard packaging for the postharvest handling of horticultural produce. Biosystems Engineering, vol. 171, pp. 220–244.

[2] Van Hung, D., Nakano, Y., Tanaka, F., Hamanaka, D. and Uchino, T. and (2010). Preserving the strength of corrugated cardboard under high humidity condition using nano-sized mists. Composites Science and Technology, vol. 70, no. 14, pp. 2123–2127.

[3] Erasmus, N.J. (2024 March). Defining citrus pallet stability and modes of failure: Developing tools to make stronger and cheaper cartons. Master thesis, Department of Mechanical and Mechatronic Engineering, Stellenbosch University, Stellenbosch.

[4] Homewood Disposal Service (2022). Your guide to cardboard recycling and types of cardboard. Available at: https://mydisposal.com/cardboard-recycling-and-types-ofcardboard

[5] Huysamer, M. (2021). Carton integrity in air- and sea freight export. Available at: https://www.capeflorasa.co.za/wp-content/uploads/docs/2021/12/Carton-issues-3-Dec-2021.pdf

[6] Department of Agriculture, Forestry and Fisheries (1990). Agricultural product standards act, 1990 (act no. 119 of 1990): Standards and requirements regarding control of the export of fresh cut flowers and fresh ornamental foliage. Available at: https://www.capeflorasa.co.za/export - standards/

[7] Huysamer, M. (2023 July 3). Extra-Ordinary appointed at Department of Horticulture, Stellenbosch University. Email conversation.

[8] BS EN ISO 187:2022 (2022 December). Paper, board and pulps - Standard atmosphere for conditioning and testing and procedure for monitoring the atmosphere and conditioning of samples. Standard, British Standards Institution, 389 Chiswick High Road, London, United Kingdom.

[9] BS EN ISO 12048:2001 (2001 February). Packaging - Complete, filled transport packages - Compression and stacking tests using a compression tester. Standard, British Standards Institution, 389 Chiswick High Road, London, United Kingdom.

[10] EUMOS 40509:2020 (2020 July). Test Method for Load Unit Rigidity. Standard, AISBL European Safe Logistics Association, Brussels, Belgium.

Dimensional evaluation of a scaled-down aircraft nose wheel fork produced in Ti6Al4V(ELI) through selective laser melting

Lehlohonolo Francis Monaheng[a,*], Willie Bouwer du Preez[b], Claudia Polese[c,d]

[a]*Department of Mechanical and Mechatronics Enginering, Faculty of Engineering, Built Enviroment and Information Technology, Central University of Technology, Free State, Bloemfontein 9300, South Africa*

[b]*Center for Rapid Prototyping and Manufacturing, Faculty of Engineering, Built Environment and Information Technology, Central University of Technology Free State, Bloemfontein 9300, South Africa*

[c]*School of Mechanical, Industrial and Aeronautical Engineering, University of the Witwatersrand, Johannesburg 2050, South Africa*

[d]*ARUA Centre of Excellence in Materials, Energy and Nanotechnology, University of the Witwatersrand, Johannesburg 2050, South Africa*

*Corresponding author email: lmonaheng@cut.ac.za

Abstract

Selective Laser Melting (SLM) has evolved from a technology primarily used for prototype production to one capable of manufacturing fully functional components that meet the stringent standards of the aviation industry. However, geometrical deviations resulting from the SLM process could significantly impact the dimensional accuracy of the final product. Consequently, further investigations are needed to determine these geometrical deviations in aircraft structural components produced through SLM.

In this study, a geometrical deviation analysis was conducted on a scaled-down light aircraft nose wheel fork manufactured using SLM of Ti6Al4V(ELI), utilising both voxel-based geometric modelling and experimental measurements. The SLM process chain was virtually simulated using Simufact Additive® software to assess the geometrical distortion of the component following SLM manufacturing, while deviation was measured with a Kreon scanning system. The simulation accurately captured the experimentally measured deviations. This study contributes to the aviation industry's efforts to characterise the geometrical accuracy of the SLM technique. It demonstrates that SLM, used to successfully produce an accurate scaled-down Ti6Al4V(ELI) nose wheel fork for a light aircraft, has the potential to be applied in the manufacturing of full-scale undercarriage components.

Keywords: Geometric deviation, Selective laser melting, Ti6Al4V(ELI), Nose wheel fork, Lightweight aircraft.

1. Introduction

An aircraft component should comply with the stringent standards in the aeronautical industry. Dimensional accuracy is among the important factors that have to be strictly adhered to, since a geometric deviation of a part, such as a nose wheel fork, could affect the assembly parameters and the functional characteristics of the product [1].

A geometric deviation can occur at different points in the manufacturing process. This includes deviation results from the creation of a discrete mesh usually in stl. format during the data pre-processing phase [2]. Additionally, each layer is two-dimensional (2D) and the 3D object is obtained by joining individual layers of a metal alloy, such as Ti6Al4V(ELI), layer upon layer, resulting in a staircase effect on the surfaces of the end-product. Other factors contributing to dimensional deviation are residual stresses [3] and shrinkage that results from rapid cooling from high temperatures [4].

A voxel-based volumetric representation of a part is a computational method that can be used to determine the geometric deviation. In this method, the manufacturability of the SLM part is evaluated by importing a CAD model of the part into a software package and predicting the geometric deviation upfront in the design process to guide design for manufacturing (DfAM) practice [5]. For example, with Simufact Additive® the entire SLM process chain of the part can be simulated and the geometric deviation can be determined [6].

A multiple coordinate measuring machine (CMM) with tactile, optical, or x-ray sensing capability, can be used to measure the geometric deviation of built parts [7]. The applicable method is selected based on the complexity of the part. With an optical CMM, large amounts of data can be captured at a rate faster than the tactile CMM. A triangulation (laser and light projection) active optical CMM, with an accuracy ranging from 10 to 100 μm, is more appropriate for the measurement of freeform parts used in airplanes [8]. These measurement systems are widely used in the metrology of metal AM parts due to several factors, which include SLM part surface roughness, typical part size, measurement accuracy required, as well as practical matters such as commercial availability and cost [9].

This study aimed at simulating and measuring the geometric deviation after stress-relieving heat treatment plus support structure removal and high-temperature annealing (HTA) of a scaled-down light aircraft nose wheel fork manufactured by SLM in Ti6Al4V(ELI).

2. Methodology

2.1. Redesign of the nose wheel fork for production in Ti6Al4V(ELI) through SLM

Topology optimisation technology was used to redesign the lightweight nose wheel fork of the Advanced High-performance Reconnaissance Light Aircraft (AHRLAC) for production in Ti6Al4V(ELI) through SLM. The detailed design process of the nose wheel fork was discussed by Monaheng et al. [10]. To obtain the scaled-down fork, a factor of 0.4 was applied to the full-scale fork leading to the overall dimensions of 222.4 mm and 103.6 mm for the length and width, respectively.

2.2. Simulation of the manufacturability of the nose wheel fork

Version 4.1 of the Simufact Additive® software was used to predict the SLM manufacturability of the scale-down nose wheel fork of Ti6Al4V(ELI). Firstly, the machine parameters of the EOSINT M280 direct metal laser sintering (DMLS) system were set. The most important parameters were the laser power of 170 W and layer thickness of 30 µm. Thereafter, the orientation was predicted and set. The Simufact Additive® orientation assistance was used to determine the best building orientation. A selection was made from ten possible orientations of the scaled-down fork provided by the software. In Fig. 1, more appropriate build orientations are shown, and the B orientation was selected. The selected orientation was based on simplicity of removal of the support structure and to minimise distortion. Most importantly, reducing the building time would improve energy efficiency, since lower building height results in lower energy required for the production of the scaled-down nose wheel fork. The distortion for building orientations A, B and C was recorded as 1.58 mm, 1.52 mm and 1.43 mm, respectively. This was followed by the auto-generation of the support structure using the Materialise Magics software which is incorporated in the Simufact Additive® software.

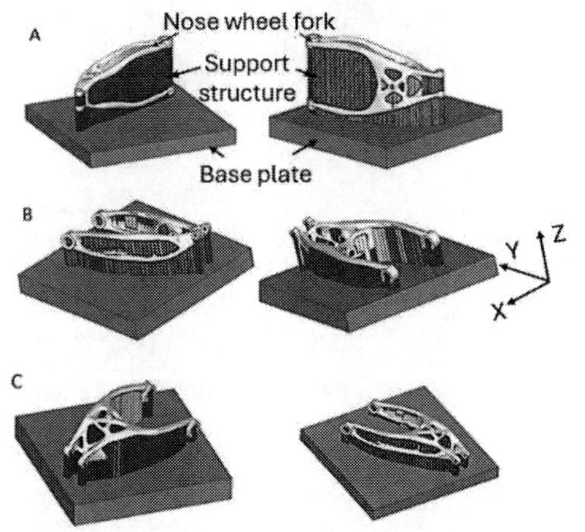

Fig. 1. Illustration of more appropriate building orientation of the scaled-down nose wheel fork according to Simufact Additive® software. The B orientation was selected.

Ti6Al4V(ELI) material properties were assigned to both the base plate and scaled-down nose wheel fork. Thereafter, a non-uniform voxel mesh was also assigned to the scaled-down nose wheel fork, with dimensions of 0.8 mm in the X and Y directions, and 1.2 mm in the Z direction. For Scan A, the geometric deviation simulation process included building the nose wheel fork using SLM, followed by stress-relieving heat treatment at 650 °C with a 3-hour soaking time, cutting off the support structure from the building base plate, and removal of the support structure from the fork. For Scan B, an HTA heat treatment at 950 °C with a 2-hour soaking time was added to the Scan A sequence simulation process chain. The results recorded from the simulation included geometric deviation and residual stresses.

2.3. Manufacturing of the nose wheel fork

Spherical Ti6Al4V(ELI) powder with particle size of less than 45 μm and chemical composition that complies with ASTM F3001-14 was used to build the nose wheel fork [11]. This powder was supplied by TLS Technik GmbH & Co. (Spezialpulver KG, Bitterfeld-Wolfen, Germany). In Fig. 2, the scaled-down nose wheel fork on the base plate is shown with various standard test specimens. The current paper focuses on the dimensional analysis.

Therefore, detailed testing procedure and results of tensile, fatigue crack growth rate, impact toughness and high cycle fatigue specimens were presented in separate papers [12][13][14].

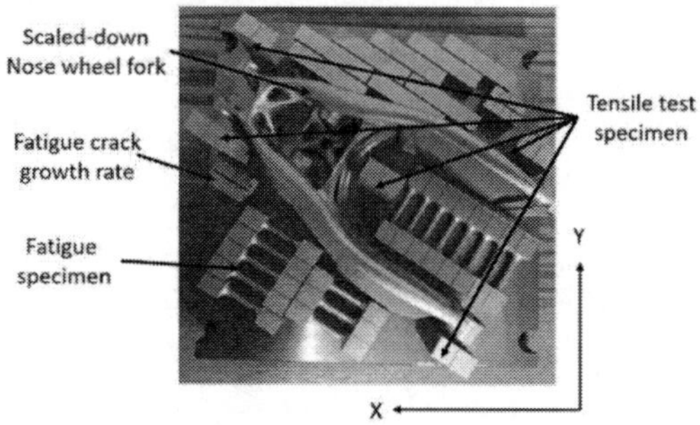

Fig. 2. The SLM Ti6Al4V(ELI) scaled-down nose wheel fork orientation and support structure on the building base plate together with various standard test specimens.

The scaled-down CAD model of the nose wheel fork was converted into stereolithography (STL) file format and was also sliced into layers to allow SLM production. An EOSINT M280 system (EOS GmbH, Munich, Germany), equipped with a 200 W Ytterbium fibre laser with a laser beam diameter of 80 μm and powder layer thickness of 30 μm, was used to build the scaled-down nose wheel fork in Ti6Al4V(ELI). The regular back-and-forth raster scanning pattern, with a shift angle of 67° for each pass, was used in the production of the fork. Importantly, the fork was anchored on the base plate using multiple pins and traditional support structures. The pin support has a cone shape with a diameter of 2 mm and 4 mm on the side of the fork and base plate, respectively. The pin support was used to minimise distortion from residual stresses as presented in section 3.1 of this paper.

After manufacturing, the nose wheel fork was submitted to a two-stage heat treatment using the Super Series Model SS12/24-13MDX vacuum furnace (Cinnaminson, NJ, USA). The heat treatment temperatures and heat treatment sequences that were applied during the simulation were carried out on the SLM Ti6Al4V(ELI) scaled-down nose wheel fork.

2.4. Measurement of the geometrical deviation of the nose wheel fork

To measure the geometric deviations from the CAD model of the nose wheel fork, two digital scans of the surface geometry of the SLM fork were performed. Scanning was done after submitting the SLM Ti6Al4V(ELI) fork to similar processes as were done during simulation, see section 2.2. A CMM triangulation Kreon 3D scanning system (Kreon GmbH Deutschland, Germany) was used to scan the nose wheel fork. This scanner uses a laser with a maximum scanning speed of 600 points per second. Its laser line resolution is 25 μm and has an accuracy of 40 μm, whereas the accepted measurement tolerance in the industry was recorded as 100 μm [15]. Moreover, the geometric deviation was calculated by auto-aligning the scan data on the CAD model using Geomagic® software. In Fig. 3, the in-plane sketch is shown to demonstrate the computation of deviation between the measured data and CAD model.

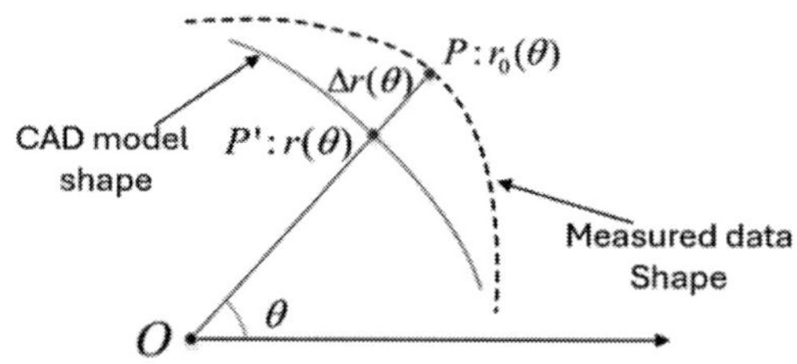

Fig. 3. An in-plane sketch demonstrating the deviation between measured data and CAD model [16].

Deviation is represented as the difference between the radii of the measured data shape ($r_0(\theta)$) and CAD model shape ($r(\theta)$). Therefore, it was presented by Eq. 1.

$$\Delta r(\theta) = r(\theta) - r_0(\theta) \quad (1)$$

Finally, the deviation patten was investigated using the software's deviation distribution ($\Delta r(\theta)$) with respect to θ. The measured data shape of the SLM Ti6Al4V(ELI) nose wheel fork resulted in the intricate deviation distribution.

3. Results and Discussions

3.1. Predicted geometric deviation results

The simulated geometric deviation distributions of the scaled-down nose wheel fork for Scan A and Scan B are presented in Fig. 4 (a) and (b), respectively. The negative values on the map represent shrinkage, whereas positive values illustrate the expansion of the nose wheel fork. The predicted geometric deviation of the nose wheel fork for both Scan A and Scan B was within 0.84 mm to -1.04 mm. Thick portions of the nose wheel fork were predicted to have higher manufacturing accuracy than the thin sections with a minimal deviation of $\pm$0.2 mm, whereas the thin sections had a maximum deviation of 0.84 mm. This variation of the predicted deviation was attributed to residual stress for thick portions and combination of residual stress as well as low structural stiffness for thin sections of the nose wheel fork. Furthermore, the highest shrinkage value was observed on the lower surface of the wheel axle bush, which was the closest to the base plate. Moreover, the inner surface of the scaled-down nose wheel fork experienced higher expansion than the external surface. This resulted from the lower rate of heat dissipation on the inner surface of the fork compared to its outer surface.

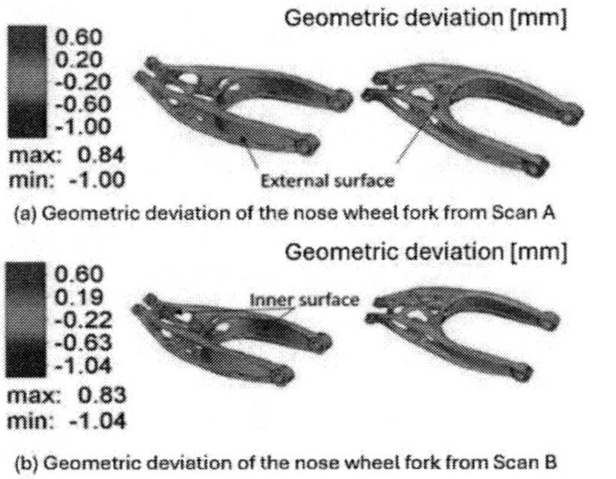

(a) Geometric deviation of the nose wheel fork from Scan A

(b) Geometric deviation of the nose wheel fork from Scan B

Fig. 4. Maps of the predicted geometric deviation of (a) Scan A and (b) Scan B of the scaled-down nose wheel fork.

In general, the residual stress was the main contributing factor towards geometric deviation. In Fig. 5, the simulated distribution of residual stress along the build Z-direction of the SLM Ti6Al4V(ELI) scaled-down nose wheel fork is illustrated.

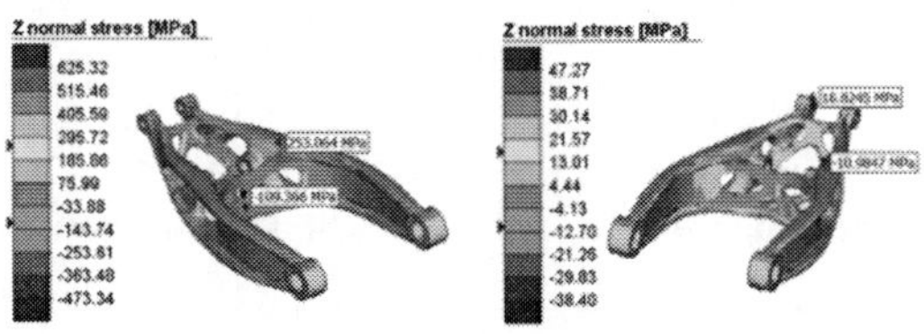

(a) (b)

Fig. 5. Predicted residual stress on the SLM Ti6Al4V(ELI) scaled-down nose wheel fork along the building Z-direction, (a) was after Scan A and (b) after Scan B

The maximum and minimum residual stress on the surface of the scaled-down nose wheel fork were predicted as 253.0 MPa and -109.4 MPa, respectively for Scan A. After HTA Scan B, residual stress was predicted to diminish by 90%. To minimise distortion due to residual stresses, the scaled-down SLM Ti6Al4V(ELI) nose wheel fork was built with multiple pins and the conventional support structure defined in section 2.3 of this paper. The stress relieving heat treatment was applied before the support structure was removed.

3.2. Measured geometric deviation results

Similar to the simulation of the nose wheel fork geometric deviation, the scanned images of the SLM produced scaled-down nose wheel forks were also superimposed on the CAD model to obtain a broad view of the shape deviations. The measured maximum and minimum geometric deviations in the scaled-down nose wheel fork for both Scan A and Scan B were 0.40 mm and -0.80 mm, respectively. Moreover, the bottom surface of the fork after Scan A depicted shape deviations of 0.3 mm to 0.4 mm, whereas after Scan B the same surface recorded -0.3 mm to -0.3 mm deviations. In addition, the experimentally measured data also confirmed expansion on the inner surface of the SLM Ti6Al4V(ELI) scaled-down nose wheel fork. For example, the inner surface of Scan A depicted no deviation, while the inner surface of the Scan B showed a deviation of 0.3 mm. Moreover, the deviation distribution for scan A and Scan B depicts compression on the outer surface of the leg of the fork and expansion on the inner surface, suggesting that the legs curved inwards. When observing lower surface deviation distribution on each scan for each leg of the SLM Ti6Al4V(ELI) fork, comparable results were identified confirming that there was an insignificant tortional deformation experienced by the fork. These were irrespective of different measured geometric deviation distributions during Scan A and Scan B, see Fig. 6. Finally, it can be suggested that the HTA can improve the geometric accuracy of the SLM Ti6Al4V(ELI) nose wheel fork by more than 50%.

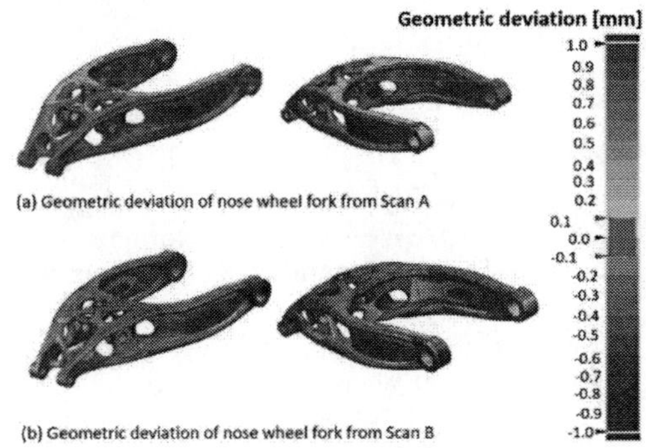

(a) Geometric deviation of nose wheel fork from Scan A

(b) Geometric deviation of nose wheel fork from Scan B

Fig. 6. Maps of the physically measured geometric deviation of (a) Scan A and (b) Scan B of the SLM Ti6Al4V(ELI) scaled-down nose wheel fork

In general, the aircraft structural component that was manufactured in Ti6Al4V(ELI) through SLM resulted in geometric deviation in various stages of building and post processing.

4. Conclusions

The study on the geometric deviation after SLM, followed by removal of support structure plus stress relieving heat treatment (Scan A), and after additional HTA (Scan B) demonstrated that a dimensionally accurate nose wheel fork of a light aircraft could be built through SLM in Ti6Al4V(ELI), if geometric deviation was factored in through simulation. An accurate prediction of the SLM

Ti6Al4V(ELI) could be achieved through voxel-based simulation with the Simufact Additive® software. The key to successfully building the fork with minimal geometric deviation was the selected build orientation, whereas the main factor contributing towards geometric deviation was the residual stress.

Future study should focus on validating the Simufact Additive® software for predicting and compensating the geometric deviation for accurate building of the topologically optimised structural aircraft component. This should be followed by investigating geometric deviation from multiple topologically optimised structural components to statistically confirm deviation distribution data that resulted from SLM Ti6Al4V(ELI) building and post processing.

Acknowledgements

This research was supported and funded by the South African Department of Science and Innovation through the Collaborative Program in Additive Manufacturing, Contract No.: CSIR-NLC-CPAM-18-MOA-CUT-01. Financial support from the Chair in Innovation and Commercialisation of Additive Manufacturing is also gratefully acknowledged.

References

[1] Federal Aviation Administration (2010). Part 23-airworthiness standards: Normal, utility, acrobatic, and commuter category airplane, pp. 189-366.

[2] Ning Y, Wong Y, Fuh JY, Loh H (2006). An approach to minimize build errors in direct metal laser sintering. IEEE Trans. Autom. Sci. Egineering, **3**: pp. 73–80.

[3] Singh UP, Swaminathan S, Phanikumar G (2022). Thermo-mechanical approach to study the residual stress evolution in part-scale component during laser additive manufacturing of alloy 718. *Mater. Des.*, **222**: pp.111048

[4] van Zyl I, Yadroitsava I, Yadroitsev I (2016). Residual stress in Ti6Al4V objects produced by direct metal laser sintering. South African J. Ind. Eng., **27**: pp. 134–141.

[5] Tedia S. Williams CB (2016). Manufacturability analysis tool for additive manufacturing using voxel-based geometric modeling. 27th Annual International Solid Freeform Fabrication Symposium - An Additive Manufacturing Conference pp. 322.

[6] Moriconi C (2019). Simufact additive: Collaborative simultaneous engineering tool for additive manufacturing. Safran additive manufacturing, IX, pp. 27–29.

[7] Rokicki P, Budzik G, Kubiak K, Dziubek T (2016). The assessment of geometric accuracy of aircraft engine blades with the use of an optical coordinate scanner. Aircr. Eng. Aerosp. Technol. **88**: pp. 374–381.

[8] Sovio E, De Chiffre L, Schmitt R (2007). Metrology of freeform shaped parts. CIRP Ann., **56**: pp. 810–835.

[9] Stavroulakis PI, Leach RK (2016). Invited Review Article: Review of post-process optical form metrology for industrial-grade metal additive. Rev. Sci. Instrum., 87: pp. 8704–1101.

[10] Monaheng LF, Preez WB, Kotze N, Vermeulen M (2020). Topology optimisation of an aircraft nose-wheel fork for production in Ti6Al4V by the Aeroswift high-speed laser powder bed fusion machine. 14th World Conf. Titan., 321, pp. 1–11.

[11] ASTM (1875). ASTM F136-08 standard specification for wrought titanium 6 aluminum 4 vanadium ELI (Extra Low Interstitial) alloy for Surgical implant applications. pp. 6–9.

[12] Miya HP, du Preez WB, Monaheng L. F (2021). High cycle fatigue performance of Ti6Al4V(ELI) specimens produced with inherent laser powder bed fusion surface roughness. *South African J. Ind. Eng.*, **32**: pp. 248–257.

[13] Monaheng LF, du Preez WB, Polese C (2021). Towards Qualification in the Aviation Industry: Impact toughness of Ti6Al4V(ELI) specimens produced through laser powder bed fusion followed by two-stage heat treatment. *Metals (Basel).*, **11**: pp. 1–12.

[14] Monaheng LF, du Preez WB, Polese C (2023). Selective laser melting process chain for development of a Ti6Al4V(ELI) nose wheel fork of light aircraft, in *Proceedings of the 15th World Conference on Titanium*, pp. 35–40.

[15] iTech standards and preview (1989). Geometric tolerance for feartures without individual tolerance indications. International Standard ISO 2768-2.

[16] Zhu Z, Anwer N, Mathieu L (2018). Shape transformation perspective for geometric deviation modeling in additive manufacturing. Procedia CIRP, **75**: pp. 75–80.

Development of an electromagnetic acoustic transducer for measurement of Lamb waves in plates

Rolaedzwa Mathoho[a], Sebutjiwane Kgwete[a], Paul Fromme[b] and Philip W. Loveday[a,*]

[a]*University of the Witwatersrand, Johannesburg, South Africa*

[b]*Department of Mechanical Engineering, University College London (UCL), London, UK*

[*] *corresponding author email: Philip.Loveday@wits.ac.za*

Abstract

Guided wave ultrasound may be used to efficiently inspect plate-like structures from a few transducer locations. The multi-modal and dispersive nature of the Lamb waves requires more sophisticated measurement equipment than what is used in conventional ultrasound. Measurements of phase and group velocity are commonly based on measuring the response at a series of points using a scanning laser vibrometer. This equipment may be unaffordable and the alternative of using only low-cost instrumentation is being explored. In this paper the development of an electromagnetic acoustic transducer (EMAT) to replace the laser vibrometer in the measurement system is considered. An EMAT consists mainly of a magnet and a coil. A previous design was found to have low signal to noise ratio and improvement of the sensitivity of the EMAT and reduction of the noise is investigated in this paper. Numerical simulation of the magnetic flux density was used to optimise magnet dimensions and to investigate the use of a magnetic field concentrator. The number of turns used in the coil is expected to increase the sensitivity and this was confirmed experimentally. Electrical shielding by a stainless steel housing and effective grounding of the housing, magnet and plate was found to significantly reduce the noise in the signals. Overall, the signal to noise ratio of the EMAT was improved by over 20 dB.

Keywords: EMAT development, Lamb waves in plates

1. Introduction

Guided wave ultrasound may be used to inspect large volumes of structures that support guided ultrasonic waves. Plate-like structures are the topic of extensive research due to the potential for industrial applications such as corrosion detection in tanks and crack detection in aircraft wings and fuselages. The multi-modal and dispersive nature of the Lamb waves, which propagate in plates, requires more sophisticated equipment than what is used in conventional ultrasonic testing. Measurements of the phase and group velocities in plates may be performed by exciting waves using an attached piezoelectric transducer, driven with a toneburst signal, and measuring the response at several equally spaced distances from the transmitting transducer, using a scanning laser vibrometer. This equipment is expensive, and the authors have previously investigated what can be achieved with only low-cost equipment [1]. In this case a piezoelectric transducer was manually positioned at the measurement points and coupling gel was reapplied before each measurement. It was found that the repeatability of measurements performed with a piezoelectric transducer was limiting the accuracy of the method. Measurements with a custom built electromagnetic acoustic transducer (EMAT) were performed [2] to investigate if these transducers could provide more repeatable measurements. Unfortunately, the signal to noise ratio (SNR) achieved with the EMAT was poor.

An EMAT consists of a permanent magnet and a copper wire coil offering the potential for a very low-cost transducer [3]. The design tested in [2] used an axially polarised cylindrical magnet placed in the centre of an axially polarised ring magnet with the polarisation of the two magnets being in opposite directions. The spiral coil was positioned in the region of the interface between the two magnets where the radial component of the magnetic field density is maximum. A similar concept was used by Sun et al. [4] to increase the sensitivity of the EMAT. In addition, Sun et al. [4] used a second, larger spiral coil to utilise the radial component of the magnetic field density at the outer edge of the ring magnet. In the application being developed in this paper the wavelength of the wave in the plate is less than 20 mm and this limits the diameter of the coil to 10 mm, making the use of a second coil impractical. Standard magnet sizes were used in [2] even though it was realised that these would not be optimal. The orthogonal test method is a simple to implement optimisation method, which has been used for EMAT design [5]. In this method the design parameters are set at different levels, and the design is analysed with different combinations of levels covering the design space. The method produces results similar to those from extensive full-scale analyses but with significantly fewer analyses [5].

Apart from the magnetic flux density it is also known that the signal strength is proportional to the number of turns in the coil, so an increase in the number of turns should result in higher signal and improved SNR. The influence of electrical shielding and grounding was investigated in [6]. It was determined experimentally that including a stainless steel housing around the transducer and connecting this to the EMAT ground and the specimen improved the SNR by a factor of four.

The objective of this investigation was to improve the SNR of the EMAT. The optimisation of the magnetic field was investigated using numerical modelling. The optimal dimensions of the magnets were determined and tested experimentally. The use of a concentrator to guide the magnetic field from a larger magnet to the region of the coil was investigated numerically. The influence of the number of turns in the coil and electrical shielding and grounding were investigated experimentally.

2. Magnetic Field Modelling & Optimisation

The sensitivity of the EMAT is proportional to the magnetic field density in the plate in the area under the coil. In this section numerical modelling is used to compute the magnetic flux density and to optimise the dimensions to maximise the magnetic flux density. Firstly, the EMAT configuration used in [2] is optimised and secondly the use of a magnetic field concentrator between the magnet and the plate is investigated and different concentrator geometries are compared.

2.1. Modelling & Optimisation of Magnet Dimensions

In [2] standard size magnets were used. Here we relax this constraint and perform optimal sizing of the magnets. Numerical modelling was shown to be effective for predicting the sensitivity of the EMAT due to changes in the geometry in [2] and was used again here. The free software FEMM 4.2 was used and due to the simple geometry of the EMAT it was possible to use an axisymmetric model, which was computationally very efficient.

Only brief details of the modelling method can be provided here. The interested reader is referred to the FEMM User's Manual [7], which is freely available. Magnetostatic problems are governed by equations 1-3 [7],

$$\nabla \times H = J \tag{1}$$

$$\nabla \cdot B = 0 \tag{2}$$

$$B = \mu H \tag{3}$$

where, the magnetic flux density (B) is related to the magnetic field intensity (H) by the permeability μ. FEMM models the permanent magnet as a thin sheet of current surrounding the solid ferromagnetic material, represented by the current density J. The material type of the magnets used was N42 and this material model was available in FEMM. A linear B-H relationship was selected and the relative permeability for the N42 material was 1.05. The coercivity was 994529 A/m and the electrical conductivity was 0.667 MS/m. A finite spherical volume of air around the permanent magnets and aluminium plate was modelled and a seven-layer impedance matching boundary was applied to represent the unbounded air domain.

It is possible to run the FEMM software from an Octave or Matlab script and to extract results. This makes it possible to automate the optimisation. The orthogonal test method was adopted in this study.

The EMAT comprises a concentric cylinder and ring magnet and the four design variables (dimensions) that were selected for optimisation are shown in Fig. 1. The lift-off distance between the magnet and plate was fixed at 0.5 mm and the plate thickness was fixed at 5 mm. Four levels of each design variable dimension were analysed resulting in a total of 16 analyses. The magnetic flux density for the original design (standard sized magnets) and the optimised design are shown in Fig. 2a. The sensitivity is determined by the radial component of the magnetic flux density in the plate, and this is plotted in Fig. 2b for the original and optimal configurations. It is seen that the maximum radial magnetic flux density is significantly greater for the optimised design.

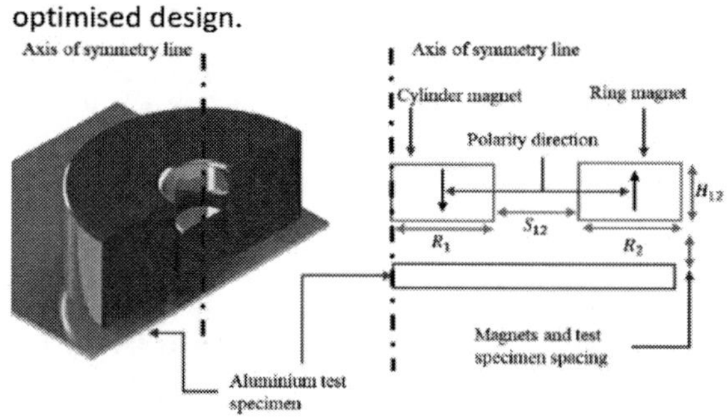

Fig. 1. EMAT configuration and design variables.

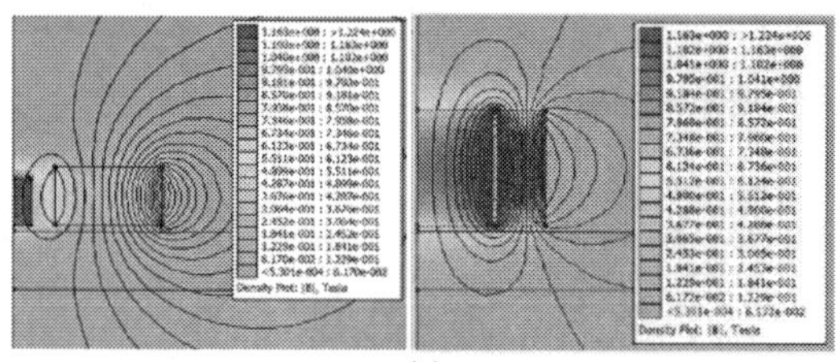

(a)

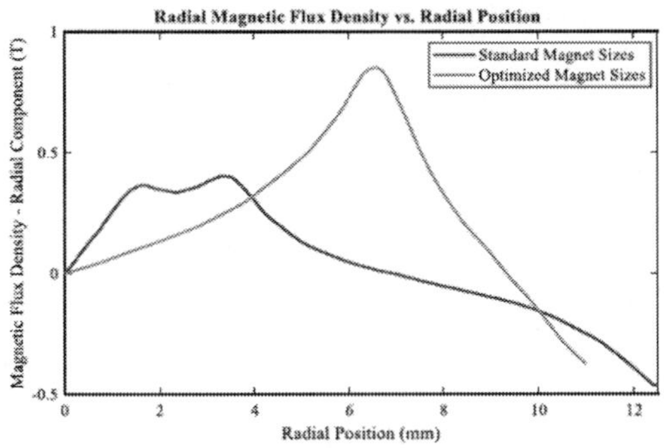

(b)

Fig. 2. (a) magnetic flux density of initial and optimised design; (b) radial components of magnetic flux density for the two designs.

2.2. Magnetic field concentrator investigation

The second technique for increasing the magnetic field was to use a large magnet along with a concentrator component to guide the magnetic field to the region of importance in the plate (below the coil). The concept is illustrated in Fig. 3a where the coil would be placed between the concentrator and the plate. The geometries for the concentrator, which were analysed, are shown in Fig. 3b. The smaller area end would be directed towards the plate.

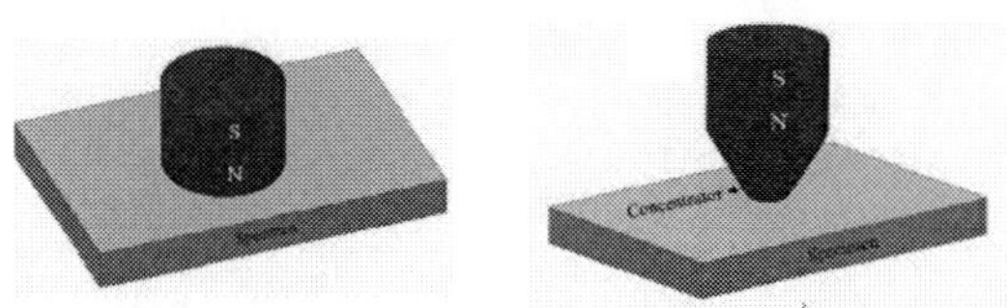

(a)

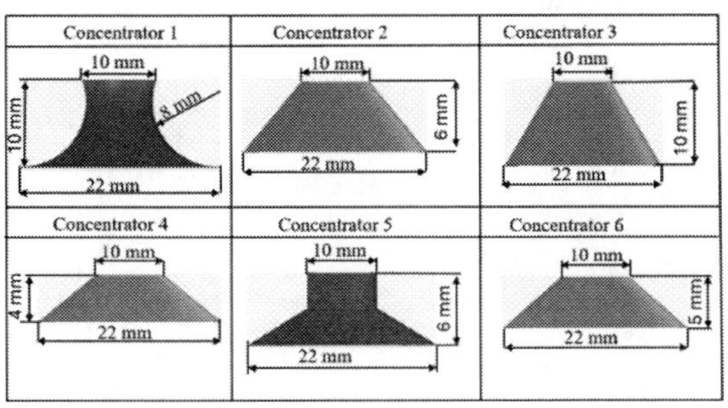

(b)

Fig. 3. (a) concentrator concept; (b) concentrator geometries analysed.

Axisymetric analyses of the magnet-concentrator-plate combinations were performed. FEMM was used for simple geometries, while concentrator No. 1 was analysed using Ansys Maxwell. It was found that the two software packages produced practically identical results if the same geometry was analysed. Initially, the concentrators were modelled as ferrite material but when this material could not be obtained for experimentation the material was changed to mild steel in the analyses. The results of the analyses are compared to experimental results in table 1.

3. Experimental Investigation

3.1. Magnet dimension and number of turns investigation

Two new EMATs were constructed. These used the optimised magnet dimensions. The magnets were produced by wire cutting (EDM) a 22 mm diameter, 10 mm thick neodymium N42 magnet into a smaller cylinder of 13 mm diameter and an outer ring using a molybdenum wire of 0.18 mm diameter. The smaller cylinder was flipped over and placed in the ring so that the magnetic polarisation directions of the two magnets were opposite. The ring magnets cracked during wire cutting but this is not expected to degrade the performance significantly. Two copper wire coils were wound by hand using 0.18 mm and 0.1 mm diameter enamel-coated copper wire. The number of turns in each coil was, unfortunately not counted, but it is expected that the coil made from the thinner wire would have approximately double the number of turns in the coil made from the thicker wire. The outer diameter of both coils was approximately 10 mm. Two stainless steel housings were manufactured, and these included a panel-mount BNC connector. One end of the coil was connected to the housing, to the magnets and to the outside of the BNC connector, while the other end of the coil was connected to the centre pin of the BNC connector. The assembled EMATs are shown in Fig. 4.

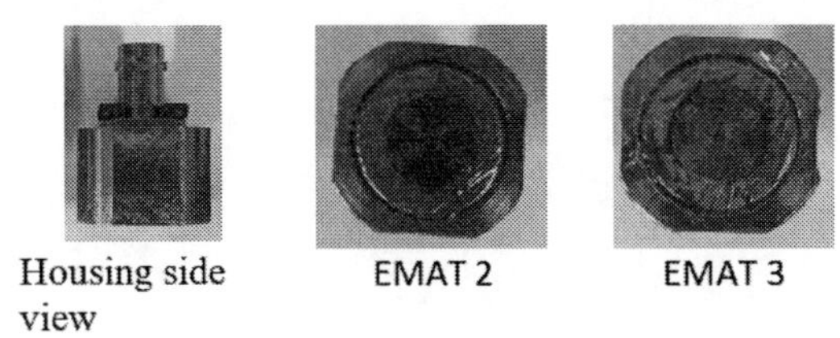

Fig. 4. Assembled EMATs with different thickness copper wire.

Measurements of the sensitivity of the EMATs were performed by using a piezoelectric transducer to transmit Lamb waves in an aluminium plate and capturing the signals generated by the EMATs at a specified distance from the transmitter. A Digilent AD2 USB oscilloscope and setup similar to that in [8] was used for the measurements. A Sonemat SAA100 pre-amplifier was used to amplify the signals from the EMATs. The EMAT constructed using standard sized magnets reported in [2] was also housed in a stainless steel housing and tested for comparison. The detected signals after averaging and filtering are shown in Fig. 5. It is seen that the EMAT with the optimised magnet dimensions is more sensitive than the EMAT based on standard sized magnets and that increasing the number of turns in the coil (by reducing the diameter of the copper wire) increased the sensitivity of the EMAT.

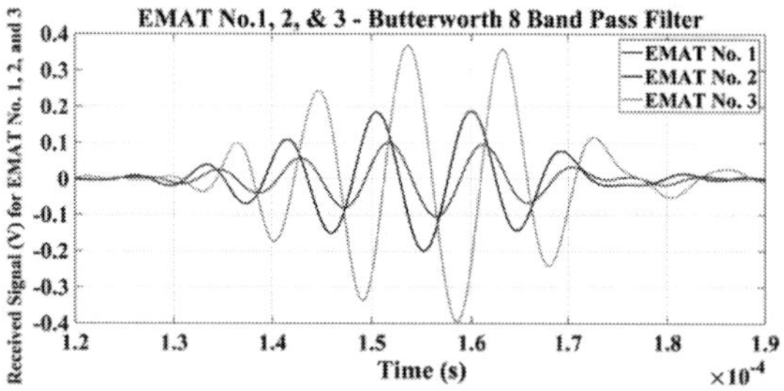

Fig. 5. Signals received from EMAT 1 (standard magnets), EMAT 2 (optimised magnet dimensions) and EMAT 3 (optimised magnet dimensions and more turns in coil).

The effectiveness of the electrical shielding provided by the stainless steel housing and electrical grounding was investigated by performing measurements before and after the EMAT was housed and grounded. These measurements are shown in Fig. 6 where it is observed that cross-coupling of the drive signal and noise in general is significantly reduced by the housing and grounding. In this case the SNR increased from 1.62 (without housing) to 345 (with housing and grounding).

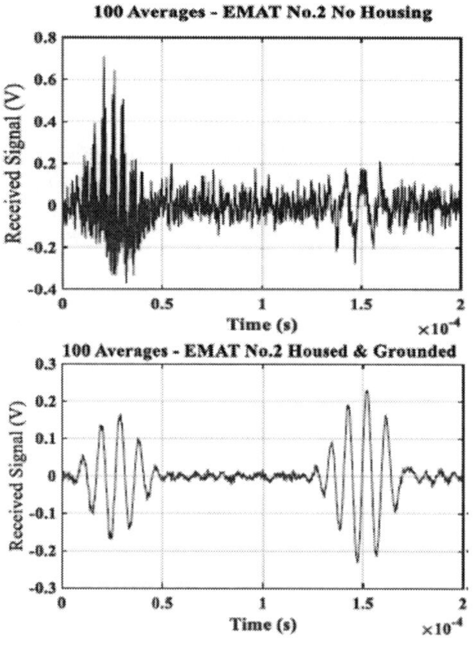

Fig. 6. Influence of shielding and grounding on Lamb wave signals received using EMAT.

3.2. Concentrator Investigation

The six concentrators illustrated in Fig. 3 were manufactured from mild steel and tested by placing them between a housed 22 mm diameter, 10 mm thick magnet and the coil as shown in Fig. 7. The amplitude of the received signal was determined for each concentrator, and this is listed in Table 1 along with numerical modelling predictions of the magnetic flux density for the concentrators made from ferrite material and from mild steel. There appears to be a correlation between the predictions and the measurements of the sensitivity with concentrator No. 1 having the best performance.

Fig. 7. Concentrator No. 1 placed between magnet and coil.

Table 1. Comparison of concentrator predictions and measurements

Concentrator No.	1	2	3	4	5	6
Predicted magnetic flux density for ferrite material (T)	6.60	3.04	2.92	5.11	6.03	3.50
Predicted magnetic flux density for mild steel material (T)	2.05	1.47	1.46	1.52	1.86	1.50
Measured signal amplitude (mV)	204	124	104	148	173	132

The predictions show that a ferrite material would produce superior performance to the mild steel that was tested. This should be investigated in future. SNR measurements were not performed as a housing covering the concentrator and coil would be required and this should be investigated.

4. Conclusions and Recommendations

Numerical modelling was used to optimise the dimensions of two concentric magnets used in the EMAT and the increased magnetic flux density produced a higher sensitivity, which was confirmed experimentally. Further research could be performed on the optimisation of the magnets by considering more levels of the design variables or using a gradient based optimisation algorithm. The use of a single magnet with magnetic field concentrators was predicted to increase the magnetic flux density in the plate and experimental measurements of EMAT sensitivity appear to confirm the numerical predictions. Experiments should be performed with concentrators made from ferrite material incorporated in properly housed and grounded EMATs to determine the performance that can be achieved with this concept.

A coil made using a smaller diameter copper wire and hence more turns further increased the sensitivity of constructed EMATs. It was found that adding a stainless steel housing and grounding the housing, magnet and aluminium plate test specimen produced a significant reduction in the noise. Overall, a substantial improvement of the order of 20 dB in the SNR was achieved through this investigation while the size of the EMAT was reduced.

Acknowledgements

The third and fourth authors acknowledge the financial support from the Wits-UCL Research/Teaching Collaborative Activity Seed Fund.

References

[1] Fromme P, Loveday PW (2024). *Guided ultrasonic wave measurement in plates using low-cost equipment.* In: Health Monitoring of Structural and Biological Systems XVIII, SPIE.

[2] Loveday PW, Fromme P. (2024). *Measurement of Lamb wave velocity using low-cost equipment.* In: Proceeding of the 13th South African Conference on Computational and Applied Mechanics, SACAM 2024.

[3] Hirao M, Ogi H (2017). *Electromagnetic Acoustic Transducers.* Tokyo: Springer Japan.

[4] Sun W, Liu G, Xia H, Xia Z (2018). *A modified design of the omnidirectional EMAT for antisymmetric Lamb wave generation*, Sensors & Actuators A Phys, 282, p. 251–258.

[5] Lan J, Zhang J, Jia X, Gao R (2022). *Optimization Design of Surface Wave Electromagnetic Acoustic Transducers Based on Simulation Analysis and Orthogonal Test Method*, Sensors, 22(2).

[6] Kogia M, et al., (2015). *Electromagnetic acoustic transducers applied to high temperature plates for potential use in the solar thermal industry*, Applied Sciences (Switzerland), 5(4), p. 1715–1734.

[7] Meeker D (2020) *Finite Element Method Magnetics Version 4.2 User's Manual*, https://www.femm.info/Archives/doc/manual42.pdf.

[8] Loveday PW, Fromme P, (2025) *Low-cost instrumentation for high frequency ultrasonic guided wave laboratory research in free rock bolts*, Applied Acoustics, 227.

Measurement of acoustic properties of slurries using low-cost laboratory equipment

Meshack Mabandla[a,b], Philip W. Loveday[b,*]

[a]Measurement and Control Division, Mintek,
Johannesburg, South Africa
[b]University of the Witwatersrand,
Johannesburg, South Africa
[*] corresponding author email: Philip.Loveday@wits.ac.za

Abstract

When an acoustic wave propagates through a fluid containing particles, the wave may be scattered and absorbed causing a measurable change in the sound speed and attenuation of the wave. Measurements of acoustic sound speed and attenuation may therefore be used to estimate the particle size distribution and volume fraction of solids in slurries. While commercial characterization systems are available, these are expensive and may not always be suitable for the investigations to be performed to support mineral processing applications in the mining industry. A low-cost measurement system was developed based in a USB oscilloscope and inexpensive 1 MHz acoustic transducers. When measurements of acoustic wave transmission in water were performed the system achieved better than 80 dB signal to noise level. Measurements were performed on a typical slurry containing small particles with good results. The addition of small quantities of larger particles could be detected although some variations were observed, which require further investigation.

Keywords: acoustic properties of slurries, low-cost measurement.

1. Introduction

When an acoustic wave propagates through a fluid containing particles, some of the energy may be absorbed or scattered causing a measurable change in the velocity and attenuation of the wave. Measurements of acoustic velocity and attenuation may therefore be used to estimate the particle size distribution and volume fraction of solids in slurries by solving the inverse problem [1]. Monitoring of slurry properties is important for optimising gold processing operations. Mintek has developed instruments for on-line measurement of slurry properties including solid volume fraction of larger concentrated carbon particles. It has been observed that other factors may influence the acoustic properties and therefore reduce the accuracy of estimates made by the instruments. Laboratory scale instrumentation is required to perform specific investigations aimed at understanding these errors, such as studying the influence of mineralogy on the acoustic properties [2].

While a commercial instrument was developed, flexible laboratory equipment is not available at Mintek. Sophisticated systems for laboratory testing are available internationally. These systems can perform acoustic measurements on a sample over a wide range of frequencies and with different distances between transmit and receive transducers. Unfortunately, these instruments are expensive and proved to be unaffordable. In addition, the purchase of an instrument does little to further the capability to develop custom online monitoring systems, which is the desired capability required to transfer this technology to gold processing plants. It was therefore decided to investigate if a low-cost measurement system could be developed to produce satisfactory performance over a limited but useful range of parameters.

Acoustic or ultrasonic methods are applicable to slurries with high volume fractions of particles where other techniques fail [3]. Extensive literature is available on the use of acoustic measurements to determine the properties of particle suspensions. If the attenuation of a slurry (as a function of frequency and particle size distribution) is available from a model, and the particle size distribution is known, then the attenuation may be computed [4]. This is the forward problem. Alternatively, if the attenuation of the slurry is measured over a wide frequency range, and the attenuation (as a function of frequency and particle size distribution) is available, the inverse problem may be solved to determine the particle size distribution in the slurry [4]. This is valid when the particles are small compared to the wavelength and viscous effects are the dominant attenuation mechanism [1].

The relative scale between the wavelength and the particle radius has a strong influence on the attenuation of acoustic waves in the slurry and three regimes may be identified [5]. When the particle is much smaller than the wavelength, viscous losses dominate. When the particle size increases there is a transition region followed by the dominance of inertial forces/losses. When the wavenumber (2π/wavelength) multiplied by the particle radius increases past unity then multiple scattering losses become the dominant source of attenuation. If a frequency of 1 MHz is used for the acoustic signal, then based on a graph in [5] particle radii below 1 micron fall in the viscous regime, particles with radii of tens of microns will fall in the inertial regime while particles with radii larger than 1 mm will fall in the multiple scattering regime causing high attenuation.

The measurement of ultrasonic attenuation spectra is detailed in an international standard [6]. Two different techniques for performing the measurement are described. The first technique uses measurements at different distances between the transducers and the attenuation as a function of distance is obtained from these measurements. A minimum of two distances is required but many distance steps can be used with linear regression analysis to improve accuracy. This technique is used in the Dispersion Technology DT100 acoustic spectrometer described in [1] where accurate alignment of the transducers and very accurate measurement of the displacement steps are performed. The second technique uses a measurement in a reference medium (degassed water) followed by a measurement in the slurry. The attenuation caused by the slurry (relative to water) is then determined. This technique requires measurements at only one fixed distance between the transducers.

The objective is to investigate the development of a measurement system for performing slurry measurements, in the laboratory, using only low-cost components. The performance of the system will be quantified by measuring signal-to-noise ratios and showing that it is suitable for measurements in high volume fraction slurries typical in gold processing.

2. Measurement Requirements and Equipment

A low-cost laboratory measurement system, based on technology that could be used in mining operations, is required. The use of a fixed distance between the two transducers was selected as this does not need accurate and reliable positioning of the transducers. Using two transducers, in pitch-catch mode, instead of a single transducer and a reflector, in pulse-echo mode, is easier to implement and is preferred if the transducers are relatively inexpensive.

The receive transducer must detect the acoustic wave, which was transmitted by the transmitter transducer and then attenuated by propagation through the medium. The minimum amplitude of signals that can be detected is dependent on the background noise. Therefore an important performance parameter for the system is the signal-to-noise ratio (SNR). The SNR is generally expressed in dB and is computed as $SNR = 20 \log {}^{v_s}/_{v_n}$, where v_s is the signal voltage and v_n is the noise voltage. The procedure described in [1] involves increasing the drive amplitude until a SNR of 40 dB is achieved before the measurement is performed. It is expected that the slurries of interest would have an attenuation of up to 20 dB/cm more than degassed water. Therefore, if the transducer spacing is 1 cm we require a 60 dB SNR in degassed water.

The monitoring instrument developed by Mintek works at 1 MHz. It is believed that this frequency was selected because it is low enough to have acceptable attenuation, in the slurry with particles of the order of tens of microns, and yet be sensitive to scattering by the larger concentrated carbon particles. It was decided to start by using this frequency and possibly add other frequencies of operation later, if necessary.

It has been shown that a low-cost ultrasonic measurement system for measurement of guided wave ultrasound in rods could be developed for less than USD 1000 [7]. The system to be investigated for slurry measurements is based on this system although the transducers used are different. It is believed that a cost of USD 1000 should be achievable for this system.

The equipment used in this investigation consisted of a USB oscilloscope, which included an arbitrary waveform generator, a drive amplifier and two acoustic transducers.

Inexpensive 1 MHz acoustic transducers were purchased online from AliExpress. No technical information was supplied but the applications listed included ultrasonic flow sensors and level sensors. The transducers were significantly smaller than those previously used by Mintek, which is an advantage for laboratory use as they can fit in a smaller container of slurry. It is believed that the transducer comprises a piezoelectric ceramic disc with a backing material and is cast in polyurethane. The maximum outer diameter was 21 mm but the front face was smaller with a diameter of 17 mm. It is expected that the piezoelectric disc may be 15 mm in diameter or have a radius R of 7.5 mm. At 1 MHz the wavelength, λ, in water would be 1.5 mm and the extent of the near field acoustic region would be less than 20 mm. This was calculated from ${}^{R^2}/_{2\lambda}$[1]. In the near-field region, acoustic spreading can be ignored, and the attenuation is only due to absorption and scattering in the medium. It was therefore decided to experiment with approximately 15 mm between the transmit and receive transducers. A transducer holder was designed and 3D printed. The USB oscilloscope, amplifier, stirrer and beaker, holder and transducers are shown in figure 1.

Fig. 1. USB Oscilloscope, amplifier, transducer and holder, and stirrer in beaker

In previous work [7] the Digilent AD2 USB Oscilloscope was used and performed well. This has been replaced by the Digilent AD3, which was used in this investigation. A Juntek DPA-2698 amplifier was used to amplify the signal from the waveform generator in the USB oscilloscope before applying it to the transmit transducer. This is a low-cost amplifier with a gain of 6 dB and an output power of 5 W per channel.

The total cost of the system was approximately USD 500 or half of the target of USD 1000.

3. Results and discussion

Measurements were initially performed with the transducers held in the holder and submerged in water. This setup was used to determine the performance of the system. The system was then used to perform measurements on a slurry.

3.1. Measurements in water

The performance was first measured by exciting the transmit transducer with a ten cycle Hann windowed tone burst signal, having a 1 MHz centre frequency and measuring the signal from the receive transducer. The distance between the transducers was 14.27 mm and the signals are shown in Fig. 2.

The SNR for this case was determined by computing the RMS noise level during a time before the excitation signal was applied and estimating the peak amplitude of the received signal. The peak amplitude was divided by $\sqrt{2}$ and then divided by the RMS noise and this ratio was expressed in dB in table 1. The noise level can be reduced by averaging and by bandpass filtering. Averaging decreases the noise level by the square root of the number of averages so taking 100 averages decreased the noise by ten times resulting in a 20 dB higher SNR. An eighth order Butterworth filter, with corner frequencies of 0.8 MHz and 1.2 MHz, was then applied to the averaged signal. It was found that this further improved the SNR by approximately 14 dB. The corner frequencies were selected because the bandwidth of a Hann windowed tone burst may be estimated from the centre frequency multiplied by 4 and divided by the number of cycles in the tone burst [8].

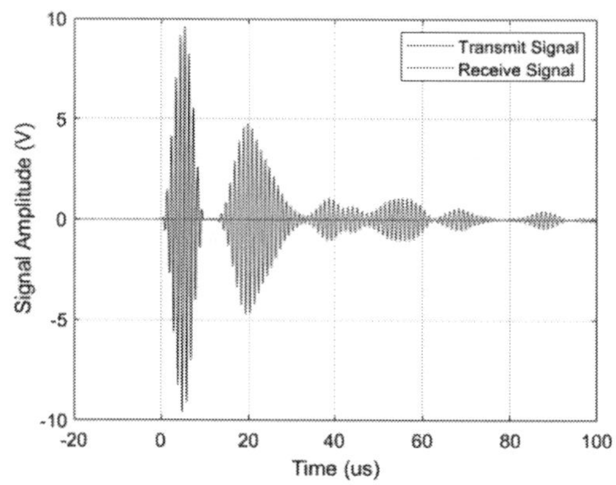

Fig. 2. Transmit and receive signals in water

Table 1. SNR for various measurement parameters

Measurement parameters	SNR (dB)
	A (t)
1 average	48.9
100 averages	68.9
100 averages and bandpass filter 0.8-1.2 MHz	82.9

The transducers were sold as 1 MHz transducers but no information about the bandwidth was provided. The frequency response of the transducer pair was measured by stepping the centre frequency of the excitation signal from 0.5 MHz to 4 MHz and estimating the maximum excitation voltage and the maximum received voltage at each step. The maximum receive voltage was divided by the maximum transmit voltage and this was plotted in Fig. 3. Two peaks are seen in the graph corresponding to the fundamental half-wavelength thickness resonance of the piezoelectric disc and the three half-wavelength second harmonic.

The noise level was also estimated at each frequency, and this was used to plot the SNR at each frequency in Fig. 4. This graph shows that the maximum SNR of approximately 85 dB occurs at 1 MHz and that the SNR is greater than 60 dB for the frequency range of 0.67 MHz to 1.29 MHz.

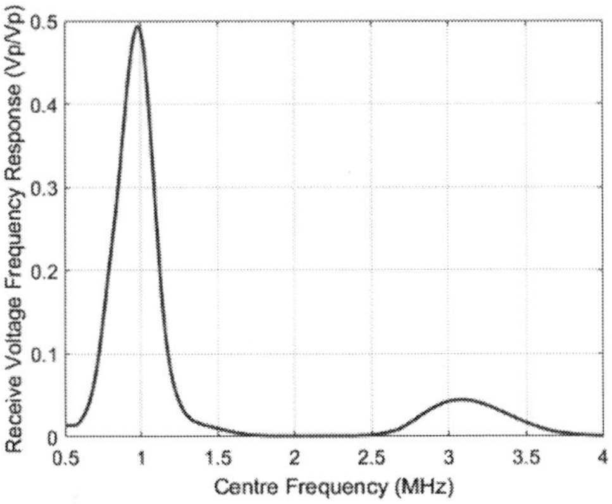

Fig. 3. Frequency response of transducer pair

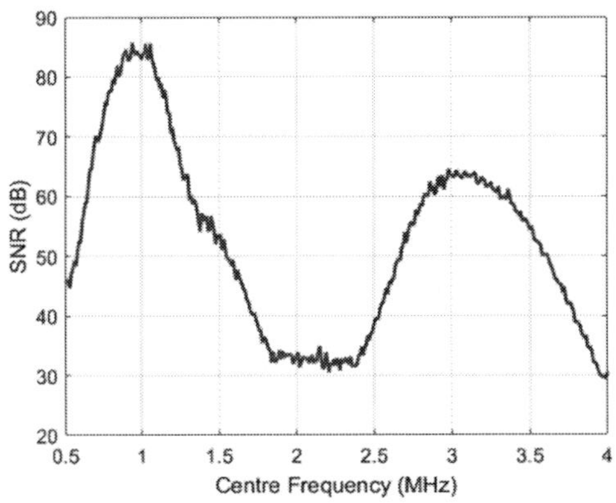

Fig. 4. Signal to noise ratio for varying centre frequency

3.2. Measurements in slurry

A slurry was prepared by combining 4 l of water and 2 kg of mineral particles with size less than 53 microns to produce approximately 5 l of slurry. A stirrer was used to continuously agitate the slurry to prevent the settling of the particles. Measurements across the frequency range were performed on the slurry and captured. Concentrated carbon particles with a size of approximately 3 mm were then added in 20 g increments and the measurements were repeated. The first measurements were performed with 100 averages at each centre frequency, and it was observed that the received signal fluctuated over time. It was speculated that this could be due to there being a different discrete number of large particles between the transducers at any time. Increasing the number of averages to 1000 produced more stable results as the number of particles between the transducers was averaged over a longer time. The measurements, shown in Fig. 5, performed on slurries containing the larger concentrated carbon particles were all performed using 1000 averages to reduce the variations. Fig. 5 shows that the addition of the 2 kg of fine particles increased the attenuation by over 6 dB. The addition of relatively small amounts of the larger concentrated carbon particles then further increased the attenuation as more particles were added. It is noted that the curves became less smooth (there are variations with frequency) when more particles were added. It is believed that the SNR is still very high and electronic noise is not the cause of the variation. This requires further investigation and using more averages could be attempted to decrease the variability.

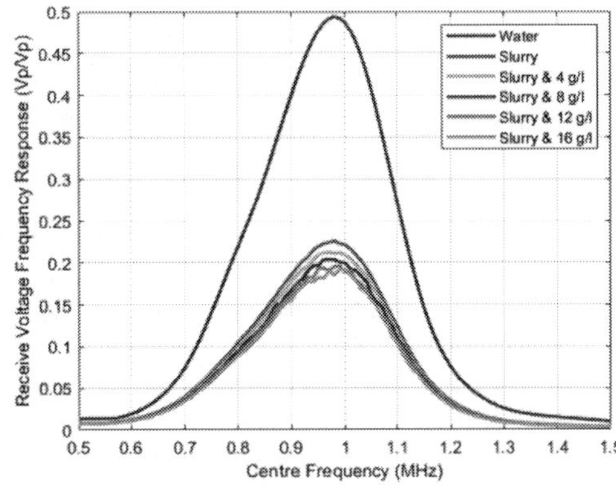

Fig. 5. Influence of addition of particles on signal strength

4. Conclusions and Recommendations

A low-cost measurement system was developed and tested. When measuring water, the system had a maximum SNR of better than 80 dB and the SNR was greater than 60 dB between 0.67 MHz and 1.29 MHz. This represents a fairly narrow frequency range compared to that of commercial systems. Good signals were obtained when measuring a slurry containing fine particles. The low-cost system would be useful for measuring high volume-fraction slurries in the lab and could be modified to perform measurements in processing tanks.

A comparison of the performance of the low-cost system against that of a commercial system should be performed in future if access to such a system can be obtained.

The addition of a relatively low number of larger particles produced some variations in the measurements. These were the only significant variations observed, and it is believed to be due to the small and discrete number of particles present in the measurement region at any time. Longer averaging times may be required, not to reduce electrical noise as initially thought but to average out the number of larger particles that are present in the measurement region. This requires further investigation, which could be performed by manipulating the position of a single particle between the transducers.

Commercial systems perform measurements over a range of accurately measured distances between the transducers to determine the attenuation per mm of propagation. Further development of the low-cost system could include the use of a holder that allows two or three distances between the transducers to be tested.

Measurements could be performed around 3 MHz with this setup to determine if this is possible and if this frequency is more sensitive to the larger particles.

References

[1] Dukhin AS, Goetz PJ (2010). *Characterization of Liquids, Nano- and Microparticulates, and Porous Bodies Using Ultrasound*, Elsevier, 2010

[2] Mabandla SM, et al. (2024). *Investigating the Influence of Ore Mineralogical Composition on Ultrasonic Attenuation and Speed Measurement Techniques*, In: 2024 IEEE Ultrasonics, Ferroelectrics, and Frequency Control Joint Symposium (UFFC-JS), IEEE.

[3] Scott DM (2022). *Recent advances in in-process characterization of suspensions and slurries*, Powder Technology, 399.

[4] Parker D, Lec RM, Pendse HP, Vetelino JF (1990). *Ultrasonic sensor for the characterization of colloidal slurries.* In: Ultrasonics Symposium Proceedings, IEEE, 1990, p. 295–298.

[5] Kytomaa HK (1995). *Theory of sound propagation in suspensions: a guide to size and concentration characterization*. Powder Technology, 82, p. 115–121.

[6] BS ISO 20998-1:2006 *Measurement and characterization of particles by acoustic methods-Part 1: Concepts and procedures in ultrasonic attenuation spectroscopy*, 2006.

[7] Loveday PW, Fromme P (2025). *Low-cost instrumentation for high frequency ultrasonic guided wave laboratory research in free rock bolts*, Applied Acoustics, 227.

[8] Fromme P (2028), *Guided Wave Testing*, In: Handbook of Advanced Non-Destructive Evaluation, N. Ida and N. Meyendorf, Eds., Springer International Publishing.

Decoding Reflection Patterns: A Damage Evolution – Reverberation Interaction Study in Welded Rail Tracks for Ultrasonic Guided Wave Inspections

Dineo A. Ramatlo[a,*], Philip W. Loveday[b,*], Daniel N. Wilke[a,b]

[a]*Department of Mechanical and Aeronautical Engineering, University of Pretoria, Pretoria, 0002, South Africa*

[b]*School of Mechanical, Industrial and Aeronautical Engineering, University of the Witwatersrand, Johannesburg, 2000, South Africa*

**dineo.ramatlo@up.ac.za*

Abstract

The presence of multiple discontinuities, such as welds and damage in welded rail tracks, leads to complex reverberations of ultrasonic guided waves, making the interpretation of inspection signals difficult. In the case of growing damage, the complexity of these signals increases significantly. Growing damage introduces additional propagation paths and more intricate reverberations. Moreover, the growing damage can overlap with reflections from the welds, further complicating signal interpretation. These overlapping signals can obscure the true nature of the damage, making it challenging to assess its size, location, and progression accurately. Understanding how these signal complexities evolve with damage growth is crucial for improving detection methods. This paper simulates ultrasonic inspection signals with a growing crack in the head section of the rail using two numerical modelling methodologies for critical interpretation. The methods account for the excitation, propagation and scattering of guided waves from discontinuities by respectively employing a hybrid model combining a 3D Finite Element (FE) model of a transducer with a 2D Semi-Analytical Finite Element (SAFE) model of the rail and another hybrid method integrating a 3D FE model of the arbitrary discontinuity (weld/damage) with two SAFE models for incoming and outgoing rails. The simulation results explore the synergy between these two methods, where one decodes reflection complexities for understanding and interpreting damage growth – reverberation interaction in the inspection signals. The other approach provides accurate simulations, accounting for all reflections in the waveguide. These insights can help to improve the utility and reliability of guided wave-based monitoring systems.

Keywords: Guided Wave Inspection, Welded Rail Tracks, Damage Evolution, Ultrasound Reverberations.

1. Introduction

Detecting and assessing damage in welded rail tracks is critical for maintaining railway safety. The use of guided wave ultrasound (GWU) facilitates the inspection of long rail lengths from a single transducer location, making it a valuable tool for monitoring. However, one important factor intensifying the complexity of ultrasonic inspections is the presence of multiple reflectors, such as welds and damage. These complexities make interpreting the measurement signals taken on operational rail tracks very challenging, which may make damage detection impossible. Modelling and simulation of GWU are thus essential for decoding the intricate nature of wave interactions caused by multiple discontinuities and improving the design and reliability of monitoring systems. They enable examining scenarios that may not be observable in measurements from operational rail tracks.

Careful selection of approaches in physics-based modelling could enhance or reduce the interpretability of inspection signals. The traditional 3D Finite Element Method (FEM) modelling can provide detailed insight but is typically constrained to short sections of rail. Various numerical approaches exist, with the Semi-Analytical Finite Element (SAFE) method frequently employed for 1D waveguide simulations, allowing for effective long-range wave propagation analyses in rails [1]. However, the method is constrained to waveguides with constant cross-sections. When the guided waves encounter various discontinuities, such as welds and damage, they interact with them, leading to both transmission and reflection of wave energy. This scattering effect can be accounted for by combining a 3D FEM model of arbitrary discontinuity and two SAFE models to represent the semi-infinite incoming and outgoing waveguides [2].

The reflection of wave energy can be direct, implying that the excited guided waves are reflected from a single discontinuity, where the wave energy is redirected back to the pulse-echo transducer along its original path. In addition to these direct reflections, there are other complex reflections where a guided wave reflects multiple times between two or more discontinuities before being redirected back to the transducer. This phenomenon of reverberating waves caused by a network of multiple reflectors is also common in other fields where researchers model and simulate the propagation of waves in multilayer structures. For example, the propagation of seismic waves in multiple layers of the earth, ray propagation modelling to characterise different types of materials, wave propagation in multi-segment transmission lines and non-destructive evaluation of adhesively bonded structures. Fig 1 shows a schematic representation of a series of propagation paths resulting from direct reflections (plotted in black arrows), where the guided waves reflect (R) off a single weld. An example of a multiple reflection between two welds is shown using red arrows. The excited guided waves propagate forward to weld A, reflect (R) from A, propagate backwards, transmit (T) through the transducer and propagate to weld B, where the second reflection happens, causing the waves to propagate forward again back to the transducer. This multiple reflection is called a double reflection.

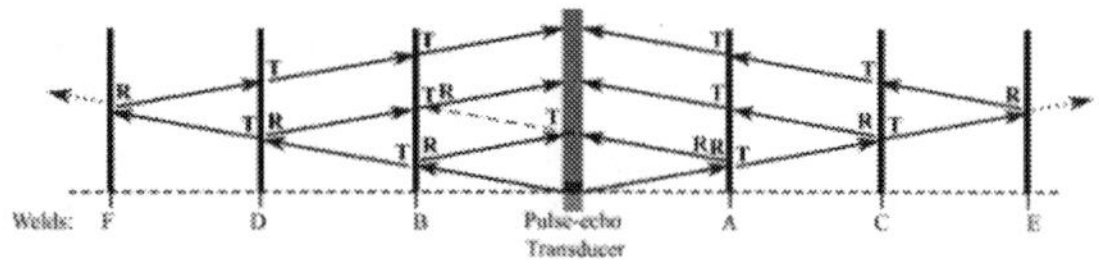

Fig. 1. Schematic representation of direct reflections (in black) and a double reflection between welds A and B (in red). R (Reflection) and T (Transmission).

Two physics-based finite element modelling frameworks have recently been introduced to simulate comprehensive configurations of GWU-based inspections in rail tracks featuring multiple discontinuities, as referenced in [3,4]. Both methods effectively account for the excitation, propagation, and scattering of guided waves interacting with these discontinuities by utilising a resonant model of a piezoelectric transducer [5] and hybrid models of the discontinuities [2]. The key distinction between the two frameworks lies in the first method's approach, which considers a finite number of reflections by simulating each reflection path individually [3]. This facilitates the interpretation and isolation of overlapping reverberations from the various reflectors. In contrast, the second method encompasses all infinite reverberations within the rail, leading to enhanced accuracy [4].

The interaction of ultrasonic-guided waves with discontinuities can cause changes in the amplitude, phase, or shape of the reflected waves, which varies based on the characteristics of each discontinuity. This dynamic further complicates the intricacies of reverberations. When one (or more) of these discontinuities is in the form of growing damage, the complexity of these signals increases significantly. Damage evolution introduces additional propagation paths and creates more intricate reverberations. Furthermore, this ongoing damage can overlap with reflections from welds, adding further challenges to signal interpretation. Such overlapping reflections can obscure the true nature of the damage, making it difficult to assess its size, location, and progression accurately. Therefore, understanding how these signal complexities evolve alongside damage growth is crucial for enhancing detection methods, as they introduce variations that can restrict the capabilities of detecting potential defects.

In this study, we investigate the role of reflection patterns in understanding damage progression over time, particularly in railway lines. By analysing the interaction of complex reverberations from various discontinuities, which are the aluminothermic welds and a crack in the head of the rail, we aim to uncover insights into how these factors influence damage evolution. To enhance our understanding, we utilise the two physics-based modelling frameworks in [3,4] that have proven effective in simulating complex reverberations, ensuring accuracy and facilitating critical interpretation. Our results highlight the synergy between the two methodologies: one decodes the complexities of reflections to interpret damage growth, while the other provides precise simulations that account for all waveguide reflections, offering a comprehensive approach to damage assessment in rail systems.

We begin by briefly presenting the application's inspection setup and simulation procedure. We then offer some preliminary results and present concluding remarks.

2. Inspection setup and the simulation procedure

We consider a section of the UIC60 rail with several reflectors, specifically four aluminothermic welds and a crack in the rail head, as shown in Fig 2. The welds are spaced 240 meters apart. A pulse-echo piezoelectric transducer is placed between welds A and B to excite guided waves through the rail head. The crack is located 108 meters to the right of the transducer. A field experiment of this setup (excluding the crack) was previously conducted on an operational heavy-haul track; measurement details are provided in [6]. This section outlines the simulation procedure for replicating this field experiment using two modelling frameworks that account for finite [3] and infinite reflections [4]. Only key aspects of these methodologies are discussed; full modelling details, including the empirical equations and validation, are provided in [3,4]. The focus here is not on the modelling procedures, but on using these frameworks to decode and interpret complex reflection patterns.

The finite element models for individual components within the rail are depicted in Fig 3. These components encompass excitation through a transducer, wave propagation in regions of consistent cross-section, and scattering due to discontinuities. For the excitation process, a hybrid model integrating a 3D finite element model (FEM) of the piezoelectric transducer with a 2D SAFE model of the rail was used. A comprehensive explanation of this excitation model is presented and validated in reference [5]. The SAFE method for modelling wave propagation is detailed in reference [1]. In contrast, the hybrid approach, which combines a 3D FEM model of arbitrary discontinuities (such as welds or damage) with two SAFE models for incoming and outgoing rails, is thoroughly discussed in reference [2]. Mesh convergence studies were conducted for the individual elements used in both modelling frameworks to ensure accuracy and mesh independence of the simulation results.

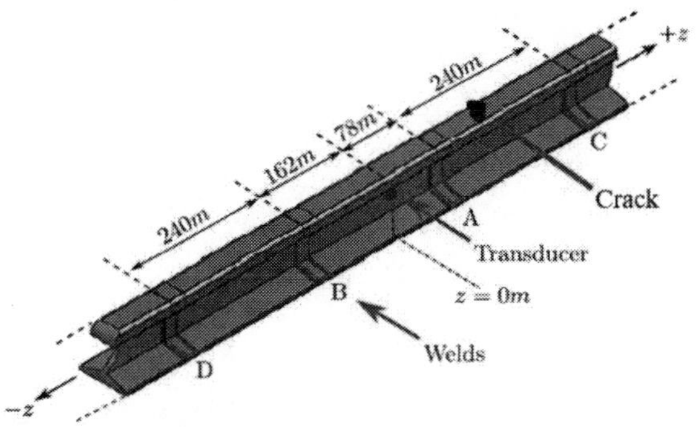

Fig. 2. Inspection setup with four welds (A, B, C and D) and a crack in the rail head.

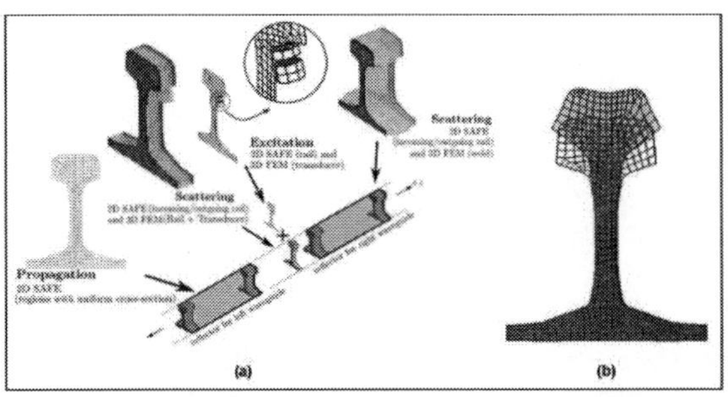

Fig. 3. Models of individual elements in the rail employed in the modelling framework for finite reflections and (a) the modelling framework for infinite reflections; and (b) the prominently excited mode.

Using the first modelling framework in reference [3], the finite reflections are manually simulated by propagating the modes excited by the transducer. The simulation includes all the direct reflections and selected complex reverberations, which are summed together to obtain the resultant response. The second modelling framework for infinite reflections employs the models of individual elements according to Fig 3(a). It aims to represent several local reflectors by introducing a single scattering matrix, as explained in detail in reference [4]. For both methods, the transducer excited several modes. The most prominently excited mode is shown in Fig 3(b).

Damage was modelled using a hybrid model consisting of two SAFE cross-sectional meshes on either side of a 3D mesh enclosing a transverse crack originating at the rail head. Long and Loveday [7] validated the correctness of our implementation by looking at several geometries with defects of different sizes across multiple regions in a UIC60 rail. Fig 6 in Appendix A illustrates the defect geometry of a radial size *r*.

3. Results

Fig 4(a) compares the baseline signal (without damage) to the simulated response signals with damage, obtained using the two modelling frameworks for finite and infinite reflections. The finite reflection results included 52 propagation paths. The strong agreement between both frameworks confirms that 52 selected paths were sufficient to capture the dominant features of the response. The direct reflections associated with each weld are labelled "A-D", while the crack reflections are labelled "cR". The crack reflection "cR path 2", is a direct reflection, while the overlapping reflections (cR paths 17 & 18) are double reflections between the crack and weld B. In Fig 5(a), the crack reflection is decomposed into interpretable components for each path using the method for finite reflections. A schematic representation of the overlapping paths 17 & 18 is shown in Fig 7 in Appendix B.

Fig 4(b) highlights additional crack reflections that emerge as the damage size increases. The crack reflections are decomposed in Fig 5(b). The crack reflection from path 2 displays additional reflection peaks due to crack growth, and new overlapping crack reflections from paths 9-11 are also observed. These complex overlapping reverberations result from waves reflecting five times between the crack, weld A, and the transducer, as illustrated in Fig 8.

Fig 4(c) presents further reflections resulting from additional crack growth. Paths 6 & 7 reflections overlap with a direct reflection from weld B.

4. Conclusions

This paper investigated the role of reflection patterns in understanding damage progression in railway lines. The results showed that as damage grows, the complexity of these signals increases significantly. Growing damage introduces additional propagation paths and more intricate reverberations. Moreover, the growing damage can overlap with reflections from aluminothermic welds, further complicating signal interpretation. These overlapping signals can obscure the true nature of the damage, making it challenging to assess its size, location, and progression accurately.

Being able to interpret these signals, as demonstrated in this study, makes it possible to uncover the true nature of the damage despite overlapping reverberations. This understanding can inform the development of inspection systems and signal processing strategies that better isolate damage-related features and reduce false alarms. Ultimately, this can lead to more accurate, robust damage detection in welded rail systems.

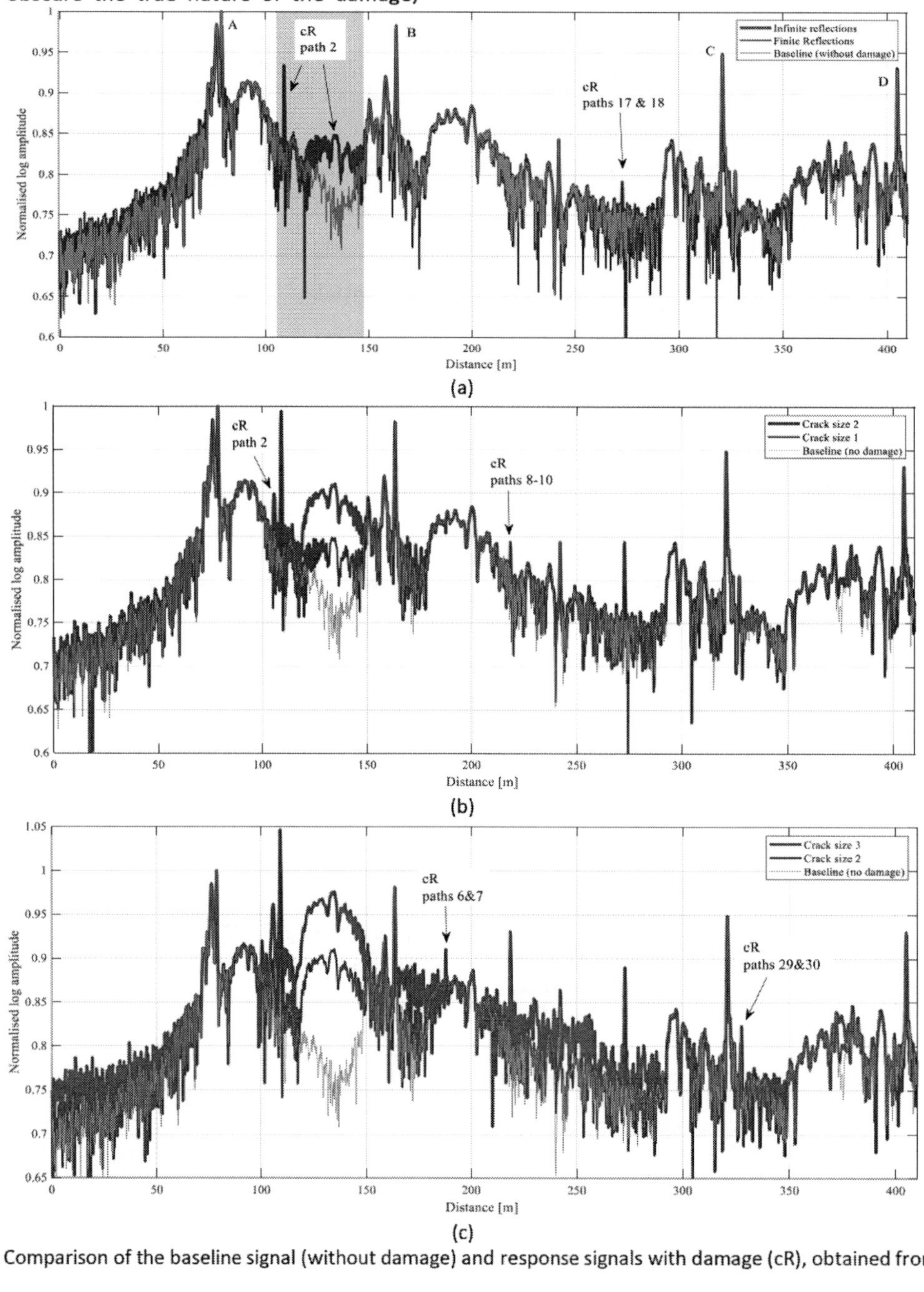

Fig. 4. (a) Comparison of the baseline signal (without damage) and response signals with damage (cR), obtained from the two

modelling frameworks; (b-c) Comparison of the baseline signal and response signals for different crack sizes. The results show that damage growth introduces more complexity and intricate reverberations.

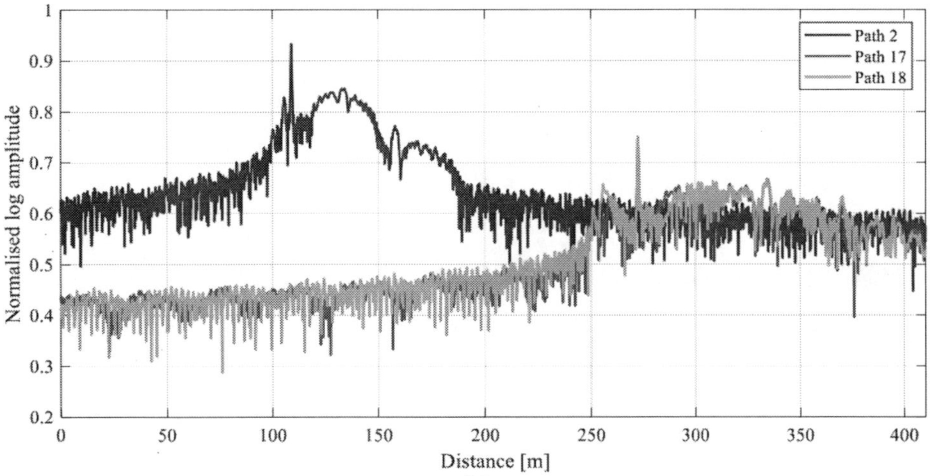

(a)

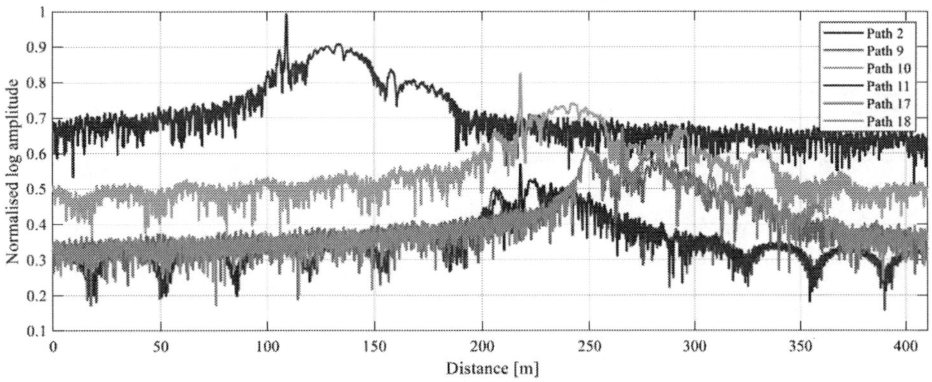

(b)

Fig. 5. (a) Crack reflections decomposed into interpretable components for (a) crack size 1 and (b) crack size 2.

Acknowledgements

AI tools (Grammarly and ChatGPT) were used to improve the spelling and grammar of this document.

References

[1] Gavric L (1995). *Computation of propagative waves*. Journal of Sound and Vibration, **185** (3) 531–543.

[2] Benmeddour F, Treyssède F, Laguerre L (2011). *Numerical modelling of guided wave interaction with non-axisymmetric cracks in elastic cylinders*, International Journal of Solids and Structures, **48** (5) 764–774.

[3] Ramatlo DA, Long CL, Loveday PW, Wilke DN (2020). *A modelling framework for simulation of ultrasonic guided wave-based inspection of welded rail tracks*. Ultrasonics, **108** (2020) 106215.

[4] Ramatlo DA, Long CL, Loveday PW, Wilke DN (2022). *Physics-based modelling and simulation of reverberating reflections in ultrasonic guided wave inspections applied to welded rail tracks*. Journal of Sound and Vibration, **530** (2022) 116914.

[5] Long CS, Loveday PW, Ramatlo DA, Andhavarapu EV (2018). *Numerical verification of an efficient coupled SAFE-3D FE analysis for guided wave ultrasound excitation, Finite Element Analysis & Design,* **149**, 45–56.

[6] Loveday PW, Long CS, Ramatlo DA (2020). *Ultrasonic guided wave monitoring of an operational rail track*. Structural Health Monitoring, **19**(6) 1666–1684.

[7] Long CS, Loveday PW (2013). *Prediction of guided wave scattering by defects in rails using numerical modelling*, In: Proceedings of the 40th Annual Review of Progress in Quantitative Nondestructive Evaluation, **1581**, 240–247.

Appendix A.

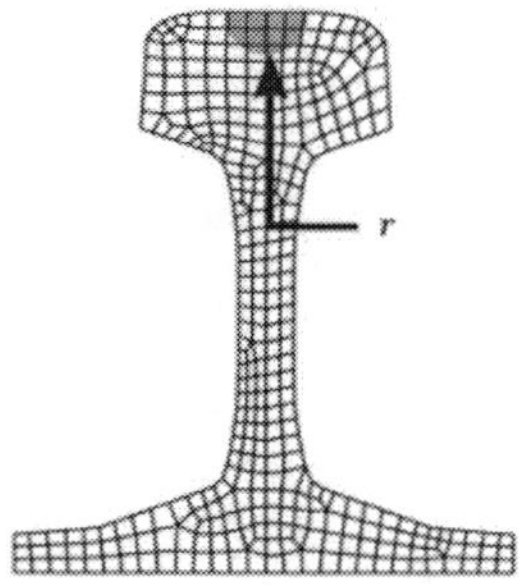

Fig. 6. Crack geometry and sizes considered.

Appendix B.

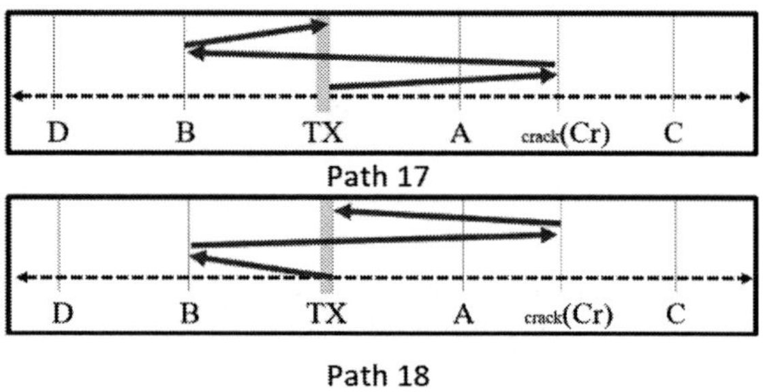

Fig. 7. Propagation paths 17 & 18.

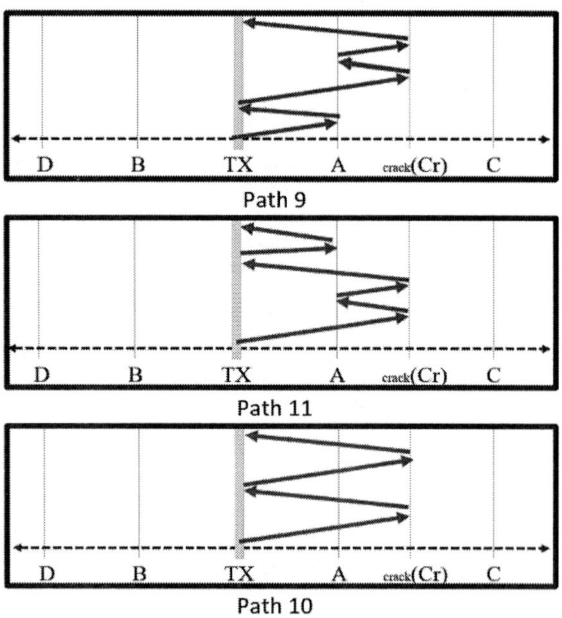

Fig. 8. Propagation paths 9-11.

Signal Analysis of Ultrasonic Testing in 3D Printed MSLA UV-Cured Resin Samples

Z. Luthuli[a*], D.N. Wilke[a], N.R. Mathe[b] and D.A. Ramatlo[a]

[a]*University of Pretoria, Faculty of Engineering, Built Environment and IT, PO Box 20, Hartfield, 0028, South Africa*

[b]*CSIR, National Laser Centre, PO Box 395, Pretoria, 0001, South Africa*

**corresponding author email: z.luthuli@up.ac.za*

Abstract

The masked stereolithography (MSLA) 3D printing method and ultrasonic testing (UT) in samples printed via this method are demonstrated and refined for high-quality sample printing and reliable UT measurements. For repeatable and meaningful results, machine evaluation was successfully performed on the 3D printer, and an uncertainty budget was established for the UT setup to ensure traceability of the measurements. Using this setup, the measured pore sensitivity for a steel material was 1.4176 ± 0.00889 mm, with 97.2% of the uncertainty coming from the measured time delay. Due to the nature of the printing process, the printed samples are anisotropic and exhibit high attenuation high frequencies, which is also demonstrated in this paper. Additionally, the tests reveal beam skewing and wave splitting phenomena inherent in 3D printed materials. These tests were conducted on both defect-free and defective samples to observe the effect of defects on the received ultrasonic signal. This work highlights the importance of gaining deeper insight into the received ultrasonic signal rather than focusing on amplitude and time delay.

Keywords: Ultrasonic Testing, Machine Evaluation, Traceability, Masked Stereolithography, Additive Manufacturing

1. Background and Motivation

Ultrasonic testing (UT) is optimised for isotropic materials with well-researched properties, such as sound propagation velocity derived from modulus and density. The pulse-echo method is suitable and very effective for detecting internal defects in these materials. However, UT in MSLA additive-manufactured (AM) ultraviolet (UV) cured photopolymer resin components is challenging due to anisotropy caused by layered printing. Anisotropy can result in direction-dependent attenuation and wave speeds, affecting signal strength and detection sensitivity differently across the material [1]. In addition to the anisotropic nature, these components can potentially exhibit inhomogeneity if the printing conditions are not controlled, like using a badly mixed resin and non-uniform post-curing of the element [2]. In these components, beam skewing, where the direction of energy propagation may not match the direction of wave propagation, causes the ultrasonic beam to deviate from its expected path [3]. Additionally, wave splitting (birefringence), where a single wave splits into multiple waves travelling at different velocities [4], results in multiple signal paths that lead to complex echo patterns. Furthermore, the nature of these components already discussed above results in higher attenuation at higher frequencies, which makes the pulse-echo method unsuitable for such materials and necessitates the use of lower frequencies for in-depth penetration [4]. Unfortunately, spatial resolution is compromised at these lower frequencies, which limits UT's sensitivity to small defects common in AM parts. Although UT has been demonstrated in some AM processes [5]. Limited research is available on applying UT technology to UV resin-cured materials. In other AM processes like laser powder bed fusion (LPBF), laser-based ultrasonic testing methods have been successfully demonstrated to offer precision advantages [6]. However, compared to piezoelectric transducer-based UT testing, their cost makes them impractical for widespread use. Piezoelectric transducers, primarily used for isotropic materials, have only been applied in fused deposition modelling AM processes [7]. This paper presents the initial experimental application of UT methods to MSLA printed samples, with the aim of establishing a deeper understanding of ultrasonic signal characteristics beyond traditional metrics like time delay and amplitude. The main objectives of this research are to investigate the influence of printer settings on the acoustic response of the printed samples, to characterise the acoustic properties of the printed samples, and to identify signal features that may correlate with the presence of defects or anisotropy in the printed samples.

2. Theory

2.1. Machine Tool Evaluation

During the MSLA 3D printing process shown in Figure 1, a photopolymer resin contained in a tank is exposed to UV light that is selectively masked to cure only specific areas corresponding to a single layer of the desired shape. This process is repeated until the whole sample is printed [8].

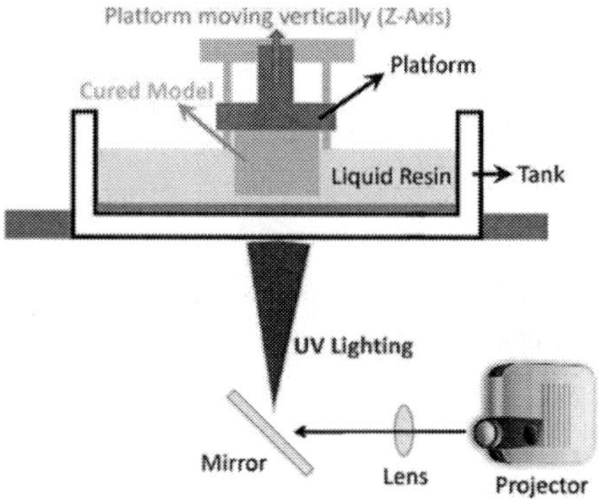

Figure 1: MSLA additive manufacturing process[8]

Machine tool evaluation is a process of evaluating whether a machine is performing as specified by the manufacturer [9]. The goal is to ensure that machines are within the required accuracy specified by the manufacturer for their entire operational lifespan. For 3D printers, this evaluation can be done based on the ISO 230 standard, which examines linear position accuracy, straightness, rotation, and alignment errors [9]. MSLA resin 3D printer manufacturers generally do not perform formal machine tool evaluations. Instead, they recommend using simple methods, such as inserting paper between the vat and printing bed, to mitigate any bed misalignments. Additionally, most resin material or 3D printer manufacturers provide specific procedures and physical standards, such as *"cones of calibration"* shown on the left of Figure 2, for fine-tuning the printer settings to achieve quality results [10].

Figure 2: (left) Cones of Calibration [11], (right) Tetrahedron artefact [12]

These manufacturer-recommended procedures are not calibration procedures but correction methods applied after a calibration. A helpful way to conceptualise this is to imagine a range of printers: depending on your selected settings, you configure a specific resin printer through the recommended manufacturer's procedure. Machine evaluation is then required on the configured printer, and adjustments will be made to achieve the optimum configuration for your photopolymer resin material and sample.

2.2. Traceability of UT measurements

The quality of any measurement is affected by uncertainty, which is important in establishing the traceability of a measurement instrument. Traceability, as defined in the international vocabulary of metrology, is *"the property of the result of a measurement or the value of a standard whereby it can be related to stated references, usually national or international standards, through an unbroken chain of comparisons, all having stated uncertainties"*[13]. Traceability is established by first performing a verification, followed by a task-specific calibration, before conducting the actual measurements [13]. The measurement uncertainty in UT is influenced by several factors, which will not be discussed further in this paper. The reader can refer to [14] for more on the development of an uncertainty budget for any measurement instrument.

3. Methodology

The quality of the measurements in this research depends on both the quality of the 3D-printed samples and the accuracy of the ultrasonic testing setup. In this section, we thoroughly examine the methodology used to print high-quality samples and the approach followed to ensure the accuracy of the UT setup.

3.1. Machine Evaluation in MSLA Machines

In this study, a tetrahedron artefact, shown on the right of Figure 2, was printed for machine evaluation of the 3D printer. The evaluation included an assessment of squareness, misalignments, and shrinkage at different positions within the printing volume. The printed tetrahedron artefact was then measured using a CMM to ensure that misalignments and squareness were within an acceptable tolerance of less than 0.05mm, corresponding to the print layer height. After the successful machine evaluation of the printer, calibration cones shown on the left of Figure 2 were printed alongside every print as a quality assurance measure.

3.2. Ultrasonic Testing Setup

A conventional experimental configuration for acquiring the ultrasonic signal is shown in Figure 3 [15].

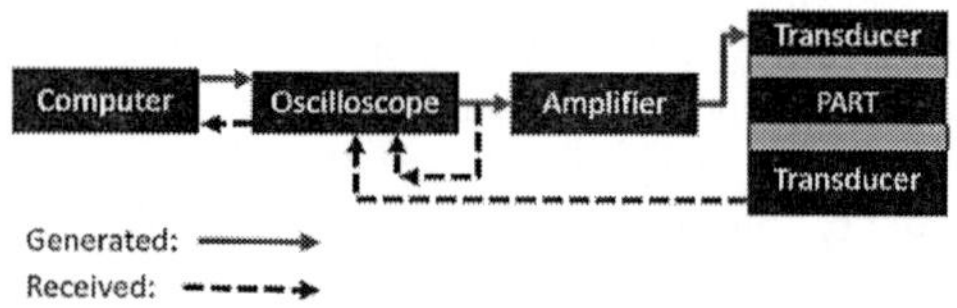

Figure 3: Ultrasonic Testing Setup

The ultrasonic testing (UT) setup is designed to analyse wave propagation characteristics in MSLA printed samples, specifically evaluating signal attenuation and velocity parameters. The experimental configuration consists of a computer with waveform software that generates and processes signal data, an oscilloscope for real-time signal visualisation, an amplifier to enhance signal strength, and piezoelectric transducers that convert electrical signals into ultrasonic waves and vice versa. The transducers are coupled to the sample surface using coupling gel to ensure efficient wave transmission. Several parameters can be controlled to optimise signal acquisition and ensure reliable measurements. These include the excitation signal's amplitude and frequency, transducer placement (opposite sides for through-transmission or the same face for pulse-echo in high-attenuation cases), amplifier gain to maintain signal clarity, and coupling gel application to minimise transmission loss. Adjusting these factors allows for accurate characterisation of the material's ultrasonic response and ensures optimal signal acquisition for further analysis.

In this study, the verification of the UT setup, including the development of an uncertainty budget, was carried out using a 5x6x30mm steel gauge block as a transfer standard. This gauge block was calibrated using an interferometer. For the UT verification measurements, complex 2MHz longitudinal piezoelectric transducers with a 10mm diameter were used in a transmission through configuration. The signal was generated by a laptop as shown in Figure 4, amplified twice (x2) by the amplifier, and transmitted to the transducer at 10V. However, due to uncertainties in the laptop supplied voltage, wiring losses, and transducer torque (tightness applied when pressing the transducer against the sample surface)[16], the actual transmitted signal was consistently slightly below 10V on average.

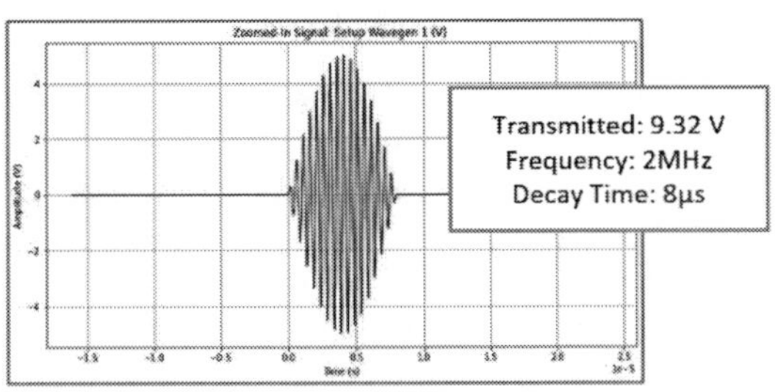

Figure 4: Generated Ultrasonic Signal

From the received signal data, both time delay and amplitude were calculated and used to estimate the ultrasonic sound propagation longitudinal velocity of the steel gauge block and compare it to the known sound velocity of steel [17], confirming conformance and accuracy of the measurements.

This experiment was repeated several times in the uncertainty budget to assess repeatability. Using this information, the Rayleigh diffraction limit [18] and Monte Carlo simulation, the axial and lateral sensitivity of the setup to defects in steel were calculated. Both sensitivities were calculated to be 1.338 ± 0.035 mm at a depth of 3.975 ± 0.0286 mm, with all the measurements reported at a confidence level of more than 95%. Notably, more than 99% of the uncertainty comes from the time delay measured using a computer. At the length of the gauge block (35mm), the lateral sensitivity has decreased to 11.77 ± 0.796 mm, which is relatively too big for defect detection in 3D printed components. Additionally, the setup achieved a signal-to-noise ratio (SNR) of 29.4dB. All human error was mitigated in the setup, and signal averaging as well as removal of electrical crosstalk was performed to ensure accurate measurements. The same uncertainty budget and methodology were used to perform task-specific calibration of the UT setup with 3D printed reference samples that were printed based on the manufacturer's recommended optimum conditions for the Nylon-like black photopolymer resin material [19].

3.3. Key Results and Discussion

After the machine evaluation of the 3D printer and establishment of traceability of the UT setup, extensive printing was done for ultrasonic testing. The ultrasonic signal shown in Figure 4 was transmitted through the sample shown in Figure 5. The sample was printed with a 3mm cubic defect running through its thickness to evaluate the effect of the defect on the received ultrasonic signal. Following a methodology similar to that used for the steel gauge block, the longitudinal wave velocity in the material was estimated to be 2510.47m/s. Applying the Rayleigh diffraction limit, the spatial resolution was estimated at 0.641mm. Based on this resolution, the system is theoretically capable of detecting the fabricated 3mm defect.

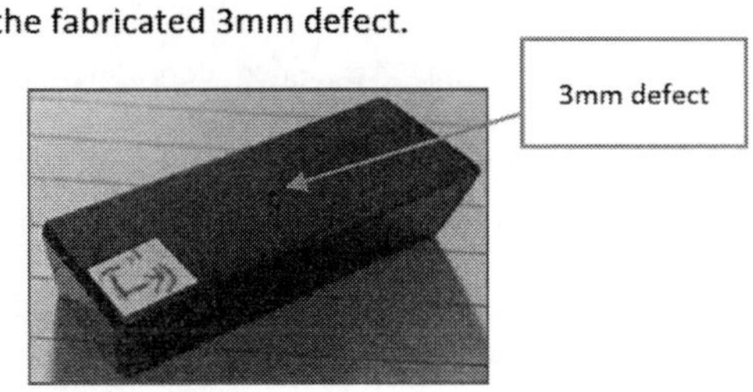

Figure 5: Black Like nylon resin MSLA Printed Sample (11x11x50mm)

Initially, the signal was transmitted through the defect-free section of the sample. The received signal is shown in Figure 6, along with its Fast Fourier Transform (FFT).

FFTs are generally used to provide more insight into the effect of the sample on the signal frequencies. The same generated signal was then transmitted over the defect, and the corresponding received signal and its FFT are also included in Figure 6.

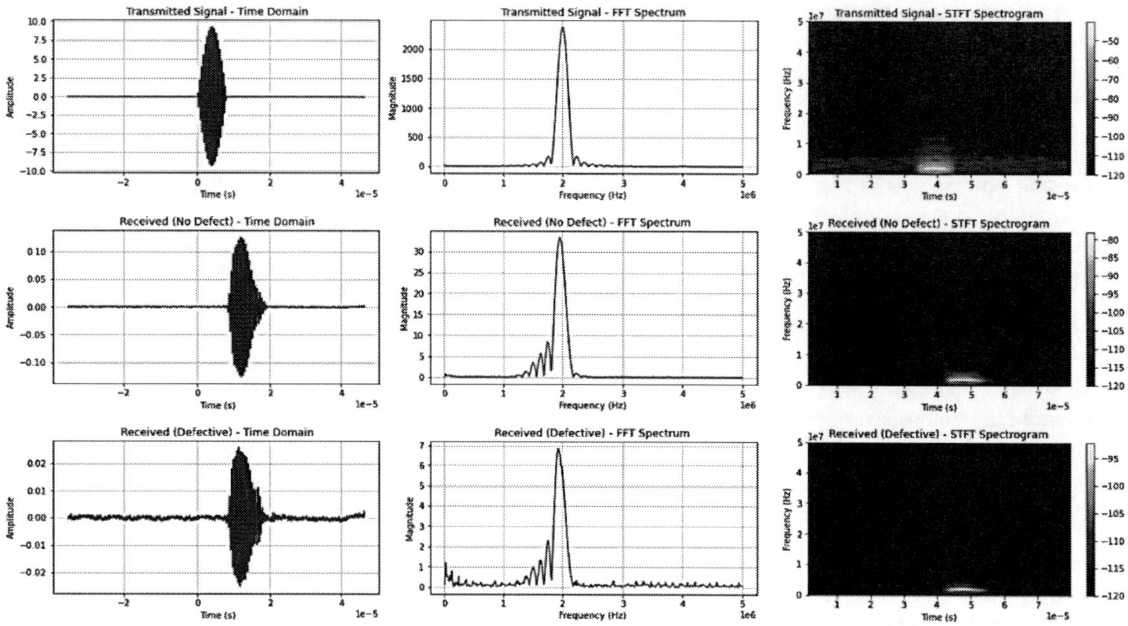

Figure 6: Time domain, FFT, and Short Time Fourier transform (STFT) of the transmitted signal, received defect-free signal, and received defective signal

Comparing the received signal in the section without a defect to the received signal from the section with a defect, clear differences can be seen in both the amplitude and frequency. When the signal was transmitted through the section without a defect, the received signal amplitude was approximately 0.15V, which is significantly lower than the transmitted amplitude of 9.32V. On the other hand, when the signal passed through a defect, the received amplitude dropped further to 0.025V. The FFT shows that although the dominant frequency remains close to 2MHz in both cases, all frequencies are attenuated, with the higher frequency components attenuated more significantly, especially in the defective sample. The short-time Fourier transform (STFT) further confirms this with the energy concentrated around the dominant frequency. In the defective signal, the higher frequencies decay more rapidly and are less pronounced overall.

Additionally, both received signally took more than 9.0 μs to decay fully, compared to the 8.0 μs of the transmitted signal. This increase in decay time can be attributed to wave splitting and birefringence. A two-sample t-test comparing the filtered raw data of the received signals yielded a p-value of 0.95, indicating that the difference in the mean amplitudes of the two received signals is not statistically significant at common confidence levels [20]. Although the above t-test analysis on time domain averages does not clearly indicate a detection of the defect, amplitude decay and frequency domain analysis provide a more reliable statistical basis that the flaw has been detected.

3.4. Conclusion and Future Work

In addition to evaluating the MSLA resin printer to produce quality printed samples, and establishing traceability of ultrasonic testing measurements. We have demonstrated that MSLA additively manufactured samples exhibit significant attenuation of high frequencies and indicated the importance of an in-depth analysis of the received ultrasonic signal during ultrasonic testing of such samples.

Acknowledgements

AI tools such as Grammarly and ChatGPT were used for the improvement of spelling and grammar in this document, but were not used to generate any ideas or content.

References

[1] Putkis, O., R.P. Dalton, and A.J. Croxford, *The anisotropic propagation of ultrasonic guided waves in composite materials and implications for practical applications.* Ultrasonics, 2016. **65**: p. 390-399.

[2] Alshamrani, A.A., R. Raju, and A. Ellakwa, *Effect of Printing Layer Thickness and Postprinting Conditions on the Flexural Strength and Hardness of a 3D-Printed Resin.* BioMed Research International, 2022. **2022**(1): p. 1-9.

[3] Hervin, F. and P. Fromme, *Guided wave propagation and skew effects in anisotropic carbon fiber reinforced laminates.* The Journal of the Acoustical Society of America, 2023. **153**(4): p. 2049-2049.

[4] Solodov, I., Y. Bernhardt, L. Littner, and M. Kreutzbruck, *Ultrasonic anisotropy in composites: Effects and applications.* Journal of Composites Science, 2022. **6**(3): p. 1-18.

[5] Cerniglia, D., M. Scafidi, A. Pantano, and J. Rudlin, *Inspection of additive-manufactured layered components.* Ultrasonics, 2015. **62**: p. 292-298.

[6] Wang, F., H. Mao, D. Zhang, X. Zhao, and Y. Shen, *Online study of cracks during laser cladding process based on acoustic emission technique and finite element analysis.* Applied surface science, 2008. **255**(5): p. 3267-3275.

[7] Bhavsar, P., B. Sharma, W. Moscoso-Kingsley, and V. Madhavan, *Detecting first layer bond quality during FDM 3D printing using a discrete wavelet energy approach.* Procedia Manufacturing, 2020. **48**: p. 718-724.

[8] Zol, S.M., M.S. Alauddin, Z. Said, M.I. Mohd Ghazali, L. Hao-Ern, D.A. Mohd Farid, N.A.f.H. Zahari, A.H.A. Al-Khadim, and A.H. Abdul Aziz, *Description of poly (aryl-ether-ketone) materials (PAEKs), polyetheretherketone (PEEK) and polyetherketoneketone (PEKK) for application as a dental material: a materials science review.* Polymers, 2023. **15**(9): p. 1-22.

[9] *ISO 230 - Test Code for Machine Tools* International Organisation for Standardisation [accessed 2024 December]; Available from: https://www.iso.org/search.html?PROD_isoorg_en%5Bquery%5D=ISO%20230.

[10] *TableFlip Foundry*. The Cones of Calibration version 3 [accessed 2024 December]; Available from: https://www.tableflipfoundry.com/3d-printing/the-cones-of-calibration-v3/.

[11] All3DP. *How to calibrate your 3D printer: step by step guide*. [accessed 2024 December]; Available from: https://all3dp.com/2/how-to-calibrate-a-3d-printer-simply-explained/.

[12] Acko, B., M. McCarthy, F. Haertig, and B. Buchmeister, *Standards for testing freeform measurement capability of optical and tactile coordinate measuring machines.* Measurement science and technology, 2012. **23**(9): p. 1-13.

[13] Luthuli, Z., *Traceability of measurements in optical coordinate measuring machines*, in *Department of Mechanical Engineering*. 2020, Stellenbosch University: South Africa.

[14] Kirkup, L., *A guide to GUM.* European journal of physics, 2002. **23**(5): p. 483-487.

[15] Gaydecki, P., F. Burdekin, W. Damaj, and D. John, *The propagation and attenuation of medium-frequency ultrasonic waves in concrete: a signal analytical approach.* Measurement Science and Technology, 1992. **3**(1): p. 126 -131.

[16] Jiang, X., K. Wang, and D. Zhang, *Determining the optimal pre-tightening force of a sandwich transducer by measuring resonance resistance.* Applied Acoustics, 2017. **118**: p. 8-14.

[17] *Engineering Toolbox*. Acoustic Properties [accessed 2024 November]; Available from: https://www.engineeringtoolbox.com/.

[18] Chandran, L., M.S. Syed Akbar Ali, B. Bobbs, C. Dutta, J. Jd, E. Bhattacharya, and P. Rajagopal, *Micron-scale imaging using bulk ultrasonics.* Scientific Reports, 2024. **14**(1): p. 1 - 16.

[19] *Sunlu*. Nylon like black resin [accessed 2025 July]; Available from: https://www.diyelectronics.co.za/store/specialised/5132-sunlu-nylon-like-resin-black-1-litre.html?srsltid=AfmBOopKAayQlpnk7FaiQy

vKnZmteG1N84W04M15Xxst5Gl9xWpzx5Kl.

[20] Witt, J.K., *Insights into criteria for statistical significance from signal detection analysis.* Meta-Psychology, 2019. **3**.

Estimating Epistemic and Aleatoric Uncertainties in Experimental Rolling Element Bearing Condition Monitoring Data

Ayomide D. Osetoba[1]*, Stephan Schmidt[1], Daniel N. Wilke[1,2]

1. Centre for Asset Integrity Management, Department of Mechanical and Aeronautical Engineering, University of Pretoria, South Africa. Private bag X20, Hatfield 0028, South Africa.
2. Department of Mechanical, Industrial and Aeronautical Engineering, University of Witswatersrand, Braamfontein, Johannesburg, South Africa.

ayomide.osetoba@tuks.co.za*, stephan.schmidt@up.ac.za, daniel.wilke@wits.ac.za

Abstract

Rolling Element Bearings (REBs) are prone to failure, so Condition Monitoring (CM) is often performed to enable more effective maintenance. The measured CM data are used to diagnose REB faults and to predict the remaining useful life of the bearings. However, complex epistemic and aleatoric uncertainties are often present in the measured data due to routine inspections, maintenance, and replacements of parts and changes in the operating and environmental conditions of the system. These uncertainties affect the performance of CM methods.

This study investigates these uncertainties by analysing vibration measurements from a bearing test bench to quantify the impact of aleatoric and epistemic uncertainties in REB CM. The results highlight the importance of considering these uncertainties when performing experiments and developing and using CM methods.

Keywords: Rolling Element Bearings, Condition Monitoring, Epistemic Uncertainty, Aleatoric Uncertainty, Experimental Data.

1. Introduction

Rolling Element Bearings (REBs) are pervasive and critical machine elements, often associated with most rotating machinery, where their failures can result in catastrophic consequences [1]. REBs have relatively low costs and operational simplicity, and combined with the risks, make them sensible candidates for Condition-Based Maintenance (CBM).

To enhance the reliability of mechanical equipment and to reduce downtime, several CBM techniques have been developed and implemented, such as vibration analysis, temperature-based analysis, and acoustic emission [2]. Over the past decade, several data-driven techniques have been developed, such as machine learning and Deep Learning (DL) techniques [3]-[7]. Mantani et al. [8] conducted a detailed literature review on DL for rotating machinery and concluded that these data-driven techniques have yet to be widely implemented in real-life applications.

To ensure the reliability and efficiency of these CBM techniques, experimental data is needed. This measured data is used in developing and validating CBM techniques for diagnostics and prognostics of REBs. However, the data can be incomplete (e.g., missing key attributes of interest), have low signal-to-noise ratios, and contain errors or extraneous components that disrupt the overall pattern. This can introduce both epistemic and aleatoric uncertainties. Epistemic uncertainty arises from a lack of knowledge or data about the system under consideration, such as operational conditions and characteristics of fault behaviour [9]. Aleatoric uncertainty arises due to the inherent randomness in the system under consideration [10]. In general, epistemic uncertainty arises from a lack of knowledge or insufficient data about the system under consideration. In the context of CBM, it often arises from unseen faults that lead to out-of-distribution data, as well as from physical factors such as variations in installation techniques. Aleatoric uncertainty, on the other hand, stems from noise in the data [11]. These uncertainties, if not captured, affect the performance of CBM techniques [10].

To the authors' knowledge, researchers have not fully quantified or investigated the influence of these experimental uncertainties during data collection.

One commonly used open-access dataset is the one from Case Western Reserve University (CWRU) [12]. Smith & Randall [13] noted that in many cases, for the CWRU dataset, the test bench assembly influenced the diagnosis of bearing faults more significantly than the faults themselves. Using the CWRU datasets to develop a multi-label approach for DL models, Rosa et al. [14] realised during error analysis that their proposed model struggled to distinguish between outer and inner race faults, thereby attributing the cause to issues previously identified by Smith & Randall [13]. Hendriks et al. [15] urged researchers to revisit this issue and identify how it affects the practical applicability of data-driven methods.

Consequently, in this study, epistemic uncertainty is focused on the assembly and disassembly of a bearing in a bearing block of an experimental bearing test bench. Aleatoric uncertainty is focused on the variability in the measured data, which results from sensor readings, noise, etc.

This paper presents a newly developed method to quantify and investigate isolated epistemic and aleatoric uncertainties using a bearing test bench and develops a statistical approach to quantify them. The Squared Envelope Spectrum (SES), a widely recognised vibration analysis technique [16] and one of the most effective tools for bearing condition monitoring in industrial applications [17], is employed for the analysis in this study.

In the following sections, Section 2 describes the experimental setup and data collection processes and processing, followed by the presentation and discussion of results in Section 3. Finally, Section 4 concludes the paper.

2. Experimental Analysis

2.1 Experimental Setup

In this study, the bearing test bench in the Centre for Asset Integrity Management (C-AIM), in the Department of Mechanical and Aeronautical Engineering, University of Pretoria, is used for this experiment. The bearing test bench is shown in Fig. 1. It is powered by a 3-phase induction motor with a power rating of 0.55kW and a maximum rotational speed of 2850rpm. It contains two Tapered Rolling Bearings (TRB), one as the support bearing and the other as the target bearing. Three industrial piezoelectric ICP accelerometers were positioned along the x, y, and z axes to record the vibration data. A four-channel QuantumX Data Acquisition System (DAQ) stores the measured data. The shaft has a radial bearing load in the form of a disk attached to it, which could be used to introduce unbalances.

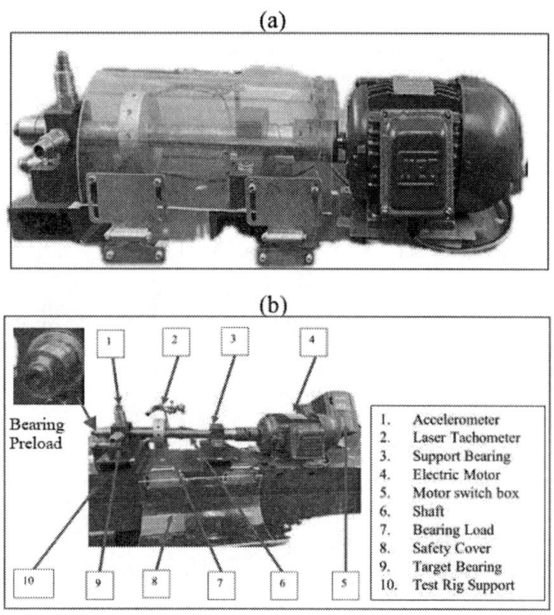

Figure 1: Bearing Test Bench. (a) Test bench with safety cover closed; (b) Test bench when the safety cover is opened with the identification of the components.

TRBs, a type of REB, comprise an outer race, an inner race, and tapered rollers. One bearing condition is considered: a damaged outer race state. A seeded fault is introduced into the outer race to induce damage. The fault has a depth of 0.6mm, a width of 1.43mm, and was applied at an angle of 76° as shown in Fig. 2.

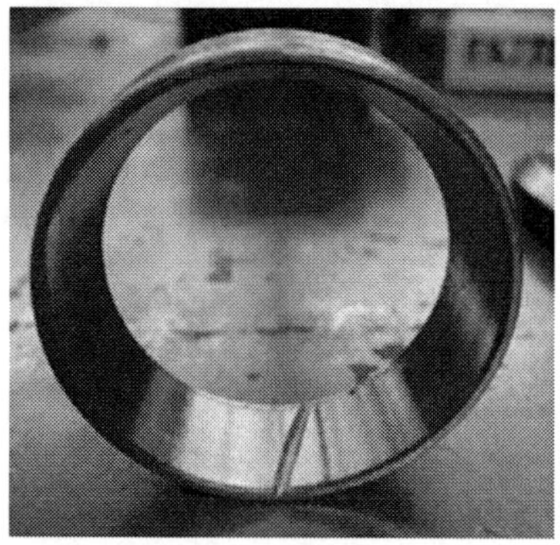

Figure 2: Damaged rolling element bearing outer race.

2.2 Vibration Data Collection and Analysis

In the data collection process, two cases are considered:

- Case 1: Initial vibration data recorded with the test bench in its baseline condition, immediately after it was set up.
- Case 2: Vibration data recorded after only the bearing preload (i.e., the axial tightening force applied during mounting, as shown in Fig. 1) was removed, followed by a visual inspection and reinstallation.

The data is recorded five times for each case at an interval of three minutes, with each recording lasting 10seconds. The five recordings are labelled R1-R5. To capture detailed bearing operation and prevent aliasing, a high sampling frequency of 48kHz is chosen, exceeding the 40kHz threshold recommended by Smith & Randall [18], who noted that bearing faults often occur at high frequencies. This high sampling rate also allows for a thorough understanding and quantification of the data variation every second. A constant high speed of 2000rpm is selected for this data collection experiment. The fault characteristics frequencies corresponding to TRB are given in Table 1. These frequencies are calculated using the formulae presented by Randall & Antoni [16]. The values for the TRB physical characteristics are obtained from the SKF website [19]. The above data collection cases and procedures aim to demonstrate how epistemic and aleatoric uncertainties influence CM techniques, particularly in this study, SES. Specifically, the cases were designed to introduce epistemic uncertainty, while the measurements within each case aim to introduce aleatoric uncertainty.

Table 1: The shaft's speed and the corresponding tapered REB fault frequencies.

Frequencies	Value [Hz]
Shaft Speed	33.33
Ball Pass Frequency Outer race (BPFO)	276.27
Ball Pass Frequency Inner race (BPFI)	357.07
Fundamental Train Frequency (FTF)	14.54

It should be noted that prior investigations have been carried out, and temperature effects that could contribute to epistemic uncertainty have been investigated through these multiple tests carried out under controlled conditions. The disassembly and assembly of the bearing in the bearing block were found to be more dominant than any residual effect due to running temperature.

The flowchart for processing and analysing this vibration signal is shown in Fig. 3. The collected 10seconds signal, referred to as the long signal, is further divided into 1 second segments, termed segmented signals. This segmentation enables detailed analysis and quantification within each dataset and facilitates the observation of the influence of aleatoric uncertainty in these datasets.

The segmentation length of 1 second is selected based on the detailed investigation carried out using simulated signals. It was observed that aligning segmentation with the bearing impulse period or its multiples does not consistently reduce the uncertainty introduced by segmentation, due to varying bearing behaviours and the DFT's assumption of periodicity. Consequently, a 1-second segmentation length is arbitrarily chosen without significantly affecting the overall analysis.

Using SES techniques, a significant peak at any of the fault frequencies listed in Table 1, or their harmonics, indicates the presence of a fault in the bearing. Since the seeded fault is located in the outer race of the bearing, the peak of the SES corresponding to the Ball Pass Frequency Outer race (BPFO) was identified within a ± 5Hz band. This range was chosen to account for possible deviation of the BPFO from its theoretical value, effectively highlighting the bearing fault. Additionally, the band accounts for potential frequency shifts caused by variations in motor speed, as fault frequencies depend on rotational speed. The frequency domain signals and SES are normalised by scaling the Fast Fourier Transform (FFT) output according to the total number of samples, as expressed in the SES formula in Equation 1. The conventional envelope analysis method is used in this study; however, it is important to note that the signal was not bandpass filtered, as the fault could be readily identified in the SES without filtering.

$$\mathbf{SES_{norm}[k]} = \left| \frac{1}{N} \sum_{n=0}^{N-1} |\mathcal{A}(n)|^2 \, e^{-j\frac{2\pi kn}{N}} \right| \quad (1)$$

where $\mathcal{A}(n)$ is the analytic signal obtained using the Hilbert Transform, $|\mathcal{A}(n)|^2$ is the squared envelope of the original signal, N is the number of samples in the signal, and k is the frequency bin index.

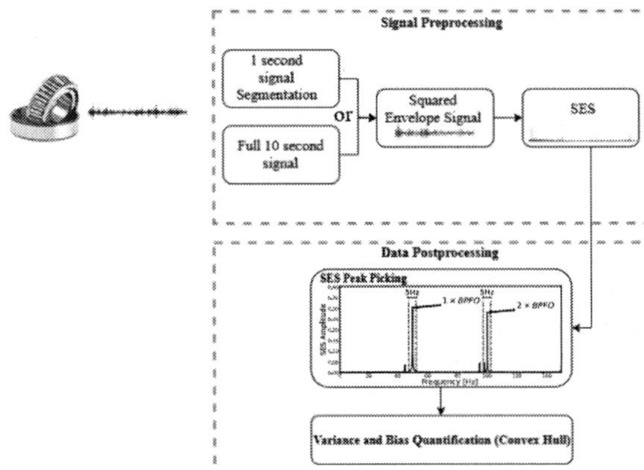

Figure 3: Vibration signal processing flow chart.

Using SES, the amplitude of the first two harmonics of the BPFO are extracted and subsequently their convex hull is calculated. The convex hull is a non-parametric method used to estimate the support of the distribution of the observations and their bivariate dispersion.

3. Results and Discussion

In Fig. 4, the long signal of Case 1's first recording (R1) is shown with its spectrum and normalised SES. From Fig. 4c, a significant peak at the BPFO and its harmonics can be observed, indicating the presence of an outer race fault on the REB in this study.

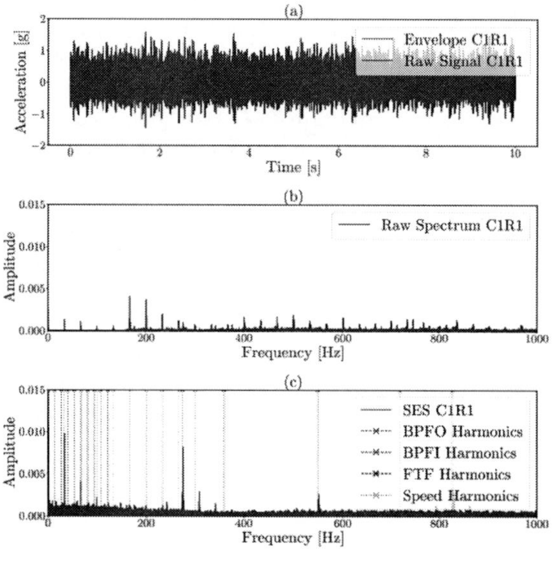

Figure 4: Case 1 Signals. (a) Envelope signal superimposed on the vibration raw signal (b) Normalised frequency domain of the raw vibration signal (c) Normalised SES of the raw vibration signal.

Fig. 5 shows the extracted BPFO peaks for the long signal (big markers) and the 1 second segmented signals (small markers) for both Case 1 (Fig. 5a) and Case 2 (Fig. 5b). The different colours highlight the different recordings (R1-R5) for each case. The segmented signal is enclosed within a convex hull, encompassing 80% of the maximum distance from the mean (indicated by a bolded diamond marker) of each recording. In contrast, the remaining points outside the hull are treated as outliers. This percentage was selected as its variance better represents the data for both cases. In this figure, the aleatoric uncertainty (variance) is observed for the segmented signal by visualising the 2 × BPFO against the 1 × BPFO. Due to noise and the system's operating conditions, the BPFO peak for each recording fluctuates, and it can be seen to be decreasing progressively from recording one to five in Case 1. In Case 2, no distinct pattern is observed;

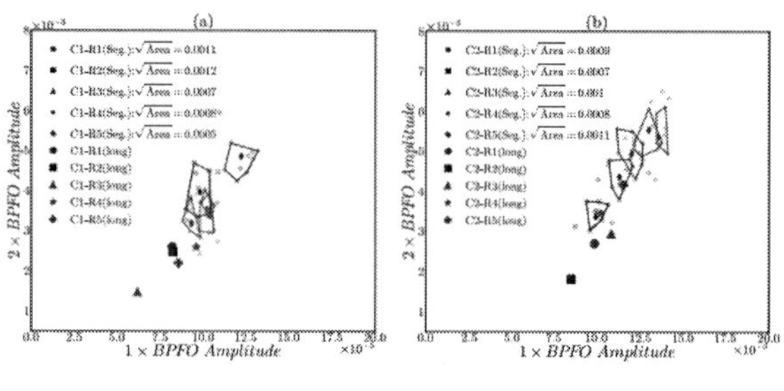

Figure 5: Quantification within each dataset. (a) Case 1. (b) Case 2. Note that small marker represents segmented signal while big marker represents long signal.

however, significant variations across all recordings highlight the presence of aleatoric uncertainty (variance) in the data.

Fig. 6 presents the quantification across each case for all recordings, including segmented and long signals. Both cases are enclosed within separate convex hulls, capturing 80% of the maximum distance from the mean (indicated by a bolded marker). This figure shows a significant offset in Case 2 compared to Case 1 in both segmented and long signals, indicating the presence of epistemic uncertainty (bias) in the data. This uncertainty arises from disassembling and reassembling the bearing preload within the bearing block.

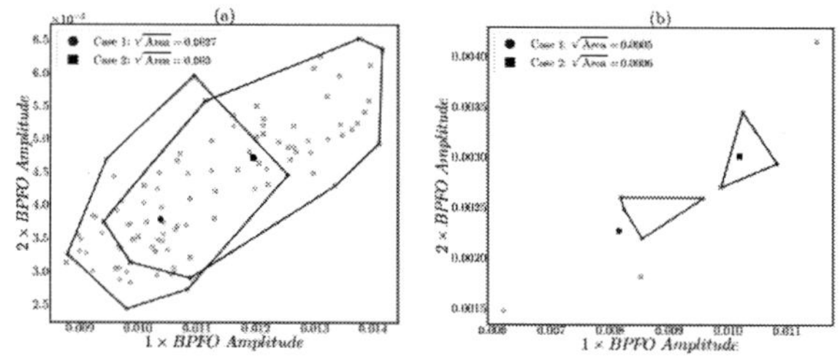

Figure 6: Quantification across each case. (a) Segmented signal. (b) Long signal.

The square root of the area of each polygon in Figures 5 and 6 is calculated, as indicated in the figure legends. These values serve as the measure of the spread of the extracted peaks within each recording and across the different cases. In Fig.5, these values alternate between increasing and decreasing in an oscillatory manner within the recordings for both Case 1 and Case 2. In Fig. 6, there is an increase in the square root of the area from Case 1 to Case 2 for both segmented and long signals. The inherent variability in these values further highlights the presence of aleatoric uncertainty within each dataset and epistemic uncertainty across the two cases.

In general, during experimental bearing testing, the test bench is repeatedly disassembled and reassembled and can introduce a bias (epistemic uncertainty) into both the collected and future data This effect is exemplified in this study, as shown in Fig. 6. If this bias, along with the variance (aleatoric uncertainty), is not accounted for, it can impact the reliability and robustness of CM methods.

Figures 5 and 6 demonstrate the significant influence and importance of quantifying these uncertainties. It is crucial to account for these experimental uncertainties during the development of CM techniques to achieve robust and reliable methods and ensure their practical applicability.

4. Conclusion

This study provides valuable insights into epistemic (bias) and aleatoric (variance) uncertainties in REB condition monitoring experimental data involving a damaged outer race. A novel approach to REB experimental data collection and processing was developed to quantify the datasets' BPFO peak in the SES. This includes computing the convex hull as a non-parametric way to estimate the support of the observed distributions and as an estimate of their bivariate dispersion. The findings illustrate the influence of these experimental uncertainties, emphasising their significance and underscoring the need for meticulous data collection practices when experimenting.

Acknowledgements

A. Osetoba gratefully acknowledges the Mastercard Foundation Scholars Program for supporting his studies.

References

[1] F. Xu et al., "A review of bearing failure Modes, mechanisms and causes," Engineering Failure Analysis, vol. 152 Oct. 2023, doi: 10.1016/j.engfailanal.2023.107518.

[2] E. Yadav and V. K. Chawla, "An explicit literature review on bearing materials and their defect detection techniques," in Materials Today: Proceedings, Elsevier Ltd, 2021, pp. 1637–1643. doi: 10.1016/j.matpr.2021.09.132.

[3] B. Ran, Y. Peng, and Y. Wang, "Bearing degradation prediction based on deep latent variable state space model with differential transformation," Mech Syst Signal Process, vol. 220, Nov. 2024, doi: 10.1016/j.ymssp.2024.111636.

[4] M. M. Manjurul Islam, A. E. Prosvirin, and J. M. Kim, "Data-driven prognostic scheme for rolling-element bearings using a new health index and variants of least-square support vector machines," Mech Syst Signal Process, vol. 160, Nov. 2021, doi: 10.1016/j.ymssp.2021.107853.

[5] S. Du, J. Liu, and L. Zhao, "Research on Rolling Bearing Condition Monitoring Method Based on Deep Learning," in Proceedings - 2020 Chinese Automation Congress, CAC 2020, Institute of Electrical and Electronics Engineers Inc., Nov. 2020, pp. 74–78. doi: 10.1109/CAC51589.2020.9327515.

[6] P. S. Jadhav et al., "Identification and Fault Diagnosis of Rolling Element Bearings Using Dimension Theory and Machine Learning Techniques," J Tribol, vol. 146, no. 9, Sep. 2024, doi: 10.1115/1.4065335.

[7] M. Kahr, G. Kovács, M. Loinig, and H. Brückl, "Condition Monitoring of Ball Bearings Based on Machine Learning with Synthetically Generated Data," Sensors, vol. 22, no. 7, Apr. 2022, doi: 10.3390/s22072490.

[8] O. Matania, I. Dattner, J. Bortman, R. S. Kenett, and Y. Parmet, "A systematic literature review of deep learning for vibration-based fault diagnosis of critical rotating machinery: Limitations and challenges," Nov. 10, 2024, Academic Press. doi: 10.1016/j.jsv.2024.118562.

[9] T. Zhou, T. Han, and E. L. Droguett, "Towards trustworthy machine fault diagnosis: A probabilistic Bayesian deep learning framework," Reliab Eng Syst Saf, vol. 224, Aug. 2022, doi: 10.1016/j.ress.2022.108525.

[10] G. Li, L. Yang, C. G. Lee, X. Wang, and M. Rong, "A Bayesian Deep Learning RUL Framework Integrating Epistemic and Aleatoric Uncertainties," IEEE Transactions on Industrial Electronics, vol. 68, no. 9, pp. 8829–8841, Sep. 2021, doi: 10.1109/TIE.2020.3009593.

[11] R. Jalayer, M. Jalayer, A. Mor, C. Orsenigo, and C. Vercellis, "Evaluating deep learning models for fault diagnosis of a rotating machinery with epistemic and aleatoric uncertainty," Dec. 2024, [Online]. Available: http://arxiv.org/abs/2412.18980

[12] Case Western Reserve University, "Bearing Data Center." Accessed: Jan. 04, 2025. [Online].Available:https://engineering.case.edu/bearingdatacenter

[13] W. A. Smith and R. B. Randall, "Rolling element bearing diagnostics using the Case Western Reserve University data: A benchmark study," Dec. 01, 2015, Academic Press. doi: 10.1016/j.ymssp.2015.04.021.

[14] R. K. Rosa, D. Braga, and D. Silva, "Benchmarking deep learning models for bearing fault diagnosis using the CWRU dataset: A multi-label approach," Jul. 2024, [Online]. Available: http://arxiv.org/abs/2407.14625

[15] J. Hendriks, P. Dumond, and D. A. Knox, "Towards better benchmarking using the CWRU bearing fault dataset," Mech Syst Signal Process, vol. 169, Apr. 2022, doi: 10.1016/j.ymssp.2021.108732.

[16] R. B. Randall and J. Antoni, "Rolling element bearing diagnostics-A tutorial," 2011, Academic

Press. doi: 10.1016/j.ymssp.2010.07.017.

[17] P. Borghesani, P. Pennacchi, R. B. Randall, N. Sawalhi, and R. Ricci, "Application of cepstrum pre-whitening for the diagnosis of bearing faults under variable speed conditions," Mech Syst Signal Process, vol. 36, no. 2, pp. 370–384, Apr. 2013, doi: 10.1016/j.ymssp.2012.11.001.

[18] W. A. Smith and R. B. Randall, "Rolling element bearing diagnostics using the Case Western Reserve University data: A benchmark study," Dec. 01, 2015, Academic Press. doi: 10.1016/j.ymssp.2015.04.021.

[19] SKF Group, "SKF Product Selection Tool," https://productselect.skf.com/#/bearing-selection-start.

Optimizing Input Matrix Construction for Dimensionality Reduction in Unsupervised Fault Detection.

Wish Z. Mundondo[a,*], Daniel N. Wilke[b], Dineo A. Ramatlo[a]

[a]Department of Mechanical and Aeronautical Engineering, University of Pretoria, Pretoria, 0002, South Africa

[b]School of Mechanical, Industrial and Aeronautical Engineering, University of the Witwatersrand, Johannesburg, 2000, South Africa.

* u20473380@tuks.co.za.

Abstract.

Mechanical faults in vibrating systems often manifest as subtle deviations in dynamic behavior, making early detection crucial for predictive maintenance. One way to implement predictive maintenance is through unsupervised machine learning. The most common unsupervised machine learning technique is dimensionality reduction, and in this context, it is Principal Component Analysis (PCA).

This study examined the best way to develop an input matrix usable for PCA. Using vibration data from a quarter-car suspension model with six channels (displacement, velocity, and acceleration of sprung and unsprung masses), we studied both multichannel and single-channel data configurations. Single-channel augmentation techniques, such as window sampling, cumulative sums, and feature engineering, are used, and the effect on PCA's latent space representation as well as the reconstruction accuracy are studied. The findings highlight the importance of the input X-matrix design in improving PCA's latent space quality, fault detection accuracy, and computational efficiency of PCA.

Results show that capturing a single channel (displacement for the sprung mass) and applying augmentation techniques, particularly time lagging and consecutive window sampling, produce an optimal input matrix (X-Matrix) resulting in improved PCA latent space quality. These findings offer practical guidelines to enhance predictive maintenance in vibrating systems by optimizing X-Matrix construction for unsupervised learning.

Keywords: **PCA, latent space, data augmentation.**

1. Introduction

1.1. Background and motivation

Fault detection in mechanical systems relies on analyzing system vibrations, which indicate wear, misalignments, or imbalance. Predictive maintenance using unsupervised learning methods, such as Principal Component Analysis (PCA), requires the construction of optimal input matrices (X-Matrices) from sensor data to compress high-dimensional data into informative latent spaces. In this study, the quarter car suspension model serves as a case study to generate high-dimensional data, which is used to evaluate various X-Matrix construction approaches.

1.2. Research Problem Statement

Traditional fault detection methods struggle to process the high-dimensional data generated by multiple sensors. This challenge is exacerbated when the number of sensors is large, and the system's behavior needs to be monitored over extended periods.

The primary problem addressed in this study is to find the best way to use the collected data to construct an input matrix (X matrix) that captures the system dynamics while enabling effective dimensionality reduction.

1.3. Research Objectives

1. To evaluate the effect of using multiple channels (sprung and unsprung mass' displacements, velocities, and accelerations) directly as the input matrix.
2. To explore alternative data augmentation techniques on a single channel, such as consecutive window sampling, random window sampling, cumulative sum, and time lagging, on a single channel to create a higher-dimensional input matrix.
3. PCA was applied to each constructed input matrix to evaluate its performance in terms of latent space quality and reconstruction accuracy.
4. The best matrix construction method for capturing the system's behavior while enabling effective anomaly detection was determined.

In this paper, we first describe the suite of data-augmentation and windowing methods used to construct high-dimensional input matrices from the quarter-car suspension signals. We then apply Principal Component Analysis (PCA) to each configuration to extract dominant latent component (PCs) and evaluate both the explained variance and the reconstruction error distributions. The results demonstrate which augmentation strategies yield the most informative latent representations and provide practical guidelines for real-world maintenance applications.

2. Key Literature

2.1. Fault Detection and Predictive Maintenance in Mechanical Systems

Fault detection in mechanical systems, which is critical for predictive maintenance, can leverage unsupervised machine learning to identify anomalies early, preventing major breakdowns [4]. Vibration data, a key indicator of a mechanical system's health, can reveal faults such as imbalance, misalignment, or wear through subtle changes in vibration patterns [11]. Real-time sensor data, processed into suitable formats (X-Matrix) for machine learning, offer a wealth of information that can be processed and analyzed using unsupervised learning techniques.

In this context, the ability to preprocess raw data into an appropriate input format (X-matrix) usable for machine learning algorithms is paramount.

2.2. Machine Learning Approaches in Fault Detection

Unsupervised Machine learning is vital for the predictive maintenance of mechanical systems, enabling anomaly detection from unlabeled vibration data based on deviations from normal system behavior [15]. Dimensionality reduction methods, such as PCA, compress high-dimensional sensor data to low dimensions for improved visualization, computational efficiency, and model performance, thereby enhancing time-series data analysis and fault detection [2]. This study emphasizes the growing use of data-driven approaches in industrial systems and the resulting importance of dimensionality reduction methods, such as PCA.

2.3. Dimensionality Reduction Techniques

[1] provides a thorough survey of dimensionality reduction techniques, with Principal Component Analysis (PCA) being among the most used methods. PCA is a statistical technique that reduces the dimensionality of the input data while preserving the variance of the data, which is crucial for detecting subtle changes in the system vibration patterns that may indicate faults. [9] presents PCA as an optimal technique for dimensionality reduction in linear systems. In this study, PCA is evaluated for its effectiveness in reducing the dimensionality of vibration data while retaining critical system dynamics for fault detection.

PCA transforms high-dimensional data into a new coordinate system defined by orthogonal axes (principal components), which are linear combinations of the original features [9]. The latent space is a reduced-dimensional representation of the data in this coordinate system, where each dimension corresponds to a principal component that encodes the most important patterns or structures in the data [7].

Designing an effective input matrix is a critical step in applying dimensionality-reduction techniques to time-series data. Techniques can be used to transform a single channel of sensor data (e.g., displacement, velocity, or acceleration) into a high-dimensional matrix, allowing for a more detailed analysis of the system's dynamic behavior.

By augmenting the data using techniques such as windowing, cumulative sum, feature expansion, or time lagging, it is possible to increase the number of samples (rows) and features (columns) in the input matrix, potentially enhancing the model's ability to learn a latent space that accurately represents the system's behavior.

3. Methodology

3.1. Sensor data generation: Quarter-Car Model

In practice, sensors (accelerometers, velocity, or displacement transducers) are mounted at certain locations on a car to capture vibration time-series data for analysis. This vibration data feedback is the basis for monitoring the suspension health and designing control or diagnostic algorithms. The equations of motion to represent a quarter car model can be derived from Newton's second law, accounting for the forces exerted by the suspension and tire on the masses representing the car parts (see Appendix A).

Healthy and faulty states can be simulated to obtain sprung and unsprung mass displacement, velocity, and acceleration vibration signals. The fault data types explored included reduced spring stiffness, altered damping, and noisy sensors applied to the quarter-car model.

3.2. Input Matrix Construction

Windowing (consecutive and random).

The consecutive window sampling method segments time-series data from a single channel into a series of continuous fixed-length windows. Each window captures the sequential evolution of the signal, preserving the time dependencies within the sampled data. This method aids in detecting gradual deviations in the system behavior by monitoring data trends over time.

[16] used sliding window techniques for outlier detection in hydrological time series. He demonstrated how windowing helps identify abnormal patterns, which is analogous to fault detection in mechanical systems, by analyzing data in segments to detect deviations.

For a signal $x[n]$ of length N, consecutive windowing is performed as follows: Windowed Segment: For a window starting at index k, the segment is:

$x_k[n] = x[n]$ for $n = k, k+1, \ldots, k+W-1$

Shift: The next window starts at $k + S$, where

$S \leq W$ for overlapping windows

$S = W$ for nonoverlapping windows.

The number of windows: $M = \text{floor}((N-W) / S) + 1$. where S is the step size (or hop size), and N is the signal length.

Random window sampling creates a non-sequential set of data windows from a single channel, breaking the continuous dependencies and providing a diverse set of system states. By selecting windows without time-order constraints, this method allows the model to capture a broader range of operational behaviors, which can enhance its ability to detect atypical patterns within the dataset. This approach reduces temporal correlations, which may result in more robust representations for fault detection. For random windowing. Windowed Segment: For random starting index k, where k is chosen uniformly between 0 and N - W, the segment is

$x_k[n] = x[n]$ for $n = k, k+1, \ldots, k+W-1$

Random Indices: Generate M random starting indices k_i, each uniformly distributed between 0 and N-W.

where M is the number of windows generated, and N and W are defined above.

Cumulative sum transformation.

Applying a cumulative sum transformation to the data generates cumulative trends within the time series, highlighting gradual increases or decreases over time. Once transformed, the cumulative data were segmented into windows. This approach is particularly useful for identifying long-term drifts and systematic changes in mechanical systems. [17] applied CUMSUM to vibration signals for fault detection in rolling bearings, demonstrating its effectiveness in mechanical systems by detecting small changes in the health states of bearings.

For a sequence a1, a2..., an a_1, a_2, the cumulative sum at position k, denoted as S_k, is the sum of the first k terms:

$S_1 = a_1$

$S_2 = a_1 + a_2$

$S_3 = a_1 + a_2 + a_3$

Time-lagged.

In the time-lagging technique, the original data are transformed by creating multiple lagged versions of the single-channel data, with each lagged version representing the same signal at different time shifts. The phase space is reconstructed by stacking lagged copies of the signal into an input matrix, allowing the matrix to capture temporal dependencies across successive time steps. This setup enables the PCA to analyze both short- and long-term patterns in the data, making it particularly effective for identifying periodic behaviors and gradual changes in the performance of a mechanical system over time.

For a single-channel signal $x[n]$ with lag τ and embedding dimension L, each row of the data matrix X is

$X_k = [x[k], x[k-\tau], x[k-2\tau], \ldots, x[k-(L-1)\tau]]$

for $k=(L-1)\tau, (L-1)\tau+1 \ldots N-1$.

Polynomial expansion.

Introduces non-linear combinations of the original time-series data by raising it to various powers. This feature engineering technique increases the dimensionality of the dataset, allowing it to capture complex nonlinear relationships within single-channel data. [7] presented polynomial basis expansions as a method to capture nonlinear relationships within regression and classification frameworks. Each expansion is taken as a column offering a richer representation of the system dynamics to PCA

For powers $j = 1,2,\ldots,P$, the jth polynomial feature is

$\Phi_j[n] = (x[n])^j$,

The augmented data matrix X has columns $\phi_1[n], \phi_2[n], \ldots, \phi_P[n]$

frequency domain (FFT).

[18] discusses the use of FFT for feature engineering in time-series forecasting, which can be applied to fault detection by identifying frequency-based anomalies. This includes transforming the data into the frequency domain using the Fast Fourier Transform (FFT), which reveals frequency-based characteristics that are critical for identifying periodic and oscillatory faults. By analyzing these frequency components within windows, this method enables the model to detect specific vibrational patterns that are indicative of mechanical faults. This frequency-domain transformation, followed by windowing, can capture both time-independent and frequency-specific anomalies in the system.

Multi-Channel Configurations.

Averaging two related channels (e.g., the displacement of sprung and unsprung masses) provides a composite view of the system response, smoothing out noise and capturing overall trends. The averaged data were then windowed to create a new input matrix.

Considering the difference between two related channels (e.g., the displacement of sprung and unsprung masses) emphasizes the relative motion between them, which can be crucial for detecting misalignments or imbalances. This difference data, segmented into windows, highlights the deviations in the synchronized behavior, aiding in the detection of faults that manifest as relative motion anomalies. This technique is beneficial for systems in which relative movements are significant indicators of performance issues..

The last method explored was combining all six channels to form a single X Matrix. Here, each signal becomes a column, giving an X-matrix with six features/columns. This results in a matrix rich in every modality but also with noise.

4. Evaluation Metrics

4.1. Latent Space Clustering

As explained in the literature, PCA transforms the data into a new coordinate system defined by orthogonal axes (principal components), which are linear combinations of the original features [9], and the latent space is the reduced dimensional representation of the data in this coordinate system, where each dimension corresponds to a principal component, encoding the most important patterns or structures in the data [7]. It offers:

A qualitative way of seeing the difference between healthy and faulty states with respect to the difference in their respective scatter plot clustering.

A quantitative method to deduce how healthy and faulty states differ within the latent space by examining the difference in the cluster compactness (density) of clustering.

The scatter plot of the scores of one PC on one axis and the scores of another PC on the other axis was produced..

4.2. Data Preprocessing

Centering and normalizing the generated data is effective for standardizing them. Centering means subtracting the average value (mean) from each data point so that the data have a center at zero. This is helpful when features have different starting points. In contrast, normalization scales the data to fit within a specific range, such as 0 to 1, making all features comparable.

A common approach, called standardization, does both: it centers the data and then scales it so that the spread (standard deviation) is 1.

4.3. Reconstruction Loss

Reconstruction loss is a metric used in machine learning to evaluate how well a model can recreate the original input data from its encoded or compressed representation. When the model learns to compress the input data into a lower-dimensional space, it must be reconstructed as accurately as possible [6]).

The reconstruction loss quantifies the difference between the original input and the reconstructed output, with common choices including the mean squared error (MSE) for continuous data or binary cross-entropy for binary data. By minimizing this loss during training, the model is encouraged to learn a representation that captures the essential features of the data while discarding noise or irrelevant information; hence, the ability to choose the optimal number of PCs.

5. Results and Discussion

The quarter-car suspension model exhibits oscillatory behavior owing to its mass-spring-damper dynamics. The latent space reflects these oscillations, with spirals or loops corresponding to specific periodic trajectories in the state space of the system (reducing the amplitude with time). Healthy systems may produce spiral patterns with consistent dynamics (e.g., amplitude reduces to zero in a certain time), whereas faulty systems' data (e.g., underdamped or noisy conditions) result in a shift in the latent space patterns.

5.1. Single Channel Augmentation

Windowing: Consecutive Sampling

As shown in the figures below, creating the X Matrix by consecutively sampling from one channel, that is, sprung mass displacement, and compressing into latent scores using PCA, the variance in the resulting data can be explained using two PCs for healthy data (green) as well as the other three types of unhealthy datasets. The healthy state dataset scatter plots (green)within the latent space differ greatly in their clustering compared to the latent space clustering for the unhealthy dataset, indicating that the latent space can differentiate between healthy and unhealthy states. The reconstruction loss after using only the first two PCs to reconstruct the X Matrices from the latent scores was low.

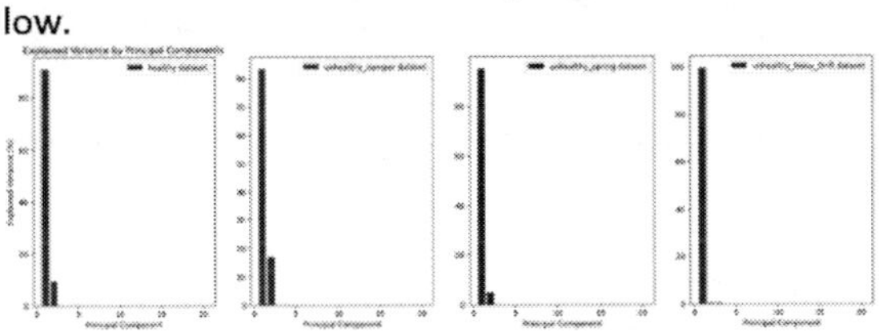

Fig 1. Sampling Consecutive Windows: Explained Variance by PCs.

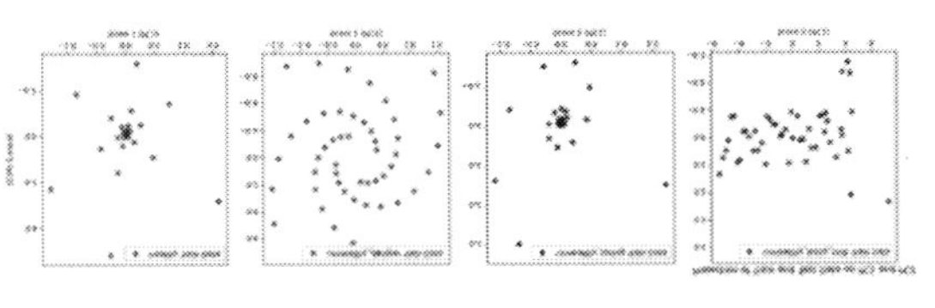

Fig 2. Sampling Consecutive Windows: Latent Space Scatter Plots.

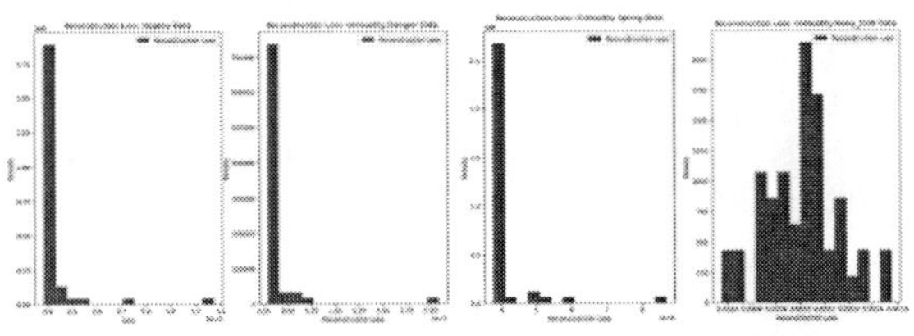

Fig 3. Sampling Consecutive Windows: Reconstruction Loss.

Consecutive windowing preserves temporal dependencies, creating structured patterns in the latent space that align with the dynamics of the oscillatory system. As mentioned above, the latent space exhibits spiral clustering, converging at (0,0). The nature of this clustering pattern differs for healthy and unhealthy data sets. This pattern reflects the inherent time-dependent structure of the data, which is preserved by sequential sampling. The reconstruction loss was small.

Windowing: Random Sampling

The variance in the data, as shown in the figure below, can be explained using two PCs for healthy datasets (green) and the other three types of unhealthy datasets. The healthy state dataset scatter plots (green) within the latent space differ greatly in their clustering compared to the latent space clustering for unhealthy states. The reconstruction loss after using only the first two PCs to reconstruct the X Matrices from the latent scores was low.

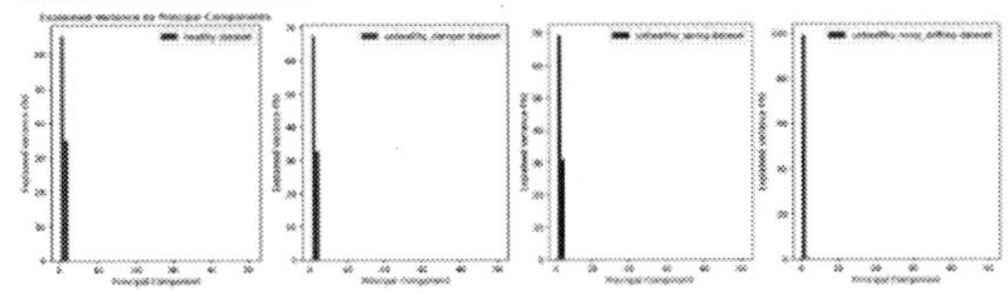

Fig 4. Random Sampling Windows: Explained Variance by PCs.

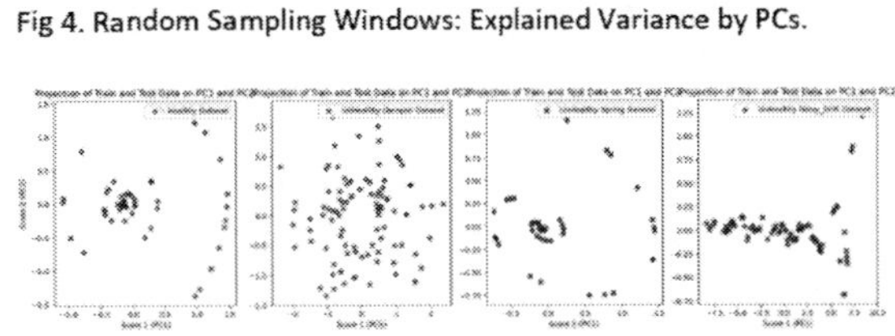

Fig 5. Random Sampling Windows: Latent Space Scatter Plots.

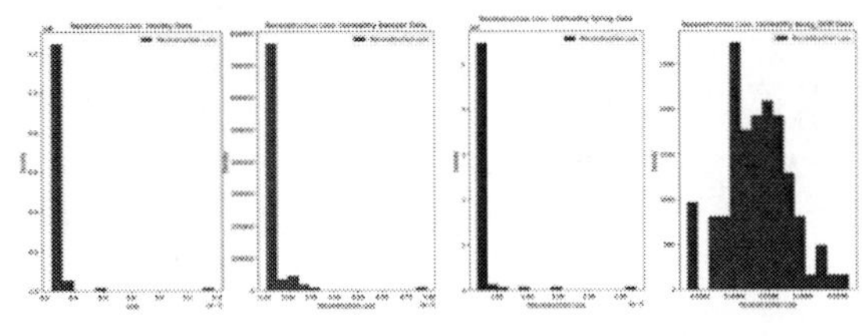

Fig 6. Random Sampling Windows: Reconstruction Loss.

Random window sampling disrupts the sequential order of the data, creating a more diverse set of windows with reduced temporal dependencies. The latent space maintained spiral clustering, although the patterns were less uniform and less continuous than those of consecutive sampling. This randomness may help the PCA generalize better across diverse system states, as reflected in the consistent clustering separation between healthy and faulty datasets. The reconstruction loss remained low for all datasets.

Cumulative Sum

The variance in the data, as shown in the figure below, can be explained using two PCs for healthy datasets (green) and the other three types of unhealthy datasets. The healthy state dataset scatter plots (green) within the latent space differ greatly in their clustering compared to the latent space clustering for unhealthy states. Cumsum suppresses short-term or local variability in the data because each data point is averaged along with all preceding values as each step aggregates more of the past data. Consequently, the variance that can be explained by only the first two PCs is reduced. The reconstruction loss was significant.

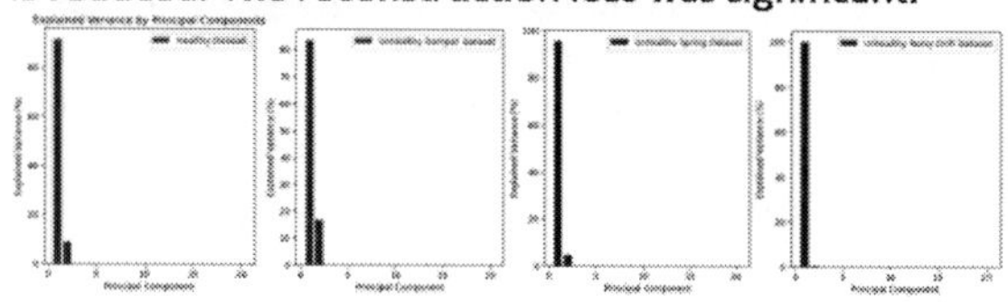

Fig 7. Cumulative Sum of Single-Sensor Data: Explained Variance by PCs.

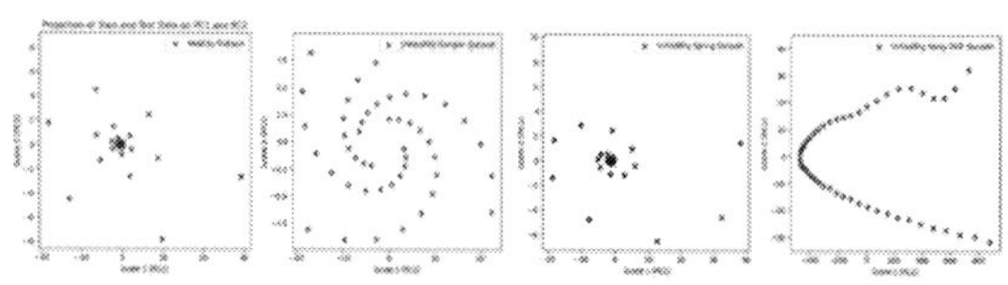

Fig 8. Cumulative Sum of Single Sensor Data: Latent Space Scatter Plots.

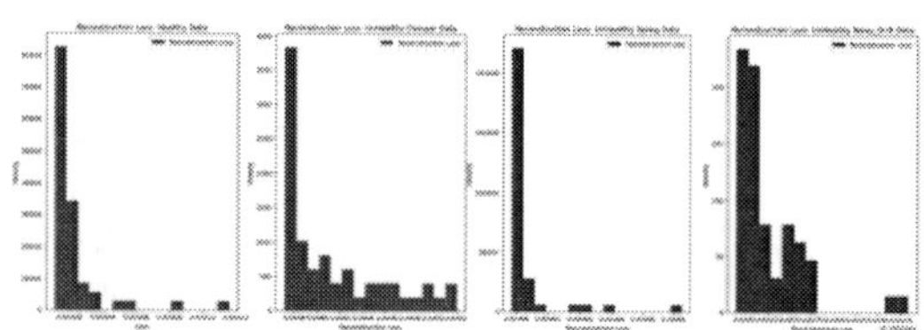

Fig 9. Cumulative sum of single-sensor data: reconstruction loss.

Applying the cumulative sum to a single channel, followed by windowing, introduces significant alterations to the latent space. As shown, the latent space clusters exhibit a distinct spiral pattern. Although this clustering effectively separates healthy and faulty states qualitatively, the high reconstruction loss indicates limitations in capturing fine-grained temporal variations, particularly for datasets with induced noise and drift.

Time Lagging

The variance in the data, as shown in the figure below, can be explained using two PCs for healthy datasets (green) and the other three types of unhealthy datasets. The first PC explained relatively more variance than the previous augmentations, with the rest of the PCs explaining very little variance. It is evident that the healthy state dataset scatter plots (green) within the latent space differ greatly in their clustering compared to the latent space clustering for unhealthy data. The reconstruction loss after using only the first two PCs to reconstruct the X Matrices from the latent scores was low.

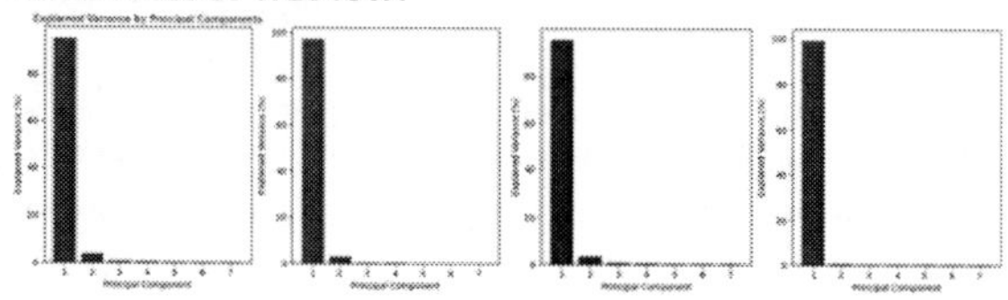

Fig 10. Time Lagging the single sensor data: Explained Variance by PCs.

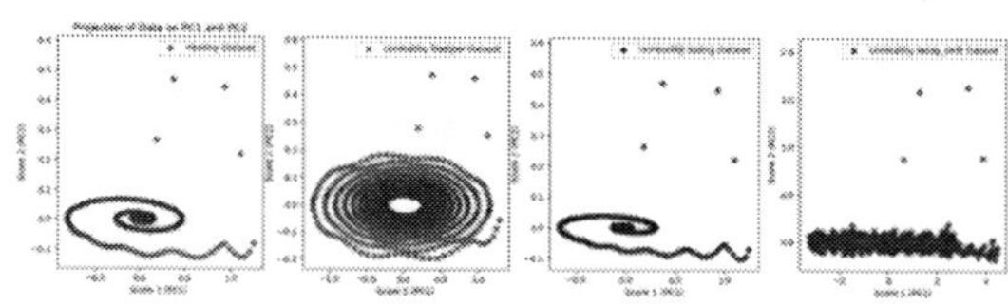

Fig 11. Time Lagging the single sensor data: Latent Space Scatter Plots.

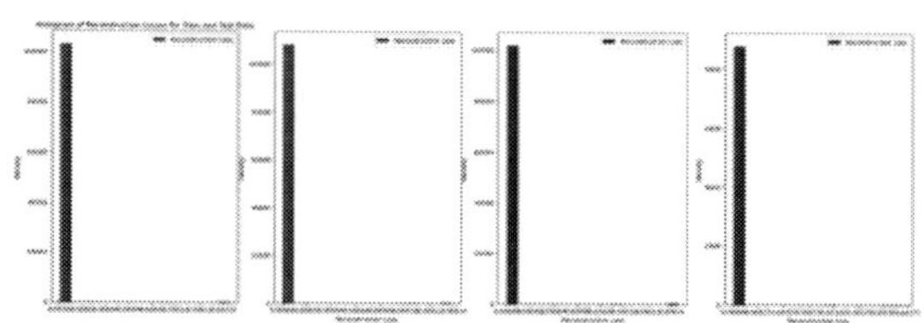

Fig 12. Time Lagging the single sensor data: Reconstruction Loss.

Time lagging creates lagged versions of the same data, introducing temporal dependencies into the X matrix. As seen , the latent space shows clear separation between healthy and faulty datasets, the clustering patterns for healthy data differ significantly from unhealthy,. Reconstruction loss is low, showing effective representation of temporal patterns with the first two PCs.

Polynomial Expansion

The variance in the data, as shown in the figure below, can be explained using two PCs for healthy datasets (green) and the other three types of unhealthy datasets. However, there are up to 3 significant PCs for X Matrix from sensor data with noise and drift needed to fully explain the variance in the data, different from previous augmentations. It is evident that the healthy state dataset (green)within latent space, does not have a significant difference in its clustering compared to the latent space clustering for unhealthy data. The reconstruction loss after using only the first two PCs to reconstruct the X Matrices from the latent scores was low.

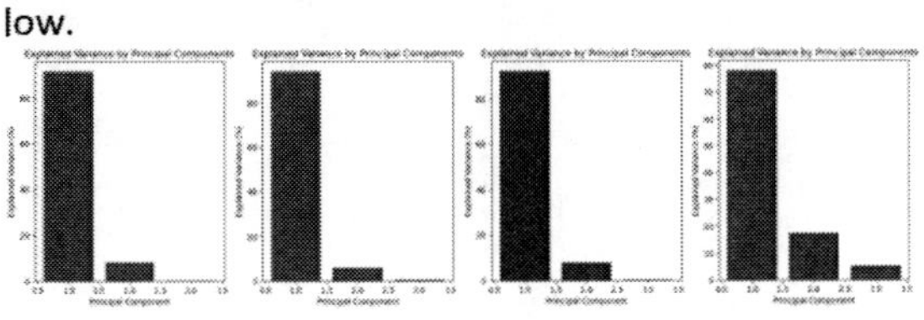

Fig 13. Polynomial Expansion : Explained Variance By PCs.

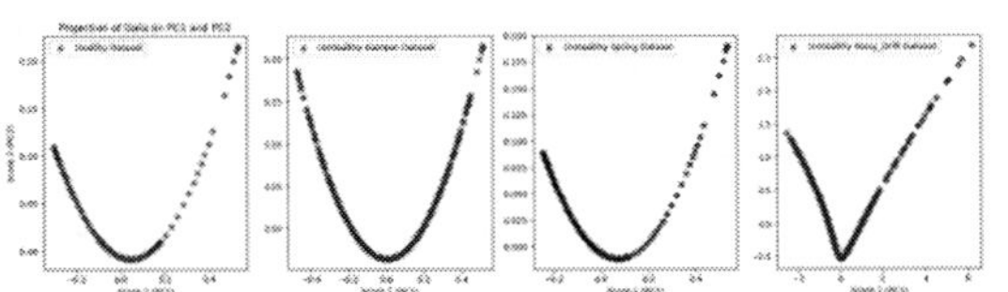

Fig 14. Polynomial Expansion: Latent Space Scatter Plots.

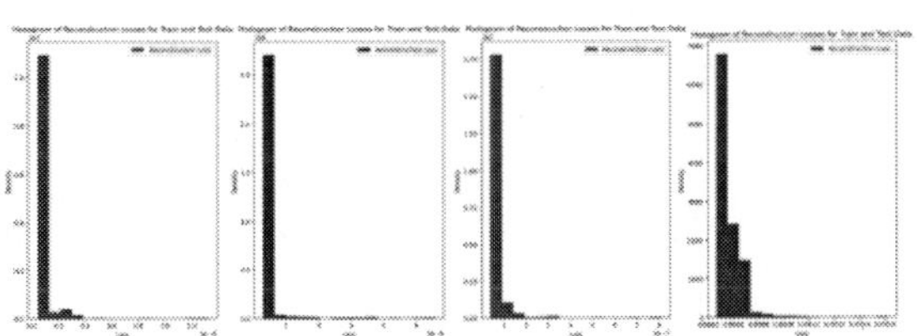

Fig 15. Polynomial Expansion: Reconstruction Loss.

Expanding data features using polynomial terms introduces non-linear relationships into the X matrix. Figures reveal that spiral patterns no longer exist in latent space clustering for this case, and the separation between healthy and faulty datasets diminishes. This suggests that higher-order expansions amplify redundancies in the data, reducing the utility of PCA for detecting subtle anomalies. Reconstruction losses remain low.

Frequency Domain (FFT)

The variance in the data, as shown in the figure below, can be explained using two PCs for healthy datasets (green) and the other three types of unhealthy datasets. However, it shows that the 3rd PCs are relatively significant in each compared to some previous augmentations. It is evident that the healthy state dataset (green)within latent space, does not have a significant difference in its clustering compared to the latent space clustering for unhealthy. The reconstruction loss was significant. This is even worse when trying to reconstruct the X Matrix for the noisy and drift induced dataset.

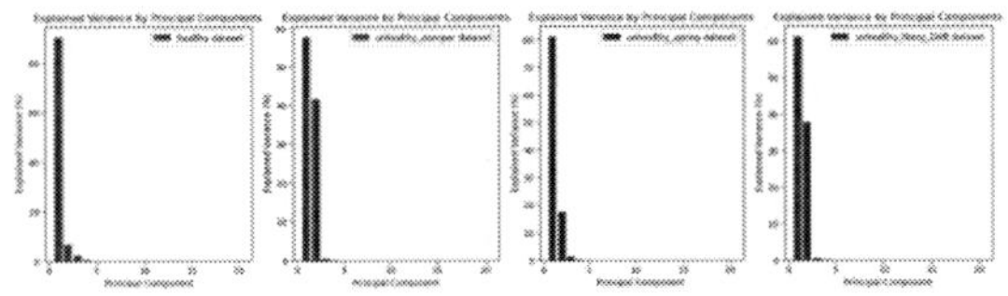

Fig 16. Frequency Domain Analysis of the Single sensors data: Explained Variance by PCs.

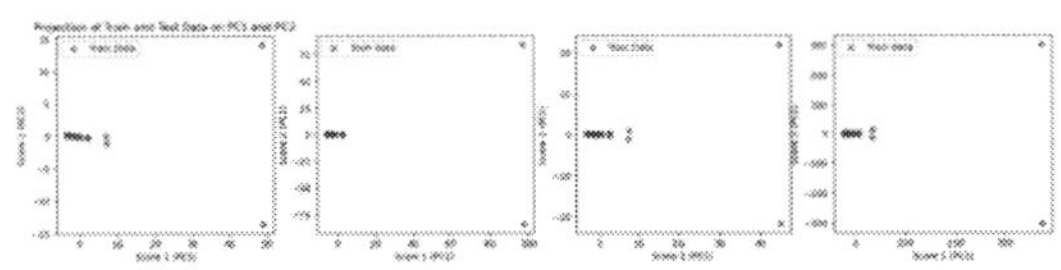

Fig 17. Frequency Domain Analysis of the Single sensors data: Latent Space Scatter Plots.

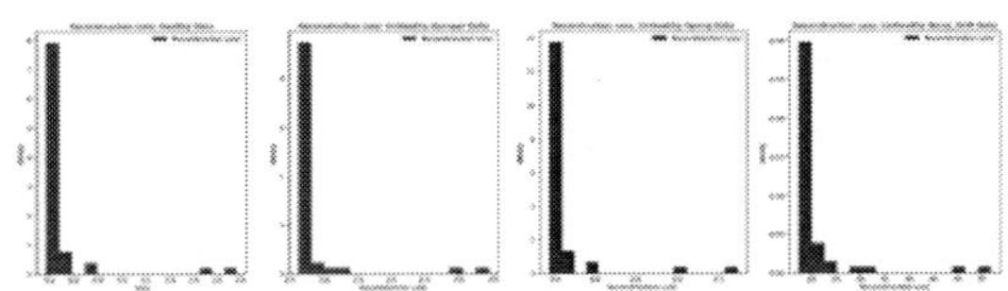

Fig 18. Frequency Domain Analysis of the Single sensors data: Reconstruction Loss.

Transforming single-channel data into the frequency domain emphasizes oscillatory and periodic behaviors. The, latent space clustering for frequency-domain data lacks distinct separation between healthy and faulty datasets and does not have the spiraling characteristics within the latent space. Reconstruction losses are significantly higher, especially for noisy datasets, due to PCA's inability to capture frequency-specific information efficiently in its linear framework.

5.2. Multi-Channel Configurations: Aggregated Channels

In each of the cases where there is use of multiple channels, either where they are aggregated or taken as they are, the spiraling patterns that were seen within single channel augmentation differ within healthy data latent space.

Averaging

The variance in the data, as shown in the figure below, can be explained using two PCs for healthy datasets (green) and the other three types of unhealthy datasets. However, the 3rd PCs are relatively significant in each case. From the figure below it is evident that the healthy state dataset (green)within latent space, differs in its clustering compared to the latent space clustering for unhealthy, and the spiralling pattern still exists. The reconstruction loss after using only the first two PCs to reconstruct the X Matrices from the latent scores was low.

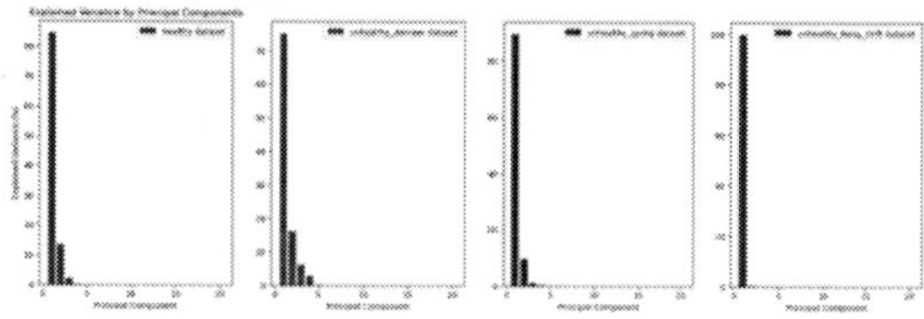

Fig 19. Aggregating the two-sensor data: Averaging: Explained Variance by PCs.

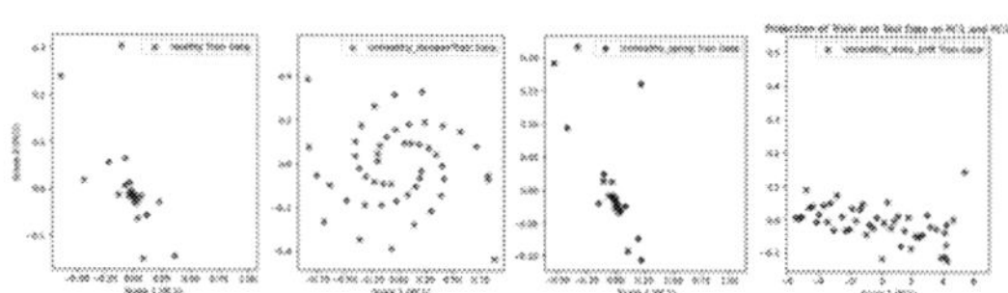

Fig 20. Aggregating the two-sensor data: Averaging: Latent Space Scatter Plots.

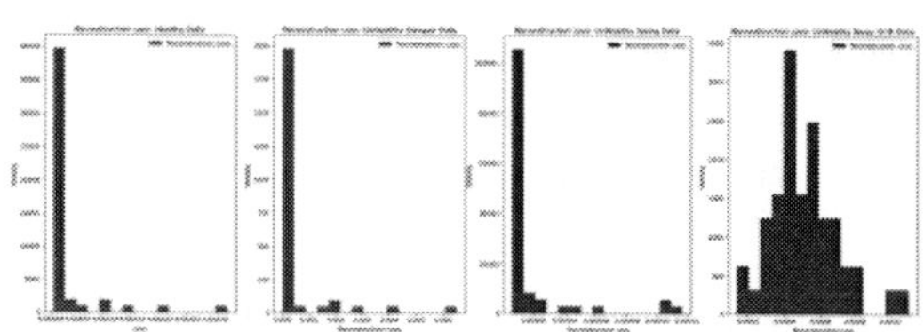

Fig 21. Aggregating the two-sensor data: Averaging: Reconstruction Loss.

Averaging two related channels smooths data, reducing noise and emphasizing overall trends. Figure demonstrates distinct clustering between healthy and faulty datasets in latent space. Reconstruction loss is low

Relative Motion (Difference)

The variance in the data, as shown in the figure below, can be explained using more than 2 PCs for healthy dataset (green) as well as the other three types of unhealthy datasets. It is evident that the healthy state dataset (green)within latent space, differs in its clustering compared to the latent space clustering for unhealthy with similar patterns as for the single channel augmentations . The reconstruction loss after using only the first two PCs to reconstruct the X Matrices from the latent scores was low.

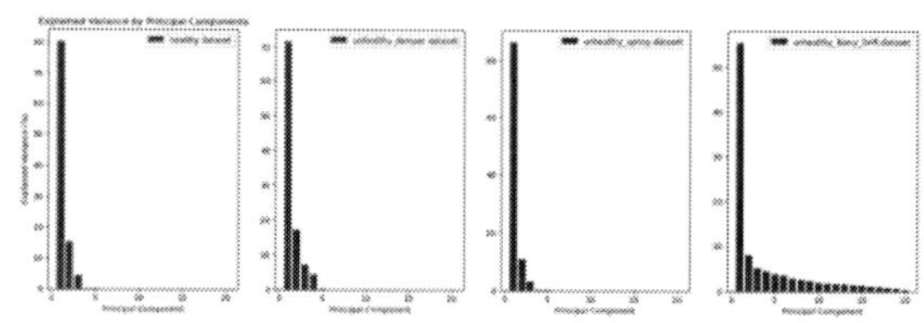

Fig 22. Aggregating the 2-sensor data: Difference (Relative Motion): Explained Variance by PCs.

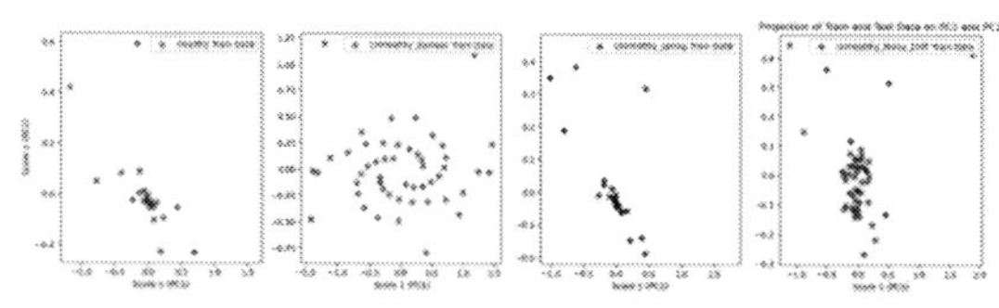

Fig 23. Aggregating the 2-sensor data: Difference (Relative Motion): Latent Space Scatter Plots.

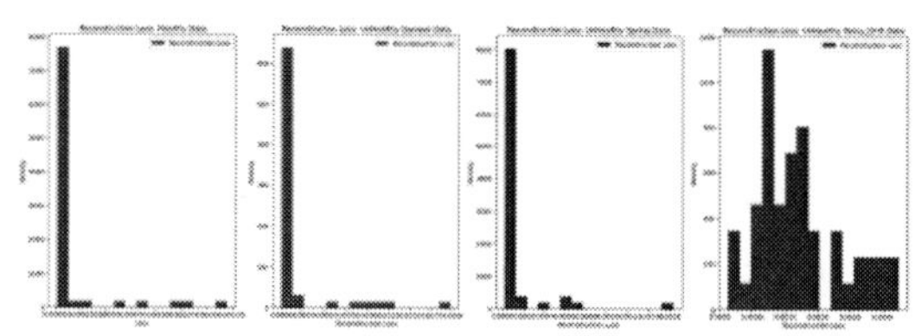

Fig 24. Aggregating the 2-sensor data: Difference (Relative Motion): Reconstruction Loss.

Difference highlights relative motion between channels, which is critical for detecting misalignment or imbalance faults. As shown in Figure, latent space clusters for faulty datasets (e.g., unhealthy damper) deviate significantly from the healthy dataset,. Reconstruction losses are comparable to averaging (order 10^{-3}), suggesting relative motion captures unique fault indicators without introducing excessive noise.

Multiplication and Ratios

Multiplying two channels emphasizes interactions between variables, introducing non-linear dependencies. Figures show that latent space clustering lacks discernible patterns, making it challenging to distinguish between healthy and faulty datasets. Reconstruction losses remain low, but the lack of clustering separation suggests this method introduces noise rather than meaningful features.

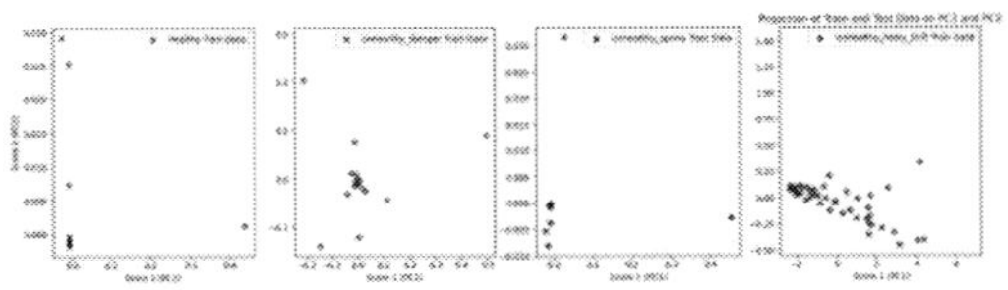

Fig 25. Aggregating the two-sensor data (Ratio of the 2 sensor data): Latent Space Scatter Plots.

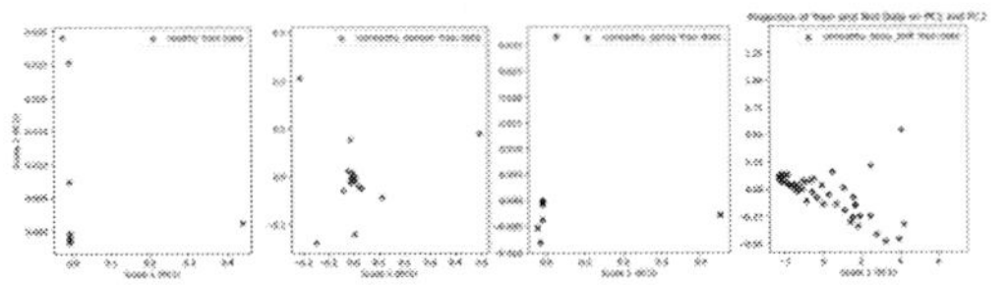

Fig 26. Aggregating the two-sensor data (Multiply the 2-sensor data): Latent Space for all datasets.

The ratio of two channels highlights proportional relationships, which are sensitive to imbalances or systematic deviations. Figures shows minor clustering separation, with noise-induced datasets showing notable divergence. Reconstruction losses are moderate (10^{-3}), suggesting ratios capture fault-related changes but lack robustness in noisy environments.

Raw Data

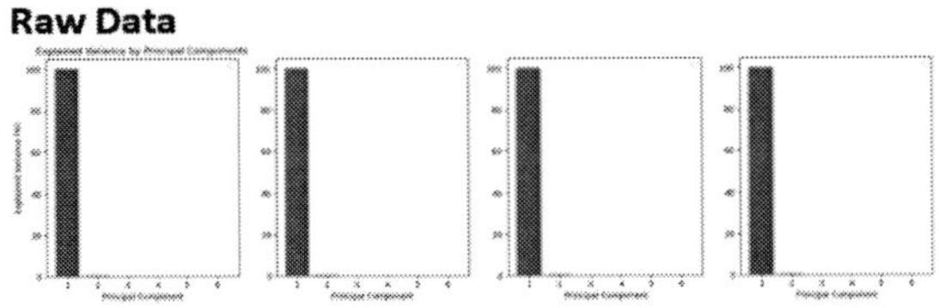

Fig 27. All Sensors: Analyze the raw data: Explained Variance By PCs.

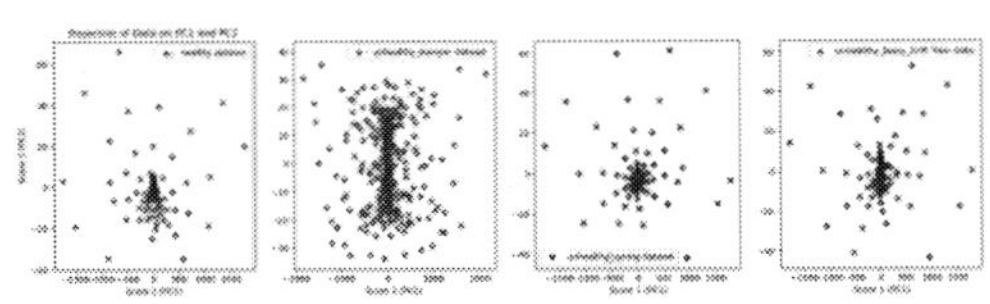

Fig 28. All Sensors: Analyze the raw data: Latent Scatter Plots.

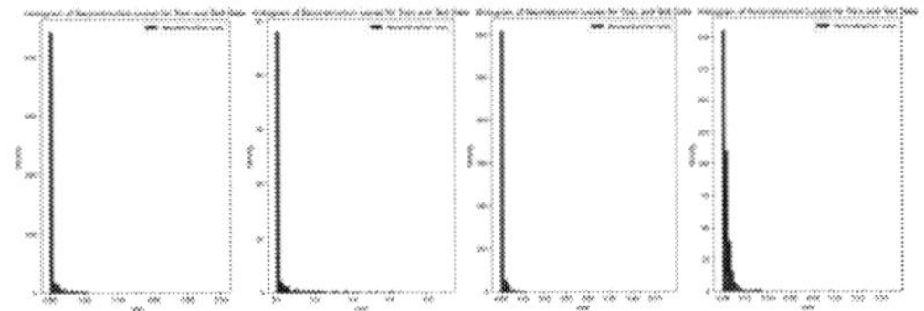

Fig 29. All Sensors: Analyze the raw data: Reconstruction Loss.

Using all six sensor channels preserves maximum information but increases matrix dimensionality. Latent space clustering qualitatively separates healthy and faulty datasets, though overlaps persist. Reconstruction losses are higher, reflecting PCA's challenges with high-dimensional inputs due to noise and redundancy the defect geometry of a radial size *r*.

6. Conclusion and Future Work

Consecutive Windowing gives X-Matrix that produces a latent space that is interpretable as one can visually separate between healthy based clusters and unhealthy data-based clusters in the scatter plots. It also results in low reconstruction loss when using only 2 PCs to reconstruct original datasets. Time lagging had similar results as consecutive windowing, only it visually had relatively better contrast due to more datapoints. Multi-channel configurations provided comprehensive data but increased noise and hence higher reconstruction loss.

As explained under methodology, time lagging setup enables PCA to analyze both short-term and long-term patterns in the data, making it particularly effective for identifying periodic behaviours and gradual changes in the mechanical system's performance over time hence can be concluded as a better augmenting method. Another feedback from this research is that, it is not always necessary to implement multiple sensors to collect data usable for machine learning purposes, a single signal can be augmented into high dimension usable for unsupervised machine learning.

Acknowledgements

AI tools (Grammarly and ChatGPT) were used to improve the spelling and grammar of this document.

Appendices.

Appendix A: Mathematical Model

Below is a detailed explanation of the quarter car mathematical model.

In a 2DOF quarter-car model the car body mass (sprung mass), denoted m_s, represents the portion of the vehicle supported by the suspension. the tire mass (unsprung mass), denoted m_t includes the tire and wheel assembly and the suspension system connects the car body to the tire, and is modelled with a spring with spring constant k_s, and a damper with damping coefficient c_s. The tire itself is modelled as a spring with spring constant k_t, connecting the tire mass to the road, accounting for its elasticity.

Displacements are:

$X_s(t)$: Displacement of the car body (y2 in the image),

$X_t(t)$: Displacement of the tire,

$X_r(t)$: Excitation of the road (an external input).

The car body interacts with the tire through the suspension, and the tire interacts with the road through its stiffness.

Equations of Motion.

Using Newton's second law, we derive the second order differential equations for each mass. For the Car Body (m_s) the forces on the car body come from the suspension:

Spring force: $k_s(x_t - x_s)$ where $x_t - x_s$ is the relative displacement. If $x_t < x_s$, the spring pulls the car body downward.

Damping force: $c_s(\dot{x}_t - \dot{x}_s)$ based on the relative velocity.

Net force equals mass times acceleration:

$$m_s\ddot{x}_s = k_s(x_t - x_s) + c_s(\dot{x}_t - \dot{x}_s)$$

For the Tire (mt).

The forces on the car body come from the suspension:

Suspension forces (acting opposite to the car body): $k_s(x_s - x_t) + c_s(\dot{x}_s - \dot{x}_t)$

Tire spring force: $k_t(x_r - x_t)$ where $x_r - x_t$ is the tire deflection from the road.

Net force equals mass times acceleration:

$$m_t\ddot{x}_t = k_s(x_s - x_t) + c_s(\dot{x}_s - \dot{x}_t) + k_t(x_r - x_t)$$

State-Space Representation.

To analyse this system, we convert these two second-order equations into a set of first-order equations using a state-space approach. Define the state vector:

$$X = \begin{bmatrix} x_1 \\ x_2 \\ x_3 \\ x_4 \end{bmatrix} = \begin{bmatrix} x_s \\ \dot{x}_s \\ x_t \\ \dot{x}_t \end{bmatrix}$$

From car body second order equations:

$$m_s\ddot{x}_s = -k_s(x_s - x_t) - c_s(\dot{x}_s - \dot{x}_t)$$

$$\ddot{x}_s = -\frac{k_s}{m_s}(x_s - x_t) - \frac{c_s}{m_s}(\dot{x}_s - \dot{x}_t)$$

$$\ddot{x}_s = -\frac{k_s}{m_s}(x_1 - x_3) - \frac{c_s}{m_s}(x_2 - x_4)$$

$$\dot{x}_2 = -\frac{k_s}{m_s}x_1 - \frac{c_s}{m_s}x_2 + \frac{k_s}{m_s}x_3 + \frac{c_s}{m_s}x_4$$

From the tire second order equation:

$$m_t\ddot{x}_t = k_s(x_s - x_t) + c_s(\dot{x}_s - \dot{x}_t) + k_t(x_r - x_t)$$

$$\ddot{x}_t = \frac{k_s}{m_t}(x_s - x_t) + \frac{c_s}{m_t}(\dot{x}_s - \dot{x}_t) - k_t(x_t - x_r)$$

$$\ddot{x}_t = \frac{k_s}{m_t}(x_1 - x_3) + \frac{c_s}{m_t}(x_2 - x_4) - k_t(x_3 - x_r)$$

$$\dot{x}_4 = \frac{k_s}{m_t}x_1 + \frac{c_s}{m_t}x_2 - \left(\frac{k_s}{m_t} + \frac{k_t}{m_t}\right)x_3 - \frac{c_s}{m_t}x_4 - \frac{k_t}{m_t}x_r$$

State First Order Equations:

$$\dot{x}_1 = x_2$$

$$\dot{x}_2 = -\frac{k_s}{m_s}x_1 - \frac{c_s}{m_s}x_2 + \frac{k_s}{m_s}x_3 + \frac{c_s}{m_s}x_4$$

$$\dot{x}_3 = x_4$$

$$\dot{x}_4 = \frac{k_s}{m_t}x_1 + \frac{c_s}{m_t}x_2 - \left(\frac{k_s}{m_t} + \frac{k_t}{m_t}\right)x_3 - \frac{c_s}{m_t}x_4 - \frac{k_t}{m_t}x_r$$

$$\therefore \dot{X} = \begin{bmatrix} \dot{x}_1 \\ \dot{x}_2 \\ \dot{x}_3 \\ \dot{x}_4 \end{bmatrix} = \begin{bmatrix} 0 & 1 & 0 & 0 \\ -\frac{k_s}{m_s} & -\frac{c_s}{m_s} & \frac{k_s}{m_s} & \frac{c_s}{m_s} \\ 0 & 0 & 0 & 1 \\ \frac{k_s}{m_t} & \frac{c_s}{m_t} & -\frac{k_s + k_t}{m_t} & -\frac{c_s}{m_t} \end{bmatrix} \begin{bmatrix} x_1 \\ x_2 \\ x_3 \\ x_4 \end{bmatrix} + \begin{bmatrix} 0 \\ 0 \\ 0 \\ -\frac{k_t}{m_t} \end{bmatrix} x_r$$

Matrix Form:

$\dot{\mathbf{x}} = \mathbf{Ax} + \mathbf{Bu}$, $u = x_r$

State Matrix for the 2dof System which includes both the car body (sprung mass) and the tire (unsprung mass), is:

$$\begin{bmatrix} 0 & 1 & 0 & 0 \\ -\frac{k_s}{m_s} & -\frac{c_s}{m_s} & \frac{k_s}{m_s} & \frac{c_s}{m_s} \\ 0 & 0 & 0 & 1 \\ \frac{k_s}{m_t} & \frac{c_s}{m_t} & -\frac{k_s + k_t}{m_t} & -\frac{c_s}{m_t} \end{bmatrix}$$

References

1. Abdi, H. and Williams, L.J., 2010. Principal component analysis. Wiley Interdisciplinary Reviews: Computational Statistics, 2(4), pp.433-459. https://doi.org/10.1002/wics.101
2. Bishop, C.M., 2006. Pattern recognition and machine learning. Springer.
3. Chen, Z. and Liu, C., 2019. Fault detection in mechanical systems using unsupervised machine learning. Journal of Vibration Engineering & Technologies
4. Fodor, I.K., 2002. A survey of dimension reduction techniques. Lawrence Livermore National Laboratory Technical Report, UCRL-ID-148494.
5. Gao, R.X. and Yan, R., 2007. Wavelets: Theory and applications for manufacturing. Springer.
6. Goodfellow, I., Bengio, Y. and Courville, A., 2016. Deep learning. MIT Press.
7. Hastie, T., Tibshirani, R. and Friedman, J., 2009. The elements of statistical learning: Data mining, inference, and prediction. 2nd ed. Springer.
8. Hyvärinen, A. and Oja, E., 2000. Independent component analysis: Algorithms and applications. Neural Networks, 13(4-5), pp.411-430.
9. Jolliffe, I.T., 2002. Principal component analysis. 2nd ed. Springer.
10. Kingma, D.P. and Welling, M., 2014. Auto-encoding variational Bayes. arXiv preprint arXiv:1312.6114.
11. Randall, R.B., 2011. Vibration-based condition monitoring: Industrial, aerospace and automotive applications. Wiley.
12. Smith, M.C., 2002. Synthesis of mechanical networks: The inerter. IEEE Transactions on Automatic Control, 47(10), pp.1648-1662.
13. Tobin, J., et al., 2020. Windowing strategies for time-series forecasting in industrial applications. Journal of Time Series Analysis.
14. Wong, J.Y., 2008. Theory of ground vehicles. 4th ed. Wiley.
15. Zhang, Y., et al., 2019. Unsupervised learning for fault detection in mechanical systems. Mechanical Systems and Signal Processing, 123, pp.456-472.
16. Yufeng Yu et al. (2014) Time Series Outlier Detection Based on Sliding Window Prediction (Wiley Online Library, DOI: 10.1155/2014/879736.
17. M. R. M. Rejab et al. (2020) Rolling Bearing Fault Detection Based on Vibration Signal Analysis and Cumulative Sum Control Chart, ResearchGate (DOI: 10.1007/s00170-020-05845-6).
18. SpringerLink (2023) An Approach to Enhance Time Series Forecasting by Fast Fourier Transform, Lecture Notes in Computer Science (DOI: 10.1007/978-3-031-42529-5_25).

Towards a phenomenological rolling element bearing outer race damage model for simulating multi-sensor vibration data under constant speed conditions

Rotondwa Ramabulana[a*], Stephan Schmidt[a]

[a]*Centre for Asset Integrity Management, Department of Mechanical and Aeronautical Engineering, University of Pretoria, Pretoria, 0028, South Africa*

**rotondwa.ramabulana@tuks.co.za, stephan.schmidt@up.ac.za*

Abstract

Phenomenological rolling element bearing models are often used in Vibration-based condition monitoring (VBCM) to rigorously assess and compare the performance of new and existing diagnostic methods. This study develops a phenomenological multi-sensor bearing vibration model to support multi-sensor VBCM (MSVBCM) research. Several model parameters are informed by experimental data from a Semi-Autogenous Grinding (SAG) mill. The model's potential to support MSVBCM research is highlighted.

Keywords: rolling element bearings, bearings models, condition monitoring.

1. Introduction

Rolling element bearings (REB) are critical elements in rotating machinery and prone to failure [1]. Vibration-based condition monitoring is used to diagnose REB faults and facilitate condition-based maintenance. Single-sensor bearing diagnosis methods are much researched, however, in recent years, multi-sensor methods have shown improved performance [2-4].

REB models are used to critically assess and compare the performance of new and existing VBCM methods. There are different types of REB models such as lumped mass models, finite element models, and phenomenological bearing models (PBMs) [5-7]. PBMs can model the complex stochastic behaviour of REB vibration signals and are the focus of this study.

Although several single-sensor PBMs (SSPBMs) exist (e.g., [7]), to the best of the authors' knowledge, there is no multi-sensor PBM (MSPBM) available to support MSVBCM research. Therefore, this study introduces a preliminary MSPBM for an outer race fault in Sec. 2. Experimental data, discussed in Sec. 3, are used to inform several important unknowns. The results of the MSPBM are shown in Sec. 4, and the work is concluded in Sec. 5.

2. Multi-sensor phenomenological model

In this study, the objective is to model the response of multiple accelerometers placed at different locations on a complex rotating system to a single outer race REB fault under constant speed conditions. The response of the sth sensor of the MSPBM is modelled as follows:

$$x_s(t) = \kappa \cdot h_s(t) \otimes F(t) + n_s(t), \qquad (1)$$

where $h_s(t)$ denotes the impulse response function (IRF) for the transmission path (TP) between the fault location and sensor s, $F(t)$ models the impulse function (IF) caused by the outer race fault, $n_s(t)$ denotes the noise component and $\otimes$ denotes the convolution operation. The impulse magnification factor κ can be used to model changes in fault severity.

The fault impulses' Time-of-Arrival (ToA) is stochastic due to rolling element slip and cage slip [7,8]. Considering this, the following experimentally validated mixed slip model for the bearing IF is used [7-9]:

$$F(t) = \sum_i A_{is}\delta(t - iT - \tau_i - \Delta T_i), \qquad (2)$$

where $A_{is} \sim \mathcal{N}(\mu_s, \sigma_A^2)$ is the amplitude of the ith impulse of the sth sensor, $\delta(t)$ is the Dirac delta function, and T is the theoretical period of the fault. The term $\tau_i \sim \mathcal{N}(0, \sigma_\tau^2)$ accounts for the rolling element slip and $\Delta T_i \sim \mathcal{N}(T, \sigma_{\Delta i}^2)$ accounts for the cage slip of the ith impulse, where $\Delta T_i = T_i - T_{i-1}$, i.e., the difference between the time-of-arrival of consecutive impulses. Note: A sample x from a Gaussian distribution with a mean μ and a variance σ^2 is denoted $x \sim \mathcal{N}(\mu, \sigma^2)$.

The displacement IRF of the SSPBMs is often modelled as follows [9]:

$$d_s(t) = \begin{cases} e^{-\zeta_s \omega_{ns} t} \sin(\omega_{ds} t) & t > T_s \\ 0 & \text{otherwise,} \end{cases} \qquad (3)$$

where T_s, ζ_s, ω_{ns}, and ω_{ns} are the response start time, the damping ratio, the undamped and damped circular natural frequencies associated with the sth sensor, respectively. The acceleration IRF $h_s(t)$ is obtained by differentiating $d_s(t)$ twice.

The rotational speed and bearing geometry can be used to find T. Sensible slip model parameters (i.e., σ_τ and $\sigma_{\Delta i}$) are available [9]. This study is focused on finding plausible models and values for T_s, $n_s(t)$, and μ_s for the different sensors in a relative sense. To isolate the different components, two sets of experimental datasets were used. First, experiments were conducted on a stationary, empty mill. Impacts were applied to the bearing housing to simulate the behaviour of a damaged bearing. This allowed the starting time T_s and the relative magnitude of the impacts μ_s to be estimated from acceleration data recorded at different positions. Second, an experiment was conducted using an empty mill rotating at 60 RPM, without any impacts applied. This was used to record acceleration data that informed the noise component $n_s(t)$ for a single rotational speed and empty mill. Experimentally informed values for ζ_s, ω_{ns}, and ω_{ds} are needed but reserved for future work.

2.1. Experimental setup and data collection

The setup, located at the University of Pretoria C-AIM laboratory, is shown in Fig. 1. The drum is driven by a motor and is supported by three bearings, namely, B1, B2, and B3. No bearings were damaged in this study. Instead, artificial impulses were introduced using a modal hammer to simulate the effect of bearing fault impacts, which enables the quantification of the impulses' relative properties. Impacts and acceleration measurements were made and taken at several locations, labelled as 'L' and 'A' in Fig. 1b, respectively. The impact locations are on B1, B2, the drum, and B3, labelled L1, L2, L3, and L4, respectively. The sensor locations are on B1, B2, the base, and B3, indicated by A1, A2, A3, and A4, respectively. All impacts were made in the vertical direction, except at L3, which were made in the axial direction. Additionally, all sensors measured in the vertical direction, except for A3 which measured in the transverse direction. At each impact location, 10 sequential impacts were made in a single measurement.

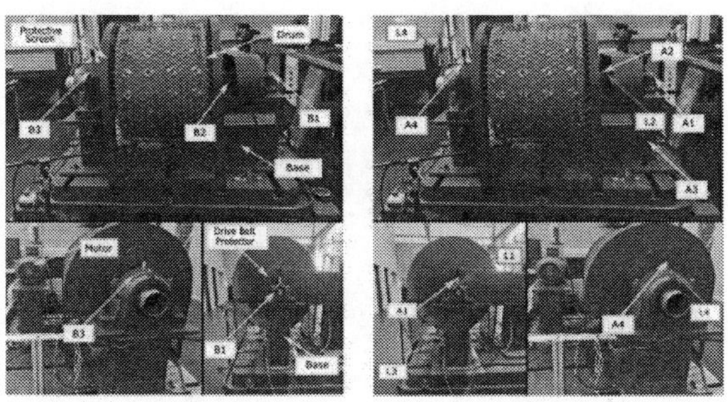

(a) (b)

Fig. 1. SAG mill experimental setup. (a) Setup showing a driving motor, a shaft, three bearings, a drum, and two protective covers. (b) The impact locations L1, L2, L3, and L4, and sensor locations A1, A2, A3, and A4 are also shown.

2.2. Experimental data analysis

In the impact experiments, 16 signals were collected, each containing 10 impacts. An example impact signal and the corresponding response are shown in Fig. 2a and Fig. 2b, respectively, for impacts made at L1 and measured at A1. The manual impacts resulted in some variability in the impacts' magnitude and time between impacts, however, the focus is on the relative magnitudes and time delays across sensors.

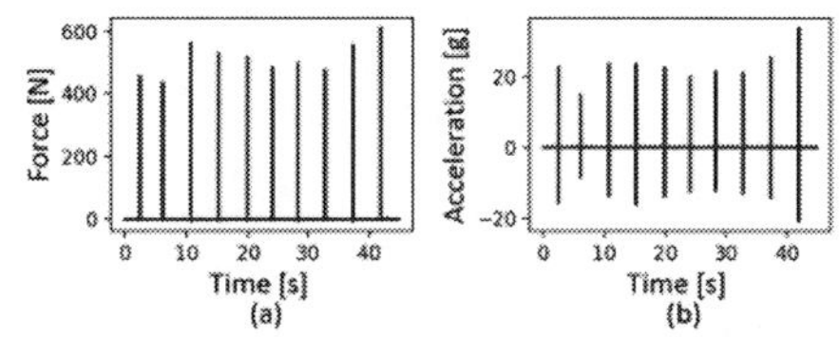

Fig. 2. (a) Impact signal for impacts at L1. (b) Response signal measured by A1 for impacts at L1.

To quantify the sensors' relative amplitudes, the 10 impulse responses in each signal were isolated, and their respective maxima calculated. Thereafter, the average of the maxima was calculated and compared between the sensors.

A time delay, whose magnitude depends on the TP between the impact location and sensor location, may occur between the time of force impact and the start of the impulse response. The time delay, T_d, is therefore equal to the response start time T_s minus the impact time. App. A explains the computation of the impact time and T_s. Fig. 3 shows the impact signal (black), response at A1 (red), and response at A4 (blue) for an impact at L1. Evidently, the starting time of A4's signal is later than that of A1's, meaning A4's time delay is larger than A1's.

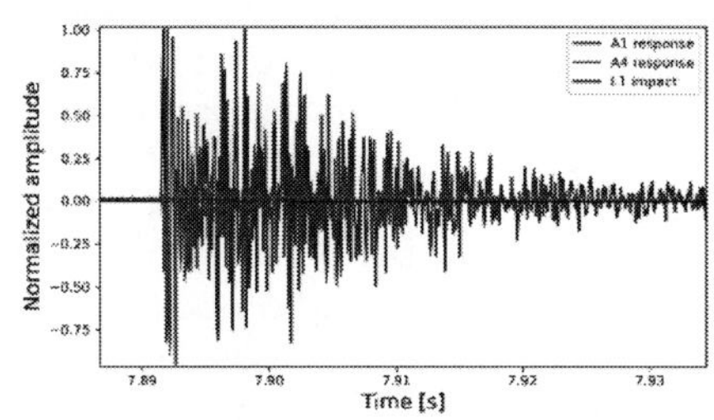

Fig. 3. Impact signal (black), impulse response signal at A1 (red), and impulse response signal at A4 (blue) for impact location L1. The signals were normalised by their respective maximum values for demonstrative purposes only.

Tab. 1 shows the IRF's average maximum magnitude for all impact and sensor locations. The trend makes sense; for an impact at a given location, the average maximum magnitude decreases as the sensors are placed further away. For example, the closest sensor to L1, L2, L3, and L4 is A1, A2, A2, and A4, respectively, meaning the sensor closest to the impact location has the highest average maximum magnitude. This information is used to inform μ_s in Eq. (2).

Tab 1. Average maximum magnitude across all impulses for all impact and sensor locations with unit g.

	L1	L2	L3	L4
A1	23.090	2.467	0.155	0.451
A2	8.402	8.707	0.371	0.749
A3	4.843	4.013	0.195	0.847
A4	0.228	0.203	0.150	27.985

Tab. 2 shows the IRF's time delay for all impact and sensor locations. The closer the sensor location to the impact location, the smaller the time delay, and vice versa. When the sensor location is near the impact location (e.g. A1 to L1), there is virtually no time delay. This information is used to inform T_d and therefore T_s in Eq. (3).

Tab 2. Average time delay across all impacts for all impact and sensor locations with unit of seconds.

	L1	L2	L3	L4
A1	0	17.1e-5	15.6e-5	108.3e-5
A2	16.7e-5	0	2.3e-5	37.5e-5
A3	27.6e-5	25.3e-5	64.2e-5	139.6e-5
A4	109.4e-5	15.9e-5	9.3e-5	5.21e-5

Data from an experiment of the empty SAG mill operating at 60 RPM was used to emulate $n_s(t)$ in Eq. (1), the noise present in a complex rotating system. No impacts were made during these experiments. A Student's t-distribution model—characterised by a mean (μ_n), standard deviation (σ_n), and degree-of-freedom (DF_n) parameter—was used to model $n_s(t)$ and was found to give a reasonable fit to the density plots, with the resulting fits show in App. B. A Student's t-distribution only captures the marginal distribution of the signal's values and not its cyclic structure or spectral content. Improving the model is reserved for future work.

Tab. 3 shows the values for μ_n, σ_n, and DF_n of the fitted Student's t-distributions to model $n_s(t)$ for all the signals. A2's signal displays Gaussian behaviour (large DF) while, A1, A3, and A4's signal display more leptokurtic behaviour (lower DF).

3. Results

Informed by the experiments, three accelerometer signals were simulated using Eq. (1). The signals are denoted S1, S2, and S3, with the distance from the fault source increasing from S1 to S3. A simulated fault frequency of 22 Hz was chosen as a plausible bearing fault frequency. The simulated signals were sampled at a frequency of 100 kHz.

Tab 3. shows the values for μ_n, σ_n, and DF_n of the fitted Student's t-distributions used to model $n_s(t)$ for all the signals with units g. This was done for a single speed case: 60 rpm.

	A1	A2	A3	A4
μ_n	-0.0020	0.0193	0.0033	0.0026
σ_n	0.0046	0.0217	0.0046	0.0033
DF_n	2.7116	7.50e6	11.0258	2.9385

The amplitude (μ_s), noise ($\mu_{ns}, \sigma_{ns}, DF_n$), and time delay ($T_d$) parameters were informed by the experimental results discussed in Sec. 3. The amplitudes were inspired by the values of A2, A3, and A4 in Tab. 1 for L1, signal noise was generated from the parameters of A2, A3, and A4 in Tab.3, and time delay parameters were taken from A1, A2, and A3 in Tab. 2 for L1. With the chosen fault frequency, the theoretical fault period, T, is 0.0455 seconds. To add effects of the rolling element and cage slip, σ_τ and $\sigma_{\Delta i}$ are 1% and 0.2% of T, respectively. Tab. 4 shows the values of the parameters of the three simulated signals. The IRF's ζ_s and ω_{ds} were selected to be distinct in order to make their effects more visible, however, further investigation is needed to assess the sensor's location's impact on these parameters. Fig. 4 superimposes the experimental and simulated noise distribution, which closely match, as expected.

Tab 4. Model parameters for the three simulated signals. The experimentally informed (EI) column highlights which quantities were informed by experiments (Y) and which quantities require further investigation (N).

Parameters	S1	S2	S3	EI
μ_s[g]	20	10	5	Y
σ_A[g]	0.01	0.01	0.01	N
ζ_s	0.1	0.15	0.045	N
$\frac{\omega_{ds}}{2\pi}$ [kHz]	1.5	2.5	2.0	N
σ_{ns}[g]	0.0217	0.0046	0.0033	Y
μ_{ns}[g]	0.0193	0.0033	0.0026	Y

T_d[s]	0	0.000167	0.001094	Y
DF_n	7.50e6	11.02	2.93	Y

The first 0.15 seconds of the simulated IF is shown in Fig. 5a when $\kappa = 1$. Fig. 5b-5d respectively show simulated signals S1, S2, and S3 with added noise. These signals show consistent behaviour with Tab. 4; sensors S1's magnitude is the largest and S3's the smallest. Furthermore, the damping behaviour is also consistent with Tab. 4.

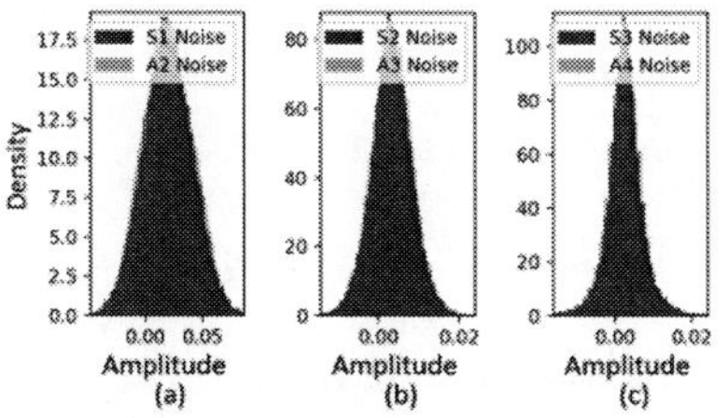

Fig. 4 shows the distribution of the noise from A2, A3, and A4 overlayed on the distribution for S1, S2, and S3.

Fig. 5e-5h shows the experimental IF, whose frequency is about 0.3 Hz, and IFRs of A2, A3, and A4, respectively, from the impact experiments. The oscillations are less visible due to the scale, but closer inspection (see Fig. 3) reveals that the simulated signals mimic the experimental signals to some degree.

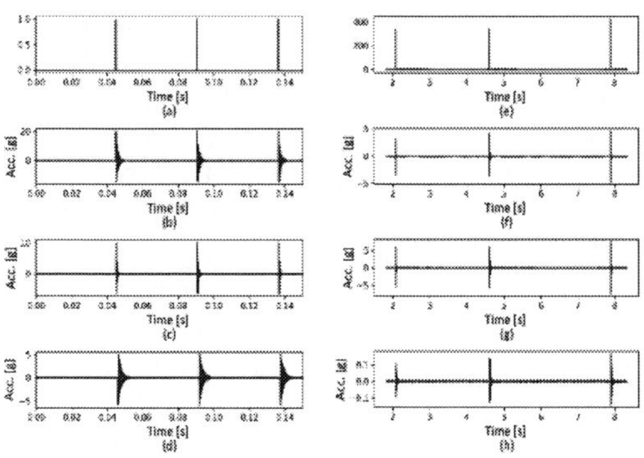

Fig. 5. Simulated and experimental signals. (a) Simulated impulse function (IF), (b) S1, (c) S2, and (d) S3 with $\kappa = 1$. (e) Experimental IF, (f) A2, (g) A3, and (h) A4.

Fig. 6a - 6c shows S1, S2, and S3, respectively, when $\kappa = 0.002$, simulating a case where the fault signal is weak in comparison to the noise—as is the case in complex, rotating machinery. Evidently, the fault information is hardly visible, unlike the case where $\kappa = 1$ in Fig. 5a-d.

The squared envelope spectrum (SES), a standard fault analysis method in VBCM (App. C) of the IF, S1, S2, and S3 is shown in Fig. 7a-7d, respectively, for $\kappa = 1$. Fig. 7e-7h shows the same for $\kappa = 0.002$. Prominent peaks are present at the fault frequency (22 Hz) and its harmonics, revealing the simulated damage. The IF's SES represents the fault information without TP effects. The SESs of S1, S2, and S3 have different structures (to each other and to that of the IF) due to TP effects, but their relative magnitudes are consistent with the Tab. 4. For $\kappa = 0.002$, the SESs of S1, S2, and especially S3, are of much lower quality as the peaks are less discernible, a pertinent research problem tackled in VBCM research. The results show that the model can be used to model different fault severities, which could be used to assess the sensitivity of the fault diagnosis methods. Furthermore, the model can be used to assess different sensor scenarios (e.g., where some sensors contain more prominent fault information than others). These capabilities are useful for critical assessment of MSVBCM methods.

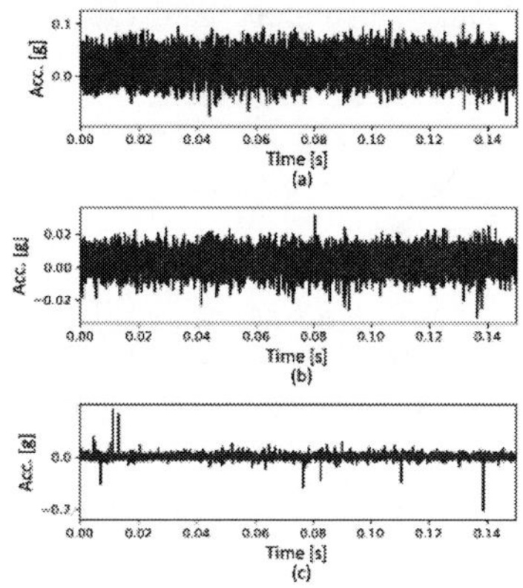

Fig. 6. Simulated signals (a) S1, (b) S2, and (c) S3 with noise added when $\kappa = 0.002$.

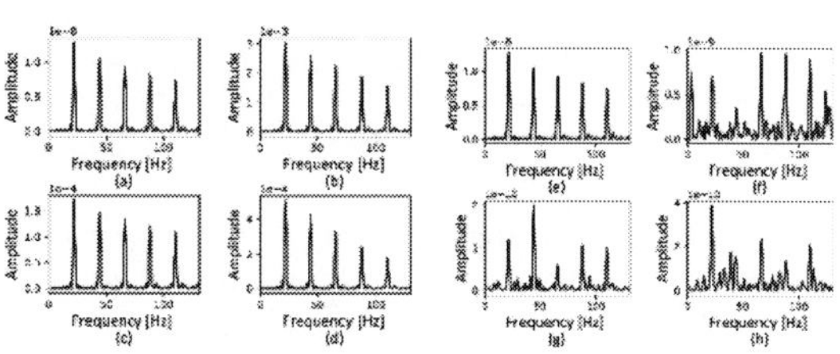

Fig. 7. SES of (a) IF, (b) S1, (c) S2, and (d) S3 for $\kappa = 1$. SES of (e) IF, (f) S1, (g) S2, and (h) S3 for $\kappa = 0.002$.

4. Conclusion and future work

The MSPBM introduced in this paper has been shown to can generate signals that mimic the experimental data presented and other characteristics relevant in VBCM such as varying fault severity. This suggests it can be used to as a tool to support MSVBCM research. Future work should involve extending the model to other damage modes and speed cases, improving the IRF model for multi-sensor configurations, and developing more accurate noise models. Additionally, validation using recordings from damaged bearings is needed.

Acknowledgements

Ms. R. Jones is gratefully acknowledged for her support during the impact test experiments and for sharing the data she collected from the operating SAG mill.

References

[1] Pandiyan M, Babu TN (2024). *Systematic review on fault diagnosis on rolling-element bearing.* Journal of Vibration Engineering & Technologies, 1-35.

[2] He Z, Shao H, Zhong, X, Zhao X (2020). *Ensemble transfer CNNs driven by multi-channel signals for fault diagnosis of rotating machinery cross working conditions.* Knowledge-Based Systems, 207, 106396.

[3] Safizadeh MS, Latifi SK (2014). *Using multi-sensor data fusion for vibration fault diagnosis of rolling element bearings by accelerometer and load cell.* Information Fusion, 18, 1-8.

[4] Saimurugan M, Ramprasad R (2018). *A dual sensor signal fusion approach for detection of faults in rotating machines.* Journal of Vibration and Control, 24(12), 2621-2630.

[5] Wang H, Zheng J, Xiang J (2023). *Online bearing fault diagnosis using numerical simulation models and machine learning classifications.* Reliability Engineering & System Safety, 234, 109142.

[6] Kiral Z, Karagülle H (2003). *Simulation and analysis of vibration signals generated by rolling element bearing with defects.* Tribology International, 36(9), 667-678.

[7] Antoni J, Randall RB (2002). *Differential diagnosis of gear and bearing faults.* Journal of Sound and Vibration, 124(2), 165-171.

[8] Antoni J, Randall RB, Chobsaard S (2001). *The relationship between spectral correlation and envelope analysis in the diagnostics of bearing faults and other cyclostationary machine signals.* Mechanical Systems and Signal Processing, 15(5), 945-962.

[9] Borghesani P, Smith WA, Randall RB, Antoni J, El Badaoui M, Peng Z (2022). *Bearing signal models and their effect on bearing diagnostics.* Mechanical Systems and Signal Processing, 174, 109077.

[10] Li H, Lian X, Guo C, Zhao P (2015). *Investigation on early fault classification for rolling element bearing based on the optimal frequency band determination.* Journal of Intelligent Manufacturing, 26, 189-198.

[11] Chen B, Cheng Y, Zhang W, Gu F (2022). *Enhanced bearing fault diagnosis using integral envelope spectrum from spectral coherence normalized with feature energy.* Measurement, 189, 110448.

[12] Luo C, Mo Z, Miao Q (2020). *Cyclic harmonic ratio defined in squared envelope spectrum and log-envelope spectrum for gearbox fault diagnosis.* IEEE Transactions on Instrumentation and Measurement, 69(12), 9568-9577.

Appendix A. Impact time and response start time

The time of the force impact is taken as the time at which the maximum force of the force impact is seen. Additionally, the start of the impulse response is taken as the time at which the magnitude of the response first reaches 20% of the maximum magnitude of the impulse response and is denoted T_s. Furthermore, the time delay, T_d, is computed by subtracting the impact time from T_s. This is done for all impacts in a signal and averaged to get the average time delay.

Appendix B. Quantification of model fit quality

The density plots (blue) of the noise signal from sensor A1, A2, A3, and A4 are shown in Fig. B1 as well as the Student's t-distribution model fits (red). The goodness of the model fits is tabulated in Tab. B1 using the Kullback-Leibler divergence (KLD) measure, a standard statistics method used to compare two distributions. A value of zero denotes that two distributions are identical. While the model's fit on data from A2, A3, and A4 are reasonable, that of A1 is very high. This is due to the small modes present to the left and right of the main mode in the distribution. This is not seen as an issue as only A2, A3, and A4's data was used further. Ideally, a different model can be used for A1's distribution, such as a Gaussian mixture model with three components to capture the main and secondary modes.

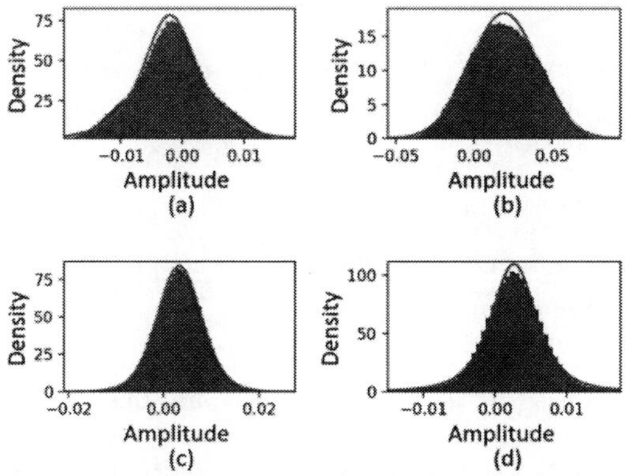

Fig. B1. Density plot (in blue) of noise of signal from (a) A1, (b) A2, (c) A3, and (d) A4. Student's t-distributions have been fitted onto the density plots (in red).

Tab. B1. The Kullback-Leibler divergence (KLD) between the histogram of the signals obtained from running the SAG mill empty at 60 RPM and the fitted Student's t-distribution.

Signal	KLD
A1	101.39
A2	7.367
A3	5.406
A4	22.949

Appendix C. The Squared Envelope Spectrum (SES)

While several formulations of the SES exist [11-12], the form used in this paper is as follows,

$$\mathrm{SES}(\alpha) = \left(\frac{1}{N} \cdot \mathrm{DFT}_{n \to \alpha}(|\tilde{x}(n)|)\right)^2, \qquad (1C)$$

where α is the frequency, DFT is the discrete Fourier Transform operator, and $\tilde{x}(n)$ is the mean-removed analytical function of the Hilbert Transform.

Appendix D. Summary of the simulation process

The process followed in the study is summarized as follows:

- Perform experiment where hammer impulses are applied to the housing of a bearing of an empty, stationary SAG mill. Record the response using several sensors.
- Take note of how the amplitudes and time delays of the response signals vary across all the sensors. Use this to inform parameters μ_s and T_d(and therefore T_s) in Eq. (2) and (3), respectively.
- Run the empty SAG mill at 60 RPM without applying impacts and record the response using several sensors—this represents noise in a complex rotating machine.
- Model the distribution of the signals obtained in Step 3 using a Student's t-distribution with parameters μ_n, σ_n, and DF_n. Use this to inform $n_s(t)$ in Eq. (1).
- Use Equation (1) to generate three signals and add noise to the signal using the model parameters from Step 4.
- Compare the generated signals to the experimental signals.
- Plot the SES of the signals to see if the fault information is present in the frequency domain.
- Change the value of κ to simulate varying fault severity.

Virtual measurement of a vibrating beam

Gerhard Durandt*, Anriëtte Bekker

Stellenbosch University, Corner of Banhoek Road and Joubert Street, Stellenbosch, 7600, South Africa

**corresponding author email: gdurandt@sun.ac.za*

Abstract

The evolution of cyber-physical connections enables benefits that can inform decision-making in terms of optimising the life span of structures in an operational setting [1]. Literature indicates that structural fatigue on maritime structures due to continuous wind and wave excitation pose a threat to structural longevity [2, 3]. Full field strain measurements that can be used for fatigue assessment is often not feasible in the operational setting. As such, virtual sensing methodologies have the potential to enable real-time monitoring of structures from a limited number of sensors.

This work entails a preliminary validation study of the virtual displacement measurement on a laboratory scale beam that is subjected to impulsive and random excitation. A finite element model is used to derive the mass, damping and stiffness matrices to formulate a state-space model of the beam. Next, a Kalman filter is applied with acceleration measurements as the input variable to predict the displacement at a specific location. The predicted displacement of the beam is validated with measured displacements at the same location.

Keywords: Virtual Sensor, Kalman Filter, Vibration, Finite Element Analysis, Experimental Modal Analysis

1. Introduction

Digital services and digital twins are gaining interest in various industrial sectors [1]. A digtial twin is a virtual representation of an asset and its operational environment, and simulates the interactions between the asset and its environment in (close-to) real time [4, 5]. This digital representation presents an opportunity to provide operational insight into the behaviour of processes, systems or structures. As a result, operators of large industrial equipment could benefit from real-time operational knowledge and decision support. Digital twins often entail the integration of cyber-physical systems in operational settings. This requires some form of statistical or physics-based modelling to enhance digital services from basic monitoring systems to useful predictive tools [6] which can aid in decision-making. In addition, it might be the case that parameters of interest cannot be directly measured due to installation cost, impractical sensor locations, or aggressive operational environments. However, the concept of virtual measurement can provide a means to bridge this gap.

The context of this paper is the development of digital solutions for the *SA Agulhas II*—a polar research and supply vessel owned by the Department of Forestry, Fisheries and the Environment (DFFE). This vessel is equipped with a network of accelerometers to measure hull vibrations. However, due to the vessel's unique hull loading phenomena, specifically referring to slamming events, concerns have been raised regarding accumulated fatigue and the remaining useful life of the structure [3, 7]. Digital solutions for monitoring and estimating full-scale hull fatigue have been proposed in the literature [8]. Consequently, this paper presents a laboratory-scale implementation of a virtual sensor methodology with the goal of eventual full-scale application to the complex hull structure of a ship.

1.1. Virtual sensing

Virtual measurement is a concept where system or process parameters are indirectly estimated with a computer algorithm based on the measurement from a physical sensor somewhere else in the system or process [9]. The methodology further entails the combination of information obtained from various sources, such as sensors and system models, to indirectly predict a parameter of interest [10]. Modal expansion [11, 12] and Kalman filters [2, 13, 14] are commonly listed response estimator processes in literature.

1.2. Scope and objectives

The aim of this paper is to investigate a virtual sensing application for structural monitoring using a laboratory-scale beam by incorporating a response estimator algorithm. Specifically, the objective is to develop and validate a Kalman filter-based virtual sensor capable of predicting the displacement of a laboratory-scale beam subjected to impulsive and random excitations. The scope of the study is limited to validating the virtual sensor under simply supported boundary conditions. This work addresses the shortcomings in the implementation and validation of such methods at the laboratory scale, serving as a first step toward full-scale deployment for structural strain response prediction.The input to the Kalman filter is the measured acceleration along the length of the beam. Finally, the predicted displacement is compared to the measured displacement at the same location.

Four objectives are identified to achieve the aim of this study, namely:

1. Represent the simply supported beam in the state-space formulation.
2. Derive the mass, damping and stiffness matrices from a finite element (FE) model that closely simulates the modal frequencies as measured on the physical beam.
3. Apply the response estimator algorithm (Kalman filter) to predict the displacement of the beam for the given set of acceleration measurements.
4. Compare the predicted displacement with the measured displacement at the same location.

2. Methodology and Computational Approach

A framework for a virtual sensor is presented in Figure 1. A FE model of the physical beam is created in order to derive system matrices. Additionally, vibration measurements of the physical beam is performed to obtain the damping ratios and modal frequencies with experimental modal analysis (EMA) or operational modal analysis (OMA) techniques.

Next, the FE model is calibrated according to the measured damping ratios, ξ, and modal frequencies, ω_n. This ensures the simulated dynamic response correlates with the physical response. The calibrated mass, damping and stiffness matrices ($\boldsymbol{M}, \boldsymbol{C}$ and $\boldsymbol{K}$) are then utilised to derive state-space equations that model the dynamic response of the structure. Finally, a response estimator algorithm is applied with the state-space model and measured input data which act as the virtual sensor.

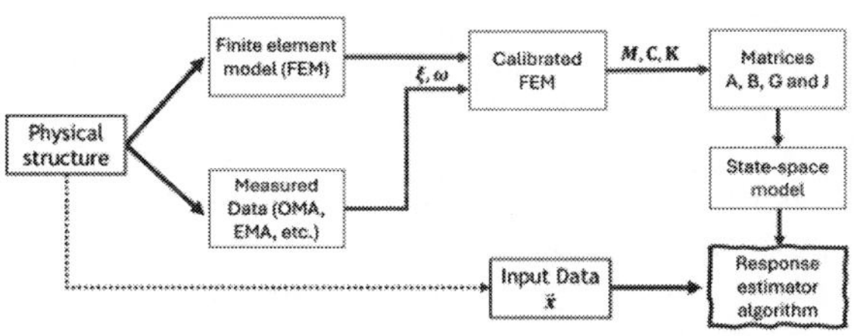

Figure 1: Framework for the development of a virtual sensor.

2.1. State-space model

Consider the equation of motion for a linear discretized system

$$\boldsymbol{M}\ddot{\boldsymbol{u}}(t) + \boldsymbol{C}\dot{\boldsymbol{u}}(t) + \boldsymbol{K}\boldsymbol{u}(t) = \boldsymbol{f}(t) \tag{1}$$

where $\boldsymbol{u}(t) \in \mathbb{R}^{n_{DOF}}$ is the displacement vector, $\boldsymbol{M}, \boldsymbol{C}$ and $\boldsymbol{K}$ refer to the mass, damping and stiffness matrix respectively. The term $\boldsymbol{f}(t)$ refers to the discretised excitation vector. The othogonality conditions according to the mass-normalized eigenvectors are applied. As a result, the decoupled governing equation of motion in modal coordinates, $\boldsymbol{u}(t) = \boldsymbol{\Phi}\boldsymbol{z}(t)$, takes the form

$$\ddot{\boldsymbol{z}}(t) + \boldsymbol{\Gamma}\dot{\boldsymbol{z}}(t) + \boldsymbol{\Omega}^2\boldsymbol{z}(t) = \boldsymbol{\Phi}^T\boldsymbol{f}(t) \tag{2}$$

Next, the measurement vector $\boldsymbol{d}(t)$ is defined as

$$\boldsymbol{d}(t) = \boldsymbol{S}_a\ddot{\boldsymbol{u}}(t) \tag{3}$$

where the selection matrix $\boldsymbol{S}_a$ is populated to describe the spacial distribution of the acceleration sensor locations.

The discretized state-space formulation of the system is given by Equations (4), (5) and (6). The interested reader is referred to [15] for more detail on the derivation of the state-space matrices $\boldsymbol{A}$ and $\boldsymbol{G}$.

$$\boldsymbol{x} = \begin{pmatrix} \boldsymbol{\Phi}^{-1}\boldsymbol{u}(k) \\ \boldsymbol{\Phi}^{-1}\dot{\boldsymbol{u}}(k) \end{pmatrix} \tag{4}$$

$$\boldsymbol{x}_{[k+1]} = \boldsymbol{A}\boldsymbol{x}_{[k]} + \boldsymbol{w}_{[k]} \tag{5}$$

$$\boldsymbol{d}_{[k]} = \boldsymbol{G}\boldsymbol{x}_{[k]} + \boldsymbol{v}_{[k]} \tag{6}$$

2.2. Kalman filter

The Kalman filter equations are derived from the state space equations. The algorithm consists of two stages: *update* and *prediction*. The interested reader is referred to [16] for detailed discussion on the Kalman filter algorithm.

Update parameters

$$\widetilde{\boldsymbol{R}}_k = \boldsymbol{G}\boldsymbol{P}_k\boldsymbol{G}^T + \boldsymbol{R} \tag{7}$$

$$\boldsymbol{L}_k = \boldsymbol{P}_{k|k-1}\boldsymbol{G}^T\widetilde{\boldsymbol{R}}_k^{-1} \tag{8}$$

$$\widehat{\boldsymbol{x}}_{k|k} = \widehat{\boldsymbol{x}}_{k|k-1} + \boldsymbol{L}_k(\boldsymbol{d}_k - \boldsymbol{G}\widehat{\boldsymbol{x}}_{k|k-1}) \tag{9}$$

$$\boldsymbol{P}_{k|k} = \boldsymbol{P}_{k|k-1} - \boldsymbol{L}_k\boldsymbol{G}\boldsymbol{P}_{k|k-1} \tag{10}$$

Predict state-variables

$$\hat{\boldsymbol{x}}_{k+1|k} = \boldsymbol{A}\hat{\boldsymbol{x}}_{k|k} \quad (11)$$

$$\boldsymbol{P}_{k+1|k} = \boldsymbol{A}\boldsymbol{P}_{k|k}\boldsymbol{A}^T + \boldsymbol{Q} \quad (12)$$

The input vector that contains the measured acceleration data is incorperated in Equation (9) as $\boldsymbol{d}_k$. In addition, the variables $\boldsymbol{R}$ and $\boldsymbol{Q}$ refer to the measurement and stochastic process noise which can be adjusted to improve the filter's performance. Lastly, the co-variance matrix $\boldsymbol{P}_{k|k-1}$ is continuously updated and evaluates whether the filter should emphasize the modelled or measured response when predicting the output response.

3. Experimental Setup

An experimental setup of a beam was constructed to measure the acceleration and displacement response for impulsive and random excitations. A steel beam (1600 mm x 40 mm x 4.5 mm) was placed on two simple supports at either end.

Figure 2 illustrates the 7 measurement positions for the beam. Acceleration responses were measured at positions 1 to 7, while vertical displacement was measured at positions 4 and 5. The sensors were installed with consideration for eliminating unintended measurement of lateral and torsional motion. The physical setup is shown in Figure 3. Impulsive forces were applied with a modal hammer at each sensor location. Similary, pseudo-random excitation was manually applied along the length of the beam for a period of 60 seconds. All measurement channels were sampled at 4.8 kHz

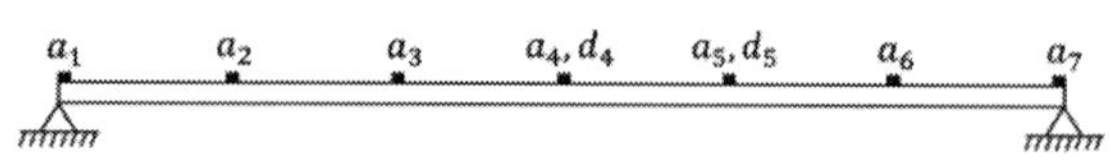

Figure 2: Sensor arrangement on a simply supported beam.

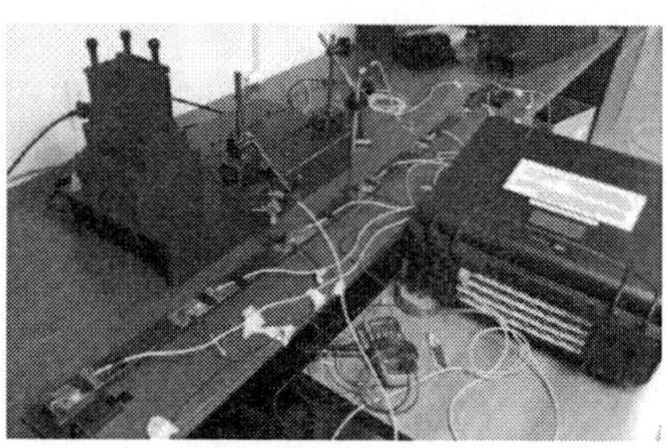

Figure 3: Experimental setup and instrumentation of the simply supported beam.

A one-dimensional FE model of the beam was constructed with 40 CBEAM elements. Translations were constrained in the transverse and axial directions. Similarly, rotations were constrained around the x and y axis. As a result, the oscillations in vertical bending plane could be modelled accurately with a reduced number of degrees of freedom, which shortened the computation time of the Kalman filter algorithm.

An experimental modal analysis was performed to identify the modes and modal frequencies. Consequently, the measured modal frequencies were used to verify whether the dynamic response of the FE model was representative of the physical beam. Table 1 lists the first 8 modal frequencies identified by EMA and from the finite element analysis (FEA). The difference between the first 6 natural frequencies for the measured and simulated beam are small, but the error seem the grow for higher modes. However, it is assumed that the first 3 modes dominate the displacement response and that the higher order modes contribute significantly less to the total response. This assumption is verified by the number of peaks visible in the frequency response shown in Figure 4. As such, the FE modelling error is deemed neglegible.

Table 1: Comparison between measured and simulated modal frequencies for the first 8 modes.

Mode number	EMA (Hz)	FEM (Hz)
1	4.5	4.0
2	16.4	16.1
3	64.8	64.4
4	101.7	100.6
5	148.6	144.8
6	197.9	197.1
7	261.9	257.4
8	332.5	325.7

4. Results and discussion

The mass, damping and stiffness matrices ($\boldsymbol{M}, \boldsymbol{C}$ and $\boldsymbol{K}$) were extracted from the FE model and used to construct the state-space model of the beam according to Equations (4), (5) and (6). Next, the state-space formulation was used with a Kalman filter to predict the displacement response of the beam following impulsive and random excitations. The measured acceleration response was the input variable to the Kalman filter. Furthermore, a damping ratio of $\zeta = 0.05$ yielded comparably good results and was subsequently used to describe the damping characteristics.

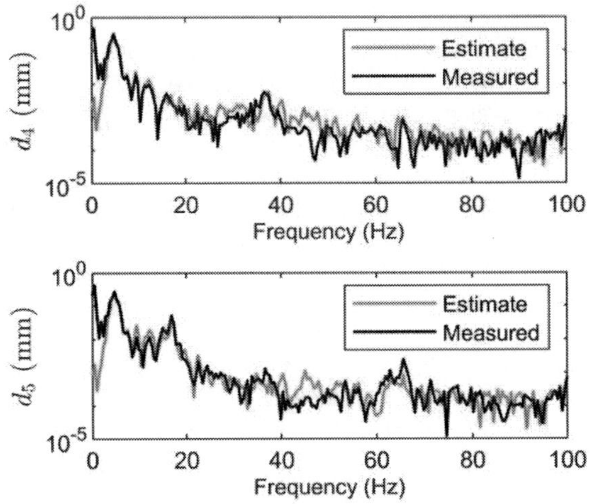

Figure 4: FFT of the measured and estimated displacement for the simply supported beam for d_4 (top) and d_5 (bottom)

4.1. Frequency response

The frequency response of the measured and estimated displacements at locations d_4 and d_5 are shown in Figure 4. The Kalman filter can estimate the first two modes with reasonable accuracy. However, at frequencies higher that 40 Hz some error starts to occur. This can be specifically observed between 40 and 50 Hz, where the filter predicts higher amplitudes while the measured response remains small. In spite of this error, the general comparison between the estimated and measured responses are similar over the frequency range of interest.

4.2. Time response

The Kalman filter calculates the predicted state variables in the time domain. As such, the impulse and random responses were evaluated to verify the filter's performance for transient and random excitations.

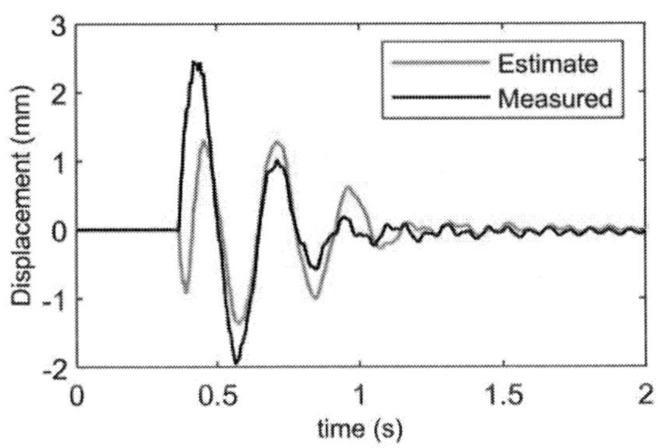

Figure 5: Measured and estimated time response from an impulsive excitation

The time-domain response from the impulsive excitation is shown in Figure 5. The Kalman filter was able to predict the sudden change of motion at $t = 0.4$, despite it initially being in the incorrect direction. Nontheless, the mode quickly corrects the error and predicts the decaying oscillation response of the beam.

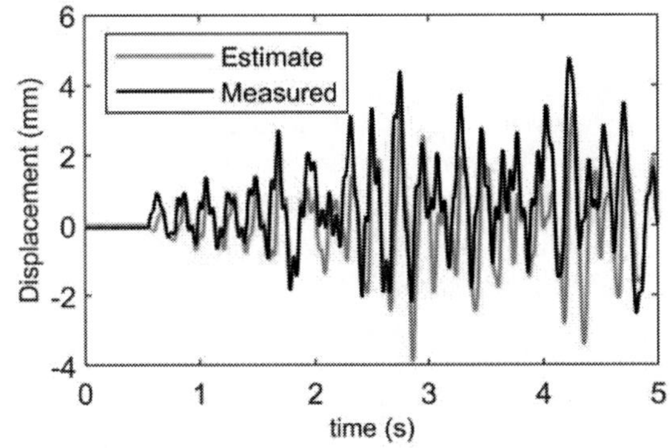

Figure 6: Measured and estimated time response from a random excitation

Next, the time response of the beam under random excitation is presented. Figure 6 shows the first 5 seconds of the measured and estimated time response of the beam, starting from rest. The estimated response compares well with the measured response. However, there are some amplitude errors that can be observed from $t = 3$ s. This observation implies that the filter initially struggles to model an instantaneous change in the response and requires a few iterations to recover and converge to the true response.

Nontheless, the Kalman filter is able to predict the frequency of oscillations and how the signal changes over time. The root mean square error (RMSE) and mean absolute error (MAE) are the two metrics used to evaluate the performance of the estimator algorithm and serves as a benchmark for future work. Table 2 shows the RMSE and MAE for the random and impulsive excitation.

Table 2: Root mean square error (RMSE) and mean absolute error (MAE) of the predicted displacement for random and impulsive excitations.

Excitation type	RMSE	MAE
Random	1.075	0.727
Impulse	0.188	0.06

5. Conclusion

The work presented entails the validation of a Kalman filter based virtual sensor that estimates the displacement of a laboratory scale beam from accelerometer measurement. A simply supported beam was instrumented and subjected to impulsive and random excitation.

A FE model of the beam was created from which the system matrices were derived. As a result, the physical beam could be modelled in the state-space format. A Kalman filter algorithm was applied with the acceleration measurements as the input parameter.

The Kalman filter was able to predict the dynamic response of the beam for both impulsive and random excitation. The frequency response of the estimated displacement aligned with the measured response. In addition, the time domain estimates correlation well with the measured displacement.

Future work entails the enhancement of the virtual sensor to improve the estimate of peak values and to accurately model higher order modes in the frequency domain. Furthermore, the development of the virtual sensor to include virtual strain measurement is critical for a structural digital twin applications where fatigue analysis and remaining useful life of a structure is modelled. Finally, the implementation of the methodology on a complex full-scale structure should be investigated and analysed.

6. References

[1] N. S. Madusanka, Y. Fan, S. Yang and X. Xiang, "Digital twin in the maritime domain: A review and emerging trends," *Journal of Marine Science and Engineering,* vol. 11, p. 1021, 2023.

[2] K. Maes, A. Iliopoulos, W. Weijtjens, C. Devriendt and G. Lombaert, "Dynamic strain estimation for fatigue assessment of an offshore monopile wind turbine using filtering and modal expansion algorithms," *Mechanical Systems and Signal Processing,* vol. 2016, pp. 592-611, 2016.

[3] K.-H. Pherdekamper and A. Bekker, "Investigation of vessel slamming and fatigue using a full-scale test sequence," *Applied Ocean Research,* vol. 103883, p. 144, 2024.

[4] S. O. Erikstad, "Merging physics, big data analytics and simulation for the next-generation digital twins," *High-performance marine vehicles,* pp. 141-151, 2017.

[5] Í. A. Fonseca and H. M. Gaspar, "Challenges when creating a cohesive digital twin ship: a data modelling perspective," *Ship Technology Research,* vol. 68, pp. 70-83, 2021.

[6] S. O. Erikstad, "Designing ship digital services," in *Proceedings of the Conference on Computer and IT Applications in the Maritime Industries*, Tullamore, Ireland, 2019.

[7] H. Omer and A. Bekker, "Detection of wave slamming sites from ship deflections," *R&D Journal,* vol. 32, pp. 50-57, 2016.

[8] A. Bekker, M. Suominen, P. Kujala, R. J. O. De Waal and K. I. Soal, "From data to insight for a polar supply and research vessel," *Ship technology research,* vol. 66, pp. 57-73, 2019.

[9] C. H. Guzmán, J. L. Carrera, H. A. Durán, J. Berumen, A. A. Ortiz, O. A. Guirette, A. Arroyo, J. A. Brizuela, F. Gómez, A. Blanco, H. R. Azcaray and M. Hernández, "Implementation of Virtual Sensors for Monitoring Temperature in Greenhouses Using CFD and Control," *Sensors,* vol. 19, p. 60, 2018.

[10] M. Tarpø, T. Friis, C. Georgakis and R. Brincker, "Full-field strain estimation of subsystems within time-varying and nonlinear systems using modal expansion," *Mechanical Systems and Signal Processing,* vol. 153, p. 107505, 2021.

[11] A. Skafte, J. Kristoffersen, J. Vestermark, U. T. Tygersen and R. Brincker, "Experimental study of strain prediction on wave induced structures using modal decomposition and quasi static Ritz vectors," *Engineering Structures,* vol. 136, pp. 261-276, 2017.

[12] M. Tarpø, T. Friis, B. Nabuco, S. Amador, E. Katsanos and R. Brincker, "Operational modal analysis based stress estimation in friction systems," in *Nonlinear Dynamics, Volume 1: Proceedings of the 36th IMAC, A Conference and Exposition on Structural Dynamics 2018*, Springer, 2019, pp. 143-153.

[13] S. E. Azam, E. Chatzi and C. Papadimitriou, "A dual Kalman filter approach for state estimation via output-only acceleration measurements," *Mechanical systems and signal processing,* vol. 60, pp. 866-886, 2015.

[14] C. Papadimitriou, C.-P. Fritzen, P. Kraemer and E. Ntotsios, "Fatigue predictions in entire body of metallic structures from a limited number of vibration sensors using Kalman filtering," *Structural Control and Health Monitoring,* vol. 18, pp. 554-573, 2011.

[15] E. Lourens, C. Papadimitriou, S. Gillijns, E. Reynders, G. De Roeck and G. Lombaert, "Joint input-response estimation for structural systems based on reduced-order models and vibration data from a limited number of sensors," *Mechanical Systems and Signal Processing,* vol. 29, pp. 310-327, 2012.

[16] Y. Kim and H. Bang, "Introduction to Kalman filter and its applications," in *Introduction and implementations of the Kalman filter*, London, UK, IntechOpen, 2019, pp. 7-22.

Identifiability of rolling element bearing damage in vibration-based condition monitoring using virtual calibration

Stephan Schmidt[a,*], Daniel N. Wilke[a,b]

[a] *Department of Mechanical and Aeronautical Engineering, University of Pretoria, South Africa*

[b] *School of Mechanical, Industrial and Aeronautical Engineering, University of of the Witwatersrand of Johannesburg, South Africa*

**stephan.schmidt@up.ac.za*

Abstract

Rolling Element Bearings (REBs) are critical components in rotating systems. Vibration-Based Condition Monitoring (VBCM) methods are developed for improved diagnosis and prognosis. However, Fault Parameter Identifiability (FPI) is not widely adopted for assessing methods. Hence, this work introduces FPI and presents a Virtual Calibration (VC) methodology to determine the FPI of REB parameters. VC is used on two signals, where the importance of using FPI to assess the performance of VBCM methods is highlighted. Lastly, we show that classification models do not necessarily solve the FPI problem, which could have detrimental consequences when applied without critical evaluation.

Keywords: Fault Parameter Identifiability, Bearings, Condition Monitoring

1. Introduction

In Vibration-Based Condition Monitoring (VBCM) of critical assets, performance metrics are used to assess existing and new methods. The performance metrics are often calculated using processed signals (e.g., signal-to-noise ratio metrics [1]), or VBCM metrics appropriate to the specific method (e.g., time of detection [2], classification accuracy-related metrics [3]). It is challenging to infer detailed fault information (e.g., the location and severity of the faults) due to complex transmission paths, dominant extraneous components, and noise. Hence, classification methods, trained using supervised learning, are frequently used to solve this challenging problem in the literature [3].

Fault diagnosis requires solving an inverse problem, and therefore, Fault Parameter Identifiability (FPI) is important for critically assessing and comparing VBCM methods. However, FPI is not widely adopted in the VBCM literature. If a representative physics-based model is available, the FPI can be quantified using Virtual Calibration (VC) [4].

This study uses a non-linear REB vibration model and VC to investigate the FPI of outer race bearing damage in a REB. The FPI of two signal representations is shown. Subsequently, we show that solving the fault diagnosis problem as a classification problem does not address the FPI problem despite obtaining perfect classification.

This document is structured as follows: In Sec. 2, parameter identifiability is discussed, whereafter the VC methodology, the results, and conclusion are presented in Sec. 3-5, respectively. Appendices A, B, and C include more information about the physics-based model, the data standardisation process, and grid-based optimisation, respectively.

2. Parameter identifiability

The concept of ill-posed problems is often misconstrued as a binary property—either a model is ill-posed or not. However, this binary perspective is misleading. Instead, it is more appropriate to view parameter identification and uncertainty as a continuous spectrum, dependent on the data quality, model form, and the relationship between the parameters and the data [4].

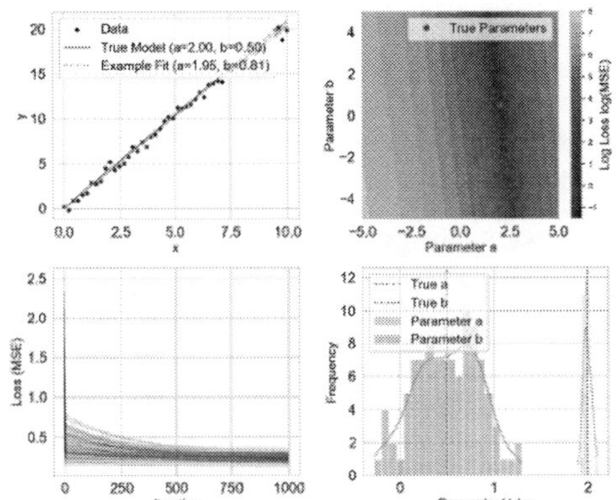

Fig. 1. shows the identification and uncertainty of parameters $\boldsymbol{\theta} = [a, b]$ in the linear regression model $y = a \cdot x + b$. **Top-left:** Data and model fits. **Top-right:** Contour plot of the log. loss (log(MSE)) over parameter space. **Bottom-left:** Convergence of the log. loss. **Bottom-right:** Shows how parameter uncertainty arises in the context of a linear regression model, where the true parameters are $\boldsymbol{\theta} = [a, b] = [2, 0.5]$.

The **top-left** panel in Fig. 1 shows the observed data (blue points) alongside the true model (red line) and an example fit (green dashed line), demonstrating the influence of parameter uncertainty on the resulting fit.

The **top-right** contour plot in Fig. 1 reveals the log Mean-Squared Error (MSE), i.e., log(MSE), over the parameter space, with the true parameters marked as a red dot. The contours' elongated shape indicates stronger identifiability for a and weaker identifiability for b, as the model is more sensitive to changes in variable a.

This is further substantiated by the **bottom-right** panel in Fig. 1, where the histogram for a is narrower and more concentrated, reflecting low uncertainty. In contrast, the broader distribution for b highlights higher uncertainty.

The convergence of the loss (MSE) during optimisation, shown in the **bottom-left** panel, stabilises over iterations, illustrating how optimisation can narrow uncertainty but does not eliminate it [4].

Fig. 1 shows that parameter uncertainty is not an all-or-nothing concept but a continuous property influenced by the interplay between model structure, data quality, and noise [4]. This example highlights that problems should not be viewed as "ill-posed" or "well-posed". Instead, the degree of identifiability, indicated by the variance in parameter estimates, provides a more nuanced understanding of how well the observed data constrain the model's parameters and the associated predictions [4].

To conclude, identifiability refers to how well the parameters can be determined from the observed data. Parameters that can be estimated with low variance are considered more identifiable, while those with high variance are less identifiable.

3. Virtual Calibration (VC) methodology

This section defines VC, the physics-based model, the signal representations, and the parameter estimation procedure in Sec. 3.1 – 3.4.

3.1. Virtual Calibration (VC)

VC solves the inverse problem in a virtual environment. In this process, a parameter vector is selected $\boldsymbol{\theta}^{act}$ and a Virtual Experimental Measurement (VEM) is simulated $\boldsymbol{x}(\boldsymbol{\theta}^{act})$ using a representative model of the system. The objective is to recover the parameters $\boldsymbol{\theta}^{act}$ by only using the VEM, i.e., $\boldsymbol{x}(\boldsymbol{\theta}^{act})$, and an assumed model form. The assumed model form can be the same as the virtual experimental model, or different to purposefully introduce epistemic modelling errors to assess their impact on identifiability. Subsequently, the inverse problem is solved to find equivalent solutions. Finally, the bias and co-variance of the estimated parameters can be used to quantify the identifiability of $\boldsymbol{\theta}^{act}$.

3.2. Physics-based model

Many advanced REB are available in the literature, e.g., [5-10]. To demonstrate the VC concept, a simple six Degree-of-Freedom (DoF) REB model, adapted from [11] and presented in Fig. 2, is used. The REB model consists of a rigid inner ring supported by a shaft and a rigid outer race supported by a pedestal. The inner and outer rings have x- and y-DOFs. The rolling elements are represented as massless non-linear springs that follow a Hertzian contact model, with the ith rolling element's linearised stiffness denoted $k_{co,i}$. A horizontal and vertical measurement location is modelled as single DoFs, with the parameters denoted by the subscript a and b of the respective horizontal and vertical DoFs. The system's mass, damping, and stiffness parameters are contained in the vector $\boldsymbol{\theta}_{sys}$. The equations of motion and Hertzian contact model are provided in App. A.

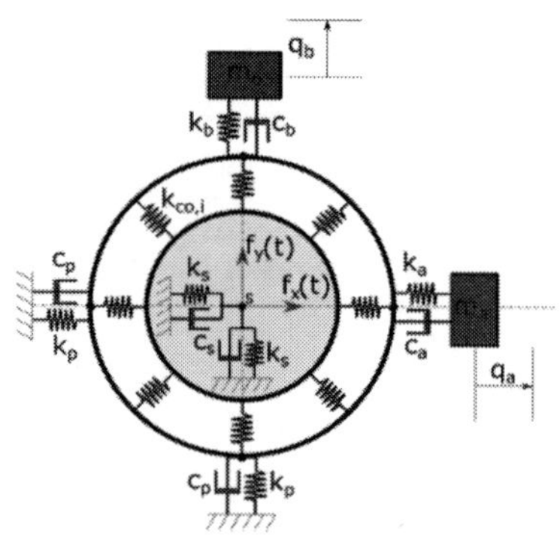

Fig. 2. The Rolling Element Bearing (REB) model adapted from [11].

In this work, outer and inner race faults are described using second-order polynomials. The fault is defined by three parameters, namely, the fault's location θ_l [rad], the fault's width θ_w [rad], and the fault's depth θ_d [mm]. The parameters are denoted $\boldsymbol{\theta}_f \in \mathbb{R}^3$ and used to alter the contact force between the rolling elements and the races as described in App. A.

The signal of the ith degree of freedom, obtained by integrating the REB's equation of motion, is denoted by $q_i(t_n;\ \boldsymbol{\theta}_{sys}, \boldsymbol{\theta}_f)$ for time t_n, where n denotes the time step number. A simulated measured signal with aleatoric uncertainty is modelled as follows:

$$x_i(t_n;\ \boldsymbol{\theta}_{sys}, \boldsymbol{\theta}_f, \sigma) = q_i(t_n;\ \boldsymbol{\theta}_{sys}, \boldsymbol{\theta}_f) + \sigma \cdot \epsilon_n, \quad (1)$$

where ϵ_n is standardised Gaussian noise and σ the standard deviation which controls the measurement noise's magnitude. The measured signal over all time steps is denoted $\boldsymbol{x}_i(\boldsymbol{\theta}_{sys}, \boldsymbol{\theta}_f, \sigma) \in \mathbb{R}^{D\times 1}$. To simplify the notation, the dependence on the parameters $\boldsymbol{\theta}_{sys}$ and σ will be implicit, i.e., the signal is denoted $\boldsymbol{x}_i(\boldsymbol{\theta}_f)$.

3.3. Signal Representation (SR)

The simulated acceleration response of mass m_b is denoted $\boldsymbol{x}(\boldsymbol{\theta}_f)$. The objective function can be calculated using some Signal Representation (SR). In VBCM, different SR (e.g., the raw signal, cyclic spectrum, and extracted features) can be used for diagnosis. Two SRs are used, with the ith M-dimensional SR of the signal denoted $\boldsymbol{\psi}^{(i)}(\boldsymbol{x})$: $\mathbb{R}^{D\times 1} \to \mathbb{R}^{M\times 1}$. SR$^{(1)}$ is the raw signal, i.e.,

$$\boldsymbol{\psi}^{(1)}(\boldsymbol{x}(\boldsymbol{\theta}_f)) = \boldsymbol{x}(\boldsymbol{\theta}_f) \tag{2}$$

and SR$^{(2)}$ is the Root-Mean-Square (RMS) and Kurtosis of the raw signal, i.e.,

$$\boldsymbol{\psi}^{(2)}(\boldsymbol{x}(\boldsymbol{\theta}_f)) = \left[\text{RMS}\left(\boldsymbol{x}(\boldsymbol{\theta}_f)\right), \text{Kurtosis}\left(\boldsymbol{x}(\boldsymbol{\theta}_f)\right)\right]^T \tag{3}$$

Each component of the SR vectors is separately standardised with the process described in App. B before evaluating the objective functions.

3.4. Parameter estimation

The Mean Squared Error (MSE) is quantified between the target solution's SR $\boldsymbol{\psi}^{(i)}\left(\boldsymbol{x}(\boldsymbol{\theta}_f^{act})\right)$ and candidate solution's SR $\boldsymbol{\psi}^{(i)}\left(\boldsymbol{x}(\boldsymbol{\theta}_f)\right)$ and calculated as follows:

$$e_i(\boldsymbol{\theta}_f; \boldsymbol{\theta}_f^{act}) = \frac{1}{M}\sum_{v=1}^{M}\left[\psi_v^{(i)}\left(\boldsymbol{x}(\boldsymbol{\theta}_f^{act})\right) - \psi_v^{(i)}\left(\boldsymbol{x}(\boldsymbol{\theta}_f)\right)\right]^2 \tag{4}$$

Since the $\boldsymbol{x}(\boldsymbol{\theta}_f)$ is random when $\sigma > 0$, the resulting SRs and the MSE are random variables, i.e., the SRs and MSE change for different realisations of $\boldsymbol{x}(\boldsymbol{\theta}_f)$. Hence, the expected error

$$\min_{\boldsymbol{\theta}_f} \mathbb{E}_{\boldsymbol{x}(\boldsymbol{\theta}_f)\sim p(\boldsymbol{x}(\boldsymbol{\theta}_f))}\{e_i(\boldsymbol{\theta}_f; \boldsymbol{\theta}_f^{act})\} \tag{5}$$

is used as an Objective Function (OF) and minimised instead. The expectation is estimated using Monte Carlo (MC) integration.

For this preliminary study, a grid-based solution enables the visualisation of the OFs and investigates and uses diverse SR. To enable this, only the fault location θ_l and depth θ_d parameters are unknown, with a fault width $\theta_w = 1$ deg is used unless stated otherwise. The OF is compared to a threshold to determine the parameters. The OF values, the estimated parameters, and resulting conclusions depend on the grid resolution and selected threshold. A more rigorous analysis is needed for future work, where gradient and Hessian information define minima [12]. The MC integration is performed with only 15 samples. However, more samples are needed to obtain a good estimate of Eq. (5). More information on the grid is provided in App. C.

4. Results

In Sec. 4.1, the identifiability of the fault location θ_l and depth θ_d parameters for an outer race fault are shown for two noise conditions. Sec. 4.2 approaches the fault diagnosis problem as a classification problem.

4.1. Fault Parameter Identifiability (FPI)

The first VEM was generated using $\theta_w = 1$ deg, $\theta_l = 270$ deg, $\theta_d = 0.6$ mm, and $\sigma = 0.0$. The OF is calculated using SR$^{(1)}$ and visualised in Fig. 3(a). A minimum is seen at $\theta_l = 270$ deg and $\theta_d = 0.6$ mm, which correspond to the parameters used in the VEM. Fig. 3(b) shows the parameter combinations that result in an error below 10^{-3}. Only a single parameter is recovered with the current grid and threshold settings.

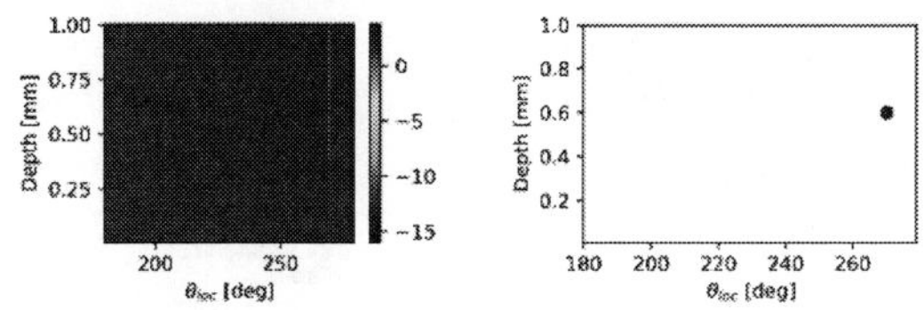

(a) (b)

Fig. 3. Logarithm of the OF (a) and the solutions (b) are shown for SR$^{(1)}$ with $\sigma = 0.0$.

The same process is used for SR$^{(2)}$, with the OF visualised in Fig. 4(a) and the candidate solutions corresponding to an error of 10^{-3} shown in Fig. 4(b), with three parameters recovered with the current grid and threshold settings.

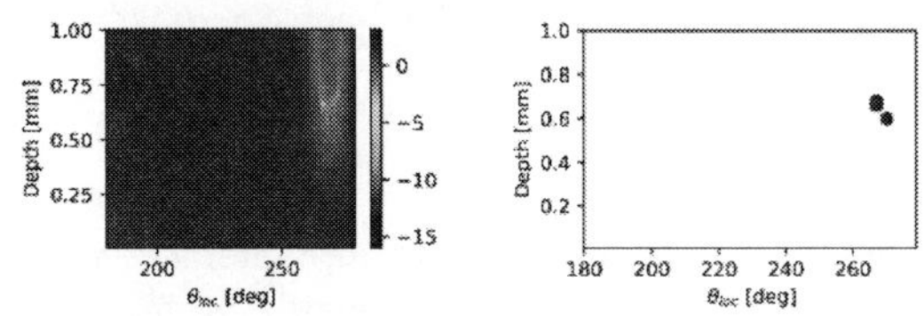

(a) (b)

Fig. 4. Logarithm of the OF (a) and the solutions (b) are shown for SR$^{(2)}$ with $\sigma = 0.0$.

For the second VEM, the process is repeated, but aleatoric noise is added with $\sigma = 10$ to the VEM and the simulations. The OF is visualised in Fig. 5(a) for SR$^{(1)}$. The minimum observed error is approximately equal to the theoretical minimum error of 200. A threshold of 205 was selected. with the samples below this threshold visualised in Fig. 5(b). With the current grid and threshold settings, the variation in the fault depth parameter is larger than the fault location, i.e., the fault depth is less identifiable than the fault location.

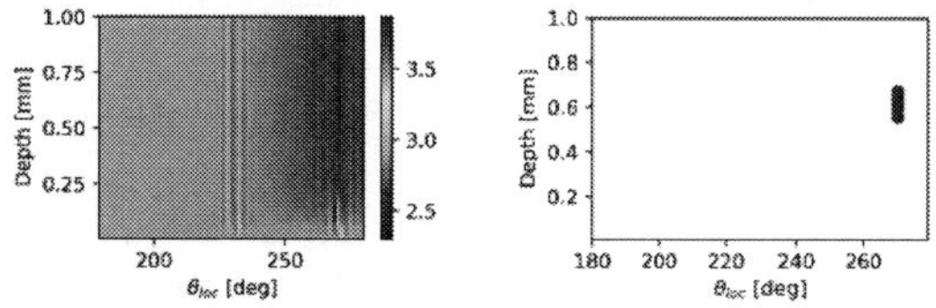

(a) (b)

Fig. 5. Logarithm of the OF (a) and the solutions (b) are shown for SR$^{(1)}$ with $\sigma = 10.0$.

SR$^{(2)}$'s results are visualised using the same VEM in Fig. 6(a). A stronger coupling between the fault depth and fault location is seen between the two parameters compared to the noiseless case in Fig. 4(a). A threshold of 10^{-3} was used to identify the candidate solutions in Fig. 6(b), with both parameters displaying variance using the current settings.

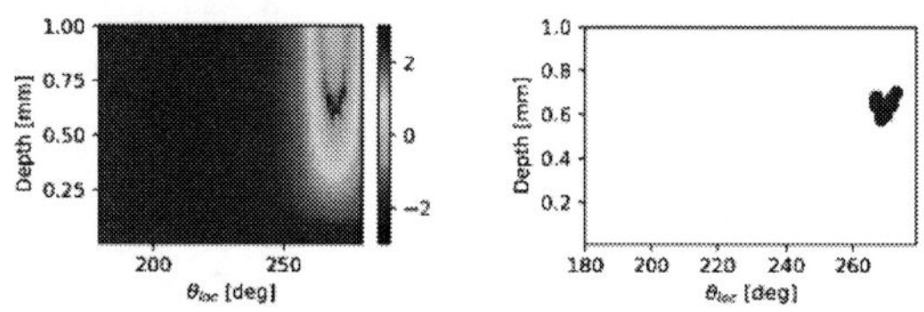

(a) (b)

Fig. 6. Logarithm of the OF (a) and the solutions (b) are shown for SR$^{(2)}$ with $\sigma = 10.0$.

Hence, using the current settings, the FPI results show that SR$^{(1)}$ is better for obtaining the fault location. However, the conclusions should be independent of the optimisation process (i.e., the grid and threshold settings), and therefore, a more rigorous approach is needed to solve this problem and to draw strong conclusions [4].

4.2. Classification

This section demonstrates feature interaction using VC when casting the fault diagnosis problem as a classification problem [13]. Only SR$^{(2)}$ is used here to visualise the data and parameter space in two dimensions. A Naïve-Bayes model [14] is employed and trained using six classes: A healthy class (H), three outer race fault classes (O1, O2, O3) and three inner race fault classes (I1, I2, I3). For the outer race faults, Class O1: $\theta_d = 0.3$ mm, O2: $\theta_d = 0.6$ mm and O3: $\theta_d = 0.9$ mm, with all classes generated using $\theta_w = 1$ deg, $\theta_l = 270$ deg. For the inner race faults, Class I1: $\theta_d = 0.15$ mm, I2: $\theta_d = 0.3$ mm and I3: $\theta_d = 0.6$ mm, with all classes generated using $\theta_w = 1$ deg, $\theta_l = 0$ deg. Forty signals were generated for each class using $\sigma = 10.0$. Thereafter, SR$^{(2)}$ was calculated for each signal, the features were standardised, and the Naïve-Bayes classifier was trained. Forty new signals were generated for each class and used to test the model, with a classification accuracy of 100% obtained. The class predictions are visualised in Fig. 7(a) for different combinations of features. The classifier successfully diagnoses the faults using classification accuracy, and the result is shown in Fig. 7(a).

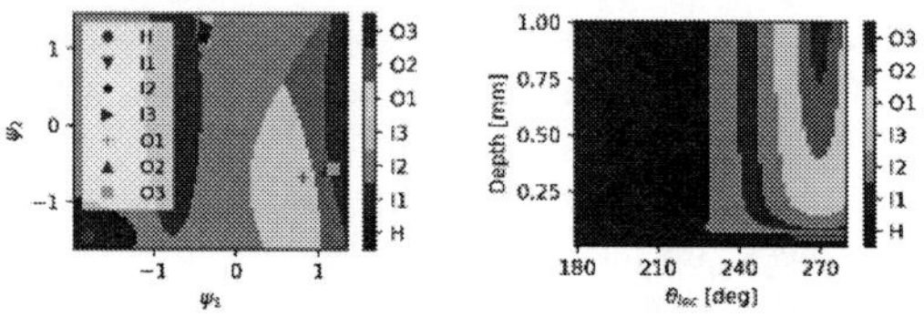

(a) (b)

Fig. 7. Classification labels in the data space (a) and the outer race's parameter space (b) are shown.

The class predictions are also visualised in the parameter space of the outer race fault's parameters in Fig. 7(b). Fig. 7(b) highlights two critical aspects. Firstly, despite the signals in Fig. 7(b) only having outer race damage, inner race faults are predicted for certain regions in the parameter space. Secondly, there is strong coupling between the fault location and fault depth parameters, a characteristic observed in SR$^{(2)}$'s OF plot in Fig. 6. Hence, non-sensible and unexpected classifications can be made by this model known as feature interactions in machine learning [13], which VC exposes. The FPI could be worse for more complex systems, which could be detrimental to the management of critical assets.

5. Conclusion

This work investigated rolling element bearings' Fault Parameter Identifiability (FPI). First, parameter identifiability was defined, and then a Virtual Calibration (VC) methodology was presented. The FPI was investigated using two signal representations, namely the raw signal and features extracted from the raw signal. The results highlight the importance of FPI in assessing vibration-based condition monitoring methods. For the Bayesian classifier, feature interaction was demonstrated using VC. However, more rigorous solution procedures are needed before methods can be critically compared. Furthermore, higher fidelity physics-based models (e.g., modelling more degrees-of-freedom and modelling the rolling element-defect interactions more accurately) are available in the literature and need to be used in future investigations. Lastly, a more in-depth investigation needs to be performed, e.g., including the defect's width as an additional unknown.

References

[1] Schmidt, S., Wilke, D.N. and Gryllias, K.C., (2025). Generalised envelope spectrum-based signal-to-noise objectives: Formulation, optimisation and application for gear fault detection under time-varying speed conditions. Mechanical Systems and Signal Processing, **224**, p.111974.

[2] Soomro, A.A., Muhammad, M.B., Mokhtar, A.A., Saad, M.H.M., Lashari, N., Hussain, M., Sarwar, U. and Palli, A.S., (2024). Insights into modern machine learning approaches for bearing fault classification: a systematic literature review. Results in Engineering, p.102700.

[3] Ali, J.B., Fnaiech, N., Saidi, L., Chebel-Morello, B. and Fnaiech, F., (2015). Application of empirical mode decomposition and artificial neural network for automatic bearing fault diagnosis based on vibration signals. Applied Acoustics, **89**, pp.16-27.

[4] Ben Turkia, S., Wilke, D.N., Pizette, P., Govender, N. and Abriak, N.E., (2019). Benefits of virtual calibration for discrete element parameter estimation from bulk experiments. Granular Matter, **21**, pp.1-16.

[5] Singh, S., Howard, C.Q. and Hansen, C.H., 2015. An extensive review of vibration modelling of rolling element bearings with localised and extended defects. Journal of Sound and Vibration, **357**, pp.300-330.

[6] Liu, J. and Shao, Y., (2015). A new dynamic model for vibration analysis of a ball bearing due to a localized surface defect considering edge topographies. Nonlinear Dynamics, **79**, pp.1329-1351.

[7] Li, X., Liu, J., Ding, S., Xu, Y., Zhang, Y. and Xia, M., 2024. Dynamic modeling and vibration analysis of double row cylindrical roller bearings with irregular-shaped defects. Nonlinear Dynamics, **112**(4), pp.2501-2521.

[8] Liu, J., Wang, L. and Shi, Z., 2022. Dynamic modelling of the defect extension and appearance in a cylindrical roller bearing. Mechanical Systems and Signal Processing, **173**, p.109040.

[9] Safian, A., Zhang, H., Liang, X. and Wu, N., 2021. Dynamic simulation of a cylindrical roller bearing with a local defect by combining finite element and lumped parameter models. Measurement Science and Technology, **32**(12), p.125111.

[10] Wang, P., Zhao, X., Xu, H., Ma, H., Yin, X., Zhang, X. and Zhao, S., (2024). Nonlinear dynamic behaviors of a shaft-bearing-pedestal system with outer ring slip and damage. Nonlinear Dynamics, pp.1-29.

[11] Sawalhi, N. and Randall, R.B., (2008). Simulating gear and bearing interactions in the presence of faults: Part I. The combined gear bearing dynamic model and the simulation of localised bearing faults. Mechanical Systems and Signal Processing, **22**(8), pp.1924-1951.

[12] Snyman, J.A. and Wilke, D.N., (2018). Practical Mathematical Optimization: Basic Optimization Theory and Gradient-Based Algorithms. Springer, Cham.

[13] Friedman, J. H., & Popescu, B. E. (2008). Predictive learning via rule ensembles. The Annals of Applied Statistics, **2**, 916-954.

[14] Pedregosa et al., (2011). Scikit-learn: Machine Learning in Python, JMLR **12**, pp. 2825-2830. Version: 1.0.2.

[15] Borghesani P, Smith WA, Randall RB, Antoni J, El Badaoui M, Peng Z (2022). Bearing signal models and their effect on bearing diagnostics. Mechanical Systems and Signal Processing, **174**, 109077.

[16] Pauli Virtanen, Ralf Gommers, Travis E. Oliphant, Matt Haberland, Tyler Reddy, David Cournapeau, Evgeni Burovski, Pearu Peterson, Warren Weckesser, Jonathan Bright, Stéfan J. van der Walt, Matthew Brett, Joshua Wilson, K. Jarrod Millman, Nikolay Mayorov, Andrew R. J. Nelson, Eric Jones, Robert Kern, Eric Larson, CJ Carey, İlhan Polat, Yu Feng, Eric W. Moore, Jake VanderPlas, Denis Laxalde, Josef Perktold, Robert Cimrman, Ian Henriksen, E.A. Quintero, Charles R Harris, Anne M. Archibald, Antônio H. Ribeiro, Fabian Pedregosa, Paul van Mulbregt, and SciPy 1.0 Contributors. (2020) SciPy 1.0: Fundamental Algorithms for Scientific Computing in Python. Nature Methods, **17**, 261-272. DOI: 10.1038/s41592-0190686-2

Appendix A. Physics-based Model

In this section, the parameters, the equations of motion, and an example signal generated with the model are shown. The model was adapted from Ref. [11]. Higher fidelity models are available in the literature (e.g., Refs. [5-10]) and need to be considered for future investigations.

A.1. Parameters

The parameters of the implemented physics-based model are provided in Tab. 1.

Table 1: Parameters of the implemented Rolling Element Bearing (REB) model. The symbols are also used in Fig. 2.

Symbol	Description	Value
m_s	Mass [kg]	1
c_s	Viscous damping [N.s/m]	500
k_s	Stiffness [N/m]	10^5

m_p	Mass [kg]	10
c_p	Viscous damping [N.s/m]	$2 \cdot 10^3$
k_p	Stiffness [N/m]	$12 \cdot 10^6$
m_a	Mass [kg]	1
c_a	Viscous damping [N.s/m]	4398.23
k_a	Stiffness [N/m]	$1.934 \cdot 10^9$
m_b	Mass [kg]	1
c_b	Viscous damping [N.s/m]	4398.23
k_b	Stiffness [N/m]	$1.934 \cdot 10^9$
k_γ	Hertzian contact stiffness parameter 1	10^{10}
n_γ	Hertzian contact stiffness parameter 2	1.5
N_B	Number of rolling elements [-]	9
$\frac{D_b}{D_p}$	Rolling element diameter / Pitch diameter of the REB [-]	0.2034
ω_s	Inner ring's rotational speed [Hz]	5.0
$f_x(t)$	Horizontal shaft force [N]	0.0
$f_y(t)$	Vertical shaft force [N]	-3000
c	Radial clearance [mm]	0

The contact stiffness parameter k_γ was selected to have a representative magnitude, however, further work is needed to derive the parameter for the exact geometrical specifications of the bearing. The horizontal and vertical pedestal stiffness and damping parameters, denoted k_p and c_p, respectively, are assumed to be equal in this work.

A.2. Equation of motion

The REB model is adapted from Ref. [11] to include damping on the shaft, radial forces on the shaft, and to include a horizontal degree-of-freedom. The equations of motion were re-arranged to include the acceleration term on the left side:

$$\ddot{x}_s = \frac{-k_s x_s - c_s \dot{x}_s - f_{con,x}(t) + f_x(t)}{m_s}$$

$$\ddot{y}_s = \frac{-k_s y_s - c_s \dot{y}_s - f_{con,y}(t) + f_y(t)}{m_s}$$

$$\ddot{x}_p = \frac{-(c_p + c_a)\dot{x}_p - (k_p + k_a)x_p}{m_p} + \frac{k_a q_a + c_a \dot{q}_a + f_{con,x}(t)}{m_p}$$

$$\ddot{y}_p = \frac{-(c_p + c_b)\dot{y}_p - (k_p + k_b)y_p}{m_p} + \frac{k_b q_b + c_b \dot{q}_b + f_{con,y}(t)}{m_p}$$

$$\ddot{q}_a = \frac{-k_a (q_a - x_p) - c_a (\dot{q}_a - \dot{x}_p)}{m_a}$$

$$\ddot{q}_b = \frac{-k_b (q_b - y_p) - c_b (\dot{q}_b - \dot{y}_p)}{m_b}$$

The stiffness, damping, and mass parameters are given in Tab. 1. The values of the external forces $f_x(t)$ and $f_y(t)$ are also provided in Tab. 1, with only a downward vertical force being applied to the bearing's inner ring. The degree-of-freedoms of the measurement locations are denoted by q_i to simplify the notation, e.g., in Eq. (1). The overall contact forces between the rolling elements and the races for a bearing with N_B rolling elements are given by [11]:

$$f_{con,x}(t) = k_\gamma \cdot \left(\sum_{j=1}^{N_B} \gamma_j \cdot \cos(\phi_j) \cdot \delta_j^{n_\gamma} \right)$$

for the x-direction and by:

$$f_{con,y}(t) = k_\gamma \cdot \left(\sum_{j=1}^{N_B} \gamma_j \cdot \sin(\phi_j) \cdot \delta_j^{n_\gamma} \right)$$

for the y-direction, with $\gamma_j = 1$ if $\delta_j > 0$ and $\gamma_j = 0$ otherwise. The angle of the jth rolling element is given by ϕ_j. The contact stiffness parameters k_γ and n_γ are dependent on the geometry of the bearing. The relative displacement between the shaft-inner ring and the pedestal-outer ring is given by:

$$\delta_j = (x_s - x_p)\cos(\phi_j) + (y_s - y_p)\sin(\phi_j) - c - B(\phi_j)$$

where c is the clearance parameter and $B(\phi_j)$ is the fault geometry function. The fault geometry function $B(\phi_j)$ is only non-zero when the jth rolling element is in contact with a fault. For simplicity's sake, we assume a quadratic profile. This means that the defect's geometry can be described by three parameters (its location θ_l, width θ_w, and depth θ_d) using this assumption. An example of an outer race defect is shown in Fig. 8. More sophisticated defect modelling approaches are available in the literature (e.g., [5-10]) and need to be used in future work.

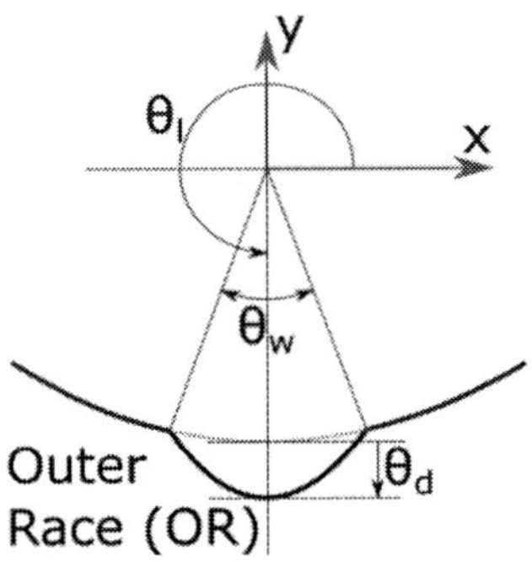

Fig. 8. Three fault parameters of an outer race bearing fault (its location θ_l, width θ_w, and depth θ_d) are shown.

The angle of the jth rolling element is given by:

$$\phi_j(t) = 2\pi \frac{j-1}{N_B} + \phi_c(t) + \phi_{slip}(t)$$

with $j = 1,2,\dots,N_B$. The rolling elements display slip, i.e., there is a deviation between the theoretical and actual angle of arrival of the elements. A detailed discussion of slip is presented in Ref. [15]. For this preliminary study, no slip is modelled in this work. The cage's phase is given by:

$$\phi_c(t) = \left(1 - \frac{D_b}{D_p}\right)\frac{\phi_s(t)}{2.0}$$

where the relationship between the shaft's speed and instantaneous phase is given by $\omega_s = \frac{d}{dt}\phi_s(t)$. The rolling element's diameter and the pitch diameter is denoted D_b and D_p respectively. The Ball-Pass Outer Race Order (BPOO), that is, the Ball-Pass Outer Race Frequency (BPFO) normalised by the reference shaft speed, is given by:

$$\text{BPOO} = \frac{N_b}{2}\cdot\left(1 - \frac{D_b}{D_p}\right)$$

A.3. Numerical integration

The equations of motion were integrated using Runge-Kutta 45 (RK45) using SciPy [16]. The integration was performed over 1.25 seconds with the initial displacements and velocities for all degrees of freedom set to 0. The first 0.25 seconds was ignored to reduce the impact of initial values and the associated transients. Fig. 9 shows a simulated measurement signal, obtained with Eq. (1) for an outer race fault. A fault location $\theta_l = 270$ rad (i.e., the defect is located in the load zone), a fault depth $\theta_d = 0.2$ mm, and $\sigma = 10$ were used. The last 1.0 seconds of the signal is used in the subsequent investigations. The periodic impulses in Fig. 9 are caused by the outer race bearing damage .

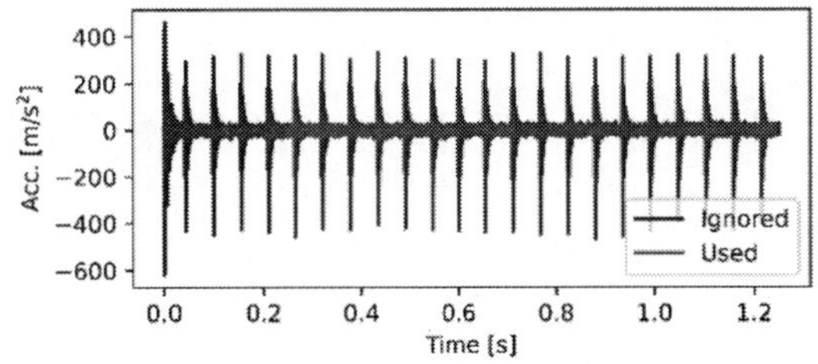

Fig. 9. An example of a 1.25 simulated measurement with the first 0.25 seconds ignored and therefore only the last 1.0 seconds used in the investigations.

A.4. Code

The code used to generate the signals in this study is available here:

https://gitlab.com/stephanschmidteng/sacam2025

Appendix B. Signal Representation (SR)

The Signal Representation (SR) is multi-dimensional and denoted $\boldsymbol{\psi}$, with its vth component denoted ψ_v. For simplicity's sake, no reference is made to the SR number (i.e., $\boldsymbol{\psi}$ is used instead of $\boldsymbol{\psi}^{(i)}$). The standardised SR's v^{th} component is defined by $\frac{\psi_v - \mu_v}{\sigma_v}$ where its mean and standard deviation are denoted μ_v and σ_v respectively. The parameters μ_v and σ_v are determined using the training data of Sec. 4.2, whereafter the parameters are fixed and re-used in Sec. 4.1 and Sec. 4.2. This ensures that consistent parameters are used between Sec. 4.1 and 4.2 and it also ensures that the data space in Sec. 4.2 is on a reasonable range (i.e., the standardised data space simplifies the grid specification) in Fig. 7.

Appendix C. Grid-based optimisation

A fixed grid was defined as follows:

- θ_l: Spacing of 1.5 deg between 180 deg and 280 deg.
- θ_w: 1.0 deg
- θ_d: Spacing of 0.1 mm between 0.0 mm and 1.0 mm

The objective function is computed for all combinations of parameters in this grid and then the parameter combinations that correspond to an objective function below a specified threshold are used and visualised, i.e., a level set is used. Hence, the results and conclusions are strongly coupled with this grid. A more rigorous optimisation process is needed before firm conclusions can be drawn.

Influence of Axle Load, Wheel Flat Length and Velocity on Impact Forces Due to Wheel Flats using Numerical Simulation

Michael Perumal[a*], Brendon Nickerson[a], Annie Bekker[a]

[a]*Stellenbosch University, Stellenbosch Central, Stellenbosch, 7602 South Africa*

**mperumal@sun.ac.za*

Abstract

This study investigates the parametric influences on wheel-rail contact forces induced by wheel flat defects, with particular emphasis on axle load, defect length, and vehicle velocity using a 2D wagon-track simulation. The investigation covered a parametric study spanning operational velocities of 4 km/hr - 120 km/hr, car body masses of 20 000 kg - 90 000 kg, and wheel flat lengths of 0 (no flats) to 40 mm. Simulation results indicate that larger flats, higher velocities and larger masses produce larger contact forces with the existence of a critical velocity that produces larger contact forces at the lower end of the velocity spectrum.

Keywords: Wheel flats, multibody physics simulations, rail

1. Introduction

Wheel rail interaction can provide insight into the condition of track infrastructure and the trainset. Measured data provides a ground truth of the interacting forces between these two systems, however, systematic investigation of the influence of individual variables is constrained by the complexity of experimental setups. Additionally, measured data introduces unknown and uncontrollable variables. A simulation can be used to circumvent these obstacles; thus, a Two-Dimensional-Wagon-Track (2DWT) numerical model is developed and used to investigate vertical train-track interactions in the XZ plane. The coordinate system is defined with the X-axis oriented along the longitudinal direction of the vehicle, the Y-axis oriented transversely across the vehicle width, and the Z-axis aligned vertically in the direction of gravitational acceleration. This paper explores an application of the 2DWT model that is developed thus far and focuses on the wheel-rail contact forces generated by wheel flats. The nomenclature is available in Appendix A1.

1.1. Background and motivation

Wheel flats (shown in Figure 1) are caused by the locking of a wheel due to sudden breaking or slipping of the wheel due to loss of adhesion between the wheel and rail surfaces. This is the most common surface defect [2] and imparts cyclical impulse forces that significantly exceed the static wheel-rail contact load. The larger impact forces damage track infrastructure, trainset components, increases noise levels, induces ground borne vibrations and results in derailment in severe cases.

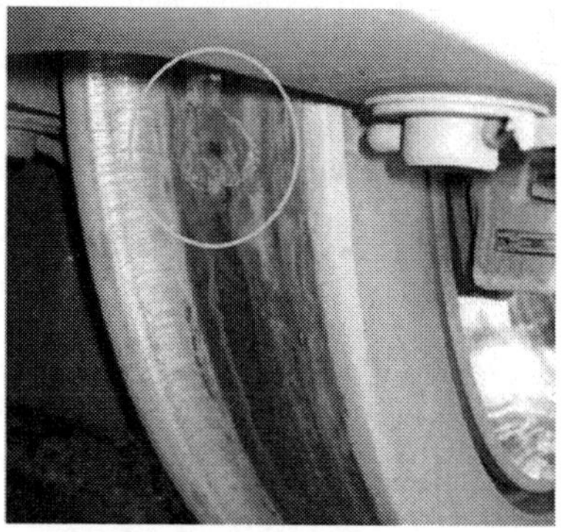

Figure 1 Wheel flat formed by sliding [1]

EN15313 provides the allowable operating conditions for a trainset with wheel flats and considers the vehicle translational speed, wheel diameter and wheel flat length. This paper aims to investigate the influence of axle weight, vehicle speed and wheel flat length on the contact forces. This can be used to provide insight on EN15313 and further assist rail operators by allowing them to mitigate structural degradation and extend the operational lifecycle of both rolling stock and track infrastructure by selecting optimal operational strategies.

1.2. Objectives and scope

The objectives of the research are:

1. Quantitatively analyze wheel-rail contact forces
    a. Implement and verify a 2DWT numerical model
    b. Quantify the relationship between wheel flat characteristics and generated contact forces
2. Examine the influence of the following variables on wheel rail contact forces:
    a. Axle load

b. Wheel flat length
c. Vehicle velocity

The scope of the research covers:

1. Investigation of contact forces in the XZ plane
2. Physical track parameters from the Datong–Qinhuangdao railway line and vehicle parameters of the Gondola Wagon C62A [2].
3. Vehicle speeds from 4 km/hr to 120 km/hr
4. Car weights from 20 000 kg to 90 000 kg.
5. Wheel flat lengths from 5mm to 40mm.

2. Methodology

2.1. Numerical Model

The system model, derived from [3], is shown in Figure 2 and consists of a 10 degree of freedom (DOF) trainset and 105 DOF track. The track is modelled as an Euler-Bernoulli beam with discrete supports. Three layers of springs and dampers represent the rail pads, ballast and subgrade while two layers of masses represent the sleepers and ballast. Interlocking between neighboring ballasts are implemented via shear springs and non-linear Hertzian contact captures the interface between the wheel and the rail.

2.2. Formulation of equations of motion

The equations of motion (EOM) are formulated (aligning with Figure 2), shown below are the equations of motion to describe the motion of wheel 1 and the rail:

Wheel 1:

$$M_w \ddot{Z}_{w1} + C_{s1}(\dot{Z}_{w1} - \dot{Z}_{t1}) + K_{s1}(Z_{w1} - Z_{t1}) - C_{s1} l_t \psi_{t1} - K_{s1} l_t \psi_{t1} + P_1(t) = 0 \tag{1}$$

Rail:

$$\ddot{q}_k(t) + \sum_{i=1}^{N} C_{pi} Y_k(x_i) \sum_{k=1}^{K} Y_k(x_i)\dot{q}_k(t) + \frac{EI}{m_r}\left(\frac{k\pi}{l}\right)^4 q_k(t)$$
$$+ \sum_{i=1}^{N} K_{pi} Y_k(x_i) \sum_{k=1}^{K} Y_k(x_i) q_k(t) - \sum_{i=1}^{N} C_{pi} Y_k(x_i)\dot{Z}_{si}(t)$$
$$- \sum_{i=1}^{N} K_{pi} Y_k(x_i) Z_{si}(t) = \sum_{j=1}^{4} P_j(t) Y_k(x_{Gj}) \tag{2}$$

The mode shape function used to describe the rail as well as the remaining EOM's are available in [3].

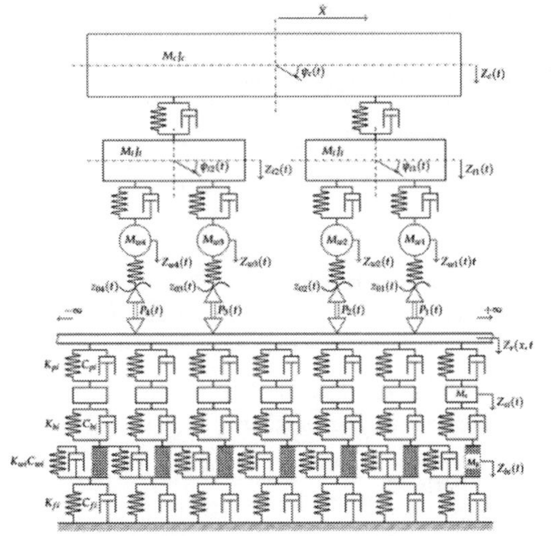

Figure 2 2DWT system model [3]

2.3. Defect implementation

The contact force is determined by the Hertzian contact law:

$$P(t) = C_h(\partial Z(t))^{3/2} \tag{3}$$

Where $\partial Z(t)$, is:

$$\partial Z(t) = z_{(w)}(t) - z_{(r)}(t) - z_{(d)}(t) \tag{4}$$

Wheel flats are categorized as new flats or rounded wheel flats (Figure 3) and are mathematically represented by a chord on the wheel circumference or a haversine function as shown in Equation 5 [4]. Figure 3 also indicates that a increases with L.

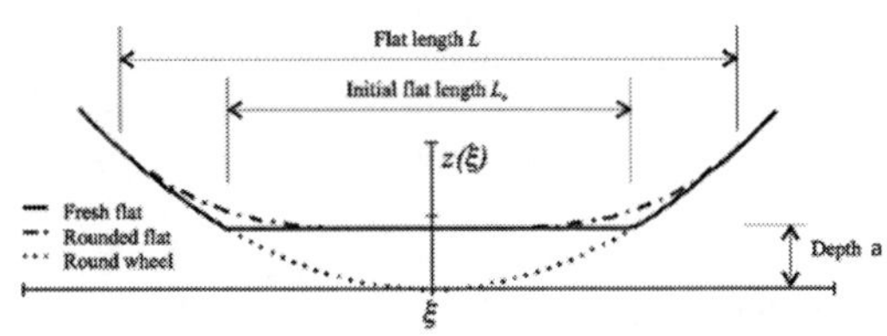

Figure 3 New wheel flat and rounded wheel flat [5]

$$z_{(defect)} = -\, {}^{a}/_{2}\left(1 - \cos\left({}^{2\pi v}/_{L}\right)\right) \tag{5}$$

2.4. Solution method

The ODE's are expressed in the standard matrix form as shown in Equation 6.

$$[M]\{\ddot{X}(t)\} + [C]\{\dot{X}(t)\} + [K]\{X(t)\} = \{F(t)\} \tag{6}$$

Large matrices deter the use of Runge-Kutta method for numerical integration [3], therefore Newmark's method, CH-alpha, Backwards Euler and JWH-alpha methods were explored [6].

3. Results and Discussion

3.1. Validation

The numerical model was validated through comparison with experimental measurements and previously published data [2] using an i7-1265U@1.80 GHz CPU with 16GB RAM. The various integration schemes performed similarly with minor deviations in processing time, as shown in in Table 1. For the verification process, parameters from the Datong–Qinhuangdao railway line and vehicle characteristics of the Gondola Wagon C62A were used. The validation case examined the contact force response to a dip defect, which can be mathematically represented using a haversine function analogous to Equation 5. As illustrated in Figure 4, the model demonstrates strong agreement with measured data and literature [2] in terms of both peak force magnitude and shape (~225 kN peak with 6.4% relative error). As the wheel rotates, exposing the defect, the wheel loses contact with the rail due to missing material, this explains the initial drop in contact force. Thereafter, the falling wheel makes impact with the rail, resulting in the peak force. This force is calculated according to equation 3, hence the vertical speed and acceleration of the wheel and rail determine the resulting force. Furthermore, the decision to use a 2D model instead of a 3D model depends on the application – while 3D models have a higher fidelity, the processing requirements are larger. 2D models tend to underpredict wheel flat impacts while 3D models tend to overpredict – but they both obtain similar results [7]. This verification establishes the model's capability to accurately simulate wheel-rail impact dynamics.

Table 1. Integration schemes performance comparison

Integration Scheme	Average run time [s]
JWH Alpha	107.947
Newmark Beta	113.202
Backward Euler	108.185
CH Alpha	111.874

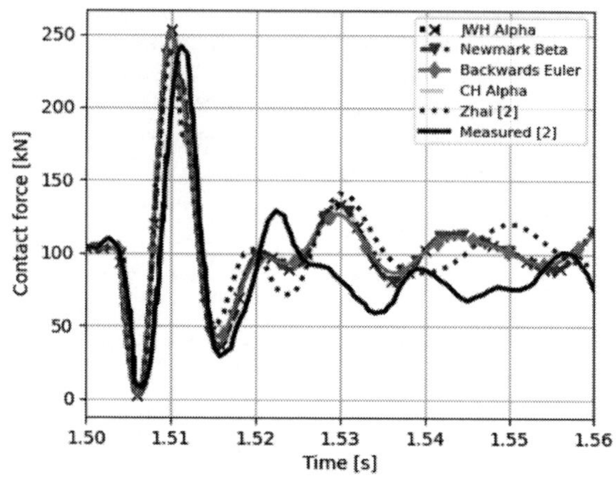

Figure 4 Contact force due to a dip in rail

3.2. Response due to wheel flats

To explore the typical response due to a wheel flat a simulation was conducted using the track and trainset parameters available in Appendix A2. A wheel flat of length 40mm was introduced via a haversine function on wheel 1 with a trainset speed of 22.6 m/s. The contact force over the duration of 3.6 seconds is shown Figure 5. The periodic nature of the contact force is evident, corresponding to each wheel revolution, with distinct force peaks occurring at the wheel flat impact locations. The delay between impacts decreases as the train accelerates, consistent with the increasing translational velocity of the train. This is indicated by the velocity profile (dotted line) which shows the transition from stationary to steady-state operation. The peak force is indicated by the dot.

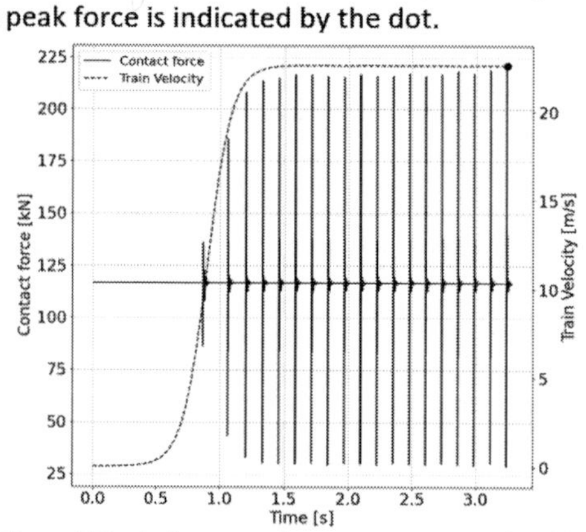

Figure 5 Typical contact force response to wheel flat

The dynamic force amplification factor (peak force divided by static force) for varying wheel flat sizes and velocities are shown in Figure 6. The contact force at wheel 1 increases with rising speeds, however, the relationship is non-linear.

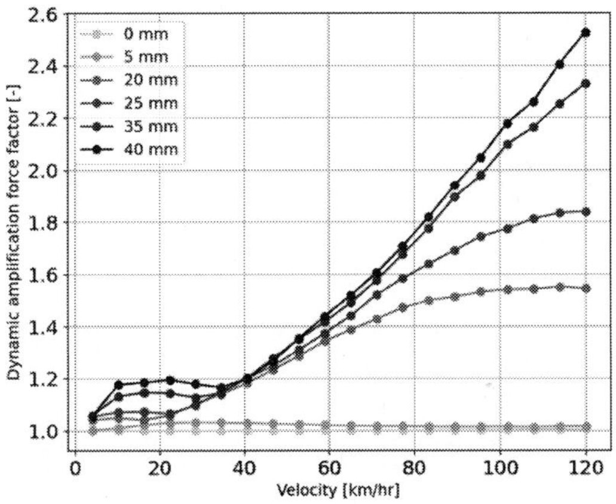

Figure 6 Dynamic force amplification factor vs velocity for flat lengths from 0 mm – 40 mm

For smaller wheel flats (5mm), the peak forces remain relatively constant across the entire velocity range, suggesting minimal velocity sensitivity for minor surface defects. This indicates that small wheel flats may not pose significant risks in terms of impact loading, even at higher operational speeds. When compared to a healthy wheel, the RMSE is 2.39 kN.

Intermediate flat lengths (20-25mm) display a transitional behavior, with a moderate increase in peak forces that begins to plateau at higher velocities. This suggests the existence of a critical velocity range where the impact mechanism transitions from quasi-static to dynamic behavior.

Larger wheel flats (35-45mm) demonstrate a markedly different behavior pattern. These defects show an approximately linear increase in peak forces at velocities above 40 km/h, with the 45mm flat generating the most severe impact forces. The divergence between different flat lengths becomes particularly pronounced at velocities exceeding 80 km/h, where the relationship between velocity and peak force becomes increasingly steep for larger defects.

It must be noted that at a velocity of ~40 km/hr, the peak forces across multiple flat lengths are similar, indicating that the influence of flat size on impact forces is less significant at lower operational speeds.

The results of varying car weight and velocity are shown in Figure 7. The peak forces follow nearly identical trajectories across the entire velocity spectrum as the car weight changes whilst keeping a 40mm flat present on wheel 1. Higher car masses produce larger forces. The peak forces between 20 km/hr and 40 km/hr remain relatively similar per car mass. Thereafter, there is exponential growth proportional to the translational velocity. It can also be seen that there exists a critical velocity at ~17 km/hr where there is a local maximum in the curve. This is caused when the flat edge of the defect is located close to the impact point. With larger flat lengths, it takes more time for the impact point to be close to the flat edge, resulting in a higher speed requirement [7][8].

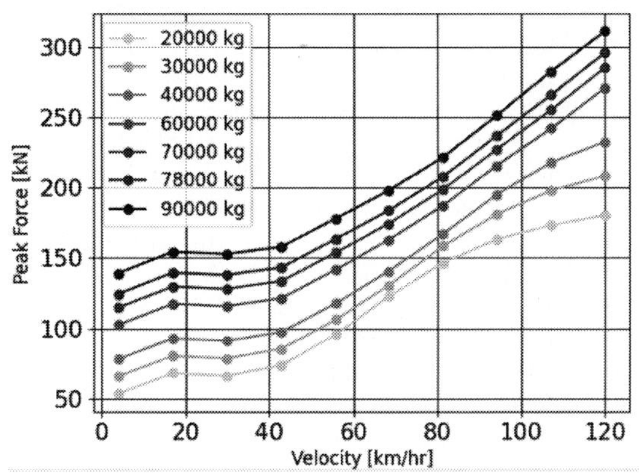

Figure 7 Peak contact force vs velocity for car weights from 20 000 kg to 90 000 kg

4. Conclusion

This study investigated the influence of wheel flat length, vehicle velocity, and axle load on wheel-rail contact forces through numerical simulation. The following results hold for the track and trainset parameters used in the model. The results demonstrated that wheel flat length has the most significant impact on force generation, with larger flats (35-45mm) showing approximately linear increases in peak forces above 40 km/h, while small flats (5mm) exhibited minimal velocity sensitivity. A critical velocity around 40 km/h was identified where the influence of flat size becomes less pronounced. Vehicle mass showed that higher masses consistently produced larger forces, with a local maximum occurring at approximately 17 km/h. These findings provide valuable insights for railway operators to optimize operational strategies and maintenance schedules, particularly for vehicles with known wheel flat defects. The verified numerical model proved effective in capturing the complex dynamics of wheel-rail interaction, offering a reliable tool for future investigations of railway vehicle dynamics. Future work to be conducted is to validate the numerical model using experimental data.

References

[1] L. Jing, "Wheel-Rail Impact by a Wheel Flat," in *Modern Railway Engineering*, InTech, 2018. doi: 10.5772/intechopen.70460.

[2] W. M. Zhai, C. B. Cai, Q. C. Wang, Z. W. Lu, and X. S. Wu, "Dynamic effects of vehicles on tracks in the case of raising train speeds," *Proc. Inst. Mech. Eng. Part F J. Rail Rapid Transit*, vol. 215, no. 2, pp. 125–135, 2001, doi: 10.1243/0954409011531459.

[3] W. Zhai and Z. Cai, "Dynamic interaction between a lumped mass vehicle and a discretely supported continuous rail track," *Comput. Struct.*, vol. 63, no. 5, pp. 987–997, 1997, doi: 10.1016/S0045-7949(96)00401-4.

[4] M. A. Spiroiu and C. I. Crăciun, "Wheel flat effect on wheel-rail dynamic interaction," *IOP Conf. Ser. Mater. Sci. Eng.*, vol. 444, no. 4, 2018, doi: 10.1088/1757-899X/444/4/042002.

[5] E. Zhou, Y., Wahab, M., Maia, N., Liu, L., Figueiredo, "Railway Wheel Out-of-Roundness and Its Effects on Vehicle–Track Dynamics: A Review," *Data Min. Struct. Dyn. Anal.*, 2019, doi: https://doi.org/10.1007/978-981-15-0501-0_3.

[6] C. Kadapa, W. G. Dettmer, and D. Perić, "On the advantages of using the first-order generalised-alpha scheme for structural dynamic problems," *Comput. Struct.*, vol. 193, pp. 226–238, 2017, doi: 10.1016/j.compstruc.2017.08.013.

[7] A. Pieringer, W. Kropp, and J. C. O. Nielsen, "The influence of contact modelling on simulated wheel/rail interaction due to wheel flats," *Wear*, vol. 314, no. 1–2, pp. 273–281, 2014, doi: 10.1016/j.wear.2013.12.005.

[8] L. Han, L. Jing, and K. Liu, "A dynamic simulation of the wheel–rail impact caused by a wheel flat using a 3-D rolling contact model," *J. Mod. Transp.*, vol. 25, no. 2, pp. 124–131, Jun. 2017, doi: 10.1007/s40534-017-0131-0.

Appendix

A.1. Nomenclature

Symbol	Description
C_b	Ballast damping
C_w	Ballast shear damping
K_w	Ballast shear stiffness
K_b	Ballast stiffness
Z_{t1}	Bogie 1 vertical displacement
Z_{t2}	Bogie 2 vertical displacement
J_t	Bogie inertia mass
M_t	Bogie mass
ψ_t	Bogie pitch
J_c	Car body inertia mass
M_c	Car body mass
ψ_c	Car pitch
Z_c	Car vertical displacement
Z_d	Defect vertical distance
l_c	Distance between bogies
l_t	Distance between wheels
M_b	Mass of ballast*
m_r	Mass of rail per unit length
M_s	Mass of sleeper*
C_{s1}	Primary suspension damping
K_{s1}	Primary suspension stiffness
EI	Rail bending stiffness
C_p	Rail pad damping
K_p	Rail pad stiffness
Z_r	Rail vertical displacement
C_{s2}	Secondary suspension damping
K_{s2}	Secondary suspension stiffness
L_s	Sleeper spacing
Z_{si}	Sleeper vertical displacement
C_f	Subgrade damping
K_f	Subgrade stiffness
M_w	Wheel mass
Z_w	Wheel vertical displacement

A.2. Track and vehicle parameters

	Value	Description
Vehicle Model Parameters		
M_c	77 000 kg	Car body mass
M_t	1 100 kg	Bogie mass
M_w	1 200 kg	Wheel mass
J_c	1.2×10^6 kg m^2	Car body inertia mass
J_t	760 kg m^2	Bogie inertia mass
K_{s1}	-	Primary suspension stiffness
C_{s1}	-	Primary suspension damping
K_{s2}	5.32×10^6 N m^{-1}	Secondary suspension stiffness
C_{s2}	7×10^4 N s m^{-1}	Secondary suspension damping
Track Model Parameters		
m_r	51.5 kg m^{-1}	Mass of rail per unit length
EI	4.2×106 N m^2	Rail bending stiffness
L_s	0.545 m	Sleeper spacing
M_s	237 kg	Mass of sleeper*
M_b	683 kg	Mass of ballast*

K_p	1.2×10^8 N m^{-1}	Rail pad stiffness
C_p	1.24×10^5 N s m^{-1}	Rail pad damping
K_b	2.4×10^8 N m^{-1}	Ballast stiffness
C_b	5.88×10^4 N s m^{-1}	Ballast damping
K_f	6.5×10^7 N m^{-1}	Subgrade stiffness
C_f	3.12×10^4 N s m^{-1}	Subgrade damping
K_w	7.84×10^7 N m^{-1}	Ballast shear stiffness
C_w	8.0×10^4 N s m^{-1}	Ballast shear damping
C_h	0.89×10^{11} N/m$^{3/2}$	Hertzian contact stiffness

*Per rail seat